# Data Assimilation for Atmospheric, Oceanic and Hydrologic Applications

Seon K. Park · Liang Xu (Eds.)

# Data Assimilation
# for Atmospheric, Oceanic
# and Hydrologic Applications

 Springer

*Editors*
Prof. Seon K. Park
Ewha Womans University
Dept. of Environmental
Science & Engineering
11-1 Daehyun-dong
Seoul, Seodaemungu-gu
Seoul 120-750
Republic of Korea
spark@ewha.ac.kr

Dr. Liang Xu
Naval Research Laboratory
7 Grace Hopper Avenue
Stop 2
Monterey CA 93943-5502
USA
liang.xu@nrlmry.navy.mil

ISBN: 978-3-540-71055-4    e-ISBN: 978-3-540-71056-1

DOI 10.1007/978-3-540-71056-1

Library of Congress Control Number: 2008935664

*Cover design:* deblik, Berlin

Printed on acid-free paper

9  8  7  6  5  4  3  2  1

springer.com

*To*
*Yoshi and Koko*
*SASAKI*

*Two views of Yoshi lecturing 4/79, JML*: Drawn by John M. Lewis, using a bamboo pen and India ink, when he attended Yoshi's lecture on variational data assimilation at University of Oklahoma in April 1979 (provided by Francois-Xavier Le Dimet).

# Preface

Data assimilation (DA) has been recognized as one of the core techniques for modern forecasting in various earth science disciplines including meteorology, oceanography, and hydrology. Since early 1990s DA has been an important session topic in many academic meetings organized by leading societies such as the American Meteorological Society, American Geophysical Union, European Geophysical Union, World Meteorological Organization, etc.

Recently, the 2$^{nd}$ Annual Meeting of the Asia Oceania Geosciences Society (AOGS), held in Singapore in June 2005, conducted a session on DA under the title of "*Data Assimilation for Atmospheric, Oceanic and Hydrologic Applications.*" This first DA session in the 2$^{nd}$ AOGS was a great success with more than 30 papers presented and many great ideas exchanged among scientists from the three different disciplines. The scientists who participated in the meeting suggested making the DA session a biennial event.

Two years later, at the 4$^{th}$ AOGS Annual Meeting, Bangkok, Thailand, the DA session was officially named "*Sasaki Symposium on Data Assimilation for Atmospheric, Oceanic and Hydrologic Applications,*" to honor Prof. Yoshi K. Sasaki of the University of Oklahoma for his life-long contributions to DA in geosciences. At the 5$^{th}$ AOGS Annual Meeting, Busan, Korea, in June 2008, two special events were hosted along with the Symposium – "*Special Lecture on Data Assimilation*" and "*Dinner with Yoshi.*" The special lecture was presented by Prof. Sasaki with a title of "Challenges in Data Assimilation." More than 50 scientists participated in the dinner and enjoyed talking with Yoshi.

Some papers of this volume are selected from the Symposium while others are invited. The first chapter of this book, titled "Sasaki's Pathway to Deterministic Data Assimilation," is contributed by John Lewis, one of Yoshi's students. I. Michael Navon provides a comprehensive review of data assimilation for numerical weather prediction. Milija Zupanski, another student of Yoshi, addresses theoretical and practical aspects of the ensemble data assimilation, especially for the maximum likelihood ensemble filter. François-Xavier Le Dimet, who worked with Yoshi as a postdoctoral scientist and served as a co-convener of the Symposium, reviews variational data assimilation in hydrology. Most notably, Yoshi himself has contributed a chapter, "Real Challenge of Data Assimilation for Tornadogenisis," which introduces a new theory based on the entropic balance.

Through prominent contributions by other scientists in DA, this book has a good mixture of collected papers in both theory and applications. Theoretical and methodological aspects span variational methods, ensemble Kalman filtering, particle filtering, the maximum likelihood ensemble filter, representer method, genetic algorithm, etc., with applications to parameter estimation, radar/satellite assimilation, data assimilation for land surface/water balance modeling, oceanic and meteorological data assimilation, adaptive (targeting) observations, and radar rainfall estimates. We hope this book will be useful to individual researchers and graduate students as a reference to the most recent research developments in the field of data assimilation. We appreciate Jeffrey Walker of the University of Melbourne, Boon Chua of Oregon State University, Lance Leslie of the University of Oklahoma, and Chun-Chieh Wu of National Taiwan University, who served as the co-convener of the Symposium previously. We are very honoured to dedicate this book to Yoshi and his lovely wife, Koko.

June 2008                                                                          Seon K. Park
                                                             *Ewha Womans University, Seoul*
                                                                                  Liang Xu
                                                      *Naval Research Laboratory, Monterey*

# Contents

Sasaki's Pathway to Deterministic Data Assimilation . . . . . . . . . . . . . . . . . 1
John M. Lewis

Data Assimilation for Numerical Weather Prediction: A Review . . . . . . . . 21
Ionel M. Navon

Theoretical and Practical Issues of Ensemble Data Assimilation
in Weather and Climate . . . . . . . . . . . . . . . . . . . . . . . . . . . . . . . . . . . . . . 67
Milija Zupanski

Information Measures in Ensemble Data Assimilation . . . . . . . . . . . . . . . . 85
Dusanka Zupanski

Real Challenge of Data Assimilation for Tornadogenesis . . . . . . . . . . . . . . 97
Yoshi K. Sasaki

Radar Rainfall Estimates for Hydrologic and Landslide Modeling . . . . . . 127
Kang-Tsung Chang, Jr-Chuan Huang, Shuh-Ji Kao and Shou-Hao Chiang

High-Resolution QPE System for Taiwan . . . . . . . . . . . . . . . . . . . . . . . . . . 147
Jian Zhang, Kenneth Howard, Pao-Liang Chang, Paul Tai-Kuang Chiu,
Chia-Rong Chen, Carrie Langston, Wenwu Xia, Brian Kaney
and Pin-Fang Lin

Assimilation of Satellite Data in Improving Numerical Simulation
of Tropical Cyclones: Progress, Challenge and Development . . . . . . . . . . . 163
Zhaoxia Pu

Diagnostics for Evaluating the Impact of Satellite Observations . . . . . . . . . 177
Nancy L. Baker and Rolf H. Langland

Impact of the CHAMP Occultation Data on the Rainfall Forecast . . . . . . . 197
Hiromu Seko, Yoshinori Shoji, Masaru Kunii and Yuichi Aoyama

**Parameter Estimation Using the Genetic Algorithm in Terms of Quantitative Precipitation Forecast** ............................... 219
Yong Hee Lee, Seon Ki Park, Dong-Eon Chang, Jong-Chul Ha and Hee-Sang Lee

**Applications of Conditional Nonlinear Optimal Perturbations to Ensemble Prediction and Adaptive Observation** .................... 231
Zhina Jiang, Hongli Wang, Feifan Zhou and Mu Mu

**Study on Adjoint-Based Targeted Observation of Mesoscale Low on Meiyu Front** .................................................. 253
Peiming Dong, Ke Zhong and Sixiong Zhao

**Ocean Data Assimilation: A Coastal Application** .................... 269
Xiaodong Hong, James A. Cummings, Paul J. Martin and James D. Doyle

**Comparison of Ensemble-Based Filters for a Simple Model of Ocean Thermohaline Circulation** ....................................... 293
Sangil Kim

**Preconditioning Representer-based Variational Data Assimilation Systems: Application to NAVDAS-AR** ........................... 307
Boon S. Chua, Liang Xu, Tom Rosmond and Edward D. Zaron

**Cycling the Representer Method with Nonlinear Models** .............. 321
Hans E. Ngodock, Scott R. Smith and Gregg A. Jacobs

**Implementation of the Ensemble Kalman Filter into a Northwest Pacific Ocean Circulation Model** ........................................ 341
Gwang-Ho Seo, Sangil Kim, Byoung-Ju Choi, Yang-Ki Cho and Young-Ho Kim

**Particle Filtering in Data Assimilation and Its Application to Estimation of Boundary Condition of Tsunami Simulation Model** ................ 353
Kazuyuki Nakamura, Naoki Hirose, Byung Ho Choi and Tomoyuki Higuchi

**Data Assimilation in Hydrology: Variational Approach** ................ 367
François-Xavier Le Dimet, William Castaings, Pierre Ngnepieba and Baxter Vieux

**Recent Advances in Land Data Assimilation at the NASA Global Modeling and Assimilation Office** ................................... 407
Rolf H. Reichle, Michael G. Bosilovich, Wade T. Crow, Randal D. Koster, Sujay V. Kumar, Sarith P. P. Mahanama and Benjamin F. Zaitchik

**Assimilation of Soil Moisture and Temperature Data into Land Surface Models: A Survey** .............................................. 429
Nasim Alavi, Jon S. Warland and Aaron A. Berg

**Assimilation of a Satellite-Based Soil Moisture Product into a Two-Layer
Water Balance Model for a Global Crop Production Decision Support
System** . . . . . . . . . . . . . . . . . . . . . . . . . . . . . . . . . . . . . . . . . . . . . . . . . . . . . . . 449
John D. Bolten, Wade T. Crow, Xiwu Zhan, Curt A. Reynolds
and Thomas J. Jackson

**Index** . . . . . . . . . . . . . . . . . . . . . . . . . . . . . . . . . . . . . . . . . . . . . . . . . . . . . 465

# List of Contributors

**Nasim Alavi** Land Resource Science, University of Guelph, Guelph, ON, Canada, N1G 2W1, nalavi@uoguelph.ca

**Yuichi Aoyama** National Institute of Polar Research, 1-9-10 Kaga, Itabashi, Tokyo 173-8515, Japan

**Nancy L. Baker** Marine Meteorology Division, Naval Research Laboratory, Monterey, CA 93943-5502, USA, nancy.baker@nrlmry.navy.mil

**Aaron A. Berg** Department of Geography, University of Guelph, Guelph, ON, Canada, N1G 2W1, aberg@uoguelph.ca

**John D. Bolten** Hydrology and Remotes Sensing Lab, U.S. Department of Agriculture, Agricultural Research Service, Beltsville, MD 20705 USA, John.Bolten@ars.usda.gov

**Michael G. Bosilovich** Global Modeling and Assimilation Office, NASA Goddard Space Flight Center, Greenbelt, MD, USA, Michael.G.Bosilovich@nasa.gov

**Williams Castaings** Joint Research Center, Ispara, Italy, william.castaings@jrc.it

**Dong-Eon Chang** Numerical Model Development Division, Korea Meteorological Administration, Seoul 156-720, Republic of Korea, dechang@kma.go.kr

**Kang-Tsung Chang** Kainan University, Taoyuan, Taiwan, chang@uidaho.edu

**Pao-Liang Chang** Meteologogical Satellite Center, Central Weather Bureau, Taiwan

**Chia-Rong Chen** Meteologogical Satellite Center, Central Weather Bureau, Taiwan

**Shou-Hao Chiang** Department of Geography, National Taiwan University, Taipei, Taiwan

**Paul Tai-Kuang Chiu** Meteologogical Satellite Center, Central Weather Bureau, Taiwan

**Yang-Ki Cho** Department of Oceanography, Chonnam National University, Yong bong-dong, Buk-gu, Gwangju, Korea, 500-757, ykcho@chonnam.ac.kr

**Byoung-Ju Choi** Department of Oceanography, Kunsan National University, Miryong-dong, Gunsan, Korea, 573-701, bjchoi@kunsan.ac.kr

**Byung Ho Choi** Department of Civil and Environmental Engineering, Sungkyunkwan University, Suwon, Gyeonggi 440-746, Korea, bhchoi@skku.ac.kr

**Boon S. Chua** SAIC, Monterey, CA, USA; Marine Meteorology Division, Naval Research Laboratory, Monterey, CA, USA

**Wade T. Crow** Hydrology and Remotes Sensing Lab U.S. Department of Agriculture, Agricultural Research Service, Beltsville, MD 20705 USA, Wade.Crow@ars.usda.gov

**James A. Cummings** Oceanography Division, Naval Research Laboratory, Stennis Space Center, MS 39529, USA

**Peiming Dong** Institute of Atmospheric Physics, Chinese Academy of Sciences, Beijing, 100029, China, dongpm@cams.cma.gov.cn

**James D. Doyle** Marine Meteorology Division, Naval Research Laboratory, Monterey, CA 93943, USA

**Jong-Chul Ha** Forecast Research Laboratory, National Institute of Meteorological Research, Korea Meteorological Administration, Seoul 156-720, Republic of Korea, bellfe@metri.re.kr

**Tomoyuki Higuchi** The Institute of Statistical Mathematics, Minato, Tokyo, 106-8569, Japan; Japan Science and Technology Agency, Japan, higuchi@ism.ac.jp

**Naoki Hirose** Research Institute for Applied Mechanics, Kyushu University, Kasuga, Fukuoka 816-8580, Japan, hirose@riam.kyushu.u.ac.jp

**Xiaodong Hong** Marine Meteorology Division, Naval Research Laboratory, Monterey, CA 93943, USA

**Kenneth Howard** NOAA/OAR National Severe Storms Laboratory, Norman, OK, USA

**Jr-Chuan Huang** Research Center for Environmental Changes, Academia Sinica, Taipei, Taiwan

**Thomas J. Jackson** Hydrology and Remotes Sensing Lab, U.S. Department of Agriculture, Agricultural Research Service, Beltsville, MD 20705 USA, Tom.Jackson@ars.usda.gov

**Gregg A. Jacobs** The Naval Research Laboratory, Stennis Space Center, MS 39529, USA, Gregg.Jacobs@nrlssc.navy.mil

**Zhina Jiang** State Key Laboratory of Severe Weather (LaSW), Chinese Academy of Meteorological Sciences, Beijing 100081, China, jzn@cams.cma.gov.cn

**Brian Kaney** Northeastern State University, Tahlequah, OK, USA

**Shuh-Ji Kao** Research Center for Environmental Changes, Academia Sinica, Taipei, Taiwan

**Sangil Kim** The College of Oceanic and Atmospheric Sciences, Oregon State University, Corvallis, OR 97331-5503, USA, skim@coas.oregonstate.edu

**Young-H. Kim** Korea Ocean Research & Development Institute, Ansan, Korea, 426-744, yhkim@kordi.re.kr

**Randal D. Koster** Global Modeling and Assimilation Office, NASA Goddard Space Flight Center, Greenbelt, MD, USA, Randal.D.Koster@nasa.gov

**Sujay V. Kumar** Science Applications International Corporation, Beltsville, Maryland, USA; Hydrological Sciences Branch, NASA Goddard Space Flight Center, Greenbelt, MD, USA, Sujay.V.Kumar@nasa.gov

**Masaru Kunii** Meteorological Research Institute, 1-1 Nagamine, Tsukuba, Ibaraki 305-0052, Japan

**Rolf H. Langland** Marine Meteorology Division, Naval Research Laboratory, Monterey, CA 93943-5503, USA, rolf.langland@nrlmry.navy.mil

**Carrie Langston** NOAA/OAR National Severe Storms Laboratory, Cooperative Institute for Mesoscale Meteorological Studies, The University of Oklahoma, Norman, OK, USA

**François-Xavier Le Dimet** Laboratoire Jean-Kuntzmann, Université de Grenoble and INRIA, Grenoble, France, Francois-Xavier.Le-Dimet@imag.fr

**Hee-Sang Lee** Forecast Research Laboratory, National Institute of Meteorological Research, Korea Meteorological Administration, Seoul 156-720, Republic of Korea, heesanglee@kma.go.kr

**Yong Hee Lee** Forecast Research Laboratory, National Institute of Meteorological Research, Korea Meteorological Administration, Seoul 156-720, Republic of Korea, yhlee@metri.re.kr

**John M. Lewis** National Severe Storms Laboratory and Desert Research Institute, 2215 Raggio Prky, Reno, NV 89512-1095, USA, jlewis@dri.edu

**Pin-Fang Lin** Meteologogical Satellite Center, Central Weather Bureau, Taiwan

**Sarith P.P. Mahanama** Global Modeling and Assimilation Office, NASA Goddard Space Flight Center, Greenbelt, MD, USA; Goddard Earth Sciences and Technology Center, University of Maryland, Baltimore County, Baltimore, MD, USA, Sarith.P.Mahanama@nasa.gov

**Paul J. Martin** Oceanography Division, Naval Research Laboratory, Stennis Space Center, MS 39529, USA

**Mu Mu** State Key Laboratory of Numerical Modeling for Atmospheric Sciences and Geophysical Fluid Dynamics (LASG), Institute of Atmospheric Physics, Chinese Academy of Sciences, Beijing 100029, China, mumu@lasg.iap.ac.cn

**Kazuyuki Nakamura** The Institute of Statistical Mathematics, Minato, Tokyo 106-8569, Japan, nakakazu@ism.ac.jp

**Ionel M. Navon** Department of Scientific Computing, The Florida State University, Tallahassee, Florida 32306-4120, USA, inavon@fsu.edu

**Pierre Ngnepieba** Department of Mathematics, Florida A&M University, Tallahassee, Florida 32307, USA, Pierre.Ngnepieba@famu.edu

**Hans E. Ngodock** The Naval Research Laboratory, Stennis Space Center, MS 39529, USA, Hans.Ngodock@nrlssc.navy.mil

**Seon Ki Park** Department of Environmental Science and Engineering, Ewha Womans University, Seoul 120-750, Republic of Korea, spark@ewha.ac.kr

**Zhaoxia Pu** Department of Meteorology, University of Utah, 135 S 1460 E, Rm. 819, Salt Lake City, UT, USA, Zhaoxia.Pu@utah.edu

**Rolf H. Reichle** Global Modeling and Assimilation Office, NASA Goddard Space Flight Center, Greenbelt, MD, USA, rolf.reichle@nasa.gov

**Curt A. Reynolds** International Production Assessment Division, U.S. Department of Agriculture, Foreign Agricultural Service, Office of Global Analysis, Washington, DC 20250 USA, Curt.Reynolds@fas.usda.gov

**Tom Rosmond** SAIC, Monterey, CA, USA; Marine Meteorology Division, Naval Research Laboratory, Monterey, CA, USA

**Yoshi K. Sasaki** School of Meteorology, National Weather Center, The University of Oklahoma, Norman, OK 73072-7307, USA, yks@ou.edu

**Hiromu Seko** Meteorological Research Institute, 1-1 Nagamine, Tsukuba, Ibaraki 305-0052, Japan

**Gwang-Ho Seo** Department of Oceanography, Chonnam National University, Yong bong-dong, Buk-gu, Gwangju, Korea, 500-757, bbinzil1324@hanmail.net

**Yoshinori Shoji** Meteorological Research Institute, 1-1 Nagamine, Tsukuba, Ibaraki 305-0052, Japan

**Scott R. Smith** The Naval Research Laboratory, Stennis Space Center, MS 39529, USA, Scott.Smith@nrlssc.navy.mil

**Baxter Vieux** University of Oklahoma, Norman, OK, USA, bvieux@ou.edu

**Hongli Wang** State Key Laboratory of Numerical Modeling for Atmospheric Sciences and Geophysical Fluid Dynamics (LASG), Institute of Atmospheric Physics, Chinese Academy of Sciences, Beijing 100029, China, hungli.wong@gmail.com

**Jon S. Warland** Land Resource Science, University of Guelph, Guelph, ON, Canada, N1G 2W1, jwarland@uoguelph.ca

**Wenwu Xia** NOAA/OAR National Severe Storms Laboratory, Cooperative Institute for Mesoscale Meteorological Studies, The University of Oklahoma, Norman, OK, USA

**Liang Xu** Marine Meteorology Division, Naval Research Laboratory, Monterey, CA, USA

**Benjamin F. Zaitchik** Hydrological Sciences Branch, NASA Goddard Space Flight Center, Greenbelt, MD, USA, Department of Earth and Planetary Sciences, Johns Hopkins University, Baltimore, MD, USA, zaitchik@jhu.edu

**Edward D. Zaron** Department of Civil and Environmental Engineering, Portland State University, Portland, OR, USA

**Xiwu Zhan** Center for Satellite Applications and Research, National Oceanic and Atmospheric Administration, National Environmental Satellite, Data, and Information Service, Camp Springs, MD 20746 USA, Xiwu.Zhan@noaa.gov

**Jian Zhang** NOAA/OAR National Severe Storms Laboratory and Cooperative Institute for Mesoscale Meteorological Studies, The University of Oklahoma, Norman, OK 73072, USA, jian.zhang@noaa.gov

**Sixiong Zhao** Institute of Atmospheric Physics, Chinese Academy of Sciences, Beijing, 100029, China, zhaosx@mail.iap.ac.cn

**Ke Zhong** Institute of Atmospheric Physics, Chinese Academy of Sciences, Beijing, 100029, China, zhongke5805@163.com

**Feifan Zhou** State Key Laboratory of Numerical Modeling for Atmospheric Sciences and Geophysical Fluid Dynamics (LASG), Institute of Atmospheric Physics, Chinese Academy of Sciences, Beijing 100029, China, zhouff04@163.com

**Dusanka Zupanski** Cooperative Institute for Research in the Atmosphere, Colorado State University, 1375 Campus Delivery, Fort Collins, CO 80523-1375, USA, Zupanski@cira.colostate.edu

**Milija Zupanski** Cooperative Institute for Research in the Atmosphere, Colorado State University, Fort Collins, CO 80523-1375, USA, ZupanskiM@cira.colostate.edu

# Sasaki's Pathway to Deterministic Data Assimilation

John M. Lewis

**Abstract** Yoshikazu Sasaki developed the variational method of data assimilation, a cornerstone of modern-day analysis and prediction in meteorology. The generation of this idea is tracked by analyzing his education at the University of Tokyo in the immediate post-WWII period. Despite austere circumstances—including limited financial support for education, poor living conditions, and a lack of educational resources—Sasaki was highly motivated and overcame these obstacles on his path to developing this innovative method of weather map analysis. We follow the stages of his intellectual development where information comes from access to his early publications, oral histories, letters of reminiscence, and biographical data from the University of Tokyo and the University of Oklahoma. Based on this information, key steps in the development of his idea were: (1) a passion for science in his youth, (2) an intellectually stimulating undergraduate education in physics, mathematics, and geophysics, (3) a fascination with the theory of variational mechanics, and (4) a "bridge to America" and the exciting new developments in numerical weather prediction (NWP).

A comparison is made between Sasaki's method and Optimal Interpolation (OI), a contemporary data assimilation strategy based on the work of Eliassen and Gandin. Finally, a biographical sketch of Sasaki including his scientific genealogy is found in the Appendix.

## 1 View from Afar

In April 1947, Yoshikazu ("Yoshi") Sasaki enrolled as an undergraduate in geophysics at the University of Tokyo.[1] Among every group of 1000 students who passed through the rigorous gymnasium education in Japan, only one gained entry into the hallowed institution (Schoppa 1991).

---

J.M. Lewis (✉)
National Severe Storms Laboratory and Desert Research Institute 2215 Raggio Prky, Reno, NV 89512-1095, USA, e-mail: jlewis@dri.edu

[1] Precise dates related to Sasaki's life and career are found in the résumé (Table 1).

S.K. Park, L. Xu, *Data Assimilation for Atmospheric, Oceanic and Hydrologic Applications*, DOI 10.1007/978-3-540-71056-1_1,
© Springer-Verlag Berlin Heidelberg 2009

**Table 1** Résumé of Yoshikazu Sasaki

| | |
|---|---|
| January 2, 1927: | Born in Akita City, Akita Prefecture<br>Parents: Kosuke Sasaki and Itsu (Kosaka) Sasaki<br>Siblings: Toshifumi and Masahiro |
| April 1947: | Entered University of Tokyo (U. Tokyo) |
| March 1950: | Bachelors Degree (Geophysics), U. Tokyo |
| March 1955: | D. Sc., U. Tokyo (Meteorology) |
| September 1956: | Emigration to USA |
| 1956–1960: | Research Scientist, Texas A & M College |
| 1960–present: | University of Oklahoma (OU)<br>    1960 – Research Scientist/Adjunct Associate Professor<br>    1965 – Associate Professor<br>    1967 – Professor (Tenured)<br>    1974 – George Lynn Cross Research Professor<br>    1994 – George Lynn Cross Research Professor Emeritus |
| 1973–1974: | Research Director<br>Naval Environmental Prediction Research Facility (NEPRF)<br>Monterey, CA (Sabbatical Leave) |
| 1980–1986: | Director<br>Cooperative Institute for Mesoscale Meteorological Studies<br>OU and National Oceanic and Atmospheric Administration |
| Awards: | Fujiwara Award (May 2000)<br>Order of Sacred Treasure (May 2004)<br>Oklahoma Hall of Fame for Higher Education (October 2004) |

At the time of naturalization (U. S. citizenship) in 1973, Yoshikazu Sasaki changed his first name to Yoshi Kazu.

At this time, less than two years after V-J day (Victory Over Japan; September 2, 1945), the country lay in ruin and the American Occupation had commenced. The occupation would last for nearly seven years. Conditions for the students were not unlike those for the general population—an absence of adequate housing and a limited supply of food. As recalled by Yoshimitzu Ogura, a fellow student at University of Tokyo:

> First of all, I could not find a place to live, since so many houses were burned down. I brought my futon to the lab, and cleaned up the top of a big table every night to spread my futon and slept there. There were half a dozen homeless fellow young scientists in the Geophysical Institute. And I was hungry all the time. With an empty stomach, you cannot think of science well.

> (Y. Ogura 1992, personal communication)

Since geophysics was a sub-specialty within the physics department, Sasaki's first three years of instruction followed the pattern for the typical physics student. During the fourth year, the geophysics student was given courses in seismology, geomagnetism, oceanography, and meteorology (usually one lecture hour per week). Sasaki especially enjoyed mathematics and has fond memories of the course in group theory, that important mathematical subject that is pervasive throughout physics, chemistry, and applied mathematics. Kunihiko Kodaira, a young assistant professor taught the course. As recalled by Yoshi:

> Professor Kodaira often came to class late and sometimes the students would leave [before he arrived][2] you know. It was kind of strange to have so few students. But it was just amazing how great a mathematician he was. I especially remember the first day of class when he came in late, bowed and then went to the blackboard and filled it with equations. It was amazing.
>
> (Y. Sasaki 1990, personal communication)

Several years after Sasaki took this course, Kodaira received the Fields Medal.[3]

After obtaining his bachelors degree in 1950, Sasaki joined the group of graduate students at the university's Geophysical Institute. There was little financial support for these students and Sasaki earned money by tutoring high school/gymnasium students as they prepared for the demanding university entrance exams. Classmate Kikuro Miyakoda taught physics and geology-geophysics part-time at two high schools, but also tried to make money by dabbling in the stock market. Akio Arakawa and Katsuyuki Ooyama served as upper-air observers aboard two Japan Meteorological Agency (JMA) ships off the east coast of Japan. The Japanese government could not cut funding for these two destroyer-sized ships, the "X-RAY" and "TANGO", since the observations were needed by the occupational forces (K. Ooyama 1992, personal communication). These jobs offered steady employment, but the work assignments were demanding, especially in wintertime when rough seas were the rule.

In the immediate post-war period, the meteorological community outside Japan was making notable advances. Upper-air observations were being routinely collected over Europe and the United States, and this inspired research on the jet stream and hemispheric circulation. And, of course, a momentous change in the scientific environment was produced by the computer—labeled the "great mutation" by Edward Lorenz (1996).

The work that most stimulated students at the Geophysical Institute was Jule Charney's paper on baroclinic instability (Charney 1947)—the dynamic underpinning for mid-latitude cyclone development. Although published in 1947, the students and chair professor of meteorology Shigekata Syono only became aware of it two years later.[4] In those early years of occupation, a two-year delay in the transmission of information from the West was typical. Journals, if available at all, were housed at the Hibiya Library in the heart of Tokyo near General MacArthur's Headquarters. And even if the journal could be located, obtaining photocopies of an article was a nontrivial task.

---

[2] Information in brackets, [...], has been inserted by the author.

[3] The Fields Medal, the so-called "Nobel Prize" for mathematicians, was awarded to Kodaira in 1954 for his major contribution to the theory of harmonic integrals. It was the first Fields Medal given to a mathematician in Japan.

[4] A scientific biography of Syono is found in Lewis (1993a).

## 2 Connection to America

Inspired by Charney's work, Kanzaburo Gambo, a research associate at the University of Tokyo, began his own study of baroclinic disturbances and established modified criterion for the development of cyclones (Gambo 1950). This note was sent to Charney at Princeton's Institute for Advanced Study. Gambo received a swift and encouraging reply with an offer to join the celebrated group of researchers at Princeton. These researchers had recently made two 24-h numerical weather predictions of transient features of the large-scale flow over the United States (Charney et al. 1950; Platzman 1979; Nebeker 1995).

During his stay at Princeton from October 1952 to January 1954, Gambo frequently sent detailed reports to the staff and students at the Geophysical Institute. These reports stimulated the formation of a numerical weather prediction (NWP) group under the leadership of Syono (and Gambo after he returned to Japan in 1954). There were approximately 25 members of this group that included JMA employees as well as the graduate students (See Fig. 1). They were, of course, without the benefit of computational power. In the absence of high-speed computation, the motivated students made use of desk calculators; but more often, they performed the integration of the governing equations via graphical methods. Science historian Frederik Nebeker has explored the history of these graphical methods in a comprehensive manner. We quote from Nebeker (1995, p. 167):

**Fig. 1** Members of the Numerical Weather Prediction Group in Tokyo, Japan (ca. 1955). Listed from left to right: *Front Row:* Akio Arakawa (1), Kanzaburo Gambo (3), Akira Kasahara (7) *Back Row:* Syukuro Manabe (1), Katsuyuki Ooyama (4), Kikuro Miyakoda (11), Yoshio Kurahara (14), and Takio Murakami (17). Sasaki was absent when this picture was taken (Courtesy of A. Kasahara)

**Fig. 2** This photograph shows faculty, students, and staff of the Geophysical Institute, University of Tokyo in the early 1950s. Yoshi Sasaki is crouched (second from the right), while Kanzaburo Gambo and Professor Syono are standing directly behind him (Gambo to the left and Syono to the right) (Courtesy of A. Kasahara)

> One of the most important graphical techniques of the 1950s was the method devised by Ragnar Fjørtoft (1952) for integrating the barotropic vorticity equation. As described in Chapter 8 [of Nebeker (1995)], Fjørtoft was working in the Norwegian tradition of the graphical calculus, and his 1952 paper may be said to have fulfilled Bjerknes's program of calculating the weather graphically.

The students who were at the Geophysical Institute in the early 1950s view Gambo as a scientific hero. His altruistic efforts to build a bridge to the West are always at the forefront of their thoughts as they retrospectively examine their own careers (Lewis 1993b). A photo of Gambo, Syono, and Sasaki is found in Fig. 2.


# 3 Track Prediction of Typhoons

Among the first contributions that came from the NWP group in Tokyo was a paper by Sasaki and Miyakoda on the track forecasting of typhoons (Sasaki and Miyakoda 1954). It is the first known publication on numerical prediction of ty-phoon/hurricane tracks.[5] Twenty-four hour track predictions were made by par-titioning the geopotential field at 700 mb into a steering current (synoptic-scale

---

[5] A historical review of hurricane track forecasting in the Atlantic Basin is found in DeMaria (1996).

flow component) and a smaller-scale circular vortex to fit the geopotential in the immediate vicinity of the typhoon. The structure of the circular vortex followed Ted Fujita's empirical formula (Fujita 1952).[6] By extracting the large-scale component from the analyzed geopotential and combining this with the 24-h prediction of the geopotential tendency (found from the graphical solution to the barotropic vorticity equation), the typhoon's motion could be determined.

Sasaki and Miyakoda presented their paper at the UNESCO Symposium on Typhoons—held in Tokyo on November 9–12, 1954. Miyakoda has a vivid memory of events surrounding this presentation. As he recalled:

> Mr. Sasaki was very creative, and original thinker. We worked together in the university on developing a method of typhoon tracks and presented our work on the international meeting in Tokyo. In that meeting, the scientists attending from US were H. [Herbert] Riehl, [Father Charles] Deppermann [,S. J.], Bob Simpson, [Leon] Sherman, possibly J. [Jerome] Namias. This is the first international meeting after the Pacific War. This method had been used in Japan Meteorological Agency for more than 10 years since then. Much later, Dr. Y. [Yoshio] Kurihara developed a hurricane forecasting system in GFDL [Geophysical Fluid Dynamics Laboratory]. In this method, he first eliminates the typhoon and replaces it with a clean circular systematic hurricane. In other words, the GFDL method of hurricane forecasts has something in common in this basic idea.
>
> (Miyakoda 2007, personal communication)

Beyond its influence on operations at JMA, the work of Sasaki and Miyakoda stimulated research in the United States—first at the University of Chicago and later at the U. S. Weather Bureau (USWB). Akira Kasahara, then a postdoctoral fellow at University of Chicago, recalls this work:

> When I joined the Platzman's research project at University of Chicago in 1956, I decided to extend the steering flow method of Sasaki and Miyakoda (1954) to suit for the numerical prediction of hurricanes using the electronic computer at the Joint Numerical Weather Prediction Unit in Suitland, MD [See Kasahara 1957]. Working with George Platzman, Gene Birchfield, and Robert Jones, we developed a series of numerical prediction models of hurricane tracks. Lester Hubert of the U. S. Weather Bureau took one of our prediction schemes, tested [it] operationally and published an article (Hubert 1959).
>
> (A. Kasahara 2007, personal communication)

## 4 Sasaki's Dissertation

Syono did not entrain students into his current interest as was the case with his contemporary C.-G. Rossby. George Platzman, a protégé of Rossby's, remembers his mentor as follows:

> Rossby had a way of engaging you on an intellectual basis, to draw you out and to co-opt you into his current interest, not by using you but by transferring his enthusiasm to you.
>
> (Platzman 1990, personal communication)

Syono expected the student to identify and find his own natural interest. He did not discourage the students from working on problems that he had investigated,

---

[6] At that time, Fujita was a professor of physics at Kyushu Institute of Technology.

most notably those related to typhoons, but such problems were deemed no more appropriate than a problem independently formulated by the student.

As an undergraduate in physics, Sasaki was captivated by the subject of variational mechanics, that post-Newtonian branch of analytical mechanics developed by Leonhard Euler, Joseph Louis Lagrange, and William Rowan Hamilton. Historically, this subject has been found so esthetically pleasing and utilitarian that words like "poetic", "Shakespearian", and "conspicuously practical" are used to describe it (Bell 1937; Lanczos 1970).[7] Sasaki had seen its application in quantum mechanics and the theory of relativity, but as he said, "I was fascinated to find whether or not there exists variational principle for meteorological phenomenon" (Sasaki 2007, personal communication). This research topic was far afield from other efforts at the Geophysical Institute, but "Professor Syono encouraged my "lone-wolf" approach and asked me to make informal presentations to him" (Sasaki 2007, personal communication). As elaborated upon by Sasaki:

> There was minimal literature on the use of variational methods in fluids—some in Horace Lamb's book on hydrodynamics [Lamb 1932] and some applications to irrotational fluids in Bateman's book [Harry Bateman, Caltech aerodynamicist; Bateman (1932)]. I had been working with Miyakoda on typhoon tracking and wanted to apply these variational ideas to a typhoon that would be idealized as a vortex. The Clebsch transformation [Clebsch 1857] is the key to this problem, and *Syōno* wanted me to discuss the physical basis for the transformation. I learned a lot from these discussions, and I made good progress. The results from the work formed the major part of my dissertation research, which was published in 1955 [Sasaki 1955].
>
> (Sasaki 1990, personal communication)

Sasaki's development of these ideas for geophysical fluid dynamics was a valuable extension of the work accomplished by Bateman (1932, Sect. 2.52)—that is, Bateman's case of irrotational flow in fixed coordinates was expanded to include the generalized hydro-thermodynamic equations on a rotating earth with application to the problem of tracking the vorticity associated with a typhoon. Again, in the absence of digital computers, Sasaki integrated the governing equations via graphical methods. Several years later, without knowledge of Sasaki's work, the noted Scripps physicist/oceanographer Carl Eckart developed similar equations for the ocean-atmosphere system (Eckart 1960a, b). Eckart and Sasaki never exchanged thoughts on these nearly equivalent developments (Sasaki 2007, personal communication).

## 5 Variational Analysis: Sasaki (1958)[8]

In March 1950, the team at Princeton succeeded in producing the two 24-h numerical forecasts mentioned earlier. Forecasts were made using the quasi-geostrophic vorticity equation at 500 mb [See Platzman (1979) for a first-hand account of these

---

[7] In Eric Temple Bell's history, these superlative forms are found in his discussion of Lagrange's work. This lofty language is found in Chap. 10 of Cornelius Lanczos's book.

[8] Although this key paper was published in 1958, two years after Sasaki emigrated to the USA, the development of this idea was generated in Japan (Sasaki 2007, personal communication).

events and Nebeker (1995) for a retrospective account]. This feat set the meteorological world abuzz; yet there were skeptics—those who expressed doubt that NWP would replace operational subjective forecasting. Nevertheless, by March 1954, a team of meteorologists in Sweden produced the first operational NWP forecast for Europe and Scandinavia with the same dynamical equations used by the Princeton group (Wiin-Nielsen 1991). A little more than a year later, on 15 May 1955, operational NWP began in the USA, this time accounting for baroclinic structure by using a 3-level quasi-geostrophic model [reviewed in Thompson (1983) and Nebeker (1995)].

Since upper-air observations worldwide are simultaneously collected every 12-h, the operational NWP models are initialized at these times. In the case of filtered models such as the quasi-geostrophic model, the initial conditions consist of analyses of the geopotential height on a network of grid points at one or more pressure levels. Time constraints at the two major centers dictated development of computer-generated initial conditions for the expansive areas of North America on the one hand and Europe on the other [respectively, Gilchrist and Cressman (1955)/Cressman (1959) and Bergthorsson and Döös (1955)]. These computer-produced analyses were labeled "objective analyses" or "numerical weather map analyses".

Into this milieu came Sasaki—fresh from his dissertation based on variational mechanics. He appreciated the need for efficient computer-generated analyses in the operational environment, yet from a more-dynamically based perspective, he realized that there was no guarantee that the operationally produced analyses would satisfy the zeroeth-order constraint for the quasi-geostrophic model—i.e., the geostrophic wind law. Generally there would be a discrepancy between the analyzed horizontal gradient of geopotential and the horizontal gradient derived from the geostrophic wind law and the observed wind. How to use both the wind and geopotential observations in accord with their accuracy to guarantee a dynamically consistent set of analyses was at the heart of his thought and contribution (Sasaki 1958).

From his dissertation work, he knew that the governing equations for atmospheric motion stemmed from minimizing the action integral, the time integrated difference between the kinetic and potential energy of the system (the Lagrangian). Now, instead of minimizing the Lagrangian, he could minimize the squared discrepancies between the desired fields (the final analyses of wind and geopotential) and the observations. Instead of integrating over time, integration over space was appropriate in this case. The approach contained the rudiments of several of the operational methods based on least squares [e.g., Panofsky (1949) and Gilchrist and Cressman (1955)], yet it was more general—allowing for differential weighting of observations while constraining the final analyzed fields to satisfy the constraint. He was most familiar with the mechanics of minimization under constraint. For this problem, governing analysis equations reduced to: (1) the solution of an elliptic equation in the geopotential forced by the vorticity (from the observed winds) and observed geopotential, and (2) the direct substitution of this geopotential into the geostrophic wind relations. Again, in the absence of computing machinery, Sasaki solved the equations via graphical methods.

Sasaki's approach seems so straightforward in retrospect, yet the magnitude of the contribution is ever so impressive in light of the fact that such methodology had never before been used in continuum mechanics. Most fundamentally, Sasaki's method is tied to Gauss's monumental work of 1801 when Gauss developed the method of least squares under constraint. Gauss's work was driven by his desire to forecast the time and reappearance of the planetoid Ceres that came into conjunction with the sun after only 42 days of observation. He needed to find the initial conditions, the celestial position and velocity of the planetoid at some point during the 42 day period of observation in order to predict its place and time of reappearance. Gauss's work, however, was based on the gravitational mechanics of discrete objects: the planetoid, the earth, and the sun. Sasaki was unaware of Gauss's work on this subject (Gauss 1809, 1857) (Sasaki 2007, personal communication).

There was a strength and beauty in Sasaki's 1958 contribution. He demonstrated its versatility not only with the geostrophic constaint, but also with the more-general wind law called the "balance" equation, an equation that couples wind and geopotential for the larger-scale circulation of the atmosphere and is applicable in the tropics as well as in mid-latitudes (see Thompson 1980). Despite the solid theoretical foundation upon which the variational method rested, the computational demands in the presence of time constraints at the operational centers made it untenable for daily production of numerical map analyses in the 1950s–1960s.[9]

With the hope and promise of longer range weather prediction (beyond several days) that stemmed from Phillips's successful numerical experiment linked to hemispheric circulation (Phillips 1956), advances in general circulation modeling using more-realistic models, improved high-speed computation [e.g., Smagorinsky (1963)], and with the politico-scientific thrust that was aimed at acquisition of global observations [addressed in Smagorinsky (1978)], the environment for application of Sasaki's method was improving by the late-1960s.

In this environment, Sasaki rekindled his interest in objective analysis and expanded his view to encompass the temporal distribution of observations—in effect, he developed a four dimensional view of data assimilation that has come to be called 4DVAR (4-dimensional variational analysis). His expansive approach to variational data assimilation is found in a trilogy of papers published in the December 1970 issue of *Monthly Weather Review* (Sasaki 1970a, b, c). A photo of Sasaki shortly after completing this work is shown in Fig. 3.

# 6 Pathways to Operations

Sasaki's 1958 contribution augmented by his 1970 papers established a solid dynamical framework for data assimilation. With increased computational power that led to the inclusion of more physical processes in the models, there were

---

[9] In the late 1960s, Odd Haug of Norwegian Weather Service made an effort to analyze surface wind and pressure via Sasaki's method. The author remembers reading the report, but despite diligent efforts by the Norwegian Meteorological Institute's librarian, Greta Krogvold, the report cannot be located.

**Fig. 3** With the National Severe Storms Laboratory in the background, Sasaki is shown standing on the site of Meteorology Department's first home on the North Campus of the University of Oklahoma (1970) (Courtesy of Y. Sasaki)

attendant demands on the data assimilation—in essence, there was a need to improve the algorithms that searched for the optimal atmospheric state [See LeDimet and Talagrand (1986) for a review of these algorithms]. With these improvements, Sasaki's variational approach became operational at the U. S. Navy's Fleet Numerical Weather Center (FNWC) in 1971 [labeled the Global Band Analysis, covering the globe from 40S to 60N] (Lewis 1972). The generalized nonlinear thermal wind equation, applicable over the tropics and mid-latitudes, was used as a weak constraint. It remained operational until the mid-1990s.

Another form of variational analysis became operational at the National Meteorological Center (NMC) in the mid-1970s (Bergman 1975; McPherson 1975; and a review by Schlatter et al. 1976). This was labeled Optimum Interpolation (OI) and stemmed from the pioneering work of Arnt Eliassen (1954) and Lev Gandin (1965). Eliassen appears to be the originator of this idea in meteorology, but in his oral history, he gives Gandin credit for developing the idea further (Eliassen 1990). They were both members of the World Meteorological Organization's (WMO) committee on network design in the late-1950s and that is where the discussion of this methodology commenced.

The common ground for OI and Sasaki's method rests on fundamental least squares problem formulation. The primary difference is that Sasaki's approach is deterministic whereas OI is stochastic. Determinism's philosophy assumes that the future state of the system is completely determined by the present state. The evolution of the state is governed by causal relationships such as the Newtonian equations of motion. The stochastic philosophy rests on the assumption that there are intrinsic uncertainties in the governing dynamics.[10]

It is safe to say that both of these variational approaches play a prominent role in the practice of data assimilation at the operational centers. A stimulating and enlightening discussion of current practice at these centers is found in Rabier and Liu (2003). The relative merits of the these strategies including factors that favor one method over the other are explored in Daley (1991), Tseng (2006), and Lewis et al. (2006).

# 7 Epilogue

Data assimilation, especially as viewed in the context of initializing dynamical prediction models, continues to present challenging problems. Compared to the dynamical models of the 1950s–1960s, modern-day models exhibit a level or degree of complexity that rivals L. F. Richardson's phantasmagoric view of the atmosphere and its governing laws (Richardson 1922)[11]. Matching the ever-complex model output with measurements from conventional and new observational tools demands excellence in theory and practice. Rossby realized that it would take a team with expertise in both arenas to produce an objective weather map, a map untouched by human hand. Pall Bergthórsson and Bo Döös were his choices, the practitioner and theorist, respectively, and they delivered in splendid fashion. Nevertheless, "purists" like Bergthórsson's mentor, the celebrated analyst Tor Bergeron, found it difficult to accept the objective maps. Bergeron's maps were not only artistic, they contained information that tacitly drew upon his years of experience as an observer and analyst (Saucier 1992, personal communication). Bergeron came to accept the NWP product, but he never accepted the objective map that provided the initial condition for the prediction model (Döös 2004, personal communication). To incorporate all of the "Bergeron-like" nuances into an objective map is probably impossible; yet recent efforts by data assimilators like Jim Purser of National Center for Environmental Prediction (NCEP) hold promise for getting the "Bergeron-like" structures into analyses in the vicinity of discontinuities like fronts (Purser 2005).

The observational tools that now include the instruments aboard satellites in space, the network of surveillance radars, and wind profilers, measure "surrogate" variables like spectral radiance, radar reflectivity, and turbulence, rather than the fundamental model variables—pressure, temperature, wind, and moisture. Thus, "model counterparts"—oft times complicated nonlinear functions of the model variables— must be used to make comparisons. And we must not forget that these

---

[10] See Lewis and Lakshmivarahan (2008) for a historical review of data assimilation methods in the context and time frame of Sasaki's career.

[11] In the introduction of Dover's reprint of Richardson (1922), Sidney Chapman said, "He [Richardson] told me that he had put all that he knew into this book."

observations stream into the system at variable times and places. To consistently connect the model dynamics with these observations is a problem of Herculean dimension.

As computational power has increased over the last decades of the twentieth century, our ambitious demands on the data assimilation system have been met to a certain degree. Yet, pragmatism must be balanced by equally ambitious efforts to more-fundamentally link dynamical law with observations. It was to this problem that Yoshi Sasaki directed his energy in the 1950s. The austerity of the situation in Japan, essentially a lack of resources, drove him to examine the problem theoretically. In short, there was an unexpected advantage from the *virtues of adversity* to use historian Arnold Toynbee's phrase (Toynbee 1939). Further, he had a "lone wolf" approach to his research, a pathway or angle of attack that veered from the standard. He celebrated the alternate view in both his teaching and research. Like Gauss, he viewed the data assimilation /initialization problem deterministically, and he set the stage for 4DVAR, four-dimensional variational data assimilation, the methodology that is currently used at five operational weather prediction centers worldwide. It will become the standard for all centers in the future.

**Acknowledgements** I'm grateful to Professor Sasaki for his oral history in May 1990 that provided the foundation for this story.

Oral histories and letters of reminiscence were received from the following individuals where $^O$ indicates oral history and $^L$ a letter of reminiscence (date shown in parentheses).

Akio Arakawa[L] (June 1992)
Páll Bergthórsson[L] (June 2004; December 2007)
Bo Döös[L] (June 2004)
Kanziburo Gambo[L] (June, September 1992)
John Hovermale [Conversation] (March 1970)
Akira Kasahara [o;L] (June 1991; June 1992, May 2007)
Kikuro Miyakoda [L] (June 1992, May 2007)
Yoshimitsu Ogura [L] (June 1992)
Katsuyuki Ooyama [o;L] (March 1992; June, September 1992)
George Platzman [o] (May 1990)
Yoshi Sasaki [L] (June 2007)
Walter Saucier [L] (September 1992; June 2007)

Information from these meteorologists facilitated reconstruction of events related to Sasaki's career; accordingly, I extend my heartfelt thanks to each of them.

Comments on an early version of the paper by Norman Phillips and George Platzman were valuable. Akira Kasahara carefully checked the manuscript for accuracy of the recorded events in Japan (spelling of names, dates, times, and places).

For help in locating key historical papers, I commend Hiroshi Niino, Mashahito Ishihara, Frode Stordal, Trond Iversen, Greta Krogvold, and Fred Carr. Finally, Vicki Hall and Domagoj Podnar helped me with the electronic submission process.

## Appendix: Biographical Sketch

Sasaki's résumé is found in Table 1. Elaboration on several elements in the table follows and a discussion of his scientific genealogy is also included.

## *Youthful Experiences*

Since Yoshi Sasaki attended grade school, middle school, and high school from the mid-1930s through the end of WWII, there was disruption in his education. As he said, "...because of the irregular schedules, the study of science especially depended much upon individual study in Japan during these years" (Sasaki 2007, personal communication). His stimulation for science/mathematics came from three quarters: his father (an engineer), a 6th grade teacher, and a physics professor who was a family friend.

Yoshi's father gave him complete freedom to pursue the career of his choice. By the time he was in the 6th grade, his gift for mathematical thinking was apparent to his grade-school teacher. This teacher asked Yoshi to serve as a "teaching assistant" in mathematics and "I was stimulated by this experience" (Sasaki 2007, personal communication).

A key event occurred in the early 1940s when the physics professor mentioned above gave Yoshi a translation of Albert Einstein's 1905 paper on special relativity. It was the philosophical ideas presented by Einstein—the majesty of observation and measurement in the various frames of reference—that touched him. It indeed was a paper filled with "...text and little in the line of formal manipulations. In the end the startling conclusion was there, obtained apparently with the greatest of ease, and by reasoning which could not be refuted" (Lanczos 1965, p. 7). To this day, Sasaki exhibits excitement as he recollects this experience.

By his own admission, Yoshi was a romanticist—a child who reveled in the "...beauty of cloud patterns, floating and varying in the sky...trying to find the principles of nature, and the elegant theories and models especially in physics" (Sasaki 2007, personal communication). With his youthful passion for science that was complemented by diligent study, he not only gained entrance into the University of Tokyo, but he was awarded a national scholarship for his undergraduate education and he completed the bachelors degree in three years.

## *Emigration to the USA*

Following in Gambo's footsteps, about ten Japanese meteorologists from the University of Tokyo permanently moved to the United States during the period from the mid-1950s through the mid-1960s (Lewis 1993b). Yoshi Sasaki, his wife Koko, and their infant son Okko came to America in 1956 on the *Hikawa-maru*. This ship, known as "Queen of the Pacific", was the only mainstream Japanese passenger liner to survive WWII. Yoshi and Koko are shown aboard this passenger liner, three days out from Yokohama on route to Seattle (Fig. 4).

Sasaki appeared to be heading for Johns Hopkins University or University of Chicago, but he eventually accepted a position at Texas A&M College. The decision was not entirely made on academic grounds; Okko's poor health was central to the decision. As remembered by Yoshi:

**Fig. 4** Sasaki and his wife Koko are shown aboard the trans-Pacific liner *Mikawa-maru* on their way to America (1956) (Courtesy of Y. Sasaki)

Koko and I chose to go to Texas because of our first baby, born disabled...We received some excellent advice from the Vice Chief of Staff of the Prime Minister. He recommended that it could be easier to live in Texas because we could take Okko to the Houston Medical Center. He also told us that the Texas-born Nisei had great respect because of heroic actions in Italy during WWII and that made Texas a good place for Japanese immigrant.

(Sasaki 2007, personal communication)

Walter Saucier, then head of the meteorology program in A&M's Department of Oceanography and Meteorology, recalls details of Sasaki's academic appointment at A&M:

John Freeman (at TAM [Texas A&M] 1952–1955) had brought Kasahara from Japan into his research at TAM. With Freeman's departure, G. W. Platzman [GWP] enticed Kasahara to U. Chicago. In that era, I was visiting U. Chicago and GWP almost annually. On one occasion, at GWP's home, I suggested he help me obtain a replacement for Kasahara. Almost simultaneously he responded with info that Johns Hopkins U. had sought to obtain Sasaki (evidently for the GFD [Geophysical Fluid Dynamics] research) but could not obtain such research funding for same. GWP suggested I write to Dr. Syono at U. Tokyo about Sasaki's availability. Syono promptly responded in positive. So I initiated request for his appointment as Research Scientist on my USAF-sponsored jet stream research...After doing research on turbulence data in jet streams, he began USWB-sponsored research on subject of squall lines.

(Saucier 2007, personal communication)

Sasaki's first contribution to meteorology that had the benefit of computers/ computational power was his numerical study of squall lines (Sasaki 1960). This paper displayed his versatility with numerical experimentation in the spirit of Phillips' pioneering work (Phillips 1956). Sasaki's experiment explored the feedback of precipitation on the development of the squall line. This line of attack was inspirational to NWP researchers who were contemplating design of mesoscale models (J. Hovermale 1970, personal communication).[12]

In the summer of 1960, Saucier left Texas A&M to start the meteorology program at the University of Oklahoma (OU). As stated by Saucier:

Mesoscale weather and severe storms, of interest to the Oklahoma public and to federal sponsorship of research and to Sasaki and also to me, were key factors in drawing Sasaki to OU in 1960...[He] arrived Oct 1960 at OU, employed by OURI [Oklahoma University Research Institute] on my NSF research funding. Quickly, Sasaki obtained NSF funding for severe storms research...

(Saucier, personal communication, 2007)

Sasaki steadily advanced up the academic ladder at OU as shown by the entries in Table 1, and was awarded the prestigious George Lynn Cross Research Professorship in 1974. As remembered by the host of graduate students who studied under him, he was a most stimulating lecturer, not given to great organization, but exemplary in his manner of introducing students to a "way of thinking" about research, a gift in the tradition of outstanding mentors (Zuckerman 1977).

---

[12] Revealed in conversation between the author and John Hovermale, then an operational modeler in the Development Division of the National Meteorological Center. The conversation took place at the AMS Conference on Motion and Dynamics of the Atmosphere, Houston, TX, March 1970.

## *Scientific Genealogy*

As shown in Fig. 5, Yoshi Sasaki's mentor was Shigekata Syono, chair professor of meteorology at University of Tokyo from 1944 until his death in 1969. Syono received his doctoral degree from University of Tokyo in 1934 and then served as a research meteorologist at the Central Meteorological Observatory (CMO) under the tutelage of Takematsu Okada, a theoretician, and the Director of the CMO and professor at University of Tokyo, Sakuhei Fujiwara (Yamamoto 1969; Sasaki, 1990, personal communication). Fujiwara's name became internationally known through his study of mutually interacting vortices—later named the "Fujiwara Effect" when applied to interacting typhoons.

Sasaki's scientific progeny are listed below his name in the schematic diagram (Fig. 5). Within each decade, the students are listed chronologically (in accord with the date of their PhD degrees). Ed Barker and Rex Inman received their PhD's from the Naval Postgraduate School and Texas A&M, respectively. All others were recipients of their doctoral degrees from OU. While Sasaki served as director of research at NEPRF in 1973–1974, Barker studied under Sasaki's supervision. Inman, listed on the far right of the diagram, left Texas A&M without his doctoral degree in 1960 to join the faculty at OU. He returned to A&M for one year (academic year 1967–1968) to complete his dissertation. Professor Robert Reid served as Inman's

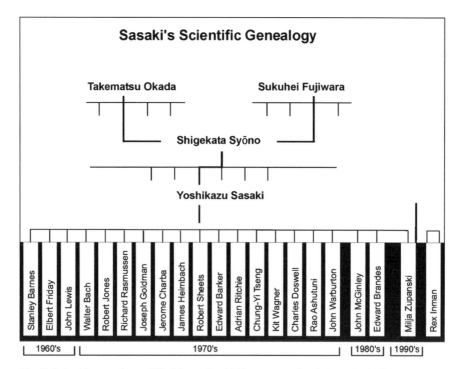

**Fig. 5** Scientific genealogy of Yoshikazu Sasaki. Progeny are listed chronologically

*de jure* advisor, but Inman was essentially a doctoral student under Sasaki and Saucier at A&M and continued to work with Sasaki after he assumed the faculty position at OU. Thus, it seems appropriate to list Inman among the protégés.

In addition to the doctoral students listed in Fig. 5 (some of whom obtained masters degrees under Sasaki), another 40 students received their masters degrees under Sasaki's supervision.

# References

Bateman H (1932) Partial differential equations of mathematical physics. Cambridge Univ. Press, pp 522

Bell E (1937) Men of Mathematics. Simon and Schuster, pp 592

Bergman K (1975) Multivariate objective analysis of temperature and wind fields. Office Note 116, Nat Meteor Cent Dev Div, NOAA, pp 29

Bergthórsson P, Döös B (1955) Numerical weather map analysis. Tellus 7: 329–340

Charney J (1947) The dynamics of long waves in a baroclinic westerly current. J Meteor 4: 135–162

Charney J, Fjørtoft R, von Neumann J (1950) Numerical integration of the barotropic vorticity equation. Tellus 2: 237–254

Clebsch A (1857) Über ein algemeine Transformation der Hydrodynamischen Gleichungen Crelle. J für Math 54: 293–312

Cressman, G (1959) An operational objective analysis system. Mon Wea Rev 87: 367–374

Daley R (1991) Atmospheric data analysis. Cambridge Univ. Press, pp 457

DeMaria M (1996) A history of hurricane forecasting for the Atlantic basin, 1920–1995. In: Fleming H (ed) Historical essays on meteorology 1919–1995. Amer Meteor Soc pp 263–305

Eckart C (1960a) Hydrodynamics of the oceans and atmospheres. Pergamon Press Ltd., UK, pp 290

Eckart C (1960b) Variation principles of hydrodynamics. Phys Fluids 3: 421–427

Eliassen A (1954) Provisional Report on Calculation of Spatial Covariance and Autocorrelation of the Pressure Field. Appendix to Report No. 5, Videnskaps-Akademiets Institutt for Vaer-Og Klimaforskning, Oslo, Norway, pp 12

Eliassen A (1990) Oral history interview (6 Nov 1990) (Interviewer: J. Green). Royal Meteorological Society, pp 43

Fjørtoft R (1952) On the numerical method of integrating the barotropic vorticity equation. Tellus 4: 179–194

Fujita T (1952) Pressure distribution in typhoon. Geophys Mag 23:437–451

Gandin L (1965) Objective Analysis of Meteorological Fields. Hydromet Press. Trans. from Russian by IPST (Israel Prog. for Sci. Trans.) in 1963, Hardin R (ed), pp 242

Gambo K (1950) On criteria for stability of the westerlies. Geophys. Notes, Tokyo Univ. 3: 29

Gauss C (1809) Theoria Motus Corporum Coelestium in Sectionibus Conicus Solem Ambientium (Theory of the Motion of Heavenly Bodies Moving about the Sun in Conic Sections). An English translation (by Chas. H. Davis) published by Little, Brown, and Co. in 1857. The publication was reissued by Dover in 1963, pp 326

Gilchrist B, Cressman G (1955) An experiment in objective analysis. Tellus 6: 309–318

Hubert L (1959) An operational test of a numerical prediction method for hurricanes. Mon Wea Rev 87: 222–230

Kasahara A (1957) The numerical prediction of hurricane movement with the barotropic model. J Meteor 14: 386–402

Lamb H (1932) Hydrodynamics (6th ed). Cambridge Univ. Press, pp 738 Reprinted by Dover, 1954.

Lanczos C (1965) Albert Einstein and the Cosmic World Order. Interscience, pp 139

Lanczos C (1970) The Variational Principles of Mechanics. Univ. Toronto Press, pp 418

LeDimet F, Talagrand O (1986) Variational algorithms for analysis and assimilation of meteorological observations, Theoretical aspects. Tellus 38A: 97–110

Lewis J (1972) An operational upper air analysis using the variational method. Tellus 24: 514–530

Lewis J (1993a) Syono: The uncelebrated teacher. Tenki (In Japanese) 40: 3–12

Lewis J (1993b) Meteorologists from the University of Tokyo: Their exodus to the United States following World War II. Bull. Amer Meteor Soc 74: 1351–1360

Lewis J, Lakshmivarahan S, Dhall S (2006) Dynamic Data Assimilation: A Least Squares Approach. Cambridge Univ. Press. pp 654

Lewis J, Lakshmivarahan S (2008) Sasaki's pivotal contribution: Calculus of variations applied to weather map analysis. Mon Wea Rev 136: 3553–3567

Lorenz E (1996) The evolution of dynamic meteorology. In: Fleming H (ed) Historical Essays on Meteorology 1919–1995. Amer Meteor Soc pp 3–9

McPherson R (1975) Progress, problems, and prospects in meteorological data assimilation. Mon Wea Rev 56: 1154–1166

Nebeker F (1995) Calculating the Weather. Academic Press. pp 255

Panofsky H (1949) Objective weather map analysis. J Meteor 6: 386–392

Phillips N (1956) The general circulation of the atmosphere: A numerical experiment. Quart J Roy Meteor Soc 82: 123–164

Platzman G (1979) The ENIAC computations of 1950 — Gateway to numerical weather prediction. Bull Amer Meteor Soc 60: 302–312

Purser R (2005) A geometrical approach to the synthesis of smooth anisotropic covariance operators for data assimilation. US Dept Commerce, NOAA, Maryland pp 89

Rabier F, Liu Z (2003) Variational data assimilation: theory and overview. Seminar on Recent Developments in Data Assimilation for Atmosphere and Ocean, 8–12 September 2003. European Centre for Medium-Range Weather Forecasts (ECMWF), 29–43. [Avail from ECMWF, Shinfield Park, Reading RG29AX, United Kingdom].

Richardson L (1922) Weather Prediction by Numerical Process. Cambridge Univ. Press (Reprinted by Dover 1965) pp 236

Sasaki Y (1955) A fundamental study of the numerical prediction based on the variational principle. J Meteor Soc Japan 33: 30–43

Sasaki Y (1958) An objective analysis based on the variational method. J Meteor Soc Japan 36: 1–12

Sasaki Y (1960) Effects of condensation, evaporation, and rainfall on development of mesoscale disturbances: A numerical experiment. In Syono S (ed) Proc. Int. Sym. on Num. Wea. Pred. (November 7–13, 1960). Meteor Soc Japan, Tokyo, Japan, pp 477–496

Sasaki Y (1970a) Some basic formalisms in numerical variational analysis. Mon Wea Rev 98: 875–883

Sasaki Y (1970b) Numerical variational analysis formulated under the constraints as determined by longwave equations and a low-pass filter. Mon Wea Rev 98: 884–898

Sasaki Y (1970c) Numerical variational analysis with weak constraint and application to surface analysis of severe storm gust. Mon Wea Rev 98: 899–910

Sasaki Y, Miyakoda K (1954) Prediction of typhoon tracks on the basis of numerical weather forecasting method. Proc. UNESCO Sym. on Typhoons, Tokyo, Japan (Nov. 9–12), pp 221–234

Schlatter T, Branstator G, Thiel L (1976) Testing a global multivariate statistical objective analysis scheme with observed data. Mon. Wea. Rev. 104: 765–783

Schoppa L (1991) Education Reform in Japan: A Case of Immoblist Politics. Routledge Pub. Co., pp 319

Smagorinsky J (1963) General circulation experiments with the primitive equations. I. The basic experiment. Mon Wea Rev 91: 99–164

Smagorinsky J (1978) History and progress. The Global Weather Experiment —Perspectives on Implementation and Exploitation. FGGE Advisory Panel Rep., National Academy of Sciences, Washington, DC, pp 4–12

Thompson P (1980) A short-range prediction scheme based on conservation principles and the generalized balance equation. Cont Atmos Phys 53: 256–263

Thompson P, 1983: A history of numerical weather prediction in the United States. Bull Amer Meteor Soc 64: 755–769.

Toynbee A (1934–1961) A Study of History (12 Volumes). Oxford Univ. Press

Tseng C - Y (2006) Inverse Problems in Atmospheric Science: Inversion, Analysis, and Assimilation (In Chinese). Two Volumes. Nat. Inst. for Compilation and Translation, Taipei, Taiwan, pp 1288

Wiin-Nielsen A (1991) The birth of numerical weather prediction. Tellus 43AB: 36–52

Yamamoto G (1969) In memory of the late Professor Shigekata Syono. J Meteor Soc Japan 47: 411–412

Zuckerman H (1977) Scientific Elite: Nobel Laureates in the United States. Free Press, pp 335

# Data Assimilation for Numerical Weather Prediction: A Review

Ionel M. Navon

**Abstract** During the last 20 years data assimilation has gradually reached a mature center stage position at both Numerical Weather Prediction centers as well as being at the center of activities at many federal research institutes as well as at many universities.

The research encompasses now activities which involve, beside meteorologists and oceanographers at operational centers or federal research facilities, many in the applied and computational mathematical research communities. Data assimilation or 4-D VAR extends now also to other geosciences fields such as hydrology and geology and results in the publication of an ever increasing number of books and monographs related to the topic.

In this short survey article we provide a brief introduction providing some historical perspective and background, a survey of data assimilation prior to 4-D VAR and basic concepts of data assimilation.

I first proceed to outline the early 4-D VAR stages (1980–1990) and addresses in a succinct manner the period of the 1990s that saw the major developments and the flourishing of all aspects of 4-D VAR both at operational centers and at research Universities and Federal Laboratories. Computational aspects of 4-D Var data assimilation addressing computational burdens as well as ways to alleviate them are briefly outlined.

Brief interludes are provided for each period surveyed allowing the reader to have a better perspective A brief survey of different topics related to state of the art 4-D Var today is then presented and we conclude with what we perceive to be main directions of research and the future of data assimilation and some open problems. We will strive to use the unified notation of Ide et al. (J Meteor Soc Japan 75:181–189, 1997).

**Keywords:** 4-D VAR data assimilation · 3-D VAR · parameter estimation · minimization methods

I.M. Navon (✉)
Department of Scientific Computing
The Florida State University, Tallahassee, Florida 32306-4120, USA
e-mail: inavon@fsu.edu

S.K. Park, L. Xu, *Data Assimilation for Atmospheric, Oceanic and Hydrologic Applications*, DOI 10.1007/978-3-540-71056-1_1,
© Springer-Verlag Berlin Heidelberg 2009

# 1 Introduction

Data assimilation in atmospheric sciences started from the fact that NWP is an initial value problem. This since we start at whatever constitutes the present state and use the NWP model to forecast its evolution. Early work by Richardson (1922) and Charney et al. (1950) were based on hand interpolations (Kalnay 2003). This in order to combine present and past observations of the state of the atmosphere with results from the model (also referred to as "Mathematical" model). Since this was a rather tedious procedure, efforts to obtain "automatic" objective analysis-the first methods have been developed by Panofsky (1949), Gilchrist and Cressman (1954), Cressman (1959), and Barnes (1964). Use of prior information to supplement rather insufficient data was pioneered by Bergthorsson and Döös (1955), Cressman (1959) followed by the comprehensive work of Lev Gandin (1965).

Early reviews of data assimilation whose purpose is that of "using all available information (data) to determine as accurately as possible the state of the atmospheric (or oceanic) flow" (Talagrand 1997) were provided by Le Dimet and Navon (1988), an in-depth survey of Ghil and Malonotte-Rizzoli (1991) as well as by the outstanding book of Daley "Atmospheric Data Analysis" (Daley 1991).

A collection of papers by Ghil et al. (1997) in "Data Assimilation in Meteorology and Oceanography: Theory and Practice (Ghil et al. 1997) summarizes state of the art of data assimilation for that period. See also a short survey by Zupanski and Kalnay (1999) along with the excellent book of Kalnay (2003) "Atmospheric Modeling, Data Assimilation and Predictability". An early effort linking Optimal Interpolation (O.I.) with the variational method was done by Sasaki (1955, 1958) and in more final form by Sasaki (1969, 1970a, b, c, d) which can be viewed as a 3-D VAR approach. It was Lorenc (1986) that showed that OI and 3-D VAR were equivalent provided the cost functional assumes the form:

$$J = \frac{1}{2}\left\{ \left[y^o - H(x)\right]^T R^{-1}\left[y^o - H(x)\right] + \left(x - x^b\right)^T B^{-1}\left(x - x^b\right) \right\} \tag{1}$$

The first term measures the distance of forecast field $x$ to observations $y^o$ and the second term measures the distance to background $x^b$.

The analysis $x$ is obtained by adding the innovation to the model forecast with weights $W$ based on estimated statistical error covariances of forecast and observations.

$$x = x^b + W\left[y^o - H(x^b)\right] \tag{2}$$

Related theoretical material related to the set-up that led to modern data assimilation may be found in the "Inverse Problem Theory" of Tarantola (1987), the optimal control book of Lions (1971), the "Perspectives in Flow Control and Optimization" by Max Gunzburger (2003) along with "Inverse Modeling of the Ocean and Atmosphere" by Andrew Bennett (2002) and "Dynamic Data Assimilation: A Least Squares Approach" by John Lewis et al. (2006), Cacuci (2003) and Cacuci et al. (2005). See also Haltiner and Williams (1980).

In this brief review we first provide some historical background to the data assimilation effort along with some basic concepts of data assimilation. We then proceed to survey in a section the early stages (1980–1990) of 4-D VAR data assimilation with brief interludes summarizing and providing perspectives as we go along. In the following section we address some computational aspects of data assimilation such as issues of automatic differentiation, and the incremental method which alleviated the computational burden of 4-D VAR and made it operationally viable at large operational NWP centers. A short section is dedicated to state-of the art of data assimilation at present time and we close with a short section outlining directions of development of 4-D VAR in the future.

## 2 Relationship Between OI and 3-D VAR

The terminology of 4-D VAR (4-dimensional data assimilation) was originally used in research centers in the context of using continuous data assimilation satellite data leading to the First Global Atmosphere Research Program (GARP) Global Experiment, Charney, Halem and Jastrow (1969).

Insertion of observations directly into primitive equations models excited spurious inertia-gravity oscillations in the model and required the use of damping schemes (Matsuno 1966) for damping the high-frequency components. A full-account of these techniques and the history of continuous data assimilation are provided in the seminal book of Daley (1991). This review will survey some aspects of variational data assimilation while only providing a brief outline of methodologies that prevailed prior to the 1980s. We will rely on work of Kalnay (2003), Daley (1991), Talagrand (1997), Zupanski and Kalnay (1999), Ghil et al. (Eds) (1997), works of the present author and his collaborators, the review of Ghil and Malonotte-Rizzoli (1991) and an early review that remained an unpublished technical report (Le Dimet and Navon 1988).

Panofsky (1949) is credited for pioneering the first objective analysis based on 2-D polynomial interpolation. It was followed by Gilchrist and Cressman (1954) who put forward an interpolation scheme for geopotential field as a quadratic polynomial in $x$ and $y$

$$E(x,y) = a_{00} + a_{10}x + a_{01}y + a_{20}x^2 + a_{11}xy + a_{02}y^2, \tag{3}$$

then minimizing mean square difference between polynomial and observations within a radius of influence of the closest grid point,

$$\min_{a_{ij}} E = \min_{a_{ij}} \left\{ \sum_{k=1}^{K_v} p_v \cdot \left( E_0^v - E(x_v, y_v) \right)^2 \right. \tag{4}$$
$$\left. + \sum_{k=1}^{K_v} q_v \cdot \left\{ [u_v^0 - u_g(x_v, y_v)]^2 + [v_v^0 - v_g(x_v, y_v)]^2 \right\} \right\}$$

where $p_v$ and $q_v$ were empirical weights and $u_g$ and $v_g$ the components of the geostrophic wind obtained from the gradient of geopotential height $E(x,y)$ at observation point$k$. $K$ was total number of observations within the radius of influence. The introduction of first guess estimate is credited to have been introduced by Bergthorsson and Döös (1955). Usually either climatology or a combination of it with first guess was used in the analysis cycle. See also the influential work of Gandin (1965), translated from Russian by the Israeli program of Translations in 1965.

# 3 Successive Correction Method

The first analysis method in 4DDA was the successive correction method developed by Bergthorsson and Döös (1955) and by Cressman (1959). The field of background was chosen as a blend of forecast and climatology with a first estimate given by the first guess field

$$f_i^0 = f_i^b. \tag{5}$$

$f_i^b$ background field estimated at the $i$-th grid point, $f_i^0$ being the zeroth iteration estimate of gridded field. This is hence followed by new iteration obtained by "successive corrections"

$$f_i^{n+1} = f_i^n + \sum_{k=1}^{K_i^n} w_{ij}^n \left( f_k^0 - f_k^n \right) + \sum_{k=1}^{K_i^n} w_{ik}^n + \varepsilon^2 \tag{6}$$

$f_i^n$- $n$-th iteration estimate at $i$th grid point,
$f_k^0$- $k$-th observation surrounding grid point,
$f_i^n$- value of $n$-th field estimate calculated at observation point $k$ derived by interpolation from nearest grid points,
$\varepsilon^2$- estimate of ratio of observation error variance to background error variance.

The important ingredient is constituted by the weights $w_{ik}^n$ which are related to a radius of influence. Cressman (1959) proposed the following weights in the SCM (Successive corrections method).

$$w_{ik}^n = \frac{R_n^2 - r_{ik}^2}{R_n^2 + r_{ik}^2} \quad \text{if} \quad r_{ik}^2 \leq R_n^2 \tag{7}$$

$$w_{ik}^n = 0 \quad \text{if} \quad r_{ik}^2 > R_n^2 \tag{8}$$

$r_{ik}^2$ square of distance between observation point $r_k$ and a grid point at $r_i$.
    The controlling parameter is the radius of influence $R_n$, allowed to vary between iterations while $K_i^n$ is the number of observations within a radius of $R_n$ of the grid point $i$. If one reduces the radius of influence, this results in a field reflecting large

scales after first iteration -and tends towards smaller scales after additional iterations. For additional technical details see Daley (1991), Kalnay (2003).

Cressman (1959) took the coefficient $\varepsilon^2$ to be zero. For noisy data with errors it
may lead to erroneous analysis. Taking $\varepsilon^2 > 0$ i.e. assuming observations with errors, allows some impact to the background field. Barnes (1964) defined the weights
to follow a Gaussian or normal distribution

$$w_{ij} = \begin{cases} exp - \left( \frac{r_{ik}^2}{d^2} \right) & \text{if} \quad r_{ik} \leq d \\ 0 & \text{otherwise,} \end{cases} \tag{9}$$

where $d$ is the radius of influence.

It uses an adaptive version where the radius of influence changes by a factor $\gamma$

$$0 < \gamma < 1. \tag{10}$$

It was shown by Bratseth (1986) that with an appropriate choice of weights these
SCM iterative method analysis increments can be made to be the same as those
obtained using optimal interpolation (OI). Lewis et al. (2006) quote also similar
independent work done by Franke and Gordon (1983), Franke (1988) and Seaman
(1988).

## 4 The Variational Calculus Approach

It was introduced in meteorology by Yoshi Sasaki in his PhD Thesis (1955) and later
extended by him to include dynamic model laws (Sasaki 1969, 1970a, b, c). He proposed three basic types of variational formalism in the numerical variational analysis
method. The basic formalisms are categorized into three areas: (1) "timewise localized" formalism, (2) formalism with strong constraint, and (3) a formalism with
weak constraint. Exact satisfaction of selected prognostic equations was formulated
as constraints in the functionals for the first two formalisms. This approach is now
generically referred to as 3-D VAR.

In 3-D VAR one defines a cost function proportional to the square of the distance between analysis and both background and observations, and it was showed
by Lorenc (1981, 1986) that the OI and the 3-D VAR approaches are equivalent
provided the cost function is defined as

$$J = \frac{1}{2} \left[ y^o - H(x) \right]^T R^{-1} \left[ y^o - H(x) \right] + (x - x^b) B^{-1} (x - x^b). \tag{11}$$

where

$B$ is the background error covariance,
$R$ is the observation error covariance,
$H$ is an interpolation operator (or observation operator),

$x^b$ is the first guess or background,
$y^o$ is the observation,
$y^o - H(x^b)$ are the observational increments

$$x^a = x^b + W\left[y^o - H(x^b)\right] \tag{12}$$

$W$ is a weight matrix based on statistical error covariances of forecast and observations.

## 5 Variational Methods

The start of variational methods is originally attributed to the work of Euler and Lagrange the seventeenth and eighteenth century. The Euler-Lagrange equation, developed by Leonhard Euler and Joseph-Louis Lagrange in the 1750s, is the major formula of the calculus of variations. It provides a way to solve for functions which extremize a given cost functional. It is widely used to solve optimization problems, and in conjunction with the action principle to calculate trajectories. Variational calculus has had a broad appeal due to its ability to derive behavior of an entire system without details related to system components. Broadly speaking variational calculus involves finding stationary points of functionals written as integral expressions. The general theory is rigorously explained in the work by Lanczos (1970) and Courant and Hilbert (1962).

Basic to the constrained minimization theory is the method of undetermined Lagrange multipliers where

$$\lambda = (\lambda_1, \ldots, \lambda_n)^T \tag{13}$$

is a vector of $n$ unknowns for the solution of

$$\min f(x) \in R^n \tag{14}$$

$$\text{subject to } g(x) = 0 \quad x \in R^m \tag{15}$$

and using the first-order conditions for a minimum we obtain using the first derivatives of the Lagrangian function

$$L(\lambda, x) = f(x) + \lambda^T g(x) \tag{16}$$

$$\nabla_x L(x, \lambda) = \frac{\partial f}{\partial x} + \lambda \frac{\partial g}{\partial x} \tag{17}$$

$$\nabla_\lambda L(x, \lambda) = g(x) \tag{18}$$

The Lagrange multiplier $\lambda$ can be viewed as measuring sensitivity of value of function $f$ at a stationary point to changes in the constraint (see also Nocedal and Wright 2006).

One can show formally (see any text book on variational methods) that finding in a given domain of admissible functions $u(x)$ the continuous first derivatives of a functional $I$ for which $I(u(x))$ is a stationary value (i.e. any function which extremizes the cost functional) must also satisfy the ordinary differential equation called the Euler-Lagrange equation

$$\frac{\partial F}{\partial u} - \frac{\partial}{\partial x}\frac{\partial F}{\partial u'} = 0 \tag{19}$$

where

$$I(u(x)) = \int_{x_a}^{x_b} F(u(x))dx \quad x_a \leq x \leq x_b \tag{20}$$

$$u' = \frac{\partial u}{\partial x} \tag{21}$$

As an example of a typical application of variational methods, consider work of Sasaki (1970a, b, c). Navon (1981) used it to enforce conservation of total enstrophy, total energy and total mass in one and two-dimensional shallow water equations models on a rotating place.

# 6 First Interlude

## 6.1 Situation in Data-Assimilation at Beginning of 1980s

Charney, Halem and Jastrow (1969) proposed that numerical models be used to assimilate newly available asynoptic data. The idea was to insert asynoptic temperature information obtained from satellite-born radiometers into the model at its true (asynoptic) time. Continuous data assimilation referred to frequent insertion of asynoptic data. Charney et al. (1969) experiment suggested continuous data assimilation I.G. Tadjbakhsh (1969). Problems of real data insertion soon emerged in the form of an inertia-gravity wave shock (Daley and Puri 1980) leading to essential rejection by the model of the information of real observational data. A remedy for continuous data assimilation of real data was to reduce the insertion interval to the time step of the model (Miyakoda et al. 1976). See also Talagrand and Miyakoda (1971).

Other approaches were via geostrophic wind correction outside the tropics or nudging also referred to as Newtonian relaxation (Hoke and Anthes 1976), Davis and Turner (1977). See also work of Talagrand (1981, 1987). Ghil, Halem and Atlas (1979), McPherson (1975). McPherson (1975) viewed data assimilation as "a process by which something is absorbed into something else". During 1974 Marchuk proposed application of adjoint method in meteorology (Marchuk 1974 – see also Russian article of Marchuk 1967) and in 1976 Penenko and Obratsov used these methods to study adjoint sensitivity (Penenko and Obratsov 1976).

In 1969, Thompson had already put forward the idea that incorrect analyses at two successive times may be optimally adjusted to maintain dynamical consistency with a given prediction model (Thompson 1969). This may be viewed as a precursor to variational data assimilation. Since 1958, Marchuk and collaborators used adjoint methods for linear sensitivity analysis problems. Atmospheric issues were also addressed in the same fashion (see Marchuk 1974). Adjoint operators have been introduced by Lagrange (1760) and have been used in modern times since Wigner (1945) and by many others in different domains.

The advent of optimal control theory of partial differential equations is attributed to Bellman starting in the late 1950s Bellman (1957) (the Hamilton-Jacobi-Bellman equation) and to Pontryagin et al. (1962) (Pontryagin's minimum principle).

The major impetus in this area came from the monograph of Lions (1968) on optimal control of partial differential equations. It was to be that a former doctoral student of Lions, Francois Le Dimet, introduced the concepts of optimal control to the meteorological community starting in the early 1980s.

One major work which impacted in a serious way the adjoint sensitivity analysis was the work of Cacuci et al. (1980), D.G. Cacuci (1981a, 1981b). Historically one can trace back linear adjoint sensitivity to work of Wigner (1940–1942) (see Wigner 1945). See the lecture of Cacuci (2004). Wigner (1949) was the first to interpret physically the adjoint functions (see also Lewins 1965) as importance functions. As mentioned above Cacuci (1980–1981) presented a complete rigorous theory for adjoint sensitivity of general nonlinear systems of equations.

Le Dimet (1981) was then preparing his technical report at Clermont-Ferrand introducing for the first time optimal control methodology with variational adjustment to the meteorological community that led to the seminal paper by Le Dimet and Talagrand (1986).

## 7 Emergence of Early Data Assimilation Works

Le Dimet (1982), Lewis and Derber (1985), Courtier (1985), Le Dimet and Talagrand (1986) were the first to work on adjoint data assimilation. Cacuci (1981a, 1981b) extended adjoint sensitivity analysis to the fully nonlinear case. Lagrange multiplier methods were presented in detail by Bertsekas (1982), while Navon and De Villiers (1983) exhibited the method in detail applied to enforcing conservation of integral invariants.

## 8 Optimal Interpolation (OI) Methods

Lev Gandin (1965) coined the term (OI) but the technique of statistical interpolation can be traced back to Kolmogorov (1941) and Wigner (1949) and the terminology of optimal interpolation was apparently due to Wigner (1949).

A review of the work of these two mathematicians is provided in the Yaglom (1962) book on stochastic processes (see Lewis et al. 2006). In atmospheric sciences use of statistical interpolation goes back to Eliassen (1954) while Krige (1951), used it in the mining industry.

Use of least-squares to obtain best estimate of state of the atmosphere by combining prior information which can consist of either a first guess or a background with observations which have errors. The concept of background field goes back to Gauss (1809). We wish to carry out a minimum variance estimation.

In a general form the optimal least-squares estimation is defined by the following interpolation equations

$$X^a = X_b + K(y - H[X_b]),\tag{22}$$

where $K$ is a linear operator referred to as gain or weight matrix of the analysis and is given by

$$K = BH^T(HBH^T + R)^{-1},\tag{23}$$

where $X^a$ is the analysis model state,

H- an observation operator,
B- covariance matrix of the background errors $(X_b - X)$,
X- being the time model state,
$X_b$- background model state,
R- covariance matrix of observation errors.
The analysis error covariance matrix is

$$A = (I - KH)B(I - KH)^T + KRK^{-1}\tag{24}$$

If $K$ is optimal least-squares gain, $A$ becomes

$$A = (I - KH)B\tag{25}$$

(see proof in Bouttier and Courtier 1999).

One can show that the best linear unbiased estimator (Talagrand 1997; Bouttier and Courtier 1999) may be obtained as the solution of the following variational optimization problem.

$$\begin{aligned}\min J &= (X - X_b)^T B^{-1}(X - X_b) + (y - H(X))^T R^{-1}(y - H(X))\\ &= J_b(X) + J_o(X)\end{aligned}\tag{26}$$

One notes that if the background and observation error probability functions are Gaussian then $X_a$ is also the maximum likelihood estimation of $X_t$ (time). Probability density function represents a probability distribution in terms of integrals, being non-negative everywhere with an integral from $-\infty$ to $+\infty$ being equal to 1. More exactly a probability distribution has density $f(x)$, if $f(x)$ is a non-negative Lebesgue integrable function from $R \to R$ such that the probability of the interval $[a, b]$ is given by $\int_a^b f(x)dx$ for any two numbers $a$ and $b$.

For a comprehensive examination of OI in meteorology we refer to Lorenc (1981) and Lorenc (1986). The most important advantage of using statistical interpolation schemes such as OI and 3-D VAR instead of empirical schemes such as SCM (1959) is the fact that they are taking into account the correlation between observational increments.

How to estimate the prior error covariances $B$ and $R$ and the observation operator $H$? A difficult issue with observation operator is the case of satellite products such as radiances, a piece of information which cannot be directly used. The observation operator performs both interpolation from model grid to satellite observation location and then uses physical theory (such as in the case of radiances) to convert model column of temperature to synthetic radiances. Observation error covariance matrix $R$ is obtained from instrument error estimates which, if independent mean that the covariance matrix $R$ will be diagonal. This can facilitate computations.

Assume that background and observation error (covariances) are uncorrelated, the analysis error covariance matrix is given as

$$A = (I - KH)B(I - KH)^T + KRK^T. \tag{27}$$

Solution of minimum covariance requires

$$\frac{\partial}{\partial K}(\text{trace}(A)) = 0 \tag{28}$$

$$\frac{\partial}{\partial A}(\text{trace}BAC) = B^T C^T \tag{29}$$

$$\frac{\partial}{\partial t}(\text{trace}ABA^T) = A(B + B^T) \tag{30}$$

$$\frac{\partial}{\partial K}(\text{trace}(A)) \equiv (I - KH)(B + B^T)H^T + K(R + R^T) \tag{31}$$

$$= -2(I - KH)BH^T + 2KR$$

$$= -2BH^T + 2K(HBH^T + R)$$

$$= 0$$

from which we obtain the optimal weight $K$

$$K = BH^T(HBH^T + R)^{-1}. \tag{32}$$

## 9 Estimating Background Error Covariances

The background error covariance is both the most difficult error covariance to estimate and it has a most important impact on results (Kalnay 2003; Navon et al. 2005). This since it is primarily the background error covariance that determines the spread of information as well as allowing observations of wind field to enrich information about the mass field and vice-versa.

In order to render modelling of $B$ practically feasible some compromises had to be made with respect to statistical aspects of the covariance matrix such as anisotropy, flow dependence and baroclinicity (Fisher 2003a, 2003b). The first approach by Hollingsworth and Lonnberg (1986) concerned statistics of innovations, namely observation – minus -background (in short forecasts) and rawinsonde observations. The assumption made was that observation errors are spatially uncorrelated and they assigned spatial correlations of innovations to the background error. Hidden in this method of use of innovation statistics is the implicit assumption of a dense homogeneous observing network.

For 3-D VAR the most popular and universally adopted method does not depend on measurements but rather uses differences between forecasts of different time-lengths which verify at the same time. It is known as the "NMC" (now NCEP) method having been introduced by Parrish and Derber (1992). In an operational numerical weather prediction they use

$$B \approx \alpha E\left\{ [X_f(48h) - X_f(24h)][X_f(48h) - X_f(24h)]^T \right\} \tag{33}$$

This provides a multivariate global forecast difference covariance. If this time interval is longer than the forecast used to generate background fields then the covariances of the forecast difference will be broader than those of the background error.

A new method based on ensemble of analyses to estimate the background errors is described in detail in Fisher (2003a, 2003b) who presents also modern approaches to background error covariance matrix construction.


## 10 Framework of Variational Data Assimilation

The objective of variational 4-D Var is to find the solution to a numerical forecast model that best fits a series of observational fields distributed in space over a finite time interval. We are assuming that the model of the atmosphere can be written as

$$B\frac{dX}{dt} + A(X) = 0 \tag{34}$$

with $B$ being identity for a dynamical model or the null operator for a steady state model. $A$ can be a linear or nonlinear operator. We have $U$ defined as a control variable which may consist of initial conditions, boundary conditions and/or model parameters.

$U$ should belong to a class admissible controls $U_{ad}$. We are looking for a unique solution $X(U)$ of (34). The major step consists in formulating the cost function $J$ which measures distance between model trajectory and observations as well as the background field at initial time during a finite time-interval, referred to as the time window.

Typically in meteorology (see Le Dimet and Talagrand 1986; Rabier 2005).

$$J(X_0) = \frac{1}{2}(X_0 - X_b)^T B^{-1}(X_0 - X_b) \tag{35}$$

$$+ \frac{1}{2}\sum_{i=0}^{N}(H_i(X_i) - y_i)^T R_i^{-1}(H_i(X_i) - y_i)$$

where

$X_0$ is the NWP model state as time $t_0$,
$X_b$-background state at time $t_0$, typically a $6h$ forecast from a previous analysis,
$B$-the background error covariance matrix,
$y_i$-the observation vector at time $t_i$,
$H_i$-observation operator,
$X_i = M_{i,0}(X_0)$ model state at time $t_i$,
$R_i$-observation error covariance matrix at time $t_i$.

where an alternative to writing the NWP model is

$$X_{i+1} = M_{i+1,i}(X_i) \tag{36}$$

$M_{i+1,i}$ is the nonlinear NWP model from time $t_i$ to time $t_{i+1}$.

The minimization of the cost functional can be viewed both in the perspective of finding its gradient in (a) Lagrangian approach, (b) adjoint operator approach and (c) a general synthesis of optimality conditions in the framework of optimal control theory approach. Requiring the gradient of the cost to vanish with respect to initial conditions control variable $X_0$ yields

$$\nabla_{X_0} J(X_0) = B^{-1}(X_0 - X_b) + \sum_{i=0}^{N} \mathbf{M}_{i,0}^T \mathbf{H}_i^T R_i^{-1}[H_i(X_i) - y_i] \tag{37}$$

where we substitute the dynamical constraint

$$X_{i+1} = M_{i+1,i}(X_i) \tag{38}$$

while perturbations of the atmospheric state are obtained by linearizing the nonlinear model (38) as

$$\delta X_{i+1} = \mathbf{M}_{i+1,i}(X_i)\delta X_i \tag{39}$$

yielding

$$\nabla_{X_0} J(X_0) = B^{-1}(X_0 - X_b) + \sum_{i=0}^{N} \mathbf{M}_{i,0}^T \mathbf{H}_i^T R_i^{-1}[H_i(X_i) - y_i] \tag{40}$$

where $\mathbf{H}_i$ is the tangent linear operator of the observation operator $H_i$ and $\mathbf{H}_i^T$ is the adjoint operator and

$$\mathbf{M}_{i,0}^T = \mathbf{M}_{1,0}^T \mathbf{M}_{2,1}^T \cdots \mathbf{M}_{i,i-1}^T \tag{41}$$

is the adjoint model consisting of a backward integration from time $t_i$ to time $t_0$.

The minimization of the cost functional is obtained using a gradient-based minimization algorithm. Starting from a first guess

$$X^0(t_0) = X_b(t_0) \tag{42}$$

while at each iteration step $k = 1, 2, \cdots, N$, we compute and store both first guess trajectory and the observation departures $H_i(X_i) - y_i$ by integrating forward in time the nonlinear model

$$X^k(t_i) = M(t_i, t_0)(X^k(t_0)) \tag{43}$$

Start with initializing the adjoint variable at time $t_N$

$$\delta' X^k(t_N) = 0 \tag{44}$$

integrating the adjoint model backwards in time from final time $t_N$ to initial time $t_0$. and whenever observations are encountered a forcing term

$$\mathbf{H}_i^T R_i^{-1}(H_i(X_i) - y_i) \tag{45}$$

is added to $\delta' X^k(t_i)$.

Finally one can show that

$$\delta' X^k(t_0) + B[X^k(t_0) - X_b] \tag{46}$$

is the gradient $\nabla J^k$ with respect to the control variable $X^k(t_0)$.

If

$$||\nabla_{X_0} J^{k+1}|| \leq \varepsilon \max\{1, ||X_k||\} \tag{47}$$

(where $\varepsilon$ is a predetermined adequately chosen tolerance.) If above criterion is satisfied then stop.

If the above criterion is not satisfied then use a stepsize search algorithm using, say, a cubic interpolation usually provided by the gradient based minimization algorithm.

One then updates the first guess, namely

$$X^{k+1}(t_0) = X^k(t_0) - \rho^k \nabla J^k \tag{48}$$

where $\rho$ is a step-size in the direction of descent and find the next minimization iterate using a gradient based minimization algorithm.

All the time we assume that the nonlinear cost function has a unique minimum and avoids temporarily addressing the complex issue of the presence of multiple minima.

# 11 Variational Formalism

## 11.1 The Lagrangian Approach

One can consider a model given as in Le Dimet and Talagrand (1986) by

$$F(U) = 0 \tag{49}$$

where $U$ denotes meteorological fields being considered. Suppose we have observations $\hat{U}$ occurring at an irregular set of points distributed in both space and time.

We wish to solve the problem of finding a solution that minimizes a cost function

$$J(U) = \int ||U - \hat{U}||^2 \, dxdydt \tag{50}$$

where $||,||$ is a suitable norm and $\hat{U}$ consists of discrete observations hence the integral is replaced by suitable finite sums. Here we view the model equation

$$F(U) = 0 \tag{51}$$

as a strong constraint on cost function $J$. Using classical Lagrange multiplier technique a Lagrangian of (50) subject to model strong constraint allows us to convert this constrained minimization into an unconstrained minimization problem by defining a Lagrangian (see Bertsekas 1982) as

$$L(U, \lambda) = J(U) + (\lambda, F(U)) \tag{52}$$

for an adequately defined inner product for a functional space in which $F(U)$ also belongs.

Then finding minima of $J(U)$ subject to

$$F(U) = 0 \tag{53}$$

is equivalent to finding the minima of

$$\nabla_\lambda L = 0 \quad \text{and} \tag{54}$$
$$\nabla_U L = 0 \tag{55}$$

which taking into account boundary conditions turns out to be the Euler-Lagrange equations of the problem. Since the Euler-Lagrange equations can seldom be solved directly, we are interested in practical algorithms for solving the minimization of cost functional subject to strong model constraint by transforming it into a sequence of unconstrained minimization problems.

There are many constrained minimization algorithms-but the simplest and most robust of them are the penalty and the multiplier (or duality) algorithms. These are presented in many numerical minimization text books, (Nash and Sofer 1996;

Nocedal and Wright 2006). For shortcomings of the penalty and duality algorithms see book of Bertsekas (1982) and Navon and De Villiers (1983).

In the augmented Lagrangian algorithm (where the constrained problem is converted into a sequence of unconstrained minimization problems) we have

$$L(\rho, U, \lambda) = J(U) + \{\lambda, F(U)\} + \rho |F(U)|^2 \tag{56}$$

This algorithm was initially proposed by Hestenes (1969) and independently by Powell (1969). Here $\rho > 0$ is the quadratic penalty coefficient.

## 12 Optimal Control View Point

In optimal control of partial differential equations developed by Lions (1968, 1971) the Lagrange multiplier is viewed as an adjoint variable. The adjoint method of optimal control allows computing the gradient of a cost $J$ with respect to the control variables.

Consider as in Gunzburger (2003) a second order nonlinear elliptic PDE

$$-\nabla(a\nabla\phi) + b \cdot \nabla\phi + \phi^3 = \sum_{k=1}^{K} \alpha_K f_K \tag{57}$$

in domain $\Omega$ with boundary conditions

$$\phi = 0 \quad \text{on} \quad \Gamma \tag{58}$$

$a, b$ and $f_K$ are given functions defined on $\Omega$.

We define a cost as

$$J(\phi, \alpha_1, \cdots, \alpha_K) = \frac{1}{2} \int_{\Omega} (\phi - \Phi)^2 d\Omega + \frac{\sigma}{2} \sum_{k=1}^{K} (\alpha_K)^2 \tag{59}$$

$\Phi$ is a given function and $\sigma$ a penalty parameter. We introduce a Lagrange multiplier (here adjoint variable) $\zeta$ and define a Lagrangian

$$L(\phi, g, \zeta) = J(\phi, g) - \zeta^T F(\phi, g) \tag{60}$$

We aim to find controls $g$, states $\phi$ and adjoint states $\zeta$ such that the Lagrangian is stationary and we obtain as in the Augmented Lagrangian approach

$$\frac{\partial L}{\partial \zeta} = 0, \text{ constraint} \tag{61}$$

$$\frac{\partial L}{\partial \phi} = 0, \text{ adjoint equation} \tag{62}$$

$$\frac{\partial L}{\partial g} = 0, \text{ optimality condition} \tag{63}$$

Taking a first order variation of $L$ with respect to the Lagrange multiplier, we obtain a variation in the state yielding an optimality condition

$$\left(\frac{\partial F}{\partial \phi}|_{(\phi,g)}\right)^{T} \zeta = \left(\frac{\partial J}{\partial \phi}|_{(\phi,g)}\right)^{T} \tag{64}$$

which yields the optimality condition.

## 13 Situation of Data Assimilation-the Early Period (1980–1987) of 4-D Var

Efforts in early adjoint applications following Francois Le Dimet (1981) pioneering technical report were by Lewis and Derber (1985) and Le Dimet and Talagrand (1986) as well as Courtier (1985). These research efforts started the meteorological optimal control application called "adjoint operator" approach.

Work of Navon and De Villiers (1983) on augmented Lagrangian methods is related to the same topic and is referred to in the early work of Le Dimet and Talagrand (1986).

John Lewis and John Derber (1985) were the first authors to present application of adjoint method, having read the report of Francois Le Dimet (1982) and inspired by earlier work of Thompson (1969). Lorenc (1986) presented a detailed account of state of theory in data assimilation for that period. See also Navon (1986).

It became soon apparent that size and complexity of atmospheric equations is such that enormous computational resources were required-limiting applications of 4-D VAR to undergo drastic approximations for actual operation forecast circumstances.

Penenko and Obratsov (1976) used adjoint data assimilation to perform simple experiments on a linear model (see Talagrand and Courtier 1987), while Derber (1985) used it in his Ph.D thesis to adjust analysis to a multi-level quasi-geostrophic model.

Hoffmann (1986) was the next to use 4-D VAR (even though he used a simplified primitive equation model and in order to estimate the gradient he perturbed in turn all the components of the initial state.)

Talagrand and Courtier (1987) presented a more in-depth general exposition of the theory of adjoint equations in the framework of variational assimilation and applied it to the inviscid vorticity equation and to the Haurwitz wave. Their results are presented in Courtier and Talagrand (1987).

## 14 OI, 3-D VAR and PSAS

Lorenc (1986) showed that the optimal weight matrix $W$ that minimizes the matrix of analysis error covariance solution may be posed in terms of a variational assimilation problem, namely that of finding the optimal analysis field $X_a$ that minimizes a cost function. The cost function measures the distance between the field variables $X$ and the background $X_b$ (the background term of the cost)-plus another term, namely the distance to the observations $y^o$ weighted by the inverse of the observation error covariance matrix $R$

$$J(X) = \frac{1}{2}(X - X_b)^T B^{-1}(X - X_b) + [y^o - H(X)]^T R^{-1}[y^o - H(X)] \qquad (65)$$

where $H$ is the forward observational operator. The cost function (97) can also be derived based on a Bayesian approach.

A formalism allowing viewing the assimilation algorithms of O-I, 3-D VAR, PSAS and 4-D VAR as a sequence of corrections to a model state can be derived from the work of Lorenc (1986), Kalnay (2003) and Courtier (1997). See also research work of Da Silva et al. (1995) who first proposed the physical space statistical analysis system (PSAS) (see also report of Aarnes 2004).

We are considering incrementing background model state $X_b$ with additional information from the observation $z$ where

$$X_a = X_b + K(z - HX_b). \qquad (66)$$

Here $H$ is an observation operator mapping the model state on space and time locations of the observation, $X_a$ is the analysis and $K$ is the gain matrix weighting the contributions from the new information according to the reliability of the observation relative to respective reliability of the model state. Following Kalnay (2003), Lorenc (1986) OI, 3-D VAR, 4-D VAR and PSAS are mathematically equivalent but 3-D VAR and related PSAS have the advantage w.r.t. OI by virtue of the fact that one can minimize the cost function $J$ with global unconstrained minimization algorithms for 3-D VAR hence all the approximation made in OI are not necessary. Other advantages of 3-D VAR are enumerated in Kalnay (2003).

To show equivalence of 3-D VAR and OI we start from the matrix system

$$\begin{pmatrix} R & H \\ H^T & -B^{-1} \end{pmatrix} \begin{pmatrix} W \\ X_a - X_b \end{pmatrix} = \begin{pmatrix} z - HX_b \\ 0 \end{pmatrix} \qquad (67)$$

where $R$ and $B$ are the error observation error and background error covariance matrices, respectively, assumed to be symmetric and positive-definite. The equivalence between OI and 3-D VAR statistical problems was proven by Lorenc (1986), Kalnay (2003) and using suggestion of Jim Purser (see Kalnay (2003))

$$W = K_{OI} = BH^T(R + HBH^T) \qquad (68)$$

To see the equivalence between OI and the PSAS scheme where minimization is performed in the space of observations rather than in the model space (Since the number of observation is usually much smaller than the dimension of model space-PSAS may turn out to be more efficient than 3-D VAR for obtaining similar results) we note that

$$\begin{pmatrix} R & H \\ H^T & -B^{-1} \end{pmatrix} \begin{pmatrix} W \\ X_a - X_b \end{pmatrix} = \begin{pmatrix} z - HX_b \\ 0 \end{pmatrix} \tag{69}$$

is equivalent to

$$\begin{pmatrix} W & 0 \\ H^T & -B^{-1} \end{pmatrix} \begin{pmatrix} W \\ X_a - X_b \end{pmatrix} = \begin{pmatrix} z - HX_b \\ 0 \end{pmatrix} \tag{70}$$

yielding

$$w = W^{-1}(z - HX_b) \tag{71}$$

and

$$X_a - X_b = BH^T W^{-1} \tag{72}$$

One first solves the linear system

$$W_w = z - HX_b \tag{73}$$

and then interpolates solution onto model space as

$$X_a = X_b + BH^T w \tag{74}$$

In PSAS one solves the first step by minimizing the cost functional

$$J(w) = \frac{1}{2} w^T W w - w^T (Z - HX_b) \tag{75}$$

thus allowing a better conditioning of the minimization due to smaller dimension of $W$ i.e

$$\dim(W) \leq \dim(B) \tag{76}$$

Courtier (1997) has shown that there is a duality between 3-D VAR and the physical space statistical analysis system (PSAS). He also showed that the temporal extension of 3-D VAR leads to 4-D VAR while the temporal extension of PSAS,4-D VAR PSAS is achieved using an algorithm related to the representers technique (Bennett 2002), which is a practical algorithm for decoupling the Euler-Lagrange equations associated with the variational problem with weak constraint. (see Amodei 1995)

## 15 4-D VAR Developments in Early 1990s

A comprehensive list of adjoint applications to meteorological problem is provided by Courtier et al. (1993). The early 1990s were characterized by publication of many research efforts related to extending 4-D VAR data assimilation to multilevel primitive-equation models using analyses as observations along with other work using synthetic observations. See for instance Thepaut and Courtier (1991), Navon et al. (1992a) and Zupanski (1993). Thepaut et al. (1993) used real observations while Rabier and Courtier (1992) studied the performance of 4-D VAR in the presence of baroclinic instability. Courtier et al. (1994) introduced an incremental formulation of the 4-D VAR, a major achievement allowing the 4-D VAR method to become computationally feasible on that period's computers.

It was perceived rather early by Derber (1989) that the perfect model hypothesis is a weakness of 4-D VAR. In the above seminal paper he assumed the model error to be fully correlated in time and solved the problem by including the bias in the control variable. Wergen (1992) and Miller et al. (1994) illustrated how serious the problem is.

At universities research in 4-D VAR data assimilation proceeded to address issues such as the impact of incomplete observations on 4-DVAR (see Zou et al. 1992), while at the suggestion and advice of Francois Le Dimet, Zhi Wang completed a doctoral thesis on second order adjoint methods (Wang 1993), as well as a first paper on second order adjoint data assimilation (Wang et al. 1995). Initial work on 4-D VAR data assimilation with the semi-implicit semi Lagrangian (SLSI) models in 2-D and 3-D was using both shallow-water and a NASA multilevel model. (see Li et al. 1993, 1994, Li and Droegemeier 1993) Basic work on optimization methods suitable for 4-D VAR was carried out by Zou et al. (1993) based on Navon (1992) and Navon et al. (1992b). Application of 4-D VAR to a finite-element model of the shallow-water equations was carried out by Zhu et al. (1994) while a novel Hessian preconditioning method based on an idea of Courtier et al. (1994) was written by W. Yang et al. (1996) Aspects of 4-D VAR dealing with boundary conditions as control variables were dealt amongst others in the work of Zou et al. (1995).

## 16 Model Error in 4-D VAR

Numerical weather prediction (NWP) models are imperfect, since they are discretized, dissipative and dispersion errors arise, and, moreover subgrid processes are not included. In addition, most of the physical processes and their interactions in the atmosphere are parameterized and a complete mathematical modeling of the boundary conditions and forcing terms can never be achieved. Usually all of these modeling drawbacks are collectively addressed by the term model error (ME). The model equations do not represent the system behavior exactly and model errors arise due to lack of resolution as well as inaccuracies occurring in physical parameters, boundary conditions and forcing terms. Errors also occur due to numerical discrete approximations. A way to take these errors into account is to use the weak constraint 4D-Var.

Variational data assimilation is based on the minimization of:

$$J(\mathbf{x}) = [H(\mathbf{x}) - y]^T R^{-1} [H(\mathbf{x}) - y] \tag{77}$$
$$+ (\mathbf{x}_0 - \mathbf{x}_b)^T B^{-1} (\mathbf{x}_0 - \mathbf{x}_b) + \Phi(\mathbf{x})^T C^{-1} \Phi(\mathbf{x})$$

Here $\mathbf{x}$ is the 4D state of the atmosphere over the assimilation window, $H$ is a 4D observation operator, accounting for the time dimension. $\Phi$ represents remaining theoretical knowledge after background information has been accounted for (such as balance relations or digital filtering initialization introduced by Lynch and Huang (1992)). One can see that model M verified exactly although it is not perfect.

## *16.1 Weak Constraint 4D-Var*

The model can be imposed as a constraint in the cost function, in the same way as other sources of information:

$$\Phi_i(\mathbf{x}) = \mathbf{x}_i - \mathbf{M}_{i-1} \tag{78}$$

Model error $\eta$ is defined as: $\eta_i(\mathbf{x}) = \mathbf{x}_i - \mathbf{M}_{i-1}$.

The cost function becomes:

$$J(\mathbf{x}) = \frac{1}{2} \sum_{i=1}^{n} (H(\mathbf{x}_i) - y_i)^T R_i^{-1} (H(\mathbf{x}_i) - y_i) \tag{79}$$

$$+ \frac{1}{2}(\mathbf{x}_0 - \mathbf{x}_b)^T B^{-1}(\mathbf{x}_0 - \mathbf{x}_b) + \frac{1}{2} \sum_{i=1}^{n} \eta_i^T Q_i^{-1} \eta$$

Another issue requiring attention is that model error covariance matrix $Q$ has to be defined. Strong constraint 4D-Var is obtained when $\Phi_i(\mathbf{x}) = 0$ i.e. $\eta = 0$ (perfect model).

Studies indicate that model error (ME) can severely impact forecast errors, see for instance Boer (1984), Dalcher and Kalnay (1987), Bloom and Shubert (1990) and Zupanski (1993).

For early methods on estimating modeling errors in operational NWP models see Thiébaux and Morone (1990) and Saha (1992). Thus giving up the assumption that the model is perfect, in the context of strong constraint VDA leads us to weak constraint formulation of VDA, and if we include time evolution of the variables, we could say we have a weak constraint 4D-Var (time plus three space dimensions).

Comparing the strong and weak constraint VDA, in the formulation of former, it is assumed that $\eta$ has mean and model error covariance matrix $Q = E(\eta(t)\eta^T(t')) = 0, \forall t$ and $t'$ and model error covariance matrix, $E[\cdot]$ is the mathematical expectation operator. It should be noted that if the mean and (co)variance of a random vector are prescribed to be equal to zero, then all realizations of that random vector are identically equal to zero, thus, $\eta \equiv 0$. In the weak constraint version of VDA, the mean and covariance of ME have to be specified. However exact statistical details of ME are difficult to obtain (Daley 1992a, b; Dee and Da Silva 1998; Zhu and Kamachi 2000) a fact which led researchers to suggest a variety of assumptions to approximate and parameterize the ME.

Early efforts to model the systematic component of ME were pioneered by Derber (1989). He suggested a simplified approach to model $\eta$ to be equal to $\lambda(t)\phi$.

The temporal part, $\lambda(t)$ is a specified function of time alone, while $\phi$ is a spatially dependent, control variable. Three different forms of $\lambda$ were considered, namely, parabolic, delta function and constant in time. It was observed that the parabolic variation of $\lambda$ provided results comparable to a constant in time$\lambda$. Using a similar approach (Wergen 1992; Zupanski 1997) it was shown that inclusion of ME allowed significant reduction in forecast RMSE.

For dynamically evolving systems such as discrete NWP models, ME is expected to depend on the model state and should be evolving in time (Griffith and Nichols 1996, 2000). Various simple forms of evolution of ME in time were considered by Griffith and Nichols (2000), Nichols (2003), At any time step,$t_k$, the evolution of ME is

$$\eta_k = T_k(\mathbf{e}_k) + \mathbf{q}_k \tag{80}$$

where $T_k$ describes the distribution of systematic errors in the NWP model equations, and $\mathbf{q}_k$, (stochastic component) is an unbiased, serially correlated, normally distributed random vector, with known covariance. The evolution of $\mathbf{e}_k$, is in-turn modeled by assuming that it depends on the state vector, $\mathbf{x}_k$

$$\mathbf{e}_{k+1} = g_k(\mathbf{x}_k, \mathbf{e}_k). \tag{81}$$

## 16.2 Systematic Model Error and State Augmentation

In order to take into account systematic components in the model errors, we assume that the evolution of the errors is described by the equations

$$\eta_k = T_k(\mathbf{e}_k) + \mathbf{q}_k \tag{82}$$

$$\mathbf{e}_{k+1} = \mathbf{g}_k(\mathbf{x}_k, \mathbf{e}_k) \tag{83}$$

where $\mathbf{q}_k \in R^n$ is unbiased, serially uncorrelated, normally distributed random vectors with known covariance matrices and the vectors $\mathbf{e}_k \in R^r$ represent time-varying systematic components of the model errors. The distribution of the systematic errors in the model equations is defined by the function$T_k : R^r \to R^n$. The functions $\mathbf{g}_k : R^n \times R^r \to R^r$ describing the systematic error dynamics are to be specified.

In practice little known about the form of the model errors and a simple form for the error evolution that reflects any available knowledge needs to be prescribed. Examples of simple forms of the error evolution includes:

$$\text{constant bias error}: \ \mathbf{e}_{k+1} = \mathbf{e}_k, T_k = I.$$

This choice allows for a constant vector $\mathbf{e} = \mathbf{e}_0$ of unknown parameters to be found, which can be interpreted as statistical biases in the model errors. This form is expected to be appropriate for representing the average errors in source terms or in boundary conditions.

$$\text{Evolving error}: \; \mathbf{e}_{k+1} = F_k\mathbf{e}_k, T_k = I.$$

Here $F_k \in R^{n \times n}$ represents a simplified linear model of the state evolution. This choice is appropriate, for example, for representing discretization error in models that approximate continuous dynamical processes by discrete time systems.

$$\text{Spectral form}: \mathbf{e}_{k+1} = \mathbf{e}_k, T_k = (I, \sin(k/N\tau)I, \cos(k/N\tau)I).$$

In this case the constant vector $\mathbf{e} \equiv \mathbf{e}_0$ is partitioned into three components vectors, $\mathbf{e}^T = (\mathbf{e}_1^T, \mathbf{e}_2^T, \mathbf{e}_3^T)$ and $\tau$ is a constant determined by the timescale on which the model errors are expected to vary, for example, a diurnal timescale. The choice approximates the first order terms in a spectral expansion of the model error.

The weak constraint VDA doubles the size of the optimization problem (compared to strong constraint VDA), in addition if the stochastic component is included in the ME formulation, then one would have to save every random realization at each model time step, which amounts to tripling the size of the optimization problem. The computational results in Griffith et al. (2000) were provided by neglecting $\mathbf{q}_k$, the stochastic component of ME and using the constant and evolving forms of the systematic component, see Griffith et al. (2000) for additional details. Similar approaches for modeling the systematic component of ME was considered by Martin et al. (2001) and reduction of ME control vector size by projecting it on to the subspace of eigenvectors corresponding to the leading eigenvalues of the adjoint-tangent linear operators was illustrated by Vidard et al. (2000)

Other choices can be prescribed, including piecewise constant error and linearly growing error (see Griffith 1997; Griffith et al. 2000; Martin et al. 2001). These techniques have been applied successfully in practice to estimate systematic errors in an equatorial ocean model (Martin et al. 2001) Zupanski et al. (2005) provided results obtained using the NCEP' s regional weather prediction system in weak constraint VDA framework. Akella and Navon (2007) studied in depth the nature of modeling errors and suggested a decreasing, constant and increasing in time forms of ME. Implementation of these forms in a weak constraint VDA framework yielded a further reduction in forecast errors.

When the number of observations is considerably smaller, the method of representers (Bennett 2002) provides a computationally efficient (in storage/ space requirements) formulation of VDA. Incorporation of ME in such framework was shown by Uboldi and Kamachi (2000).

Very little is known with certainty about ME spatio-temporal structure since MEs are not observable, contrary to forecast errors. The common practice is to assume that MEs are white. Daley (1992a) suggested use of a first order (in time) linear model for MEs. That approach was implemented by Zupanski (1997) in its simplest form; the inevitable simplicity is due to the absence of empirical estimates of parameters and even structural features of the ME model. DelSole and Hou (1999) considered the state-dependent part of ME and proposed a respective estimator.

Mitchell and Daley (1997) considered the discretization part of ME and its effect on data assimilation. Menemenlis and Chechelnitsky (2000) estimated the spatial structure of an ME white-noise model for an ocean circulation model. ME models rely on hypotheses that have never been checked namely the applicability of a stochastic model driven by an additive (and not, say, multiplicative) noise, Gaussianity of ME, the white-noise or red-noise hypotheses. Tools needed to use the information on ME (Tsyrulnikov 2005) structure in meteorology and oceanography are available such as ensemble forecasting, weak-constraint four-dimensional variational assimilation (4D-Var, e.g. Zupanski 1997; Xu et al. 2005), and Kalman filtering (e.g. Cohn 1997). Empirical approaches have been used only in ensemble techniques but cannot be used in the weak-constraint 4D-Var, where one must specify an ME spatio-temporal stochastic model.

# 17 Automatic Differentiation

Automatic differentiation (AD) is a set to techniques based on the mechanical application of the chain rule to obtain derivatives of a function given as a computer program adjoint equations resulting from differentiation the discretized model equation can be obtained.

Automatic differentiation exploits fact that a computer code executes a sequence of elementary arithmetic operations consisting of additions or elementary functions.

By applying the chain rule derivative repeatedly to these operations derivatives of any order can be computed automatically. Other classical methods to achieve the same goal are available but with inherent shortcomings are symbolic differentiation or use of finite-differences.

Symbolic differentiation is slow, while finite differences suffer from round-off errors in the discretization process and cancellations.

Automatic differentiation has the advantage of solving these problems.

There are essentially two modes of AD, namely forward accumulation and reverse accumulation. Forward accumulation is less useful for data assimilation while reverse accumulation allows efficient calculation of gradients.

The first powerful general purpose AD systems was developed at Oak Ridge National Laboratory (Oblow 1983), later endowed with the adjoint variant ADGEN for reverse AD by Pin et al. (1987). Later ADIFOR (Bischof et al. 1992) was developed at Argonne National Laboratory, Odyssee was developed at INRIA and TAMC by Giering and Kaminski (1997). In France the TAPENADE code is used (see Hascoet and Pascual 2004). There are many more automatic differentiation languages. Earlier books on AD are by Rall (1981) and Kagiwada et al. (1986). See also Navon and Zou (1991).

**Checkpointing** is a general trade-off technique, used in the reverse mode of AD that trades duplicate execution of a part of the program in order to save memory space employed to save intermediate results. Checkpointing a code fragment amounts to running this fragment without storage of intermediate values, thus saving memory space. At a later stage, when the intermediate value is required, the fragment is run a second time to obtain the required values.

Results and application studies of automatic differentiation have been published in proceedings of the international workshop on automatic differentiation held in Breckenridge (See Griewank and Corliss 1991). The most comprehensive book and work is that of Andreas Griewank (2000). See also Berz et al. (1996) and Griewank and Corliss (1991).

## 18 Second Order Adjoint Methods

Behind most of the methods used in meteorology such as: optimal interpolation, variational methods, statistical estimation etc., there is a variational principle, i.e. the retrieved fields are obtained through minimization of a functional depending on the various sources of information. The retrieved fields are obtained through some optimality condition which can be an Euler or Euler-Lagrange condition if regularity conditions are satisfied. Since these conditions are first order conditions, it follows that they involve the first order derivatives of the functional which is minimized. In this sense, data assimilation techniques are first order methods. But first order methods provide only necessary conditions for optimality but not sufficient ones. Sufficient conditions require second order information. By the same token, from the mathematical point of view sensitivity studies with respect to some parameter can be obtained through Gateaux derivatives with respect to this parameter. Therefore if we seek the sensitivity of fields which have already been defined through some first order conditions we will have to go to an order of derivation higher and in this sense sensitivity studies require second order information.

Early work on second order information in meteorology includes Thacker (1989) followed by work of Wang et al. (1992, 1993) stimulated by advice and expertise of F.X. Le Dimet. Wang (1993) and Wang et al. (1998) considered use of second order information for optimization purposes namely to obtain truncated -Newton and Adjoint Newton algorithms using exact Hessian/vector products. Application of these ideas was presented in Wang et al. (1997). Kalnay et al. (2000) introduced an elegant and novel pseudo-inverse approach and showed its connection to the adjoint Newton algorithm of Wang et al. (1997). (See Pu et al. 1997; Park and Kalnay 1999, 2004; Pu and Kalnay 1999; Kalnay et al. 2000). Ngodock (1996) applied second order information in his doctoral thesis in conjunction with sensitivity analysis in the presence of observations and applied it to the ocean circulation. Le Dimet et al. (1997) presented the basic theory for second order adjoint analysis related to sensitivity analysis.

A comprehensive review paper on second order adjoint methods was written by Le Dimet et al. (2002) considering all aspects of second order adjoint methods.

## 19 Computing the Second Order Information

In what follows we follow closely the presentation in Le Dimet et al. (2002). In general we will assume that the model has the general form:

$$F(\mathbf{X}, \mathbf{U}) = 0 \qquad (84)$$

where $\mathbf{X}$, the state variable, describes the state of the environment, $\mathbf{U}$ is the input of the model, i.e. an initial condition which has to be provided to the model to obtain from Eq. (84) a unique solution $\mathbf{X}(\mathbf{U})$. We will assume that $\mathbf{X}$ and $\mathbf{U}$ belong to a space equipped with an inner product.

The closure of the model is obtained through a variational principle which can be considered as the minimization of some functional:

$$J(\mathbf{X}, \mathbf{U}) \qquad (85)$$

For instance, in the case of variational data assimilation, $J$ may be viewed as representing the cost function measuring the discrepancy between the observation and the solution associated with the value $\mathbf{U}$ of the input parameter. Therefore the optimal input for the model will minimize $J$.

## 19.1 First Order Necessary Conditions

If the optimal $\mathbf{U}$ minimizes $J$ , then it satisfies the Euler equations given by

$$\nabla J(\mathbf{U}) = 0 \qquad (86)$$

where $\nabla J$ is the gradient of $J$ with respect to control variables.

The gradient of $J$ is obtained in the following way:

(i) we compute the Gateaux (directional) derivative of the model and of $F$ in some direction $\mathbf{u}$. We may write

$$\frac{\partial F}{\partial \mathbf{X}} \times \hat{\mathbf{X}} + \frac{\partial F}{\partial \mathbf{U}} \times \mathbf{u} = 0 \qquad (87)$$

where $(\hat{\ })$ stands for the Gâteaux derivative. Let $Z$ be an application from $R^n$ into $R^n$ with variable $\mathbf{U}$. We define the Gâteaux derivative of $Z$ in the direction $\mathbf{u}$ when this limit exists. For a generic function $Z$ it is given by:

$$\hat{Z}(\mathbf{U}, \mathbf{u}) = \lim_{\alpha \to 0} \frac{Z(\mathbf{U} + \alpha \mathbf{u}) - Z(\mathbf{U})}{\alpha} \qquad (88)$$

If $\hat{Z}(\mathbf{U}, \mathbf{u})$ is linear in $\mathbf{u}$ we can write

$$\hat{Z}(\mathbf{U}, \mathbf{u}) = <\nabla Z(\mathbf{U}), \mathbf{u}> \qquad (89)$$

where $\nabla Z$ is the gradient of $Z$ with respect to $\mathbf{U}$. The Gateaux derivative is also called a directional derivative. Here $\frac{\partial F}{\partial \mathbf{X}}$ (or $\frac{\partial F}{\partial \mathbf{U}}$) is the Jacobian of $F$ with respect to $\mathbf{X}$ (or $\mathbf{U}$)) and

$$\hat{J}(\mathbf{X}, \mathbf{U}, \mathbf{u}) = <\frac{\partial J}{\partial \mathbf{X}}, \hat{\mathbf{X}}> + <\frac{\partial J}{\partial \mathbf{U}}, \mathbf{u}> \qquad (90)$$

where $<>$ stands for the inner product.

The gradient of $J$ is obtained by exhibiting the linear dependence of $\hat{J}$ with respect to $\mathbf{u}$. This is done by introducing the adjoint variable $P$ (to be defined later according to convenience).

Taking the inner product between (87) and $P$ yields

$$< \frac{\partial F}{\partial \mathbf{X}} \times \hat{\mathbf{X}}, P > + < \frac{\partial F}{\partial \mathbf{U}} \times \mathbf{u}, P >= 0 \tag{91}$$

$$< \left(\frac{\partial F}{\partial \mathbf{X}}\right)^T \times P, \hat{\mathbf{X}} > + < \left(\frac{\partial F}{\partial \mathbf{U}}\right)^T \times P, \mathbf{u} >= 0 \tag{92}$$

Therefore using (90), if $P$ is defined as the solution of the adjoint model

$$\left(\frac{\partial F}{\partial \mathbf{X}}\right)^T \times P = \frac{\partial J}{\partial \mathbf{X}} \tag{93}$$

then we obtain

$$\nabla J(\mathbf{U}) = \left(\frac{\partial F}{\partial \mathbf{U}}\right)^T \times P + \frac{\partial J}{\partial \mathbf{U}} \tag{94}$$

Therefore the gradient is computed by solving Eq. (93) to obtain , then by applying Eq. (94).

## 19.2 Second Order Adjoint

To obtain second order information we look for the product of the Hessian $G(\mathbf{U})$ of $J$ with some vector $\mathbf{u}$ . As before we apply a perturbation to Eqs. (84), (93), and from Eq. (93) and (94) we obtain

$$\left(\frac{\partial^2 F}{\partial \mathbf{X}^2} \times \hat{\mathbf{X}} + \frac{\partial^2 F}{\partial \mathbf{X} \partial \mathbf{U}} \times \mathbf{u}\right)^T \times P + \left(\frac{\partial F}{\partial \mathbf{X}}\right)^T \times \hat{P} \tag{95}$$

$$= \frac{\partial^2 J}{\partial \mathbf{X}^2} \times \hat{\mathbf{X}} + \frac{\partial^2 J}{\partial \mathbf{X} \partial \mathbf{U}} \times \mathbf{u}$$

and

$$\widehat{\nabla J(\mathbf{U})} = G(\mathbf{U}) \times \mathbf{u} = - \left(\frac{\partial^2 F}{\partial \mathbf{U}^2} \times \mathbf{u} + \frac{\partial^2 F}{\partial \mathbf{U} \partial \mathbf{X}} \times \hat{\mathbf{X}}\right)^T \times P \tag{96}$$

$$- \left(\frac{\partial F}{\partial \mathbf{U}}\right)^T \times \hat{P} + \frac{\partial^2 J}{\partial \mathbf{U}^2} \times \mathbf{u} + \frac{\partial^2 J}{\partial \mathbf{X} \partial \mathbf{U}} \times \hat{\mathbf{X}}$$

We introduce here $Q$ and $R$, two additional variables. To eliminate $\hat{\mathbf{X}}$ and $P$, we will take the inner product of Eq. (87) and (95) with $Q$ and $R$ respectively, then add the results. We then obtain

$$< \hat{\mathbf{X}}, \left(\frac{\partial F}{\partial \mathbf{X}}\right)^T \times Q > + < \mathbf{u}, \left(\frac{\partial F}{\partial \mathbf{U}}\right)^T \times Q > + < P, \left(\frac{\partial^2 F}{\partial \mathbf{X}^2}\right) \times \hat{\mathbf{X}} \times R > \quad (97)$$

$$+ < P, \left(\frac{\partial^2 F}{\partial \mathbf{X} \partial \mathbf{U}}\right) \times \mathbf{u} \times R > + < \hat{P}, \left(\frac{\partial F}{\partial \mathbf{X}}\right) \times R >$$

$$= < \hat{\mathbf{X}}, \left(\frac{\partial^2 J}{\partial \mathbf{X}^2}\right)^T \times R > + < \mathbf{u}, \left(\frac{\partial^2 J}{\partial \mathbf{X} \partial \mathbf{U}}\right)^T \times R >$$

Let us take the inner product of Eq. (96) with $\mathbf{u}$, then we may write

$$< G(\mathbf{U}) \times \mathbf{u}, \mathbf{u} > = < - \left(\frac{\partial^2 F}{\partial \mathbf{U}^2} \times \mathbf{u} + \frac{\partial^2 F}{\partial \mathbf{X} \partial \mathbf{U}} \times \hat{\mathbf{X}}\right)^T \times P, \mathbf{u} > \quad (98)$$

$$+ < \hat{P}, \left(-\frac{\partial F}{\partial \mathbf{U}}\right) \times \mathbf{u} > < \frac{\partial^2 J}{\partial \mathbf{U}^2} \times \mathbf{u}, \mathbf{u} > + < \hat{\mathbf{X}}, \frac{\partial^2 J}{\partial \mathbf{X} \partial \mathbf{U}})^T \times \mathbf{u} >$$

From (98) we get

$$< \hat{\mathbf{X}}, \left(\frac{\partial F}{\partial \mathbf{X}}\right)^T \times Q + \left(\frac{\partial^2 F}{\partial \mathbf{X}^2} \times P\right) \times R - \frac{\partial^2 J}{\partial \mathbf{X}^2} \times R > + < \hat{P}, \frac{\partial F}{\partial \mathbf{X}} \times R >$$

$$= < \mathbf{u}, - \left(\frac{\partial F}{\partial \mathbf{U}}\right)^T \times Q - \left(\frac{\partial^2 F}{\partial \mathbf{X} \partial \mathbf{U}} \times P\right)^T \times R + \frac{\partial^2 J}{\partial \mathbf{X} \partial \mathbf{U}} \times R > \quad (99)$$

Therefore if $Q$ and $R$ are defined as being the solution of

$$\left(\frac{\partial F}{\partial \mathbf{X}}\right)^T < \mathbf{u}, - \left(\frac{\partial F}{\partial \mathbf{U}}\right)^T \times Q + \left(\frac{\partial^2 F}{\partial \mathbf{X}^2} < \mathbf{u}, - \left(\frac{\partial F}{\partial \mathbf{U}}\right)^T \times P\right) \times R \quad (100)$$

$$- \left(\frac{\partial^2 J}{\partial \mathbf{X}^2}\right)^T \times R = \left(\frac{\partial^2 J}{\partial \mathbf{X} \partial \mathbf{U}}\right)^T \times \mathbf{u} - \left(\frac{\partial^2 F}{\partial \mathbf{U} \partial \mathbf{X}} \mathbf{u}\right) \times P$$

$$\left(\frac{\partial F}{\partial \mathbf{X}}\right) \times R = -\frac{\partial F}{\partial \mathbf{U}} \times \mathbf{u} \quad (101)$$

then we obtain:

$$G(\mathbf{U}) \times \mathbf{u} = - \left(\frac{\partial^2 F}{\partial \mathbf{U}^2} \times \mathbf{u}\right) \times P + \frac{\partial^2 J}{\partial \mathbf{U}^2} \times \mathbf{u} - \left(\frac{\partial F}{\partial \mathbf{U}}\right)^T \times Q \quad (102)$$

$$- \left(\frac{\partial^2 F}{\partial \mathbf{X} \partial \mathbf{U}} \times P\right) \times R + \frac{\partial^2 J}{\partial \mathbf{X} \partial \mathbf{U}} \times R$$

For Eqs. (93), (94), (95), (96), (97), (98) and (99) we took into account the symmetry of the matrix of second derivative, e.g.

$$\frac{\partial^2 F}{\partial \mathbf{X}^2} = \left(\frac{\partial^2 F}{\partial \mathbf{X}^2}\right)^T \quad (103)$$

leading to some simplifications. The system (99) will be called the second order adjoint. Therefore we can obtain the product of the Hessian by a vector **u** by (i) solving the system (99). (ii) applying formula (102).

## 19.3 Remarks

(a) The system (99) which has to be solved to obtain the Hessian/vector product can be derived from the Gateaux derivative (99) which is the same as (101). In the literature, the system (99) is often called the tangent linear model, this denomination being rather inappropriate because it implies the issue of linearization and the subsequent notion of range of validity which is not relevant in the case of a derivative.
(b) In the case of an $N$-finite dimensional space the Hessian can be fully computed after $N$ integrations of vector of the canonical base. Equation (99) differs from the adjoint model by the forcing terms which will depend on **u** and $R$.
(c) The system (99), (100), (101) and (102) will yield the exact value of the Hessian/vector product. An approximation could be obtained by using the standard finite differences, i.e.,

$$G(\mathbf{U}) \times \mathbf{u} \approx \frac{1}{\alpha} \left[ \nabla J(\mathbf{U} + \alpha \mathbf{u}) - \nabla J(\mathbf{U}) \right] \tag{104}$$

where $\alpha$ is the finite-difference interval which has to be carefully chosen. In the incremental 3/4D-Var approach the Hessian/vector product can readily be obtained by differencing two gradients.

However several integrations of the model and of its adjoint model will be necessary in this case to determine the range of validity of the finite-difference approximation (Wang et al. (1995) and references therein).

## 19.4 Time Dependent Model

In the case of variational data assimilation the model $F$ is a differential system on the time interval $[0, T]$. The evolution of $\mathbf{X} \in H[C(0, T)]^n$ between 0 and $T$ is governed by the differential system,

$$\frac{\partial \mathbf{X}}{\partial t} = F(\mathbf{X}) + \mathbf{B} \times \mathbf{V} \tag{105}$$

The input variable is often the initial condition,

$$\mathbf{X}(0) = \mathbf{U} \in R^n \tag{106}$$

In this system $F$ is a nonlinear operator which describes the dynamics of the model, $\mathbf{V} \in V[C(0, T)]^m$ is a term used to represent the uncertainties of the model which we assume to be linearly coupled through the $(m, n)$ -dimensional matrix $\mathbf{B}$, $\mathbf{U}$ is the initial condition, and the criteria $J$ is the discrepancy between the solution of Eqs. (105) and (106) and observations

$$J(\mathbf{U}, \mathbf{V}) = \frac{1}{2} \int_0^T ||\mathbf{H}\mathbf{X} - \mathbf{X}_{obs}||^2 dt \qquad (107)$$

where $\mathbf{H}$ is the observation matrix, i.e., a linear operator mapping $\mathbf{X}$ into $\mathbf{X_{obs}}$. The problem consists in determining $\mathbf{U}$ and $\mathbf{V}$ that minimize $J$.

A perturbation $\mathbf{v}$ on $\mathbf{V}$ and $\mathbf{u}$ on $\mathbf{U}$ gives $\hat{\mathbf{X}}$ and $\hat{\mathbf{J}}$ the Gateaux derivatives of $\mathbf{X}$ and $\mathbf{J}$ as solution of

$$\frac{d\hat{\mathbf{X}}}{dt} = \frac{\partial F}{\partial \mathbf{X}} \times \hat{\mathbf{X}} + \mathbf{B} \times \mathbf{V} \qquad (108)$$

$$\hat{\mathbf{X}}(0) = \mathbf{u} \qquad (109)$$

$$\hat{J}(\mathbf{U}, \mathbf{V}, \mathbf{u}, \mathbf{v}) = \frac{1}{2} \int_0^T < \mathbf{H}\mathbf{X} - \mathbf{X}_{obs}, \mathbf{H}\hat{\mathbf{X}} > dt \qquad (110)$$

Let us introduce $P$ the adjoint variable, we take the product of (108) with $P$ after a summation on the interval $[0, T]$ and an integration by parts followed by identification of linearities with respect to $\mathbf{U}$ and $\mathbf{V}$ in (110), we conclude that of $P$ is defined as the solution of

$$\frac{dP}{dt} = \frac{\partial F}{\partial \mathbf{X}}^T \times P + \mathbf{H}^T \mathbf{H}(\mathbf{X} - \mathbf{X}_{obs}) \qquad (111)$$

$$P(T) = 0 \qquad (112)$$

and the components of the gradient $\nabla J$ with respect to $\mathbf{U}$ and $\mathbf{V}$ are

$$\nabla J_{\mathbf{U}} = -P(0) \qquad (113)$$

$$\nabla J_{\mathbf{V}} = -\mathbf{B}^T P \qquad (114)$$

$\mathbf{V}$ is time dependent, its associated adjoint variable $Q$ will be also time dependent. Let us remark that the gradient of $J$ with respect to $\mathbf{V}$ will depend on time. From a computational point of view the discretization of $\mathbf{V}$ will have to be carried out in such a way that the discretized variable remains in a space of "reasonable" dimension.

The second derivative will be derived after a perturbation $h$ on the control variables $\mathbf{U}$ and $\mathbf{V}$

$$h = \begin{pmatrix} h_U \\ h_V \end{pmatrix} \qquad (115)$$

The Gateaux derivatives $\hat{\mathbf{X}}$, $P$ of $\mathbf{X}$ and $P$ in the direction of $h$, are obtained as the solution of the coupled system

$$\frac{d\hat{\mathbf{X}}}{dt} = \frac{\partial F}{\partial \mathbf{X}}\hat{\mathbf{X}} + \mathbf{B}h_V \tag{116}$$

$$\hat{\mathbf{X}}(0) = h_U \tag{117}$$

$$\frac{d\hat{\mathbf{P}}}{dt} + \left(\frac{\partial^2 F}{\mathbf{X}^2} \times \hat{\mathbf{X}}\right)^T \times P + \left(\frac{\partial F}{\partial \mathbf{X}}\right)^T \times P = \mathbf{H}^T \mathbf{H}\hat{\mathbf{X}} \tag{118}$$

$$\nabla J_{\mathbf{U}} = -\hat{P}(0) \tag{119}$$

$$\nabla J_{\mathbf{V}} = -\mathbf{B}^T \hat{P} \tag{120}$$

We introduce $Q$ and $R$, second order adjoint variables. They will be defined later for ease use of presentations.

Taking the inner product of (116) with $Q$ and of (118) with $R$, integrating from 0 to $T$, then adding the resulting equations, we may write:

$$\int_0^T \left[ <\frac{d\hat{\mathbf{X}}}{dt}, Q> - <\frac{\partial F}{\partial \mathbf{X}} \times \hat{\mathbf{X}}, Q> - <\mathbf{B}h_V, Q> + <\frac{d\hat{P}}{dt}, R> \right. \tag{121}$$

$$\left. + <\left[\frac{\partial^2 F}{\partial \mathbf{X}^2} \times \hat{\mathbf{X}}\right]^T \times P, R> + <\left[\frac{\partial F}{\partial \mathbf{X}}\right]^T \hat{P}, R> - <\mathbf{H}^T \mathbf{H}\hat{\mathbf{X}}, R> \right] dt = 0$$

The terms in $\hat{\mathbf{P}}$ and $\hat{\mathbf{X}}$ are collected and after integration by parts and some additional transformations we obtain

$$\int_0^T <\hat{\mathbf{X}}, -\frac{dQ}{dt} - \left[\frac{\partial F}{\partial \mathbf{X}}\right]^T \times Q + \left[\frac{\partial^2 F}{\partial \mathbf{X}^2} \times P\right]^T \times R - \mathbf{H}^T \mathbf{H}R > dt \tag{122}$$

$$+ \int_0^T <\hat{P}, -\frac{dR}{dt} + \left(\frac{\partial F}{\partial \mathbf{X}}\right) \times R > dt - \int_0^T <h_V, \mathbf{B}^T \times Q > dt$$

$$+ <\hat{\mathbf{X}}(T), Q(T)> - <\hat{\mathbf{X}}(0), Q(0)> + <\hat{P}(T), R(T)>$$

$$- <\hat{P}(0), R(0)>= 0$$

Let $\mathbf{G}$ be the Hessian matrix of the cost $J$. We have

$$\mathbf{G} = \begin{pmatrix} G_{UU} & G_{UV} \\ G_{VU} & G_{VV} \end{pmatrix} \tag{123}$$

Therefore if we define the second order adjoint as being the solution of

$$\frac{dQ}{dt} + \left[\frac{\partial F}{\partial \mathbf{X}}\right]^T \times Q = \left[\frac{\partial^2 F}{\partial \mathbf{X}^2}P\right]^T \times R - \mathbf{H}^T \mathbf{H}R \tag{124}$$

$$\frac{dR}{dt} = \left[\frac{\partial F}{\partial \mathbf{X}}\right] \times R \tag{125}$$

and

$$Q(T) = 0 \tag{126}$$

$$R(0) = h_U \tag{127}$$

then we finally obtain

$$< -h_U, Q(0) > = < \hat{P}(0), R(0) > \tag{128}$$

$$\hat{P}(0) = -Q(0) \tag{129}$$

We would like to point out that Eq. (129) follows directly from Eq. (128) by using Eq. (127). The product of the Hessian by a vector r is obtained exactly by a direct integration of (125) and (127) followed by a backward integration in time of (124) and (126).

One can obtain **G** by $n$ integrations of the differential system:

$$\frac{dQ}{dt} + \left[\frac{\partial F}{\partial \mathbf{X}}\right]^T \times Q = \left[\frac{\partial^2 F}{\partial \mathbf{X}^2} \times P\right]^T \times R - \mathbf{H}^T \mathbf{H} R \tag{130}$$

$$\frac{dR}{dt} = \left[\frac{\partial F}{\partial \mathbf{X}}\right] R \tag{131}$$

with the conditions

$$Q(T) = 0 \tag{132}$$

$$R(0) = \mathbf{e}_i \tag{133}$$

where $\mathbf{e}_i$ are the n-vectors of $R^n$ the canonical base of thus obtaining

$$\mathbf{G}_{UU} \mathbf{e}_i = Q(0) \tag{134}$$

$$\mathbf{G}_{UV} \mathbf{e}_i = \mathbf{B}^T \times Q \tag{135}$$

One then integrates $m$ times the differential system

$$\frac{dQ}{dt} + \left[\frac{\partial F}{\partial \mathbf{X}}\right]^T \times Q = \left[\frac{\partial^2 F}{\partial \mathbf{X}^2} \times P\right]^T \times R - \mathbf{H}^T \mathbf{H} R \tag{136}$$

$$\frac{dR}{dt} - \left[\frac{\partial F}{\partial \mathbf{X}}\right] \times R = \mathbf{f}_j \tag{137}$$

with initial and terminal conditions

$$Q(T) = 0 \tag{138}$$

$$R(0) = 0 \tag{139}$$

where $\mathbf{f}_j$ are the $m$ canonical base vectors of $R^m$ obtaining

$$\mathbf{G}_{VV} \times \mathbf{f}_j = \mathbf{B}^T \times Q \tag{140}$$

The system defined by these equations is the second order adjoint model. The Hessian matrix is obtained via $n + m$ integrations of the second order adjoint. The second order adjoint is easily obtained from the first order adjoint – differing from it only by some forcing terms, in particular the second order term. The second equation is that of the linearized model (the tangent linear model).

One can also obtain the product of a vector of the control space, times the Hessian at cost of a single integration of the second order adjoint.

## 19.5 Use of Hessian of Cost Functional to Estimate Error Covariance Matrices

A relationship exists between the inverse Hessian matrix and the analysis error covariance matrix of either 3-D VAR or 4-D VAR (See Thacker 1989; Rabier and Courtier 1992; Yang et al. 1996; Le Dimet et al. 1997).

Following Courtier et al. (1994) we consider methods for estimating the Hessian in the weakly nonlinear problem when the tangent linear dynamics is a good approximation to nonlinear dynamics. As a consequence the cost function is near to being quadratic. If as Gauthier and Courtier (1992) we consider the observations as random variables and we look at variational analysis as attempting to solve the minimization problem

$$\min J(\mathbf{v}) = \frac{1}{2}(\mathbf{x} - \mathbf{x}_b)^T \mathbf{B}^{-1}(\mathbf{x} - \mathbf{x}_b) + \frac{1}{2}(\mathbf{Hx} - \mathbf{y})^T \mathbf{O}^{-1}(\mathbf{Hx} - \mathbf{y}) \tag{141}$$

where $\mathbf{x}_b$ is the unbiased background field and $\mathbf{y}$ the set of unbiased observations, both being realizations of random variables of covariances $\mathbf{B}$ and $\mathbf{O}$ respectively and where the operator $\mathbf{H}$ computes the model equivalent $\mathbf{Hx}$ of the observation $\mathbf{y}$. Then the Hessian $J''$ of the cost function $J$ at the minimum is given by

$$J'' = \mathbf{B}^{-1} + \mathbf{H}^T \mathbf{O}^{-1} \mathbf{H} \tag{142}$$

obtained by differentiating (141) twice.

Moreover the analysis error covariance matrix is the inverse of the Hessian as shown in Appendix B of Rabier and Courtier (1992). Calling $\mathbf{x_a}$ the result of the minimization (i.e. the analysis) and $\mathbf{x_t}$ the truth, one has

$$\mathbf{E}\left[(x_a - x_t)(x_a - x_t)^T\right] = (J'')^{-1} = (\mathbf{B}^{-1} + \mathbf{H}^T \mathbf{O}^{-1} \mathbf{H})^{-1} \tag{143}$$

A requirement is that the background error and the observation error are uncorrelated (Rabier and Courtier 1992; Fisher and Courtier 1995). See also work of Thepaut and Moll (1990) pointing out that the diagonal of the Hessian is optimal among all diagonal preconditioners.

## 20 Hessian Singular Vectors (HSV)

Computing HSV's uses the full Hessian of the cost function in the variational data assimilation which can be viewed as an approximation of the inverse of the analysis error covariance matrix and it is used at initial time to define a norm. The total energy norm is still used at optimization time. See work by Barkmeijer et al. (1998, 1999). The HSV's are consistent with the 3-D VAR estimates of the analysis error statistics. In practice one never knows the full 3-D VAR Hessian in its matrix form and a generalized eigenvalue problem has to be solved as described below.

The HSV's are also used in a method first proposed by Courtier et al. (1993) and tested by Rabier et al. (1997) for the development of a simplified Kalman filter fully described by Fisher (1998) and compared with a low resolution explicit extended Kalman filter by Ehrendorfer and Bouttier (1998). See also Buizza and Palmer (1995).

Let $\mathbf{M}$ be the propagator of the tangent linear model, $\mathbf{P}$ a projection operator setting a vector to zero outside a given domain. Consider positive-definite and symmetric operators including a norm at initial and optimization time respectively. Then the SV's defined by

$$\frac{< \mathbf{P}\varepsilon(t), \mathbf{E}\mathbf{P}\varepsilon(t) >}{< \varepsilon(t_0), \mathbf{C}\varepsilon(t_0) >} \tag{144}$$

under an Euclidean norm are solution of generalized eigenvalue problem.

$$\mathbf{M}^*\mathbf{P}^*\mathbf{E}\mathbf{P}\mathbf{M}\mathbf{x} = \lambda\mathbf{C}\mathbf{x} \tag{145}$$

In HSV, the operator $\mathbf{C}$ is equal to the Hessian of the 3-D Var cost function. As suggested by Barkmeijer et al. (1998), one can solve (145) by using the generalized eigenvalue algorithm (Davidson 1975). See also Sleijpen and Van der Vorst (1996). Using

$$\mathbf{C} \equiv \nabla^2 J = \mathbf{B}^{-1} + \mathbf{H}^T\mathbf{O}^{-1}\mathbf{H} \tag{146}$$

and carrying out a coordinate transformation

$$\mathbf{x} = \mathbf{L}^{-1}\mathbf{x}, \mathbf{L}^{-1}\mathbf{L} = \mathbf{B} \tag{147}$$

Then we obtain a transformed operator

$$(\mathbf{L}^{-1})^T\mathbf{C}\mathbf{L} \tag{148}$$

and the Hessian becomes equal to the sum of identity and a matrix with rank less or equal to the dimensions of the vector of observations (Fisher and Courtier 1995).

Veerse (1999) proposes to take advantage of this form of the appropriate Hessian in order to obtain approximations of the inverse analysis error covariance matrix, using the limited memory inverse BFGS minimization algorithm.

Let $\mathbf{H}$ be $(\nabla^2 J)^{-1}$ the inverse Hessian and $\mathbf{H}^+$ the updated version of the inverse Hessian.

$$\mathbf{s} = \mathbf{x}^{n+1} - \mathbf{x}^n \tag{149}$$

where $\mathbf{s}$ is the difference between the new iterate and the previous one in a limited-memory quasi-Newton minimization procedure.

$$\mathbf{y} = \mathbf{g}^{n+1} - \mathbf{g}^n \tag{150}$$

is the corresponding gradient increment. One has the formula

$$\mathbf{H}^+ = \mathbf{U}(\mathbf{H}, \mathbf{y}, \mathbf{s}) = \left( \mathbf{I} - \frac{\mathbf{s} \otimes \mathbf{y}}{<\mathbf{y}, \mathbf{s}>} \right) \frac{\mathbf{s} \otimes \mathbf{s}}{<\mathbf{y}, \mathbf{s}>} \tag{151}$$

where $<,>$ is a scalar product with respect to which the gradient is defined and $\otimes$ stands for the outer product.

The method is implemented by using the inverse Hessian matrix-vector product built in the minimization code and based on Nocedal's (1980) algorithm. These methods are useful when the second order adjoint method is not available due to either memory or CPU limitations.

## 21 4-D VAR Status Today

4-D VAR data assimilation is available and implemented today at several operational numerical weather prediction centers starting with European Centre for Medium-Range Weather Forecasts (ECMWF), (Rabier et al. 2000; Klinker et al. 2000) while a similar system was operational at Meteo-France in 2000 (Janiskova et al. 1999; Gauthier and Thepaut 2001; Desroziers et al. 2003). More recently 4-D VAR was implemented at UK Met office, Japan and Canada.

Park and Zupanski (2003) survey the status and progress of the four-dimensional variational data assimilation with emphasis on application to prediction of meso-scale/storm-scale atmospheric phenomena. See also Zupanski and Zupanski et al. (2002)

The impact of adopting 4-D VAR was qualified as a substantial, resulting in an improvement in NWP quality and accuracy (see Rabier (2005) in special Issue of QJRMS 2005).

4-D VAR combined with improvement in error specifications and with a large increase in a variety of observations has led to improvements in NWP accuracy (Simmons and Hollingsworth 2002).

Hollingsworth et al. (2005) shows how observing system improvements led to improvements of forecast scores while Bouttier and Kelly (2001) show that the improvement of forecast scores for the southern hemisphere are due to satellite data.

Also, error statistics for different sources of observation constitutes an active field of research aimed mainly at obtaining better representation of the specific observation operators.

## 22 Recent Algorithmic Developments of Note for 4-D VAR

Following an idea of Derber, Courtier et al. (1994) proposed and developed the incremental 4-D VAR algorithm, where minimization is carried out at reduced resolution in the inner iteration and on a linear model. The 4-D VAR incremental algorithm minimizes the following cost function (Rabier 2005)

$$J(\delta_{w_0}) = \frac{1}{2}\delta_{w_0}^T B^{-1}\delta_{w_0} + \frac{1}{2}\sum_{I=1}^{N}(H_i\delta_{X_i} - d_i)^T R_i^{-1}(H_i\delta_{X_i} - d_i) \tag{152}$$

with $\delta_{w_0} = s(X_0 - X_b)$.

Simplified increment at initial time $t_0$

$$d_i = y_i^o - H_i(X_i) \tag{153}$$

is the observation increment at time $t_i$. The solution resulting from minimization of the cost function is added to the background $X_b$ to obtain analysis at $t_0$ i.e

$$X_0^a = X^b - S^{-I}\delta_{w_0}^a \tag{154}$$

where $S^{-I}$ is the generalized inverse of operator $S$ which projects from high to low resolution (i.e $S^{-I}$ projects from low to high resolution). In an outer loop one updates the high resolution reference trajectory and observation departures. A refinement of the incremental 4-D VAR was proposed as a multi-incremental algorithm by Veerse and Thepaut (1998).

Physical parameterizations that have been modified to allow use in the linear models used in the incremental procedure were implemented by Janiskova et al. (2002), Lopez and Moreau (2005).

## 23 Impact of Observations

In view of high density of some observations horizontal thinning is performed on data sets, and optimal observation density is found by trial and error.

Another approach called "super-obbing", i.e. it averages neighboring observations. A new advance concerns the information content of the data. While usual method of estimating data impact in a forecasting system consists in performing observing system experiments (OSE) which turn out computationally expensive. However, another diagnostic called the "degrees of freedom for signal (DFS)" has been used by Rodgers (2000), Fisher (2003a, 2003b) and Cardinali et al. (2004).

Given an analysis $x_a$, background $x_b$ and observation $y^o$ we have

$$x_a = x_b + (B^{-1} + H^T R^{-1}H)^{-1}H^T R^{-1}(y^o - H(x_b)) \tag{155}$$

which can be written compactly as

$$x_a = x_b + Kd \tag{156}$$

$B$-being the background error covariance matrix, $R$ the observation error covariance, $H$-linearized observation operator of $H$. $K$ is called the Kalman gain matrix and $d$ innovation vector $d = y^o - H(x_b)$.

The DFS is defined as

$$DFS = Tr(HK) \tag{157}$$

where the trace of the matrix $HK$ measures the gain in information due to the observations of how an assimilation system extracts information signal from the background. One way to calculate DFS is the use of estimation the Hessian of the cost function provided. Fisher (2003a, 2003b) and Cardinali et al. (2004) used estimation of Hessian of the cost function provided by the minimization algorithm. Chapnik et al. (2006) use evaluation of trace of the $KH$ matrix, using a method put forward by Desroziers and Ivanov (2001) to evaluate trace of $KH$.

Computing sensitivity of forecast to the observations can be carried out by considering the adjoint of data assimilation together with the adjoint of the forecast model. This allows use of adaptive observations which is a topic of increased research efforts in 4-D VAR data assimilation (Berliner et al. 1999; Baker and Daley 2000; Daescu and Navon 2004; Langland and Baker 2004).

## 24 Conclusions

A condensed review of several aspects of 4-D VAR as it evolved in the last 30 or so years is presented. It aims to present both the history of 4-D VAR as well as its evolution by highlighting several topics of its application.

No attempt was made to cover here the vast ensemble Kalman filter data assimilation and its various flavors due to space and time limitations. In the same vein this review is not exhaustive as it is not covering all the issues dealing with 4-D VAR applications.

It has become amply evident that in the last 15 years major improvements in NWP are due to large extent to development of sources of observations and that 4-D VAR and sequential data assimilation can take advantage of them due to major research efforts at universities, federal laboratories and operational centers.

For new opportunities for research see the article by McLaughlin et al. (2005) that illuminates and outlines possibilities for enhanced collaboration within the data assimilation community.

It is certain that data assimilation concepts will become widely applied in all the geosciences as more geoscience scientific disciplines gain access to larger amounts of data, from satellite remote sensing and from sensor networks, and as Earth system models increase in both accuracy and sophistication.

It is hoped that this review highlights several aspects of 4-D VAR data assimilation and serves to attract interest of both atmospheric science practitioners as well as real time PDE constrained optimization research scientists.

**Acknowledgements** Prof. I.M. Navon would like to acknowledge the support of NSF grants ATM-0201808 and CCF-0635162.

# References

Aarnes JE (2004) Iterative methods for data assimilation and an application to ocean state modeling. Technical Report, SINTEF Applied Mathematics. Oslo, Norway

Akella S, Navon IM (2007) Different approaches to model error formulation in 4D-Var:a study with high resolution advection schemes. In revision process with Tellus

Amodei L (1995) Solution approchée pour un problème d'assimilation de données météorologiques avec prise en compte de l'erreur de modèle. Comptes Rendus de l'Académie des Sciences 321, série II: 1087–1094

Baker NL, Daley R (2000) Observation and background adjoint sensitivity in the adaptive observation-targeting problem. Q J R Meteorol Soc 126:1431–1454

Barkmeijer J et al (1998) Singular vectors and estimates of the analysis-error covariance metric. Q J R Meteorol Soc A:1695–1713

Barkmeijer J et al (1999) 3D-Var Hessian singular vectors and their potential use in the ECMWF ensemble prediction system. Q J R Meteorol Soc B:2333–2351

Barnes SL (1964) A technique for maximizing details in numerical weather map analysis. J Appl Meteor 3:395–409

Bellman RE (1957) Dynamic programming. Princeton University Press, Princeton

Bennett AF (2002) Inverse modeling of the ocean and atmosphere. Cambridge University Press, p 256

Bergthorsson P, Döös B (1955) Numerical weather map analysis. Tellus 7(3):329–340

Berliner LM et al (1999) Statistical design for adaptive weather observations. J Atmos Sci 56:2536–2552

Bertsekas DP (1982) Constrained optimization and Lagrange multiplier methods. Addison-Wesley, Tucson, AZ, p 491

Berz M et al (eds) (1996) Computational differentiation: techniques, applications, and tools. SIAM, Philadelphia, PA

Bischof CH et al (1992) Automatic differentiation of advanced CFD codes for multidisciplinary design. J Comput Syst Eng 3:625–638

Bloom SC, Shubert S (1990) The influence of Monte-Carlo estimates of model error growth on the GLA OI assimilation system. In: International symposium on assimilation of observations in meteorology and oceanography. WMO, Geneva, Switzerland, pp 467–470

Boer GJ (1984) A spectral analysis of predictability and error in an operational forecast system. Mon Wea Rev 112:1183–1197

Bouttier F, Courtier P (1999) Date assimilation concepts and methods. Meteorological Training Lecture Notes, ECMWF, Shinfield Park, Reading

Bouttier F, Kelly G (2001) Observing-system experiments in the ECMWF 4D-Var data assimilation system. Q J R Meteorol Soc 127:1469–1488

Bratseth A (1986) Statistical interpolation by means of successive corrections. Tellus 38A:439–447

Buizza R, Palmer TN (1995) The singular-vector structure of the atmospheric general circulation. J Atmos Sci 529:1434–1456

Cacuci DG et al (1980) Sensitivity theory for general systems of nonlinear equations. Nucl Sci Eng 75:88

Cacuci DG (1981a) Sensitivity theory for nonlinear systems. I: Nonlinear functional analysis approach. J Math Phy 12:2794–2803

Cacuci DG (1981b) Sensitivity theory for nonlinear systems. II. Extensions to additional classes of responses. J Math Phy 12:2803–2812

Cacuci DG (2003) Sensitivity & uncertainty analysis: A primer, CRC Press, Boca Raton, 1:304

Cacuci DG (2004) On the historical evolution of the use of adjoint operators in sensitivity and uncertainty analysis. Sixth Workshop on Adjoint Applications in Dynamic Meteorology, 23–28 May 2004, Acquafredda di Maratea, Basilicata, Italy

Cacuci DG et al (2005) Sensitivity and uncertainty analysis. Volume II Applications to Large-Scale Systems, CRC Press, Boca Rotan, p 368

Cardinali C et al (2004) Influence matrix diagnostic of a data assimilation system. Q J R Meteorol Soc 130:2767–2786

Chapnik B et al (2006) Diagnosis and tuning of observational error statistics in a quasi-operational data assimilation setting. Q J R Meteorol Soc 132:543–565

Charney JG et al (1950) Numerical integration of the barotropic vorticity equation. Tellus 2:237–254

Charney JG et al (1969) Use of incomplete historical data to infer the present state of the atmosphere. J Atmos Sci 2:1160–1163

Cohn SE (1997) An introduction to estimation theory. J Meteorol Soc Jpn 75:257–288

Courant R, Hilbert D (1962) Partial differential equations. Methods of Mathematical Physics. Wiley-Interscience, New York

Courtier P (1985) Experiments in data assimilation using the adjoint model technique. Workshop on High-Resolution Analysis ECMWF (UK) June

Courtier P, Talagrand O (1987) Variational assimilation of meteorological observations with the adjoint equations Part 2. Numerical results. Q J R Meteorol Soc 113:1329–1347

Courtier P, Derber J, Errico R, et al (1993) Important literature on the use of adjoint, variational-methods and the Kalman filter in meteorology Tellus A 45A(5):342–357.

Courtier P et al (1993) Progress in variational data assimilation at ECMWF. In: Boer GJ (eds) Research activities in atmospheric and oceanic modeling. Report No.181:42

Courtier P et al (1994) A strategy for operational implementation of 4D-VAR, using an incremental approach. Q J R Meteorol Soc 120 (519) B JUL:1367–1387

Courtier P (1997) Dual formulation of four-dimensional variational assimilation. Q J R Meteorol Soc 123 (544) B:2449–2461

Cressman GP (1959) An operational objective analysis system. Mon Wea Rev 87:367–374

Daescu DN, Navon IM (2004) Adaptive observations in the context of 4D-Var data assimilation. Meteor Atmos Phys 85:205–226

Dalcher A, Kalnay E (1987) Error growth and predictability in operational ECMWF forecasts. Tellus 39A:474–491

Daley R, Puri K (1980) Four-dimensional, data assimilation and the slow manifold. Mon Wea Rev 108:85–99

Daley R (1991) Atmospheric data analysis. Cambridge University Press, Cambridge

Daley R (1992a) The effect of serially correlated observation and model error on atmospheric data assimilation. Mon Wea Rev 120:164–177

Daley R (1992b) Estimating model-error covariances for application to atmospheric data assimilation. Mon Wea Rev 120:1735–1746

Da Silva A et al (1995) Assessing the effects of data selection with DAO's physical-space statistical analysis system. Proceedings of the second international Symposium on the assimilation of observations in meteorology and Oceanography. Tokyo, Japan

Davidson ER (1975) The iterative calculation of a few of the lowest eigenvalues and corresponding eigenvectors of a large real symmetric matrices. J Comput Phys 17: 87–94

Davis HE, Turner RE (1977) Updating prediction models by dynamic relaxation. An examination of the technique. Q J R Meteorol Soc 103:225–245

Dee D, Silva AD (1998) Data assimilation in the presence of forecast bias. Q J R Meteorol Soc 124:269–295

DelSole T, Hou AY (1999) Empirical correction of a dynamical model. Part I: fundamental issues. Mon Wea Rev 127: 2533–2545

Derber JC (1985) The variational four dimensional assimilation of analysis using filtered models as constraints. PhD, Thesis, Univ. of Wisconsin-Madison, p 141

Derber JC (1989) A variational continuos assimilation technique. Mon Wea Rev 117: 2437–2446

Desroziers G, Ivanov S (2001) Diagnosis and adaptive tuning of information error parameters in a variational assimilation. Q J R Meteorol Soc 127:1433–1452

Desroziers G et al (2003) A 4D-Var re-analysis of the FASTEX. Q J R Meteorol Soc 129:1301–1315

Ehrendorfer M, Bouttier F (1998) An explicit low-resolution extended Kalman filter: Implementation and preliminary experimentation. Sept, ECMWF Tech Memo No 259

Eliassen A (1954) Provisional report on calculation of spatial covariance and autocorrelation of the pressure field, Report # 5. Institute of Weather and Climate Res, Academy of Science, Oslo, p 11

Fisher M, Courtier P (1995) Estimating the covariance matrices of analysis and forecast errors in variational data assimilation. ECMWF Tech Memo 220:28

Fisher M (1998) Development of a simplified Kalman filter. ECMWF Tech Memo 260:16

Fisher M (2003a) Background error covariance modelling M Fisher seminar, Recent developments in data assimilation for atmosphere and ocean. 8–12 September European Centre for Medium-Range Weather Forecasts

Fisher M (2003b) Estimation of entropy reduction and degrees of freedom for signal for large variational analysis systems. Technical Memo, ECMWF, Reading, UK, p 397

Franke R, Gordon WJ (1983) The structure of optimal interpolation functions. Technical Report, NPS-53-83-0004. Naval Postgraduate School, Monterey, CA

Franke R (1988) Statistical interpolation by iteration. Mon Wea Rev 116:961–963

Gandin LS (1965) Objective analysis of meteorological fields. Leningrad: Gridromet (in Russian) (1963) English translation by the Israel program for scientific translations, p 242

Gauss CF (1809) Theoria motus corporum coelestium in sectionibus conicis solem ambientum Book 2, Section 2, pp 205–224. English translation by Ch. Davis, reprinted 1963 by Dover, New York, p 142

Gauthier P, Courtier P (1992) Data assimilation with an extended Kalman filter. ECMWF Workshop on variational assimilation with emphasis on three-dimensional aspects, Shinfield Park, Reading RG2 9AX, UK, p 171–190

Gauthier P, Thepaut JN (2001) Impact of the digital filter as a weak constraint in the preoperational 4DVAR assimilation system of Meteo-France. Mon Wea Rev 129:2089–2102

Ghil M et al (1979) Time-continuous assimilation of remote-sounding data and its effect on weather forecasting. Mon Wea Rev 107:140–171

Ghil M, Malonotte-Rizzoli P (1991) Data assimilation in meteorology and oceanography. Adv Geophys 33:141–266

Ghil M et al (1997) Data assimilation in meteorology and oceanography: Theory and practice. Meteorological Society of Japan and Universal Academy Press, Tokyo, p 496

Giering R, Kaminski T (1997) Recipes for adjoint code construction. Technical Report 212, Max-Planck-Institut für Meteorologie

Gilchrist B, Cressman GP (1954) An experiment in objective analysis. Tellus 6:309–318

Griewank A, Corliss GF (1991) Automatic differentiation of algorithms: theory, implementation, and application. SIAM, Philadelphia, p 353

Griewank A (2000) Evaluating derivatives principles and techniques of algorithmic differentiation. Frontiers Appl Math 19:369

Griffith AK, Nichols NK (1996) Accounting for model error in data assimilation using adjoint methods. In: Berz M et al (eds) Computational differentiation: Techniques, applications, and tools. SIAM, Philadelphia, PA, pp 195–205

Griffith AK (1997) Data assimilation for numerical weather prediction using control theory. The University of Reading, Department of Mathematics, PhD Thesis

Griffith AK, Nichols NK (2000) Adjoint methods in data assimilation for estimating model error. Flow Turbul Combust 65:469–488

Griffith AK, Martin MJ, Nichols NK (2000) Techniques for treating systematic model error in 3D and 4D data assimilation, in Proceedings of the Third WMO Int. Symposium on Assimilation of Observations in Meterology and Oceanography, World Meteorological Organization, WWRP Report Series No. 2, WMO/TD – 986, pp. 9–12.

Gunzburger MD (2003) Perspectives in flow control and optimization. Advances in design and control. SIAM, Philadelphia, USA

Haltiner GJ, Williams RT (1980) Numerical prediction and dynamic meteorology. John Wiley and Sons, New York, p 477

Hascoet L, Pascual V (2004) TAPENADE 2.1 user's guide. Technical report, INRIA, 300:81

Hestenes MR (1969) Multiplier and gradient methods. J Opt Theory Appl 4:303–320

Hoffmann RN (1986) A four dimensional analysis exactly satisfying equations of motion. Mon Wea Rev 114:388–397

Hoke JE, Anthes RA (1976) The initialization of numerical models by a dynamical initialization technique. Mon Wea Rev 194:1551–1556

Hollingsworth A, Lonnberg P (1986) The statistical structure of short-range forecast errors as determined from radiosonde data. Part I: the wind field. Tellus 38A:111–136

Hollingsworth A et al (2005) The transformation of earth-system observations into information of socio-economic value in GEOSS. Q J R Meteorol Soc 131: 3493–3512.

Ide K et al (1997) Unified notation for data assimilation: Operational, sequential and variational. J Meteor Soc Japan 75: 181–189

Janiskova M et al (1999) Simplified and regular physical parameterizations for incremental four-dimensional variational assimilation. Mon Wea Rev 127:26–45

Janiskova M et al (2002) Linearized radiation and cloud schemes in the ECMWF model: Development and evaluation. Q J R Meteorol Soc 128:1505–1527

Kagiwada H et al (1986) Numerical derivatives and nonlinear analysis, Mathematical Concepts and Methods in Science and Engineering. Plenum Press, New York, London

Kalnay E et al (2000) Application of the quasi-inverse method to data assimilation. Mon Wea Rev 128: 864–875

Kalnay E (2003) Atmospheric modeling, data assimilation and predictability. Cambridge Univ Press, Cambridge, p 341

Klinker E et al (2000) The ECMWF operational implementation of four-dimensional variational assimilation. III: Experimental results and diagnostics with operational configuration. Q J R Meteorol Soc 126:1191–1215

Krige DG (1951) A statistical approach to some mine valuation problems on the Witwatersrand. J Chem Metal Min Soc South African 52:119–139

Langland RH, Baker NL (2004) Estimation of observation impact using the NRL atmospheric variational data assimilation adjoint system. Tellus 56A:189–203

Lagrange JL (1760) Essai d'une nouvelle méthode pour déterminer les maxima et les minima des formules indéfinies. Miscellanea Taurinensia, 2 (1762) Oeuvres 1:365–468

Lanczos C (1970) The variational principles of mechanics. University of Toronto Press, Toronto

Le Dimet FX (1981) Une application des méthodes de contrôle, optimal 'a l'analyse variationnelle. Rapport Scientifique LAMP, Université Blaise-Pascal, 63170, Aubière Cedex, p 42

Le Dimet FX (1982) A general formalism of variational analysis. CIMMS Report, Norman, OK 73091 22:1–34

Le Dimet FX et al (1982) Dynamic initialization with filtering of gravity waves. CIMMS Report No.40, Norman, OK 73019, p 12

Le Dimet FX, Talagrand O (1986) Variational algorithms for analysis and assimilation of meteorological observations. Tellus 38A: 97–110

Le Dimet FX, Navon IM (1988) Variational and optimization methods in meteorology: A review. Tech. Rep. FSU-SCRI-88-144. Florida State University, Tallahassee, Florida, p 83

Le Dimet FX et al (1997) Sensitivity analysis in variational data assimilation. J Meteorol Soc Japan 1B:245–255

Le Dimet FX et al (2002) Second order information in data assimilation. Mon Wea Rev 130 (3):629–648

Lewins J (1965) Importance, the Adjoint Function., Pergamon Press, Oxford

Lopez P, Moreau E (2005) A convection scheme for data assimilation: Description and initial tests. Q J R Meteorol Soc 131: 409–436

Kolmogorov AN (1941) Interpolation and extrapolation of stationary random sequences. Translation by RAND Corporation, memorandum RM-30090-PR, April (1962). Bulletin of the Academy of Sciences, USSR series on Mathematics, p 15

Lewis JM, Derber JC (1985) The use of adjoint equations to solve a variational adjustment problem with advective constraints. Tellus 37A: 309–322

Lewis JM et al (2006) Dynamic data assimilation: a least squares approach. Cambridge University Press, Cambridge

Li Y et al (1993) Variational data assimilation with a semi-implicit semi-Lagrangian global shallow-water equation model and its adjoint. Mon Wea Rev 121:159–169

Li Y, Droegemeier KK (1993) The influence of diffusion and associated errors on the adjoint data assimilation technique. Tellus 45A:1–14

Li Y et al (1994) 4-D assimilation experiments with a multilevel semi-Lagrangian semi-implicit GCM. Mon Wea Rev.122(5):966–983

Lions JL (1968) Contrôle Optimal des Systèmes Gouvernes par des Equations aux Dérivés Partielles. Gauthiers-Villars, Paris

Lions JL (1971) Optimal control of systems governed by partial differential equations. Translated by Mitter SK, Springer-Verlag, Berlin-Heidelberg, p 404

Lorenc AC (1981) A global three-dimensional multivariate statistical interpolation scheme. Mon Wea Rev 109:701–721

Lorenc AC (1986) Analysis-methods for numerical weather prediction. Q J R Meteorol Soc 112 (474):1177–1194

Lynch P, Huang XY (1992) Initialization of the HIRLAM model using a digital filter. Mon Wea Rev 120(6):1019–1034

Marchuk GI (1958) Numerical calculation of nuclear reactors. Atomizdat, Moskwa, p 162

Marchuk GI (1967) Chislennye Metody v Prognoze Pogodi (Numerical Methods in Weather Forecasting). Gidrometeoizdat, Leningrad, p 356

Marchuk GI (1974) Numerical solution of the problem of dynamics of atmosphere and ocean (in Russian). Gidrometeoizdat, Leningrad, p 303

Martin MJ et al (2001) Estimation of systematic error in an equatorial ocean model using data assimilation. In: Baines MJ (eds) Numerical methods for fluid dynamics VII. ICFD, Oxford, pp 423–430

Matsuno T (1966) Numerical integrations of the primitive equations by simulated backward difference scheme. J Meteor Soc Japan 44:76–84

McPherson RD (1975) Progress, problems, and prospects in meteorological data assimilation. Bullet Am Meteorol Soc 56 (11):1154–1166

Menemenlis D, Chechelnitsky M (2000) Error estimates for an ocean general circulation model from altimeter and acoustic tomography data. Mon Wea Rev 128:763–778

Miller RN (1994) Advanced data assimilation in strongly nonlinear dynamical-systems. J Atmospheric Sci 51(8):1037–1056

Miller RN, Ghil M, Gauthiez F (1994) Advanced data assimilation in strongly nonlinear dynamical-systems. J Atmos Sci 51(8):1037–1056

Mitchell HL, Daley R (1997) Discretization error and signal/error correlation in atmospheric data assimilation, I:All scales resolved. Tellus 49A:32–53

Miyakoda K et al (1976) The near-real-time, global, four-dimensional analysis experiment during the GATE period, Part I. J Atmospheric Sci 33 (4):561–591

McLaughlin D (2005) Opportunities for enhanced collaboration within the data assimilation community. Q J R Meteorol Soc 131 (613):3683–3693 Part C

McLaughlin D, O'Neill A, Derber J, Kamachi M, (2005) Opportunities for enhanced collaboration within the data assimilation community. Q J R Meterol Soc 131 (613):3683–3693, Part C

Nash SG, Sofer A (1996) Linear and Nonlinear Programming. McGraw-Hill, p 692

Navon IM (1981) Implementation of a posteriori methods for enforcing conservation of potential enstrophy and mass in discretized shallow-water equations models. Mon Wea Rev 109:946–958

Navon IM, De Villiers RD (1983) Combined penalty multiplier optimization methods to enforce integral invariants conservation. Mon Wea Rev 111:1228–1243

Navon IM (1986) A review of variational and optimization methods in meteorology. In: Sasaki YK (ed) Festive volume of the international symposium on variational methods in geosciences. Elsevier Science Pub. Co. Developments in Geo-mathematics 5:29–34

Navon IM (1987) The Bayliss-Isaacson algorithm and the constraint restoration method are equivalent. Meteorol Atmospheric Phys 13:143–152

Navon IM, Zou X (1991) Application of the adjoint model in meteorology. In: Griewank A, Corliss G (eds) Proceedings of the international conference on automatic differentiation of algorithms: Theory, Implementation and Application. SIAM, Philadelphia, PA

Navon IM (1992) Numerical experience with limited memory quasi-Newton and truncated Newton methods. In: Phua KH et al (eds) Optimization techniques and applications, Vol 1, World Scientific Publishing Co, Singapore, 1232, p 33–48

Navon IM et al (1992a) Variational data assimilation with an adiabatic version of the NMC spectral model. Mon Wea Rev 122:1433–1446

Navon IM et al (1992b) Testing for reliability and robustness of optimization codes for large scale optimization problems. In: Phua KH et al (eds) Optimization Techniques and Applications Vol 1. World Scientific Publishing Co, Singapore, 1232, pp 445–480

Navon IM et al (2005) The impact of background error on incomplete observations for 4D-Var data assimilation with the FSU GSM. In: Sunderam VS et al (eds) Computational Science-ICCS 2005, LNCS 3515, 2005. Springer Verlag, Heidelberg, pp 837–844

Ngodock HE (1996) Data assimilation and sensitivity analysis. Ph D Thesis, LMC-IMAG Laboratory, University Joseph Fourier, Grenoble, France, p 213

Nichols NK (2003) Treating model error in 3-D and 4-D data assimilation. Proceedings of the NATO Advanced Study Institute on Data Assimilation for the Earth System, Maratea, Italy

Nocedal J (1980) Updating quasi-Newton matrices with limited storage. Math Comput 773–782.

Nocedal J, Wright SJ (2006) Numerical optimization. Second Edition, Springer, New York

Oblow EM (1983) An automated procedure for sensitivity analysis using computer calculus, Tech Report ORNL/TM-8776, ORNL

Panofsky HA (1949) Objective weather-map analysis. J Appl Meteor 6:386–392

Park SK, Kalnay E (1999) Application of the quasi-inverse method to storm-scale data assimilation. Abstracts, 5th SIAM Conference on Mathematical and Computational Issues in the Geosciences, 24–27 March, San Antonio, Texas, SIAM, 104

Park SK, Zupanski D (2003) Four-dimensional variational data assimilation for mesoscale and storm-scale applications. Meteor Atmos Phys 82:173–208

Park SK, Kalnay E (2004) Inverse three-dimensional variational data assimilation for an advection-diffusion problem: Impact of diffusion and hybrid application. Geophys Res Lett, 31:L04102

Parrish DF, Derber JC (1992) The National-Meteorological-Centers' spectral statistical-interpolation analysis system. Mon Wea Rev 120 (8):1747–1763

Penenko V, Obratsov NN (1976) A variational initialization method for the fields of the meteorological elements. Meteorol Gidrol 11:1–11

Pin et al (1987) ADGEN: An automated adjoin code generator for large-scale sensitivity analysis, Trans. Am Nucl Soc 55:311

Pontryagin LS et al (1962) The mathematical theory of optimal processes. John Wiley & Sons, New York

Powell MJD (1969) A method for nonlinear constraints in minimization problems. In: Fletcher R(eds) Optimization Chap 19. Academic Press, pp 283–298

Pu ZX et al (1997) Sensitivity of forecast errors to initial conditions with a quasi-inverse linear method. Mon Wea Rev Vol 125(10):2479–2503

Pu ZX, Kalnay E (1999) Targeting observations with the quasi-inverse linear and adjoint NCEP global models: Performance during FASTEX. Q J R Meteor Soc 125 (561):3329–3337

Rabier F, Courtier P (1992) Four-dimensional assimilation in the presence of baroclinic instability. Q J R Meteorol Soc 118:649–672

Rabier F et al (1997) Recent experimentation on 4D-Var and first results from a Simplified Kalman Filter. ECMWF Technical Memorandum No 240

Rabier F et al (2000) The ECMWF operational implementation of 4D variational assimilation Part I: experimental results with simplified physics. Q J R Meteorol Soc 126:143–1170

Rabier F (2005) Overview of global data assimilation developments in numerical weather-prediction centers. Q J R Meteorol Soc 131 (613):215–3233

Rall LB (1981) Automatic differentiation: Techniques and applications, Lecture Notes in Computer Science Vol 120. Springer-Verlag, Berlin

Richardson LF (1922) Weather prediction by numerical processes. Cambridge University Press. Reprinted by Dover (1965, New York). With a New Introduction by Sydney Chapman, xvi+236

Rodgers CD (2000) Inverse methods for atmospheres: theories and practice. World Scientific Publ, Singapore

Saha S (1992) Response of the NMC MRF model to systematic error correction within the integration. Mon Wea Rev 120:345–360

Sasaki YK (1955) A fundamental study of the numerical prediction based on the variational principle. J Meteor Soc Japan 33:262–275

Sasaki YK (1958) An objective analysis based on the variational method. J Meteor Soc Japan 36:77–88

Sasaki YK (1969) Proposed inclusion of time-variation terms, observational and theoretical in numerical variational objective analysis. J Meteor Soc Japan 47:115–203

Sasaki YK (1970a) Some basic formalisms in numerical variational analysis. Mon Wea Rev 98:857–883

Sasaki YK (1970b) Numerical variational analysis formulated under the constraints as determined by long-wave equations as a low-pass filter. Mon Wea Rev 98:884–898

Sasaki YK (1970c) Numerical variational analysis with weak constraint and application to the surface analysis of severe storm gust. Mon Wea Rev 98:899–910

Sasaki YK (1970d) A theoretical interpretation of anisotropically weighted smoothing on the basis of numerical variational analysis. Mon Wea Rev 99:698–707

Seaman RS (1988) Some real data tests of the interpolation accuracy of Bratseth's successive corrections method. Tellus 40A:173–176

Simmons AJ, Hollingsworth A (2002) Some aspects of the improvement in skill of numerical weather prediction. Q J R Meteorol Soc 128:647–677

Sleijpen GLG, HA Van der Vorst, (1996) A Jacobi-Davidson iteration method for linear eigenvalue problems. SIAM J Matrix Anal Vol 17, Issue 2:401–425

Tadjbakhsh IG (1969) Utilization of time-dependent data in running solution of initial value problems. J Appl Meteorol 8(3):389–391

Talagrand O, Miyakoda K (1971) Assimilation of past data in dynamical analysis. 2. Tellus 23(4–5):318–327

Talagrand O (1981) On the mathematics of data assimilation. Tellus 33 (4):321–339

Talagrand O, Courtier P (1987) Variational assimilation of meteorological observations with the adjoint vorticity equation-Part 1. Theory Q J R Meteorol Soc 113:1311–1328

Talagrand O (1997) Assimilation of observations, an introduction. J Meteorol Soc Japan 75 (1B) (1997):191–209

Tarantola A (1987) Inverse problem theory Methods for data fitting and model parameter estimation. Elsevier, Amsterdam, p 613

Thacker WC (1989) The role of Hessian matrix in fitting models to measurements. J Geophys Res 94 (C5):6177–6196

Thepaut JN, Moll P (1990) Variational inversion of simulated TOVS radiances using the adjoint technique. Q J R Meteorol Soc:1425–1448

Thepaut JN, Courtier P (1991) Four-dimensional variational assimilation using the adjoint of a multilevel primitive-equation model. Q J R Meteorol Soc 117:1225–1254

Thepaut JN et al (1993) Variational assimilation of meteorological observations with a multilevel primitive-equation model. Q J R Meteorol Soc 119:153–186

Thiébaux HJ, Morone LL (1990) Short-term systematic errors in global forecasts: Their estimation and removal. Tellus 42A:209–229

Thompson PD (1969) Reduction of Analysis Error Through Constraints of Dynamical Consistency. J Appl Meteorol 8:738–742

Tsyrulnikov MD (2005) Stochastic modelling of model errors: A simulation study. Q J R Meteorol Soc 131, pp 3345–3371

Uboldi F, Kamachi M (2000) Time-space weak-constraint data assimilation for nonlinear models. Tellus 52 A:412–421

Veerse F, Thepaut JN (1998) Multiple-truncation incremental approach for four-dimensional variational data assimilation. Q J R Meteorol Soc 124:1889–1908

Veerse F (1999) Variable-storage quasi-Newton operators as inverse forecast/analysis error covariance matrices in data assimilation. INRIA Technical Report 3685 Theme 4, p 28

Vidard PA et al (2000) 4-D variational data analysis with imperfect model. Flow Turbul Combust 65:489–504

Wang Z et al (1992) The second order adjoint analysis: Theory and application. Meteorol and Atmos Phy 50:3–20

Wang Z, (1993) Variational Data Assimilation with 2-D Shallow Water Equations and 3-D FSU Global Spectral Models, Ph.D. Dissertation, Department of Mathematics, College of Arts and Sciences, The Florida State University, 235 pp.

Wang Z et al (1993) The adjoint truncated Newton algorithm for large-scale unconstrained optimization. Tech Rep FSU-SCRI-92-170, Florida State University, Tallahassee, Florida, p 44

Wang Z et al (1995) A truncated Newton optimization algorithm in meteorology applications with analytic hessian vector products. Comput Opt Appl 4:241–262

Wang Z et al (1997) Application of a new adjoint Newton algorithm to the 3-D ARPS storm scale model using simulated data. Mon Wea Rev 125 (10):2460–2478

Wang Z et al (1998) The adjoint Newton algorithm for large-scale unconstrained optimization in meteorology applications. Comput Opt Appl 10:283–320

Wergen W (1992) Effect of model errors in variational assimilation. Tellus 44 A:297–313

Wigner N (1949) Extrapolation, interpolation and smoothing of stationary time series with engineering applications. M.I.T. Press, Cambridge, MA, p 163

Wigner EP (1945) Effects of small perturbations on pile period. Chicago Rep, CP-G 3048

Xu L et al (2005) Development of NAVDAS-AR: formulation and initial tests of the linear problem. Tellus 57A:546–559

Yaglom A (1962) An Introduction to the theory of stationary random functions. Dover, New York

Yang W et al (1996) A new Hessian preconditioning method applied to variational data assimilation experiments using adiabatic version of NASA/GEOS-1 GCM. Mon Wea Rev 124 (5):1000–1017

Zhu K et al (1994) Variational data assimilation with a variable resolution finite-element shallow-water equations model. Mon Wea Rev 122 (5):946–965

Zhu J, Kamachi M (2000) An adaptive variational method for data assimilation with imperfect models. Tellus 52 A:265–279

Zou X et al (1992) Incomplete observations and control of gravity waves in variational data assimilation. Tellus 44A:273–296

Zou X et al (1993) Numerical experience with limited-memory quasi-Newton methods and truncated Newton methods. SIAM J Num Opt 3:582–608

Zou J et al (1995) Sequential open-boundary control by data assimilation in a limited area model. Mon Wea Rev 123 No. 9:2899–2909

Zupanski M (1993) Regional four-dimensional variational data assimilation in a quasi-operational forecasting environment. Mon Wea Rev 121:2396–2408

Zupanski D (1997) A general weak constraint applicable to operational 4DVAR data assimilation systems. Mon Wea Rev 125:2274–2292

Zupanski M, Kalnay E (1999) Principles of data assimilation. In: Browning KA, Gurney RJ (eds) Global energy and water cycles. Cambridge Univ Press, Cambridge, pp 48–54

Zupanski D, Zupanski M, Rogers E, et al. (2002) Fine-resolution 4DVAR data assimilation for the Great Plains tornado outbreak of 3 May 1999 Weather and Forecasting 17(3):506–525

Zupanski M et al (2005) CIRA/CSU four- dimensional variational data assimilation system. Mon . Wea Rev 123:829–843

# Theoretical and Practical Issues of Ensemble Data Assimilation in Weather and Climate

Milija Zupanski

**Abstract** Practical and theoretical issues of ensemble data assimilation are presented and discussed. In presenting the issues, the dynamical view, rather than a typical statistical view, is emphasized. From this point of view, most problems in ensemble data assimilation, and in data assimilation in general, are seen as means of producing an optimal state that is in dynamical balance, rather than producing a state that is optimal in a statistical sense. Although in some instances these two approaches may produce the same results, in general they are different. Details of this difference are discussed.

An overview of several fundamental issues in ensemble data assimilation is presented in more detail: dynamical balance of analysis/forecast, inclusion of nonlinear operators, and handling of reduced number of degrees of freedom in realistic high-dimensional applications.

An ensemble data assimilation algorithm named the Maximum Likelihood Ensemble Filter (MLEF) is described as a prototype method that addresses the above-mentioned issues. Some results with the MLEF are shown to illustrate its performance, including the assimilation of real observations with the Weather Research and Forecasting (WRF) model.

## 1 Introduction

Data assimilation (DA) is an important component of everyday weather forecasts, as it provides the initial conditions for numerical models. Data assimilation can be defined as a methodology to combine the information from the models and from observations (e.g., measurements) in such way to produce best forecast and/or best analysis. In practice, both goals require that DA produce optimal initial conditions. In numerical weather prediction (NWP) and operational applications these initial conditions are used to predict future weather. Also, these initial conditions can be

M. Zupanski (✉)
Cooperative Institute for Research in the Atmosphere, Colorado State University,
Fort Collins, CO 80523-1375, USA,
e-mail: ZupanskiM@cira.colostate.edu

S.K. Park, L. Xu, *Data Assimilation for Atmospheric, Oceanic and Hydrologic Applications*, DOI 10.1007/978-3-540-71056-1_3,
© Springer-Verlag Berlin Heidelberg 2009

treated as an optimal analysis, and as such could be used as a verification field, or to initiate other models in order to simulate a weather event of interest. In climate applications there is less dependence of the prediction on the initial conditions, and more dependence on model errors and model empirical parameters. From the mathematical point of view, since all these variables are subject to the same probability laws, data assimilation in weather and in climate can be seen as equivalent.

As the prediction models are simulating finer and finer details, they require a high-dimensional state vector. In regard to DA, this implies more intensive calculation, and possibly some restrictions in practice due to limited computational resources. Typically the weather processes in smaller scales (i.e. finer model resolution) are more nonlinear than in larger scales, thus making the data assimilation a nonlinear problem. Many types of observations are nonlinearly related to the model variables, introducing additional nonlinearity to data assimilation. High-resolution and nonlinear models that cover a wide range of weather processes are also increasingly sensitive to small perturbation to its initial state, and can create spurious waves and processes, eventually leading to incorrect and useless forecast. In some instances, even if perfectly balanced, these small perturbations can still lead to useless forecast, which is then defined as a limit to predictability. This means that weather and climate are chaotic systems.

An analysis or forecast has little value if its uncertainty (e.g., error) is not known. Estimation of the analysis and forecast uncertainty, however, represents an additional challenge for DA. Uncertainty introduces the need for additional degrees of freedom, which then implies additional computational requirements and more careful theoretical development. All these components make data assimilation a very difficult problem that has to include estimation theory, control theory, probability and statistics, signal processing, and possibly other fields of mathematics and engineering.

Current operational data assimilation methods in weather and climate are the variational methods (Parrish and Derber 1992; Courtier et al. 1998; Rabier et al. 2000). Although powerful, these methods incorporate several theoretical limitations: (1) the forecast error covariance is modeled, (2) the uncertainty of analysis and forecast is typically not estimated, (3) an adjoint model is required, thus the development and maintenance of a DA system is demanding, and (4) good Hessian preconditioning is not possible on current super-computers, due to inability to invert matrices of large dimensions. It should be mentioned that there are wide spread efforts to improve all these aspects of variational data assimilation. The four-dimensional variational DA, which explicitly incorporates time dimension, has, to a certain degree, dynamically dependent analysis error covariance. In principle, the uncertainty can be estimated in variational methods, but that requires additional computational resources, and may be still an impossible task for high-dimensional problems. Most important limitation, however, comes from the fact that a modeled forecast error covariance is used, thus an estimate of uncertainty may not be very realistic. The Hessian preconditioning can be improved, but a typical state dimension, denoted $S$, is of the order of $S \sim O(10^6)$ in realistic applications. The inversion of a $S \times S$ matrix needed for preconditioning is practically impossible.

The mentioned limitations of variational data assimilation methods open an opportunity for development of new DA methodologies. Among them, the ensemble data assimilation methods are the most often used in applications to weather, oceanography and climatology. The ensemble data assimilation methods deal with $S \times N$ and $N \times N$ matrices, where $N$ denotes the number of ensembles. In a typical application $N \ll S$, making the manipulation of these matrices feasible on today's computers. Arguably the most important advantage of ensemble DA is a flow-dependent forecast error covariance. In principle, this allows more realistic estimate of uncertainty than variational methods, as well as it reduces the need for modeling of error covariance. However, since the estimate of uncertainty is an intrinsic component of ensemble DA, its calculation can require more computational time than variational methods. In practice, powerful super-computers are needed in order to have ensemble DA applied to realistic high-resolution models. Parallel processing is especially advantageous for use with ensemble DA due to a low communication between processors. This makes the computational overhead less of a problem for realistic ensemble DA applications. An additional advantage of ensemble DA is that an adjoint model is not needed, thus the system development and maintenance is not a critical issue. A potential problem of ensemble DA is that it has insufficient degrees of freedom that can lead to filter divergence. For that reason methods for increasing the effective number of degrees of freedom have been developed (Houtekamer and Mitchell 2001; Hamill et al. 2001; Ott et al. 2004; Hunt et al. 2007). Although all ensemble DA methods are generalizations of the Kalman filter to include nonlinear processes, the means of achieving this generalization differ. Consequently, a variety of ensemble data assimilation methods exist, although they could be roughly divided in two groups.

The Ensemble Kalman filtering (EnKF) method, originally developed by Evensen (1994) and Houtekamer and Mitchell (1998), can be viewed as an improvement of the Monte-Carlo Kalman filters. Therefore, it is a sample based statistical-dynamical filter. Numerous EnKF algorithms have been developed and used in meteorological data assimilation (Houtekamer and Mitchell 2001; Bishop et al. 2001; Anderson 2001; Whitaker and Hamill 2002; Ott et al. 2004; Hamill 2006; Mandel 2006, 2007). In the theoretical limit, the EnKF requires an infinite sample of forecasts, which is a practical impossibility. In order to deal with the problem of small sample, thus an under-estimation of the variance, the EnKF includes several tunable parameters related to covariance inflation and localization.

Another group of ensemble DA methods are developed as reduced-rank nonlinear approximations to the Kalman filter. As such, they utilize algebra and functional analysis in calculation of error covariances, rather than a sample-based statistical approach. The filters in this group are the MLEF (Maximum likelihood ensemble filter – Zupanski 2005; Zupanski and Zupanski 2006; Zupanski et al. 2008) and the SEEK (Single Evolutive Extended Kalman filter – Pham et al. 1998; Brasseur et al. 1999) filters. In order to calculate the square root forecast error covariance the SEEK filter calculates the leading Empirical Orthogonal Functions (EOF) from the ensemble of forecasts. The analysis is calculated using a formulation equivalent to the Kalman filter solution, however of reduced rank to correspond to the number of

EOFs. A recent review of the SEEK filter can be found in Rozier et al. (2007). For linear operators, both the MLEF and SEEK filters become equivalent to the Kalman filter. Although these two methods don't suffer from the problem of small statistical sample directly, they also have to deal with insufficient degrees of freedom in high-dimensional applications.

Another group of filters that are creating an interest in weather and climate applications of ensemble DA are the particle filters (van Leeuwen 2003; Xiong et al. 2006), originally developed for engineering applications (Gordon et al. 1993; Liu and Chen 1998). The idea of the particle filters is to approximate the posterior distribution of the state by a set of particles (i.e. ensembles). Different weight is assigned to each particle, such that the weight is proportional to the probability of that particle. Their advantage is that the non-Gaussian and nonlinear processes are much easier to be included, but there is still an issue of recalculating the weights in order to prevent the collapse of ensemble spread in situations with very few particles (e.g., Arulampalam et al. 2001; Doucet et al. 2001; Crisan and Doucet 2002).

In this paper we will focus on some practical and theoretical issues relevant to ensemble data assimilation. In Sect. 2 the issues related to nonlinearity will be discussed, and dynamical balance will be emphasized in Sect. 3. The issue of insufficient degrees of freedom will be addressed in Sect. 4. The results illustrating the ensemble data assimilation performance will be shown in Sect. 5, and conclusions will be drawn in Sect. 6.

## 2 Nonlinearity of Ensemble Data Assimilation

The prediction models and observation operators in meteorological applications are typically nonlinear. Of special importance are the satellite measurements, which overwhelmingly dominate in today's observation data. The prediction models are weakly nonlinear for advection processes, but are highly nonlinear for physical processes, such as the clouds and precipitation. Therefore, legitimate methods for data assimilation and prediction have to efficiently and accurately deal with nonlinear processes.

The nonlinearity is one of the main reasons for developing generalizations of the (inherently linear) Kalman filtering method (Jazwinski 1970). Ensemble data assimilation methods differ in their approach to nonlinearities. It is instructive to address these differences in the forecast and in the analysis separately.

### 2.1 Forecast

All ensemble data assimilation methods use a nonlinear prediction model to define the forecast error covariance. A nonlinear dynamical model $M$ transports the state vector $x$ in time according to

$$x_t = M(x_{t-1}) \tag{1}$$

where $t$-1 and $t$ refer to the current and the next analysis times, respectively. Note that the model error is neglected in Eq. (1) in order to simplify the derivation. To keep the notation manageable, we omit the time index whenever possible. The *fore-cast* increment resulting from the $i$-th analysis increment is typically defined as a difference between an ensemble forecast $x_i^f$ and a referent forecast state $x^f$

$$\Delta_i x^f = x_i^f - x^f = M(x_{t-1}^a + p_i^a) - x^f \tag{2}$$

where the superscripts $a$ and $f$ refer to analysis and forecast, respectively. The vectors $\{p_i^a; i = 1, \ldots, N\}$ represent the columns of the square-root analysis error covariance, at the time of previous analysis. The referent forecast state $x^f$ is typically defined as a deterministic forecast, or a linear combination of deterministic forecasts.

Based on the interpretation of the forecast increments $\Delta_i x^f$, one can distinguish two groups of ensemble data assimilation methods: (*i*) Monte-Carlo (MC) based methods, and (*ii*) Kalman filter (KF) based methods. The MC based methods are typically called the EnKF methods, while the KF based methods are often referred to as reduced-rank ensemble filters. In ensemble data assimilation both the dynamic and stochastic (e.g., random) components of the error are acknowledged. However, the mathematical formalism describing such-defined forecast error structure differs.

In MC based ensemble data assimilation methods the forecast increments are primarily interpreted as random realizations of the forecast error, thus the standard statistical formalism is applied. In order to create unbiased perturbations, the referent forecast state is defined as a sample mean of ensemble forecast perturbations

$$x^f = \frac{1}{N} \sum_{i=1}^{N} x_i^f \tag{3}$$

and the unbiased statistical estimate of the error covariance is given by

$$P_f = \frac{1}{N-1} \sum_{i=1}^{N} \left(\Delta_i x^f\right) \left(\Delta_i x^f\right)^T \tag{4}$$

where the index $N$ defines the number of ensembles.

In KF based ensemble data assimilation methods (e.g. SEEK and MLEF filters) the forecast increments are interpreted as span-vectors of the uncertainty subspace. The role of the forecast model is to transport the uncertainty subspace in time. From this definition, the referent forecast state, $x^f$, is the state with respect to which the error, or uncertainty is defined. Typically, it is defined as a forecast from the previous analysis, $x_{t-1}^a$

$$x^f = M(x_{t-1}^a) \tag{5}$$

while the forecast error covariance is defined as

$$P_f = \sum_{i=1}^{N} \left(\Delta_i x^f\right) \left(\Delta_i x^f\right)^T . \tag{6}$$

The advantages and disadvantages of mentioned ensemble data assimilation methods are still a subject of intensive research. We can only point out that by using the statistical framework, the Monte Carlo definition assumes random, by definition unpredictable forecast errors. Although similar assumptions are made in the initial step of the KF methods, the mathematical formalism employed in the remaining calculations (i.e. the forecast and analysis steps) is based on algebra and functional analysis. In order for statistics to be representative, the MC based methods require large, ideally an infinite statistical sample. On the other hand, the KF based methods need only up to $S$ ensemble members. This creates an algorithmic advantage for the KF based methods in low-dimensional applications, since error covariance localization or inflation is not needed. However, in realistic high-dimensional applications this advantage may disappear, and both approaches have to find the ways to deal with insufficient degrees of freedom given by inadequate number of ensemble perturbations.

## 2.2 Analysis

The analysis step of ensemble data assimilation is defined as a procedure in which the forecast information is blended with the observation information, eventually producing an analysis state. The observations $\mathbf{y}$ are related to the model state $\mathbf{x}$ by

$$\mathbf{y} = H(\mathbf{x}) + \varepsilon_R \tag{7}$$

where $H$ is a nonlinear observation operator and $\varepsilon_R$ is the observation error, which includes the measuring instrument error, the error of the operator $H$, and the representativeness error (e.g., Cohn 1997).

The analysis reveals a number of hidden and obvious assumptions in an ensemble data assimilation system. In standard EnKF methods (Evensen 1994; Houtekamer and Mitchell 1998) the statistical mathematical formalism is consistently used throughout the analysis procedure, leading to the use of perturbed observations. In square-root EnKF methods (Anderson 2001; Bishop et al. 2001; Whitaker and Hamill 2002; Ott et al. 2004) the statistical formalism is approximated by several assumptions, without perturbing the observations, leading to a more efficient algorithm. The differences and similarities of the standard EnKF and the square-root EnKF methods were discussed in detail in Tippet (2003). Also, the review by Hamill (2006) describes various EnKF methodologies. In this paper we focus on the analysis solution methodology, which reveals the assumptions related to linearity and differentiability of observation operators.

Before we discuss details, it is important to recall that Kalman filtering is originally defined for linear operators, thus the analysis and its uncertainty are calculated using a Gaussian PDF assumption and linear operators. In realistic applications, however, one is interested in developing the capability to use nonlinear operators. From that point of view, one can again distinguish two major groups of ensemble data assimilation methods: (i) direct solution methods, and (ii) iterative solution methods. The direct solution methods are the methods that use the direct linear solution of the Kalman filter, improved by inserting nonlinear observation operators. Practically all ensemble data assimilation methods belong to this category, including the standard EnKF, square-root EnKF, and the SEEK filter. The iterative solution methods find nonlinear analysis solution by minimizing a cost function. An ensemble data assimilation method that belongs to this category is the MLEF.

The direct solution methods utilize the linear solution of the Kalman filter

$$\mathbf{x}_t^a = \mathbf{x}^f + \mathbf{K}\left(\mathbf{y} - \mathbf{H}\mathbf{x}^f\right) \tag{8}$$

where the Kalman gain $K$ is

$$\mathbf{K} = \mathbf{P}^f \mathbf{H}^T \left(\mathbf{H}\mathbf{P}^f \mathbf{H}^T + \mathbf{R}\right)^{-1}. \tag{9}$$

The nonlinearity is introduced by calculating sample estimates of the matrices $\mathbf{H}\mathbf{P}^f$ and $\mathbf{H}\mathbf{P}^f\mathbf{H}^T$ with nonlinear $H$ operators (Evensen 1994; Houtekamer and Mitchell 1998). Several methods can be defined based on the technique used to invert the matrix in Eq. (9), as discussed in Tippett (2003). The analysis error covariance is

$$\mathbf{P}^a = (\mathbf{I} - \mathbf{K}\mathbf{H})\mathbf{P}^f(\mathbf{I} - \mathbf{K}\mathbf{H})^T. \tag{10}$$

As suggested in Whitaker and Hamill (2002) this estimate of $\mathbf{P}^a$ is systematically underestimated and additional procedures (e.g., error covariance inflation) are used in order to correct the problem.

In the iterative solution methods one needs to define a cost function to be minimized. The most natural choice is to define the posterior conditional probability, denoted $P(X|Y)$, where X relates to the model state and Y to the observations. Equivalent to searching for the state that maximize the posterior probability, one can search for the minimum of the cost function (as in variational methods)

$$J = -\ln P(X|Y). \tag{11}$$

For Gaussian PDFs, using the Bayes rule and a common assumption of independence between the observation and model state errors (Lorenc 1986), the cost function is

$$J(\mathbf{x}) = \frac{1}{2}\left(\mathbf{x} - \mathbf{x}^f\right)^T P_f^{-1}\left(\mathbf{x} - \mathbf{x}^f\right) + \frac{1}{2}\left[\mathbf{y} - H(\mathbf{x})\right]^T R^{-1}\left[\mathbf{y} - H(\mathbf{x})\right]. \tag{12}$$

The observation error is defined as $\mathbf{R} = \left\langle \varepsilon_R \varepsilon_R^T \right\rangle$, where $\langle \cdot \rangle$ denotes the mathematical expectation. The difference from variational cost function is in the definition of the forecast error covariance: instead of being modeled, in ensemble data assimilation the covariance $\mathbf{P}^f$ is calculated from ensembles as described in Sect. 2.1.

The minimizing solution can be expressed in various forms, depending on the minimization algorithm used. In the first iteration, however, one would obtain (Jazwinski 1970; Zupanski 2005) an equivalent of

$$\mathbf{x}^1 = \mathbf{x}^f + \alpha^1 \mathbf{H}\mathbf{P}^f\left(\mathbf{R} + \mathbf{H}\mathbf{P}^f\mathbf{H}^T\right)^{-1}\left(\mathbf{y} - H(\mathbf{x}^f)\right) \tag{13}$$

where the step-length is denoted $\alpha^1$. Formally, this expression is identical to the direct solution (8)–(9), with the exception of the step-length. In direct solution methods the step-length is assumed to be equal to one, which is true only for linear observation operators (i.e. quadratic cost function). This shows that the two methods are related, and that the direct solution methods could also be viewed as an iterative minimization with only one iterative step. The analysis error covariance can be represented as the inverse Hessian estimated at the minimum

$$\mathbf{P}^a = \left(\mathbf{P}^f\right)^{T/2} \left(\mathbf{I} + \mathbf{R}^{-1/2}\mathbf{H}\mathbf{P}^f\mathbf{H}^T\mathbf{R}^{-1/2}\right)^{-1} \left(\mathbf{P}^f\right)^{1/2}. \tag{14}$$

The actual calculation utilizes nonlinear operators, as shown in Zupanski et al. (2008). Note that the form of the analysis error covariance is similar to formulations in the ensemble transform Kalman filter (ETKF – Bishop et al. 2001) and in the SEEK filter (Pham et al. 1998; Rozier et al. 2007). The major difference is that in the MLEF the formula (14) is estimated at the minimum point, while in the ETKF and the SEEK filters it is calculated at the first guess. For highly nonlinear problems, the first guess and the minimum significantly differ, implying a difference between the methods.

An important advantage of iterative solution methods is in applications with nonlinear observation operators. For highly nonlinear operators the Gaussian PDF assumption may not hold any longer. This is another potential advantage of the iterative solution methods since one can define new cost function to reflect this change of PDF (Fletcher and Zupanski 2006a, b). On the other hand, the direct solution methods are derived using the Gaussian PDF assumption, and it may not be straightforward to include non-Gaussian PDFs.

The similarity between the analysis solution in direct methods, and in the *first iteration* of iterative methods also reveals the relevance of the Kalman gain matrix. It is a perfect quadratic Hessian preconditioner, which is utilized in the MLEF algorithm.

## 3 Dynamical Balances in Ensemble DA

Another aspect of ensemble DA is related to dynamical consistency of analysis. Since the analysis is typically used as initial conditions for the forecast model, it is important to have a balanced analysis. There are several techniques that could control the noise in the analysis. For example, it is possible to apply a low-pass filtering technique that would create a more balanced analysis. However, the filtering could also move the analysis away from observations, thus potentially worsening, or even nullifying, the positive impact of the original analysis. Possibly more appealing would be a method that simultaneously performs data assimilation and controls the dynamical balance.

We will examine the dynamical balance of analysis using the KF and the MC-based methods as examples. The analysis in the KF and MC-based ensemble DA can be written as (see Appendix for mathematical derivation)

$$(\mathbf{x}_t^a)_{MC} = \mathbf{u}_w + \left(1 - \sum_{i=1}^{N} w_i\right) \frac{1}{N} \sum_{i=1}^{N} M(\mathbf{x}_{t-1}^a + \mathbf{p}_i^a) \tag{15}$$

$$(\mathbf{x}_t^a)_{KF} = \mathbf{u}_w + \left(1 - \sum_{i=1}^{N} w_i\right) M(\mathbf{x}_{t-1}^a) \tag{16}$$

where $\mathbf{u}_w$ is a linear combination of ensemble forecasts (Appendix, Eq. (20)), and $w_i$ are the weights given to ensemble forecast perturbations.

Now, we make several assumptions. First, we assume that the analysis from the previous analysis cycle is in dynamical balance. This is not a very strict assumption being in agreement with a DA strategy in numerical weather prediction (NWP) to form balanced initial conditions for the forecast model. We will also assume that the ensemble (i.e. perturbed) initial conditions, corresponding to the previous analysis, are not in dynamical balance. In addition, we will assume that short-range ensemble forecast is not sufficiently long to eliminate the spurious noise. Although this may be more restrictive assumption, it is not unrealistic. Current ensemble DA schemes typically do not include a control of noise in ensemble initial conditions. Also, many ensemble DA techniques add new ensemble perturbations in order to improve the ensemble spread. In general these perturbations are not dynamically balanced.

Under the above assumptions, $\mathbf{u}_w$ denotes the unbalanced component of the analysis. The MC-based analysis also contains an extra term, unbalanced under our assumptions. However, this does not necessary imply greater imbalance in the MC-based method, since the term $\mathbf{u}_w$ can have larger magnitude in the KF-based methods. Relative magnitude of these terms will determine the imbalance of analysis. Equations (15) and (16) state that final analysis will be dynamically unbalanced unless a procedure for controlling the noise in ensemble forecasts is applied. One should also note that, due to an implicit control of the numerical noise, longer ensemble forecasts are more balanced, even with noisy initial conditions. For very short DA cycles that are anticipated in high-resolution mesoscale ensemble DA applications, this would imply additional difficulties regarding the dynamical balance of the analysis. One can also use other, computationally more expensive methods to control the balance, such as the ensemble smoother (e.g. Evensen and van Leeuwen 2000). Although this may be beneficial in situations where the balance is of extreme importance, such as for assimilation of precipitation and cloud variables, the computational cost may restrict their practicality.

## 4 Degrees of Freedom for Signal

A common problem in realistic high-dimensional applications is the insufficient number of ensembles. This directly relates to the inadequate number of degrees of freedom (DOF) for signal, which eventually may cause the filter divergence. This issue was addressed in ensemble data assimilation development (Houtekamer and Mitchell 1998; Hamill et al. 2001; Ott et al. 2004), and it has become an important component of a high-dimensional ensemble data assimilation algorithm. There are three basic strategies currently used in ensemble data assimilation: (i) Schur product, (ii) local domains, and (iii) hybrid. There are other methods as well (Hunt et al. 2007; Bishop and Hodyss 2007), that employ components of several strategies. This research is still very active.

The Schur product is an element-wise product between the ensemble error covariance and a prescribed space-limited covariance matrix. The resulting error covariance is a modified ensemble error covariance that is spatially localized. The benefit is that spurious distant correlations from ensembles are not affecting the analysis. The downside is that the prescribed error covariance has inadequate dynamical structure, and may introduce noise in the resulting error covariance, thus in the analysis solution. In addition, the Schur product approach may be computationally demanding in high-dimensional systems.

The local domains refer to finding local analysis solutions in prescribed geographical sub-domains, which are then combined to form the global solution. In the original local domains approach there is no prescribed error covariance, and the local analysis is in dynamical balance. The problem comes when forming the global solution, since the boundaries between local domains create a discontinuity in the global analysis. In order to smooth out the discontinuity between local boundaries several improvements were introduced, such as the assimilation of observations outside of local domain within a predefined distance from the boundary. Another problem of using local domains is that they are predefined and do not reflect the dynamics. As a consequence, the local domains can have a boundary cutting through a dynamical system (e.g., frontal zone, for example), causing the two local analyses to have a discontinuity. The computational advantage of local domains is due to the parallel processing of local analyses, without the need for point-wise processing of error covariances.

The hybrid methods (Hamill and Snyder 2000; Wang et al. 2006) improve the number of DOF by creating a modified error covariance by adding a prescribed error covariance to the ensemble error covariance. The relative weight given to each of the covariance components is also prescribed. The advantage is that all degrees of freedom can be accounted for, even if ensembles did not match well the true uncertainties. Remaining questions of this approach are related to improving the understanding of the interaction between the ensemble error covariance and the prescribed error covariance, including the impact of prescribed error covariance in creating the noise in the analysis.

One of main characteristics of the mentioned strategies is that the resulting forecast error covariance is spatially localized, which prevents spurious distant correlations. Another major impact of error covariance localization is that the number of DOF is increased. This greatly improves the fit to observations and reduces the risk of filter divergence. However, there are some limits to these approaches. Ideally, with all DOF available, the error covariance is reflecting the dynamical structure of uncertainties. Although one could, in principle, create a full-rank diagonal (co)variance matrix (mathematically accounting for all degrees of freedom), this would not correspond to dynamics and would result in spurious gravity waves or other numerical noise in the forecast model. In order to have a dynamically balanced analysis correction, it is important to create the localized error covariance that approximates the truth. Obviously, this is a difficult problem, and not fully satisfactory resolved in current ensemble data assimilation algorithms. Even with these limitations, the error covariance localization produces improved results and it is commonly used in ensemble data assimilation.

## 5 Results

In this section we illustrate ensemble DA results using the MLEF algorithm as a prototype ensemble data assimilation algorithm. We apply the MLEF *without* error covariance localization in small and medium size applications, and *with* error covariance localization in a high-dimensional application.

The first example is the assimilation of simulated observations using a one-dimensional soil temperature model (developed at the University of Belgrade, Serbia, after Mihailovic et al. 1999) and applied with the MLEF (Biljana Orescanin – private communication). An interesting aspect is that this version of the model has only a single grid point, implying the state dimension equal to one. According to our discussion in Sect. 2, a single ensemble member would be sufficient in this case for the KF-based methods. In order to test this proposition, several experiments are conducted, one without assimilation (NOASSIM) and two with the MLEF assimilation. The MLEF assimilation experiments include the test with 1 ensemble member, and also with 8 ensemble members. The results are illustrated in Fig. 1, where the difference from the true state is plotted. Due to the same forcing in both experiments, all solutions eventually merge, creating a zero error. One can

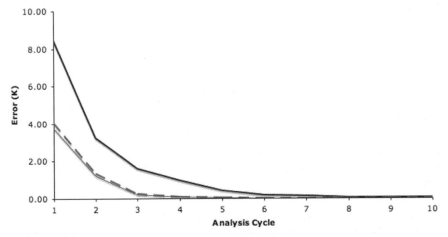

**Fig. 1** Analysis error of soil temperature (K) for 1-dimensional soil model with: (**a**) 1 ensemble (*dashed line*), and (**b**) 8 ensembles (*thin solid line*). *Thick solid line* denotes the NOASSIM experiment

see that MLEF experiments clearly outperform the NOASSIM experiment, indicating a positive impact of data assimilation. Especially interesting is the comparison between the two MLEF experiments, which produce virtually identical root-mean-squared errors. In other words, eight ensembles did not improve the results of the one ensemble experiment. This suggests that one ensemble is sufficient to represent the uncertainty of the analysis, as anticipated. For the MC based methods much larger number of ensembles would be required, since one ensemble is clearly not sufficient for a reliable statistical estimate. Although it may not be difficult to run more ensembles with such a simple model, thus creating a reliable statistical estimate, this example is illustrating the difference between the KF and the MC based methods.

The medium size example is related to the development of a super-GCM (Global Circulation Model with super-parameterization – Randall et al. 2003). In particular, the MLEF is developed to work with the two-dimensional Goddard Cumulus Ensemble (GCE) model (Tao and Simpson 1993; Tao et al. 2003) in assimilation of simulated observations. The GCE is a cloud-resolving model (CRM) and is used in Goddard multi-scale modeling framework (MMF).

The MLEF-GCE system is used in simulation of a South China Sea Monsoon Experiment (SCSMEX) convective period, prior to the monsoon development (18 May 1998). The simulated observations were created by adding a random observation error to the model forecast. The simulated observations are available in 1-hour intervals throughout the assimilation. For the experiment shown here, the potential temperature is assimilated. The GCE model used in this study has 514 points in horizontal and 43 vertical levels. The number of ensembles is 200, and there is no error covariance localization applied.

In Fig. 2 it is shown how the analysis error covariance is changing in time, from the analysis cycle 1 to the cycle 10. The striking feature is that the spatially localized structure of error covariance is developing with time, even though there is no error covariance localization procedure applied. The initially noisy error covariance (due to the inadequate initial first guess) quickly adjusts to the observations and dynamics. This is important since it indicates that the MLEF system has a natural, dynamical error covariance localization capability, that can be sufficient in some small and medium-size problems, in agreement with results for general dynamical systems (e.g., Pikovsky and Politi 1998). Note that the covariance between distant points is not identical to zero, but it has a very small value.

Another application we present is the assimilation by the MLEF with the Weather Research and Forecasting (WRF) model and real observations. The radiosonde (upper-air) and synop (surface) observations are assimilated in a hurricane Katrina example (August 25–31, 2005). Majority of observations were surface observations, with very few upper-air observations, making the recovering of the hurricane three-dimensional structure difficult. Since the hurricane had most important development over the ocean, where very few observations were available, this example also tests the robustness of the MLEF. The model horizontal resolution is 30 km, and there are 28 vertical levels. The shown results are obtained with 32 ensemble members. Since in this case the model state dimension was about 700,000 (including variables such

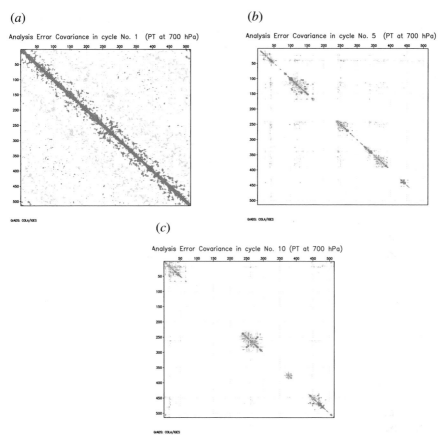

**Fig. 2** The analysis error covariance for potential temperature at 700 hPa: (**a**) in cycle 1, (**b**) in cycle 5, and (**c**) in cycle 10. The axes represent the indexes of the actual covariance matrix (514 × 514)

as surface pressure, temperature, winds, and specific humidity) an error covariance localization based on a modified local domains approach is used.

The results of the MLEF assimilation are compared with an experiment without assimilation (denoted NOASSIM – no assimilation). The difference between the 6-hour MLEF-initialized forecast and the NOASSIM forecast are shown in Fig. 3. Not only that the position and intensity of the hurricane Katrina is improved (not shown), but the impact of the analysis correction is localized. This is a desirable consequence of error covariance localization, seen in all model prognostic variables. One can see that the differences for temperature, cloud water and wind are localized in the vicinity of the hurricane. The difference in temperature and cloud water (Fig. 3a and d) indicates a correction of the position of the hurricane by the MLEF assimilation in general southwest direction, as it was observed. The wind difference (Fig. 3c) indicates an effective weakening of the cyclonic vorticity to the northeast, and strengthening to the southwest, again in dynamical agreement with

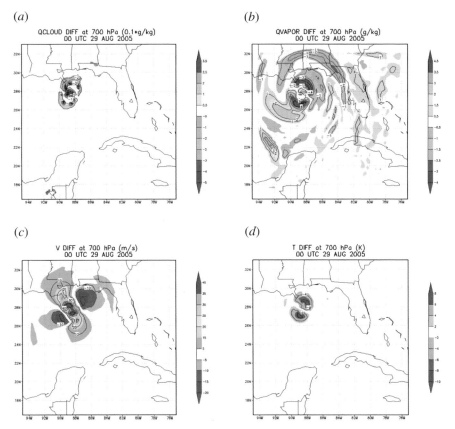

**Fig. 3** Difference between the MLEF and the NOASSIM experiments at 00 UTC on August 29, 2005, for: (**a**) cloud water, (**b**) specific humidity, (**c**) north–south wind, and (**d**) temperature

other variables. On the other hand, the specific humidity difference (Fig. 3b) is not confined to immediate vicinity of the hurricane. One can easily note the water vapor bands indicating an influx of moisture from a distance, again in agreement with general dynamics of tropical cyclones. These few examples of analysis correction are indicative of the difficulty of prescribing the error covariance localization parameters. This also suggests that a dynamical approach to data assimilation, with minimal number of prescribed values, is advantageous in weather and climate applications.

The results also illustrate that a stable ensemble data assimilation result can be obtained even in challenging high-dimensional applications with real observations. The flow-dependent character of ensemble data assimilation makes the analysis correction confined to a dominant dynamical system in the model domain, the hurricane in this case. The flow-dependence also explains why ensemble data assimilation works well with few ensembles: all degrees of freedom in ensembles are isolating unstable dynamics, and thus making most correction for the unstable dynamical system in the domain. This has an impact on reducing the error growth in the forecast, eventually producing an improved prediction.

# 6 Summary and Conclusions

An overview of ensemble data assimilation methodologies is presented, focusing on the issues related to nonlinearity and insufficient degrees of freedom. In reality data assimilation is a dynamic-stochastic method, thus it incorporates both the dynamics and the statistics. Similarity and differences between the KF-based and MC-based ensemble DA are discussed in more detail. The relevance of maintaining dynamical balance in DA is underlined. Our results indicate that a control of dynamical balance of ensemble (i.e. perturbed) initial conditions would be beneficial. This may be especially important in ensemble DA applications with shorter assimilation intervals, anticipated in high-resolution mesoscale applications.

The problem of nonlinearities in the analysis is also discussed. The numerical solution obtained by an iterative minimization is considered preferable over the more standard solution obtained by inserting nonlinear operators in the linear Kalman filter solution. The preferred iterative solution is based on the maximum posterior probability, rather than the minimum variance approach. Due to an implicit Hessian preconditioning, the iterative solution is computationally efficient. The maximum posterior probability approach has an additional advantage of efficiently including non-Gaussian PDFs, which may be important for variables belonging to skewed distributions, such as the precipitation and cloud variables.

The problems related to nonlinearities are often obscured by other outstanding problems, typically related to the insufficient number of DOF. Several basic strategies for increasing the number of DOF are discussed. In conclusion, the error covariance localization is a beneficial addition to ensemble data assimilation that provides the means for controlling the filter divergence. As a cautionary note, current error localization methods are not without limits, and an excessive amount of localization may introduce undesirable noise.

Even at the current state of development, ensemble data assimilation shows comparable results to variational data assimilation. A combination of ensemble and variational DA techniques has been pursued as well, suggesting that new development in DA can be unified. Given that there is an ample room for improvement, we believe that ensemble data assimilation will become increasingly the method of choice in future operational practice. One should also underline the systematic feedback between the ensemble data assimilation and ensemble forecasting, making the general ensemble approach to weather and climate even more appealing.

Important future development of ensemble data assimilation includes a development of new error covariance localization techniques with emphasis on satisfying the dynamical constraints. The development and applications of non-Gaussian ensemble data assimilation techniques will also become more relevant, especially in weather and climate applications to precipitation and cloud processes.

**Acknowledgements** This work was supported by the National Science Foundation Collaboration in Mathematical Geosciences Grant ATM-0327651, the National Aeronautics and Space Administration Grant NNG04GI25G, the National Oceanic and Atmospheric Administration Grant NA17RJ1228, and the DoD Center for Geosciences/Atmospheric Research at Colorado State

University under Cooperative Agreement W911NF-06-2-0015 with the Army Research Laboratory. We acknowledge computational support of NASA Ames Research Center Columbia supercomputer system. Our gratitude is also extended to the Computational and Information Systems Laboratory the National Center for Atmospheric Research, which is sponsored by the National Science Foundation, for the use of the NCAR Blueice supercomputer.

## Appendix: Analysis in Ensemble DA Methods

The square root forecast error covariance defines the space in which the analysis is corrected. One can define a general analysis update equation as

$$\mathbf{x}_t^a = \mathbf{x}^f + \mathbf{P}_f^{1/2}\boldsymbol{\gamma} = \mathbf{x}^f + \gamma_1 \mathbf{p}_1^f + \cdots + \gamma_N \mathbf{p}_N^f \tag{17}$$

where the index $t$ denotes the current analysis time, the vectors $\{\mathbf{p}_i^f \in S; i = 1, \ldots, N\}$ are the columns of the square-root forecast error covariance, and $\boldsymbol{\gamma} = (\gamma_1, \ldots, \gamma_N)^T$. In order to make the comparison between the two methods easier by using the forecast increments (e.g., Eq. (2)), we introduce a new weight $\mathbf{w} = (w_1, \ldots, w_N)^T$ such that

$$\mathbf{x}_t^a = \mathbf{x}^f + w_1 \Delta_1 \mathbf{x}^f + \cdots + w_N \Delta_N \mathbf{x}^f \tag{18}$$

For example, in the MC-based DA $\{w_i = \frac{\gamma_i}{\sqrt{N-1}}; i = 1, \ldots, N\}$, while in the KF-based DA $\{w_i = \gamma_i; i = 1, \ldots, N\}$, due to the definition of the forecast error covariance (e.g., Eqs. (4) and (6)). After substituting Eq. (2) in Eq. (17) one obtains

$$\mathbf{x}_t^a = \mathbf{u}_w + \left(1 - \sum_{i=1}^N w_i\right) \mathbf{x}^f, \tag{19}$$

where

$$\mathbf{u}_w = \sum_{i=1}^N w_i M(\mathbf{x}_{t-1}^a + \mathbf{p}_i^a). \tag{20}$$

The first term in Eq. (18) is common to all ensemble DA algorithms, and represents a linear combination of ensemble forecasts. The subscript $w$ underlines that the contribution of this term depends on the ensemble DA algorithm. For the MC-based ensemble DA, the substitution of Eq. (3) in Eq. (18) gives

$$(\mathbf{x}_t^a)_{MC} = \mathbf{u}_w + \left(1 - \sum_{i=1}^N w_i\right) \frac{1}{N} \sum_{i=1}^N M(\mathbf{x}_{t-1}^a + \mathbf{p}_i^a). \tag{21}$$

For the KF-based ensemble DA the definition (5) is substituted in Eq. (18) to give

$$(\mathbf{x}_t^a)_{KF} = \mathbf{u}_w + \left(1 - \sum_{i=1}^N w_i\right) M(\mathbf{x}_{t-1}^a). \tag{22}$$

# References

Anderson JL (2001) An ensemble adjustment Kalman filter data assimilation. Mon Wea Rev 129: 2884–2903

Arulampalam S, Maskell S, Gordon SN, Clapp T (2001) Tutorial on particle filters for on-line nonlinear/non-gaussian Bayesian tracking. IEEE Trans Signal Proc 50:174–188

Bishop CH, Etherton BJ, Majumdar SJ (2001) Adaptive sampling with the ensemble transform Kalman filter. Part I: Theoretical aspects. Mon Wea Rev 129:420–436

Bishop CH, Hodyss D (2007) Flow-adaptive moderation of spurious ensemble correlations and its use in ensemble-based data assimilation. Quart J Roy Meteor Soc 133:2029–2044

Brasseur P, Ballabrera-Poy J, Verron J (1999) Assimilation of altimetric data in the mid-latitude oceans using the Singular evolutive Extended Kalman filter with an eddy-resolving, primitive equation model. J Marine Syst 22:269–294

Cohn SE (1997) Estimation theory for data assimilation problems: Basic conceptual framework and some open questions. J Meteor Soc Japan 75: 257–288

Courtier P et al (1998) The ECMWF implementation of three-dimensional variational assimilation (3D-Var). I: Formulation. Quart J Roy Meteor Soc 124:1783–1808

Crisan D, Doucet A (2002) A survey of convergence results on particle filtering for practitioners. IEEE Trans Signal Proc 50:736–746

Doucet A, de Freitas N, Gordon N (2001) Sequential Monte Carlo methods in practice. Statistics for engineering and information science, Springer-Verlag, New York, 622pp

Evensen G (1994) Sequential data assimilation with a nonlinear quasi-geostrophic model using Monte-Carlo methods to forecast error statistics. J Geophys Res 99:10143–10162

Evensen G, van Leeuwen PJ (2000) An ensemble Kalman smoother for nonlinear dynamics. Mon Wea Rev 128:1852–1867

Fletcher SJ, Zupanski M (2006a) A data assimilation method for lognormally distributed observation errors. Quart J Roy Meteor Soc 132:2505–2520

Fletcher SJ, Zupanski M (2006b) A hybrid multivariate normal and lognormal distribution for data assimilation. Atmos Sci Let 7:43–46

Gordon NJ, Salmond DJ, Smith AFM (1993) Novel approach to nonlinear/non-gaussian Bayesian state estimation. Radar and Signal Process IEE-F 140:107–113

Hamill TM, Snyder C (2000) A hybrid ensemble Kalman filter – 3D variational analysis scheme. Mon Wea Rev 128:2905–2919

Hamill TM, Whitaker JS, Snyder C (2001) Distance-dependent filtering of background error co-variance estimates in an ensemble Kalamn filter. Mon Wea Rev 129:2776–2790

Hamill, TM (2006) Ensemble-based atmospheric data assimilation. Predictability of weather and climate. In: Palmer T, Hagedorn R (eds) Cambridge University Press, Cambridge, 718pp

Houtekamer PL, Mitchell HL (1998) Data assimilation using ensemble Kalman filter technique. Mon Wea Rev 126:796–811

Houtekamer PL, Mitchell HL (2001) A sequential ensemble Kalman filter for atmospheric data assimilation. Mon Wea Rev 129:123–137

Hunt BR, Kostelich EJ, Szunyogh I (2007) Efficient datra assimilation for spatiotemporal chaos: A local ensemble transform Kalman filter. Physica D 230: 112–126

Jazwinski AH (1970) Stochastic processes and filtering theory. Academic Press, New York.

Liu JS, Chen R (1998) Sequential Monte Carlo methods for dynamic systems. J. Amer. Stat. Soc. 93:1032–1044

Lorenc AC (1986) Analysis methods for numerical weather prediction. Quart J Roy Meteor Soc 112:1177–1194

Mandel J (2006) Efficient implementation of the ensemble Kalman filter. CCM Report 231, University of Colorado at Denver and Health Sciences Center, June 2006

Mandel J (2007) A brief tutorial on the ensemble Kalman filter. CCM Report 242, University of Colorado at Denver and Health Sciences Center, February 2007

Mihailovic DT, Kallaos G, Arsenic ID, Lalic B, Rajkovic B, Papadopoulos A (1999) Sensitivity of soil surface temperature in a force-restore equation to heat fluxes and deep soil temperature. Int J Climatol 19:1617–1632

Ott E, Hunt BR, Szunyogh I, Zimin AV, Kostelich EJ, Corazza M, Kalnay E, Patil DJ, Yorke JA (2004) A local ensemble Kalman filter for atmospheric data assimilation. Tellus 56A:273–277

Parrish DF, Derber JC (1992) The National Meteorological Center's spectral statistical interpolation analysis system. Mon Wea Rev 120:1747–1763

Pikovsky A, Politi A (1998) Dynamic localization of Lyapunov vectors in spacetime chaos. Nonlinearity 11:1049–1062

Pham DT, Verron J, Roubaud MC (1998) A singular evolutive extended Kalman filter for data assimilation in oceanography. J Marine Sys 16:323–340

Rabier F, Jarvinen H, Klinker E, Mahfouf J-F, Simmons A (2000) The ECMWF operational implementation of four-dimensional variational data assimilation. I: Experimental results with simplified physics. Quart J Roy Meteor Soc 126A:1143–1170

Randall D, Khairoutdinov M, Arakawa A, Grabowski W (2003) Breaking the cloud parameterization deadlock. Bull Amer Meteorol Soc 82: 2357–2376

Rozier D, Birol F, Cosme E, Brasseur P, Brankart JM, Verron J (2007) A reduced-order Kalman filter for data assimilation in physical oceanography. SIAM Rev 49:449–465

Tao W-K, Simpson J (1993) The Goddard cumulus ensemble model. Part I: model description. Terr Atmos Oceanic Sci 4:19–54

Tao W-K, Simpson J, Baker D, Braun S, Johnson D, Ferrier B, Khain A, Lang S, Shie C-L, Starr D, Sui C-H, Wang Y, Wetzel P (2003) Microphysics, radiation and surface processes in a nonhydrostatic model. Meteorol Atmos Phys 82:97–137

Tippett MK, Anderson JL, Bishop CH, Hamill TM, Whitaker JS (2003) Ensemble square-root filters. Mon Wea Rev 131:1485–1490

van Leeuwen PJ (2003) A variance-minimizing filter for large-scale applications. Mon Wea Rev 131:2071–2084

Xiong X, Navon IM, Uzunoglu B (2006) A note on the particle filter with posterior Gaussian resampling. Tellus 58A:456–460

Wang X, Hamill TM, Whitaker JS, Bishop CH (2006) A comparison of hybrid ensemble transform Kalman filter-optimum interpolation and ensemble square root filter analysis schemes. Mon Wea Rev 135:1055–1076

Whitaker JS, Hamill TM (2002) Ensemble data assimilation without perturbed observations. Mon Wea Rev 130:1913–1924

Zupanski D, Zupanski M (2006) Model error estimation employing ensemble data assimilation approach. Mon Wea Rev 134:1337–1354

Zupanski M (2005) Maximum likelihood ensemble filter: Theoretical aspects. Mon Wea Rev 133:1710–1726

Zupanski M, Navon IM, Zupanski D (2008) The maximum likelihood ensemble filter as a nondifferentiable minimization algorithm. Quart J Roy Meteor Soc 134:1039–1050

# Information Measures in Ensemble Data Assimilation

Dusanka Zupanski

**Abstract** In this chapter, we examine similarities and differences of the two information measures used in the current state-of-the-art data assimilation methods, the Degrees of Freedom (DOF) for signal and the E dimension. We evaluate these measures using simple arbitrary examples and a realistic ensemble data assimilation algorithm, including a complex atmospheric model and real observations.

The results indicate that the E dimension is more sensitive to the model dynamics and less sensitive to the amount and quality of the observations, while the opposite is true for the DOF for signal. These differences have to be taken into account when comparing the results of different data assimilation methods.

## 1 Introduction

Information theory (e.g., Shannon and Weaver 1949; Rodgers 2000) has been used in a number of atmospheric and oceanic studies to quantify predictability (e.g., Schneider and Griffies 1999; Kleeman 2002; Roulston and Smith 2002; DelSole 2004; Abramov et al. 2005, just to mention some).

Modern data assimilation methods employ information theory, through the use of a symmetric positive definite information matrix, to quantify and extract maximum information from the available observations. Typically, the eigenvalues of the information matrix are used to diagnose information content of assimilated observations via calculating information measures, such as Degrees of Freedom (DOF) for signal (e.g., Rodgers 2000) and Shannon entropy (e.g., Shannon and Weaver 1949; Rodgers 2000). Diagnosing information content of observations has proven especially useful in channel selections of the satellite observations (e.g., Rabier et al. 2002) and in a variety of other data assimilation applications within variational data assimilation methods (e.g., Wahba 1985; Purser and Huang 1993; Wahba et al. 1995; Fisher 2003; Johnson 2003; Engelen and Stephens 2004; L'Ecuyer et al. 2006).

D. Zupanski (✉)
Cooperative Institute for Research in the Atmosphere, Colorado State University, 1375 Campus Delivery, Fort Collins, CO 80523-1375, USA, e-mail: Zupanski@cira.colostate.edu

S.K. Park, L. Xu, *Data Assimilation for Atmospheric, Oceanic and Hydrologic Applications*, DOI 10.1007/978-3-540-71056-1_4,
© Springer-Verlag Berlin Heidelberg 2009

Evaluations of information measures within other state-of-the-art data assimilation methods, the Ensemble Kalman Filter (EnKF) methods (e.g., Evensen 1994; Houtekamer and Mitchell 1998; Anderson 2001; Bishop et al. 2001) have been limited so far. The ensemble-based information measures were first introduced to ensemble forecasting by Patil et al. (2001), where a measure called "bread vector dimension" was used, and by Oczkowski et al. (2005), where the so-called "E dimension" was defined. Both measures are essentially the same, defined following Bretherton et al. (1999). The E dimension was later examined within ensemble data assimilation (e.g., Szunyogh et al. 2005; Wei et al. 2006) and additional information measures, such as Shannon entropy and DOF for signal, were introduced to ensemble data assimilation methods (e.g., Uzunoglu et al. 2007; and Zupanski et al. 2007).

In summary, the currently available data assimilation studies on the information measures, even though valuable for specific applications, are not always directly comparable because of the use of different data assimilation methods and different information measures. It would be, therefore, useful to further examine some of these differences by (i) comparing the same observation measures within different data assimilation approaches and by (ii) comparing different information measures within the same data assimilation approach. The first comparison was performed in Zupanski et al. (2007), where the same information measure (the DOF for signal) was examined within variational and ensemble-based methods. In this chapter we compare different information measures (the DOF for signal vs. the E dimension) within the same data assimilation approach. We also discuss the results of Zupanski et al. (2007), as they apply to the results of this study.

The plan of this chapter is as follows. In Sect. 2, the information measures of interest are defined. In Sect. 3, the data assimilation experiments are shortly described. The experimental results are presented and discussed in Sect. 4. Finally, conclusions and future research directions are given in Sect. 5.

## 2 Information Measures

### 2.1 E Dimension

This measure was mainly designed to quantify degrees of freedom in atmospheric dynamics by quantifying the differences between different ensemble members. Oczkowski et al. (2005) defined E dimension as follows. "The E dimension is a local measure that varies in both space and time and quantifies the distribution of the variance between phase space directions for an ensemble of nonlinear model solutions over a geographically localized region." The E dimension is a non-negative, real number defined, under the standard Gaussian error assumption, as (e.g., Oczkowski et al. 2005)

$$E = \frac{\left(\sum\limits_{i=1}^{k} \lambda_i\right)^2}{\sum\limits_{i=1}^{k} \lambda_i^2}, \tag{1}$$

where $\lambda_i^2$ are eigenvalues of a positive definite $k \times k$ covariance matrix and $k$ is either the ensemble size (in the ensemble-based methods) or the size of the model state variable (in the variational methods). In data assimilation applications, $\lambda_i^2$ represent the eigenvalues of the information matrix $C$ defined as (e.g., Zupanski et al. 2007)

$$C = Z^T Z; \quad z^i = R^{-\frac{1}{2}} H(x + p_f^i) - R^{-\frac{1}{2}} H(x), \tag{2}$$

where index $i$ denotes an ensemble member, matrix $R$ is the observation error covariance matrix, $H$ is, in general, a non-linear observation operator, $x$ is the model state vector, vectors $p_f^i$ are columns of the square root of the forecast error covariance matrix $P_f$, and vectors $z^i$ are the columns of the matrix $Z$ of dimension $k \times k$. It is easy to see from Eq. (2) that the eigenvalues $\lambda_i^2$ depend on two main factors: the model dynamics, via the flow-dependent $P_f$, and observations, via the observation error covariance matrix $R$. Depending on how the eigenvalues are combined to define a particular information measure, one or the other of the two factors could be more dominant.

As illustrated in the original definition given in Bretherton et al. (1999), the E dimension depends on the ratio $\mu$ between the two neighboring eigenvalues (i.e., on the steepness of the eigenvalue spectrum), but not on actual eigenvalues. Assuming, as in Bretherton et al. (1999), that the eigenvalues drop geometrically with index $i$, according to the constant ratio $\mu$ defined as

$$\frac{\lambda_{i+1}}{\lambda_i} = \exp(-\mu), \quad \mu > 0, \tag{3}$$

we can see that Eq. (1) depends on the ratio $\mu$, but not on $\lambda_i^2$:

$$E = \frac{\left\{\sum\limits_{i=1}^{k} \exp\left[-(k-1)\mu\right]\right\}^2}{\sum\limits_{i=1}^{k} \exp\left[-2(k-1)\mu\right]}. \tag{4}$$

According to Eqs. (1) and (4) the E dimension varies between 1 and $k$. As explained in the earlier papers (e.g., Oczkowski et al. 2005; Szunyogh et al. 2005) the minimum value corresponds to the case where the uncertainty is confined to a single direction (the eigenvalue spectrum is steep) and the maximum value corresponds to the case where the uncertainty is equally distributed among all $k$ phase space directions (the eigenvalue spectrum is flat).

## 2.2 DOF for Signal

This measure quantifies the number of new pieces of information brought to the assimilation by the observations with respect to what was already known, as expressed by the background error covariance $P_f$. As the E dimension, the DOF for signal incorporates the impact of both factors (flow-dependent $P_f$ and $R$), however, not necessarily in the same way. The DOF for signal, denoted $d_s$, is defined as (e.g., Rodgers 2000; Zupanski et al. 2007)

$$d_s = \sum_{i=1}^{k} \frac{\lambda_i^2}{(1+\lambda_i^2)}, \tag{5}$$

where $\lambda_i^2$ and $k$ have the same meaning as before. As explained in Rodgers (2000), $\lambda_i^2 \geq 1$ corresponds to the observational signal (meaning that the observation errors are smaller than the background errors) and $\lambda_i^2 < 1$ corresponds to the observational noise (meaning that the observation errors are larger than the background errors). Note that the Eq. (5) was also derived assuming that all errors are Gaussian.

Following the same example with constant ratio $\mu$, defined in Eq. (3), we can now rewrite (5) as

$$d_s = \frac{\lambda_1^2 \sum_{i=1}^{k} \exp\left[-2\left(k-1\right)\mu\right]}{\sum_{i=1}^{k} \left\{1 + \lambda_1^2 \exp\left[-2\left(k-1\right)\mu\right]\right\}}, \tag{6}$$

and see that the DOF for signal depends, in addition to the ratio $\mu$, on the leading eigenvalue $\lambda_1^2$.

Since both information measures depend on atmospheric dynamics, as described by the forecast error covariance $P_f$, it should be advantageous to employ a flow-dependent $P_f$ (as in the ensemble based and the classical Kalman filter methods) rather than to use a prescribed $P_f$ (as in the variational methods). On the other hand, the use of a full rank forecast error covariance matrix is potentially an advantage of the variational methods compared to the ensemble-based methods. Advantages/disadvantages of the variational vs. the ensemble-based data assimilation methods were examined in Zupanski et al. (2007) and will not be addressed here.

Similarly as in Oczkowski et al. (2005), in Table 1 we illustrate the differences between the E dimension and the DOF for signal using two non-zero, arbitrarily chosen, eigenvalues $\lambda_1^2$ and $\lambda_2^2$. As the table indicates, the E dimension is higher if the variance is equally distributed between the two different eigenmodes (rows 2 and 3), and lower if one of the modes has significantly larger magnitude than the other (row 1). Thus, E dimension is sensitive to the ratios of $\lambda_i^2$, but not to their actual values. Unlike the E dimension, the DOF for signal is sensitive to the actual values of $\lambda_i^2$. We can notice that the DOF for signal is considerably smaller for $\lambda_i^2 < 1$ (row 2) than for $\lambda_i^2 > 1$ (row 3), while the E dimension is equal to the same value (2.00) in both cases. The differences between the E dimension and DOF for signal are further examined in the next subsection, within an actual data assimilation experiment.

**Table 1** E dimension and DOF for signal calculated using three arbitrary combinations of the non-zero eigenvalues $\lambda_1^2$ and $\lambda_2^2$

| First two non-zero eigenvalues | E dimension | DOF for signal |
|---|---|---|
| $\lambda_1^2 = 25.0,\ \lambda_2^2 = 1.0$ | 1.38 | 1.46 |
| $\lambda_1^2 = 0.47,\ \lambda_2^2 = 0.46$ | 2.00 | 0.63 |
| $\lambda_1^2 = 3.47,\ \lambda_2^2 = 3.46$ | 2.00 | 1.55 |

# 3 Data Assimilation Experiment

We employ an ensemble data assimilation approach referred to as the Maximum Likelihood Ensemble Filter (MLEF, Zupanski 2005; Zupanski and Zupanski 2006; Zupanski et al. 2008) and the Weather Research and Forecasting (WRF) model as examples of an ensemble-based data assimilation method and a complex atmospheric model. We assimilate conventional observations (SYNOP and radiosonde) every 6 h, and evaluate the E dimension and the DOF for signal. The observations include wind components ($u$ and $v$), surface pressure, temperature and humidity. We adopt standard observation errors for these observations, as defined in the WRF 3-dimensional variational data assimilation (WRF-var) system. The control variable includes $u$, and $v$ wind components, potential temperature perturbation, geopotential height perturbation and specific humidity. The model spatial resolution is 30 km/28 vertical levels (totaling $75 \times 70 \times 28$ grid points) and the time step is 3 min. The size of the control variable is 698277.

We employ 32 ensemble members, thus the ensemble size is several orders of magnitude smaller than the size of the control variable. In order to increase the number of degrees of freedom in the assimilation, we apply covariance localization by solving the problem within multiple local domains (e.g., Ott et al. 2004). We assume that the local domains (or sub-domains) are independent and do not overlap, however, as suggested by E. Kalnay (2008, personal communication), we smooth the analysis weights between the neighboring sub-domains, which introduces correlations between sub-domains and produces similar results as with overlapping sub-domains. Using non-overlapping sub-domains is, however, advantageous when calculating information measures since the total values of the information measures can be easily calculated by summing up the values obtained over the individual sub-domains. In the next sub-section we present the experimental results obtained for the case of hurricane Katrina (August 25–31, 2005).

# 4 Experimental Results

We have performed three different data assimilation experiments: (1) the base experiment, which employs the entire model domain (i.e., global domain, without the use of sub-domains), (2) the experiment which employs large sub-domains (the model domain is divided into 9 "independent" sub-domains of the size of $25 \times 24 \times 28$ grid

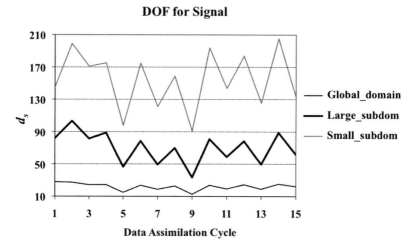

**Fig. 1** DOF for signal obtained in the experiment using global domain (base experiment) and the experiments employing large and small sub-domains, shown as a function of data assimilation cycles. The data assimilation interval is 6 h and the simulation time of 15 data assimilation cycles covers the period from 06 UTC 26 Aug 2005 to 18 UTC 29 Aug 2005. Note zig-zag behavior in all three lines, which indicates sensitivity of DOF for signal to the number of observation in different times, since more observations were used in synoptic times (00z and 12z) than in asynoptic times (06z and 18z)

points), and (3) the experiment which uses small sub-domains (total of 100 sub-domains of the size of $8 \times 7 \times 28$ grid points). In Fig. 1, the total number of DOF for signal is plotted as a function of data assimilation cycles for the three different experiments. In the base experiment, the total number of DOF for signal was obtained by combining, according to Eq. (5), the 32 values of $\lambda_i^2$, calculated over the entire model domain. In the other two experiments the total number of DOF for signal was obtained by summing-up the individual DOF for signal, calculated independently for each sub-domain. The figure indicates that the DOF for signal is generally small (and always smaller than the ensemble size) in the experiment involving the global domain. As expected, introduction of the independent sub-domains increases the DOF for signal and also improves the data assimilation results (not shown). Re-calling that similar results were also obtained in Zupanski et al. (2007), where the increased values of DOF for signal were achieved by increasing the ensemble size, we can observe that the covariance localization via independent sub-domains has a similar impact on the information measures as increasing the ensemble size.

By examining Fig. 1, we can notice a zigzag behavior in all three lines, indi-cating that the information measures obtained in all three experiments are sensitive to the larger (smaller) number of observations assimilated at synoptic (asynoptic) times. The experiment using the small sub-domains indicates most sensitivity to the number of observations, thus being able to extract the maximum amount of infor-mation from the observations. Conversely, the experiment with the global domain indicates least sensitivity to the number of observations, thus being able to extract much smaller amount of information from the observations (because the ensemble size of only 32 members was too small in this case). Nevertheless, the quality of the data assimilation results of all three experiments was acceptable, since all three experiments indicated positive impact of data assimilation on the analysis and forecast. Therefore, we consider examining information measures of these experiments worthwhile. Due to the limited scope of this chapter, we do not explicitly address the quality of the data assimilation results, however, additional discussions are provided in Chapter 3 of this book titled "Theoretical and Practical Issues of Ensemble Data Assimilation in Weather and Climate".

Let us now examine the other information measure, the E dimension, which is plotted for the same three experiments in Fig. 2. As the figure indicates, the E dimension also increases with the increasing number of sub-domains (at least in the examples examined here) and is sensitive to the number of observations (zigzag behavior is also present, though less pronounced). There is a pronounced maximum in the first data assimilation cycle in all three experiments in Fig. 2, which is not present, or at least not pronounced, in Fig. 1. This is because the uncertainty is more evenly distributed among different ensemble members in the first cycle, since the dominant directions of uncertainty have not yet fully emerged, which are exactly the properties the E dimension is sensitive to. These results indicate that the E dimension is more sensitive to the complexity of the model dynamics, and less sensitive to the number and quality of the observations, while the opposite is true for the DOF for signal.

Finally, we examine spatial distribution of the two information measures for a selected data assimilation cycle to link them to both model dynamics and observations. In Fig. 3, we plotted the DOF for signal obtained in the experiment with 9 large sub-domains (experiment denoted "Large sub-domains" in Figs. 1 and 2). The E dimension, calculated from the same experiment, is shown in Fig. 4. The results from cycle 10 (valid at 12 UTC 28 Aug 2005) are given, showing also the two most

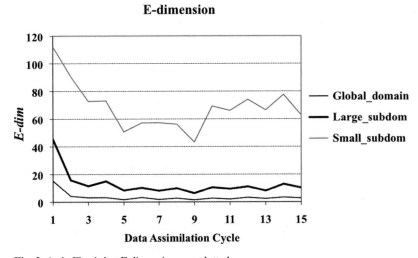

**Fig. 2** As in Fig. 1, but E dimension was plotted

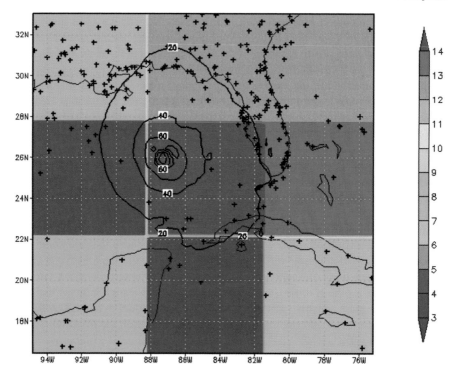

**Fig. 3** DOF for signal calculated in data assimilation cycle 10 (valid at 12 UTC 28 Aug 2005) shown, in shading, for each of the 9 sub-domains, covering the model domain. Also shown are the locations of the assimilated SYNOP and radiosonde observations (crosses) and the location of the tropical cyclone by the isolines of the magnitude of the horizontal wind (in ms$^{-1}$), as obtained by the analysis at the model level corresponding to $z = 700\,\mathrm{m}$

dominant factors influencing the information measures: the locations of the assimilated observations and the location of the tropical cyclone. By comparing the two figures we can notice that both measures have maximum values in the four northeastern sub-domains (plotted in pink, red and orange). Note that pink color indicates that the maximum values of the DOF for signal are $\geq 14$, and for the E-dimension $\geq 8$. These four sub-domains cover the areas where either the cyclone eye is located (the pink colored central sub-domain) or where the observation density is increased (the red and orange sub-domains), thus confirming that both the model dynamics and the observations are important factors in both measures. However, a very high value is obtained for the E dimension in the sub-domain located at the southeastern corner (pink color), where the number of available observations is relatively small. Not surprisingly, the value of the DOF for signal is also relatively small in this sub-domain (green color in Fig. 3).

In summary, results shown in Figs. 3 and 4, also demonstrate that the E dimension is more sensitive to the model dynamics, while the DOF for signal is more sensitive to the amount and quality of the observations. These different properties of the two information measures have to be taken into account when comparing information

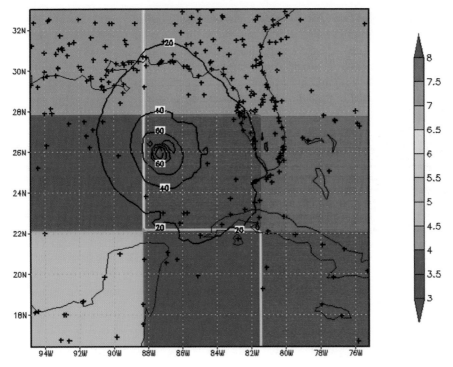

**Fig. 4** As in Fig. 3, but for the E dimension. Note that the values of E dimension are multiplied by 5 to obtain magnitudes comparable to the DOF for signal

measures obtained from different data assimilation methods, since some methods are better suitable for handling large amounts of observations (e.g., variational methods) and some for taking into account model dynamics (e.g., ensemble-based methods). It would be, therefore, helpful to evaluate multiple different information measures when comparing different data assimilation methods.

## 5 Conclusions

In this study, we have examined similarities and differences of the two information measures used in the current data assimilation methods, the DOF for signal and the E dimension. We have evaluated these measures using simple arbitrary examples (e.g., Table 1) and using a full-blown ensemble data assimilation algorithm, including a complex atmospheric model and real observations.

The results indicated that both information measures are sensitive to the following two main factors: the complexity of the model dynamics and the amount and quality of the observations. However, the E dimension indicated more sensitivity to the model dynamics and less sensitivity to the observations, while the opposite was true for the DOF for signal, which was highly sensitive to the amount and quality of the observations and less sensitive to the complexity of the model dynamics. It is, therefore, advisable to evaluate both information measures (and perhaps include new measures) when comparing the results of different data assimilation approaches. The research on this subject is currently at an early stage, thus more studies are needed in the future to fully utilize the power of the information theory within the state-of-the-art data assimilation methods.

**Acknowledgements** The author of this chapter is thankful to Seon Ki Park for the helpful discussions on the topical cyclones, which resulted in better understanding of the information measures examined in this study. The author also appreciates constructive comments made by an anonymous reviewer, which resulted in the improved final version of the manuscript. This research was supported by the NOAA/NESDIS GOES-R Risk Reduction Program under Grant No. NA17RJ1228, the NASA Precipitation Measurement Mission (PMM) Program under Grant No. NNX07AD75G, and the DoD Center for Geosciences/Atmospheric Research at Colorado State University under Cooperative Agreement W911NF-06-2-0015. We also acknowledge computational support of the NASA Ames Research Center's Columbia supercomputer and the NCAR's Blueice supercomputer.

# References

Abramov R, Majda A, Kleeman R (2005) Information theory and predictability for low-frequency variability. J Atmos Sci 62, 65–87

Anderson JL (2001) An ensemble adjustment filter for data assimilation. Mon Wea Rev 129, 2884–2903

Bishop CH, Etherton BJ, Majumjar S (2001) Adaptive sampling with the ensemble transform Kalman filter. Part 1: Theoretical aspects. Mon Wea Rev 129, 420–436

Bretherton CS, Widmann M, Dymnikov VP, Wallace JM, Blade I (1999) The effective number of spatial degrees of freedom of a time-varying field. J Climate 12, 1990–2009

DelSole T (2004) Predictability and information theory. Part I: Measures of predictability. J Atmos Sci 61, 2425–2440

Engelen RJ, Stephens GL (2004) Information content of infrared satellite sounding measurements with respect to $CO_2$. J Appl Meteor 43, 373–378

Evensen G (1994) Sequential data assimilation with a nonlinear quasi-geostrophic model using Monte Carlo methods to forecast error statistics. J Geophys Res 99, (C5), 10143–10162

Fisher M (2003) Estimation of entropy reduction and degrees of freedom for signal for large variational analysis systems. ECMWF Tech. Memo. No. 397. 18pp

Houtekamer PL, Mitchell HL (1998) Data assimilation using an ensemble Kalman filter technique. Mon Wea Rev 126, 796–811

Johnson C (2003) Information content of observations in variational data assimilation. Ph.D. thesis, Department of Meteorology, University of Reading, 218 pp. [Available from University of Reading, Whiteknights, P.O. Box 220, Reading, RG6 2AX, United Kingdom.]

Kleeman R (2002) Measuring dynamical prediction utility using relative entropy. J Atmos Sci 59, 2057–2072

L'Ecuyer TS, Gabriel P, Leesman K, Cooper SJ, Stephens GL (2006) Objective assessment of the information content of visible and infrared radiance measurements for cloud microphysical property retrievals over the global oceans. Part I: Liquid clouds. J Appl Meteor Climat 45, 20–41

Oczkowski M, Szunyogh I, Patil DJ (2005) Mechanism for the development of locally low-dimensional atmospheric dynamics. J Atmos Sci 62, 1135–1156

Ott, E, Hunt BR, Szunyogh I, Zimin AV, Kostelich EJ, Corazza M, Kalnay E, Patil DJ, Yorke JA (2004) A local ensemble Kalman filter for atmospheric data assimilation. Tellus 56A, 273–277

Patil DJ, Hunt BR, Kalnay E, Yorke JA, Ott E (2001) Local low dimensionality of atmospheric dynamics. Phys Rev Lett 86, 5878–5881

Purser RJ, Huang H-L (1993) Estimating effective data density in a satellite retrieval or an objective analysis. J Appl Meteorol 32, 1092–1107

Rabier F, Fourrie N, Djalil C, Prunet P (2002) Channel selection methods for infrared atmospheric sounding interferometer radiances. Quart J Roy Meteor Soc 128, 1011–1027

Rodgers CD (2000) Inverse methods for atmospheric sounding: Theory and practice. World Scientific, Singapore, 238 pp

Roulston M, Smith L (2002) Evaluating probabilistic forecasts using information theory. Mon Wea Rev 130, 1653–1660

Schneider T, Griffies S (1999) A conceptual framework for predictability studies. J Climate 12, 3133–3155

Shannon CE, Weaver W (1949) The mathematical theory of communication. University of Illinois Press, Champaign, 144 pp

Szunyogh I, Kostelich EJ, Gyarmati G, Patil DJ, Hunt BR, Kalnay E, Ott E, Yorke JA (2005) Assessing a local ensemble Kalman filter: Perfect model experiments with the NCEP global model. Tellus, 57A, 528–545

Uzunoglu B, Fletcher SJ, Zupanski M, Navon IM (2007) Adaptive ensemble member size reduction and inflation. Quart J Roy Meteor Soc 133, 1281–1294

Wahba G (1985) Design criteria and eigensequence plots for satellite-computed tomography. J Atmos Oceanic Technol 2, 125–132

Wahba G, Johnson DR, Gao F, Gong J (1995) Adaptive tuning of numerical weather prediction models: Randomized GCV in three- and four-dimensional data assimilation. Mon Wea Rev 123, 3358–3370

Wei M, Toth Z, Wobus R, Zhu Y, Bishop CH, Wang X (2006) Ensemble transform Kalman filter-based ensemble perturbations in an operational global prediction system at NCEP. Tellus 58A, 28–44

Zupanski M (2005) Maximum likelihood ensemble filter: Theoretical aspects. Mon Wea Rev 133, 1710–1726

Zupanski D, Zupanski M (2006) Model error estimation employing an ensemble data assimilation approach. Mon Wea Rev 134, 1337–1354

Zupanski D, Hou AY, Zhang SQ, Zupanski M, Kummerow CD, Cheung SH (2007) Applications of information theory in ensemble data assimilation. Quart J Roy Meteor Soc 133, 1533–1545

Zupanski M, Navon IM, Zupanski D (2008) The maximum likelihood ensemble filter as a non-differentiable minimization algorithm. Quart J Roy Meteor Soc 134, 1039–1050.

# Real Challenge of Data Assimilation for Tornadogenesis

Yoshi K. Sasaki

**Abstract** Successful recent numerical simulation of tornadogenesis with horizontal resolution of the order of $10\,\mathrm{m}$, $O(10\,\mathrm{m})$, and associated temporal resolution of O $(0.1\,\mathrm{s}$ and $0.01\,\mathrm{s}$ for a time-split scheme) requires a vast amount of computer time, impractical to use the simulation model for the model constraint of variational data assimilation and ensemble Kalman filter.

Also, recent advanced observations such as phased array radar have revealed spatial and temporal details of the similar high resolutions important for tornadogenesis, which should be properly reflected in the data assimilation.

To deal with them, data assimilation for operational uses requires special strategy. The author discusses, in this article, one promising strategy, especially use of the entropic balance model, which sounds computationally practical in variational data assimilation.

The entropic balance theory with entropic source and sink simplification is favorably compared to other historically proposed tornadogenesis theories and models. The theory explains the overshooting of hydrometeors against head-wind upper level westerlies and middle-level south-westerlies, mesocyclone, rear frank downdraft and tornado development. Furthermore, the theory suggests transition from dipole structure of early stage to monopole type mature stage, similar to an attractor of nonlinear system, of tornadogenesis, which explains of tilting of tornado vortex axis.

It is a real challenge to develop an operational data assimilation technology for accurate diagnosis and prediction of tornadogenesis.

Y.K. Sasaki (✉)
School of Meteorology, National Weather Center, The University of Oklahoma, Norman, OK 73072–7307, USA, yks@ou.edu

The author used the name Y. Sasaki or Yoshikazu Sasaki before 1974 for publications. The author serves also as an external member of the Board of Directors, Weathernews America Inc., USA.

S.K. Park, L. Xu, *Data Assimilation for Atmospheric, Oceanic and Hydrologic Applications*, DOI 10.1007/978-3-540-71056-1_5,
© Springer-Verlag Berlin Heidelberg 2009

# 1 Introduction

Motivation of this study will be presented by mentioning the followings; data assimilation, especially, variational data assimilation of primarily storm and precipitation, advanced numerical simulation of tornadogenesis, advanced observation, especially phased array radar, and historical review of other theories and models of tornadogenesis.

## 1.1 Data Assimilation

The data assimilation methods so far developed are well summarized in recent publication by Tsuyuki and Miyoshi (2007). The assimilation methods for mesoscale and precipitation forecast are divided primarily into two. One is the variational data assimilation (Zou et al. 1993; Zupanski 1993a, b, 1997; Zupanski and Mesinger 1995; Kuo et al. 1996; Tsuyuki 1996a, b; Tsuyuki, 1997; Mahfouf and Rabier 2000; Tsuyuki et al. 2002; Ishikawa and Koizumi 2002; Park and Zupanski 2003; Seko et al. 2004; Honda et al. 2005; Koizumi et al. 2005; Lewis et al. 2006; Kawabata et al. 2007). The other is ensemble Kalman filter (Kalney 2003) which will be discussed later.

The spatial resolution used for the both data assimilation is of the order of 1 or 10 km. As we will see in the subsequent discussion, it is too coarse.

## 1.2 Advanced Numerical Simulation

Smaller grid size was preferably chosen for better tornado simulation as the computer technology advanced. For instance, Noda (2002) and Noda and Niino (2005) used the horizontal spatial resolution of 75 m and the temporal resolution of 0.3 and 0.03 s for the time split with ARPS model (Xue et al. 1995) for their simulation.

Ideally, the same model used for Noda/Niino simulation should be used also for data assimilation. However the simulation alone for 3 hr evolution time took computation time of about 720 hrs on IBM Regatta, 16 nodes computer at Tokyo University. A similar successful simulation with high resolution was reported by Tsuboki (2007). The data assimilation will need several times more computation time. The ensemble Kalman filter method if used for data assimilation requires computer time roughly multiplied by the number of ensemble members. Therefore, it is the author's intention to develop a simpler model which still expresses the essence of tornadogenesis processes with necessary accuracy. The entropic balance model is a candidate for it.

## 1.3 Advanced Observation

In recent years, important progress has been made in observation with temporal and spatial ultra-high resolutions, the phased array/Doppler/dual polarization radar, seemingly ideal for measurement of tornado, precipitation and other mesoscale systems (Forsyth et al. 2007; Zrnic et al. 2007). In addition, presently under development is CASA dense low- level scanning adaptive radar network (Chandrasekar 2007). The phased array and CASA radars produce large volume of data. Data assimilation should be capable to handle the data selectively in time for analysis and prediction of tornado, precipitation and other mesoscale systems, together with necessary model processing.

## 1.4 Historical Review of Tornadogenesis Research

Tornado develops, most frequently in rotating supercell storm. Supercell is a special form of thunderstorm developed in the atmosphere. The first systematic investigation was undertaken in 1946–1947 by the group of the University of Chicago under Thunderstorm Project and revealed key features of thunderstorm of basically non-tornadic (Byers and Braham 1949).Their finding became the starting point of further studies including supercell. The life-cycle of thunderstorm has three stages, (a) cumulus (developing) stage with strong updraft and raindrop formation, (b) mature stage with heavy precipitation and downdraft formation, and (c) dissipating stage with rain cooled cold -downdraft spreading in the entire cloud. Since 1960, great progress was made in observation technology of thunderstorm (meso) scales of the atmosphere, especially on supercell and tornadogenesis, with development of observations such as satellite and Doppler radar.

Supercell different from the thunderstorm is characterized by storm rotation developed under certain environmental wind shear field. One of the earlier typical wind hodograph for a supercell storm is given by Chisholm and Renick (1972), also seen in Fig. 129 of Atkinson (1981), simply cited in this article as [A Fig. 129]. The typical vertical shear of horizontal wind at high, middle and low levels is discussed by Browning (1968, 1977) [A Fig. 129]. Figure 1 is shown schematically the environmental wind field of moist sir influx at the lowest level and the dry air influxes at higher levels including the surface gust fronts and tornado.

In the last several decades, question on maintenance, neither spreading nor weakening but rather intensifying, of convective storm circulation under the vertical shear of horizontal wind was investigated by several researchers (Jeffreys 1928; Asai 1970a, b). Takeda (1971) suspected cloud-physical process in storm under vertical wind shear acted as restoring force of maintaining vertical stretch of storm. Importance of the processes is further clarified in this article for the development of supercell circulation and tornadogenesis as generating source and sink of entropy based on the entropic balance theory. The numerical simulations and theoretical analyses of tornadic vortices (e.g., Klemp and Wilhelmson 1978 ; Wilhelmson and Klemp 1981; Rotunno 1981; Rotunno and Klemp 1982, 1985), and linear and semi-linear small perturbation analysis (Lilly 1982; Davies-Jones 1984; Davies-Jones and Brooks 1993) suggested the tilting and stretching mechanism of tornadogenesis representing the era of initial fruitful Doppler radar observations and computer

## Environmental Wind Field
## of
## Supercell

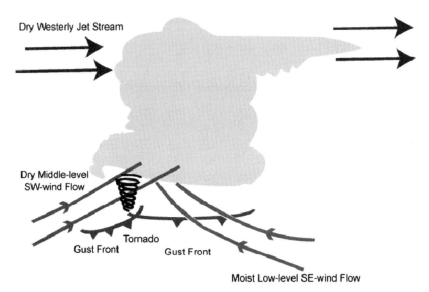

Dry Westerly Jet Stream

Dry Middle-level
SW-wind Flow

Tornado
Gust Front          Gust Front

Moist Low-level SE-wind Flow

**Fig. 1** Schematic environmental wind field of supercell. Supercell cloud, tornado, two major gust fronts, moist low-level SE-wind flow, dry middle-level SW-wind flow and dry upper-air westerly jet stream are schematically shown with surface gust fronts and tornado. All of them are observation based

simulations of supercell and tornado. The mechanisms are tilting and stretching of horizontally lying vortex tube by storm updraft and suppressing one side of vortex tube by downdraft.

However, the model has serious problems. One of them concerns with the tail end of the vortex tube which should be parallel to the ground, while it should be perpendicular to the ground according to the law of fluid mechanics as verified by observations without question. The other problem is that the upward pressure force is insufficient to tilt upwards sharply the horizontally lying vortex tube by the thunderstorm updraft (in absence of upward pressure force generated by dynamical-pipe effect (e.g., Leslie 1971; Trapp and Davies-Jones 1997)). Validity of the dynamical-pipe effect is questionable because of lack of cyclostrophic balance. On the other hand, recognized for years by tornado chasers, observations, numerical and theoretical analyses is the development of an evaporation driven rear-frank downdraft (RFD) upstream side near the ground, adjacent to the storm, forming a gust front, consequently, mesocyclogenesis due to the local baroclinicity as supported from numerous observations, numerical simulations and theoretical analyses (e.g., Simpson 1972; Lemon and Doswell III 1979; Burgess et al. 1982; Rasmussen et al. 1982;

Sasaki and Baxter 1982; Rotunno and Klemp 1982; Klemp and Rotunno 1983; Wicker and Wilhelmson 1995; Alderman, Droegemeier and Davies-Jones 1999; Noda 2002; Noda and Niino 2005).

The entropic balance model seems to provide a theoretical background, from the uniquely different view point, to the near-ground gust front type mesocyclogenesis together with tornadogenesis by a unified theoretical basis.

It is noted that development of the models of tornadogenesis may be classified into two categories, one is based on linear mathematical analysis including semi-linear and the other nonlinear. The linear model is defined with linear small pertur-bation superimposed on constant basic field and semilinear as the basic field varies with spatial coordinates. The nonlinear is the case when perturbation term includes product term of two or more perturbations. If no elimination of nonlinear term is made, it is called full nonlinear in this article. The entropic balance theory is in full nonlinear form. The linear and semilinear approximations are derived from it and all of them will be discussed in Sect. 5 and compared with the other models.

Tornadogenesis occurs in the flow characterized with super high Reynolds num-ber (Re $= 10^{8-12}$) and moderately high Rossby number (Ro $= 10^{2-4}$) and also "quasi-adiabatic" process of cloud physics. The quasi-adiabatic process assumes significant difference of the time scales between cloud-physical phase changes and supercell flow of tornadogenesis. It results in step-like discrete jump of adiabatic entropy levels between before and after phase-change of hydrometers or "water molecules variety" (water in vapor, liquid or solid form). The entropic balance was originally called Clebsch's transformation (Clebsch 1859), and extended by the au-thor (Sasaki 1999) to clarify the origin of rotations of supercell and tornado. It seems clarify also the origins of RFD, mesocyclone, hook echo and wall cloud. And it does the origin of overshooting hydrometeors against headwind westerlies and RFD.

The entropic balance is analogous to the hydrostatic balance used for shallow water system, quasi-geostrophic balance for large scale flow, and anelastic balance for convective motion (Ogura and Phillips 1962). The geostrophic balance was used by the author as the first data assimilation of large-scale flow analysis (Sasaki 1958). The entropic balance is, however, the most general form compared with the other balance form as we will be found in the subsequent discussion.

The entropic balance theory suggests development of tornadic vortex at or near the entropic sink if it locates in close proximity, about few km or less, from the en-tropic source, in supercell with certain environmental wind field. The environmental field consists of moist air influx of southeast (SE) wind at low altitudes and dry air influxes of both southwest (SW) or more effectively westerly wind at the middle altitudes and westerlies at high altitudes. These moist air and dry air influxes are in-strumental to develop intense entropic source and sinks, consequently tornado. The results are compared favorably with a vast number of observations and theoretical model development including numerical simulations made on supercell storms and tornadoes.

Under the typical environmental wind and vertical shear, as shown in Fig. 1, the rotation is anti-cyclonic, clockwise, in updraft area and cyclonic in downdraft area. The downdraft is accelerated by the dragging friction of rain droplets and ice parti-cles, like hails, heavier than the air. It coincides with the consequence derived from the entropic balance equation, that is, the flow is cyclonic around the entropic sink (or simply called as sink in this article), where downdraft develops due to evaporation cooling of the surrounding air, and anti-cyclonic around the entropic source (or simply as source), where updraft develops due to warming the air by condensation.

Together with probabilistic choice of initial entropic level and probabilistic determination of time and magnitude of phase change, the phase change of ensemble of hydrometeors is probabilistic. The phase change with small probabilistic variation creates probabilistic source and probabilistic sink (Seo and Biggerstaff 2006). In Fig. 2, marked are the one source by P1 and four sinks by N1–N4. Of course, there are more entropic sources and sinks, continuous in stead of discrete form in reality, but Fig. 2 shows only major representative ones in discrete form together with the associated trajectories of hydrometeors, marked by arrowed broken lines with the source and sinks.

The trajectory of moving up and eastwards is easily understood because of its traveling with the upper westerly jet stream but the up and westwards against the westerlies faces intense head wind and the mechanism is not clear because of weaker horizontal momentum carried upward from the moist SE wind at the lowest level.

## Supercell

## Entropic Source and Sink
## &
## Trajectories of Hydrometeors

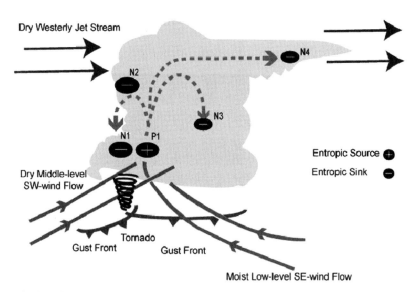

**Fig. 2** Major entropic source and sinks and trajectories of hydrometeors (water in liquid or solid form) in schematic model of supercell. Major source is one, marked by circled + and P1, and major sinks are four, marked by circled –sign and N1–N4. Here, N4 is a minor sink marked by smaller circle with –sign. The trajectory of hydrometeors is shown by broken line

Clarifying the overshooting mechanism against the headwind westeries is one of major interests of this study. Another interest is to clarify the generation mechanism of the intense RFD observed in the lower west proximity of supercell and tornado. It is shown in this article that the entropic balance equation suggests "entrotropically balanced rotation" in 3-D that explains flow patterns around entropic source and sinks, and the overshoot mechanism of hydrometeors against the westerly headwind, and intensification of the evaporation-cooled rear downdraft of supercell and tornado. From the sole diagnostic equation of the Euler-Lagrange (E-L) equations derived from the Lagrangian density of the prescribed flow leads to the entropic balance equation which describes flow around the entropic source and sinks in a supercell storm developed in the prescribed environmental wind field. Proper combinations of the entropic source and sinks simulate and explain well several major flow circulation patterns of supercell and tornadogenesis.

Development of the models of tornadogenesis may be classified into two categories, one is based on linear mathematical analysis including semilinear and the other nonlinear. The linear model is defined with linear small perturbation superimposed on constant basic field and semilinear as the basic field varies with spatial coordinates. The nonlinear is the case when perturbation is not linear, including product term of two or more perturbations. If no elimination of nonlinear term is made, it is called full nonlinear in this article. The entropic balance theory is in full nonlinear form, or in linear and semilinear approximations which can be derived from the nonlinear form and will be discussed in Sects. 4 and 5 for further details comparing the other theories and models tornadogenesis.

In the final section of this article, the entropic balance theory for detection of tornadogenesis is further discussed about the observational and modeling needs for future appropriate data assimilation.

## 2 Entropic Balance Theory

The flow is treated in the variational formalism of this study, employing the molecular approach based on the classical gas theory of Maxwell and Boltzmann (McQuarrie and Simon 1997; Atkins and de Paula 2002).

Let denote the Reynolds number by $\mathbf{R_e}$, defined as $\mathbf{VD}/v$, and the Rossby number by $\mathbf{R_o}$, $:= \mathbf{V}/(\mathbf{fD})$, where the order of magnitudes of tornadic and supercell flow velocity are represented by $\mathbf{V}$, the horizontal extent of mature tornado and supercell by $\mathbf{D}$, kinetic viscosity of the air by $v$, and the Coriolis parameter by $\mathbf{f}$. Since the appropriate values of $\mathbf{V}$ and $\mathbf{D}$ are estimated as $\mathbf{V} = 10^{1-2}$ m/s and $\mathbf{D} = 10^{2-3}$ m, and $v$ and $\mathbf{f}$ as $10^{-1}$ cm$^2$ s$^{-1}$ and $10^{-4}$ s$^{-1}$ respectively,

$$\mathbf{R_e} = 10^{8-12} \quad \text{and} \quad \mathbf{R_o} = 10^{2-4}. \tag{1}$$

The Reynolds number is super high and the flow may be treated as if it is inviscid. The Rossby number is also very high and the earth rotation, the Coriolis effect, may be neglected.

Also, very important to pay our attention is that the time scales of the phase changes in the cloud-physical processes are of the order of magnitudes different from time scales of the dynamical processes of supercell and tornado. The phase changes of water molecules variety occur almost instantaneously, the time scale of which is denoted by $\Delta t_{\text{phase change}}$, while the dynamical processes of supercell and tornado have the time scales, $\Delta t_{\text{supercell,tornado}}$, of the order of magnitude, about $10^{3-4}$ s,

$$\Delta t_{\text{phase change}} \ll \Delta t_{\text{supercell,tornado}}. \tag{2}$$

Since meteorological processes deal with ensemble of the molecules, there is a variation in the time of occurrence in the phase changes. Furthermore, before and after a phase change, the thermo-dynamical process is adiabatic and the value of entropy conserves, and the finite magnitude of entropy level jump is determined uniquely from the physical chemistry, called in this article as "Quasi-adiabatic". There is also another variation in the initial entropy level of the ensemble. These two kind of the variations are schematically represented in Fig. 3.

The Lagrangian density and the associated Lagrangian are formulated, developing the following two hypotheses based on the preceding discussion.

Quasi-Adiabatic Process : Discrete Entropy Level Change
Due to Cloud-Physical Phase Change :

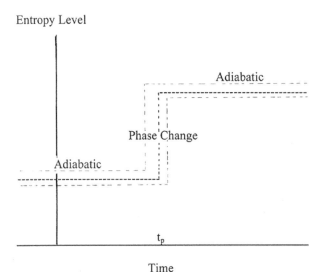

**Fig. 3** Quasi-adiabatic process. The schematic diagram shows temporal change of entropy level due to cloud-physical phase change. Two ambiguities exist, one due to initial value of entropy and other time of occurrence of phase change

The Lagrangian density is formulated on molecular level with quasi-adiabatic process. (Hypothesis 1)

The Lagrangian is temporal and spatial integration of the Lagrangian density over an appropriate integration domain, and represents ensemble of the molecules. The variations in the time of phase changes and the value of initial entropy levels are small enough to allow us to approximate them by their ensemble means. (Hypothesis 2)

Use of the conditions (1), (2), (Hypothesis 1) and (Hypothesis 2) justifies the following Lagrangian density, $\mathscr{L}$,

$$\mathscr{L} := \rho(1/2\,\mathbf{v}^2 - U - \Phi) - \alpha(\partial_t\rho + \nabla\cdot(\rho\mathbf{v})) - \beta(\partial_t(\rho S) + \nabla\cdot(\rho\mathbf{v}S)) \qquad (3)$$

where $\rho$, $U$, $\Phi$, $S$, and $\mathbf{v}$ are air density, internal energy, potential energy, entropy, and flow velocity respectively, and $\alpha$ and $\beta$ are the Lagrange multipliers to satisfy the constraints of mass continuity and entropy respectively. Then, the Lagrangian L is defined,

$$L = \int_\Omega \mathscr{L}\, dx^3\, dt. \qquad (4)$$

where $\Omega$ represents temporal and spatial integration domain and ensemble of the molecules is represented by the spatial integration.

The first variation of L leads to the E-L equations, then after mathematical manipulation, a full set of dynamical and thermo-dynamical nonlinear equations of the ideal flow (Lamb 1932 in hydrodynamics ; Bateman 1932 in mathematical physics; Sasaki 1955 in atmospheric dynamics). The E-L equations are all prognostic except only one diagnostic, so called the Clebsch's transformation (Clebsch 1859) of flow velocity,

$$\mathbf{v} = -\nabla\alpha - S\nabla\beta. \qquad (5)$$

Then, the vorticity, $\omega$, becomes

$$\omega = (1/S)\nabla S \times (-S\nabla\beta). \qquad (6)$$

Later, Dutton, John A. showed in his book published in 1976 the same equations of Eqs. (5) and (6) for an isentropic flow on absolute coordinates (Dutton 1976). The quasi-adiabatic process based on the significant time-scale difference between the microphysical phase change and mesoscale system as shown in Eq. (2) supports Hypothesis 1, and allow isentropic process before and after the phase change. It provides physical background for helicity conservation used by Lilly (1982) and demonstrated by Noda (2002) in his numerical simulation of tornadogenesis. It should be noted that the E-L equation (5) and the associated vorticity equation (6) are nonlinear similarly with the all other E-L equations (not shown) derived from the Lagrangian equation (3). Incipient stages of mesocyclone development and tornadogenesis may be described to some extent by small perturbation analysis. However, it is apparent that the magnitudes of field variables are larger if compared with those of the corresponding basic field. Instead of small perturbation analysis, full nonlinear analysis based on Eqs. (5) and (6) are interesting to describe mature stage of tornado. It is found that the perturbation analysis leads flow circulation of dipole type for incipient stage, while the nonlinear analysis does monopole type circulation for mature stage, as will be seen in Sect. 5.

The diagnostic E-L equation (5) provides the common domain in the solution space for the entire E-L equations which include solution of the diagnostic equation and those of the prognostic equations, steady, quasi-steady and non-steady (Fig. 4). The full solution is schematically shown by the heavy solid lined domain which includes non-steady solution (NSS), steady solution (SS), quasi-steady solution (QSS) and the diagnostic solution (DS) from Eq. (5). The solution in the domain covered by DS and SS has long-lasting property mathematically similar to the attracter. The schematic diagram shown in Fig. 4 suggests possibility for us of peeking into partially certain key aspects of full solution of the entire E-L equations.

The vector relation (6) is found extremely important to get clear insight into the development mechanisms of supercell and tornadogenesis as it will be discussed in the rest of this article. The diagnostic equation (6) is universal for the ideal flow, which is demonstrated in convenience by the mutually orthogonal vector relation among the variables, similar to the so called Fleming's right hand law of electro-magnetic field. The airflows around the source and sink are obtained from

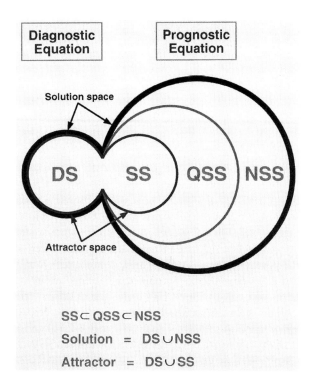

**Fig. 4** Solution space. The domain of full solution of the Euler-Lagrange (E-L) equations is schematically shown in the solution space by the heavy *solid line*. It includes non-steady state solution (NSS), quasi-steady state solution (QSS), steady state solution (SS), and the solution of diagnostic E-L equation (DS). The solution in the domain covered by DS and SS has long-lasting property mathematically similar to the attracter. The schematic diagram suggests possibility of peeking into some key aspects of full solution of the entire E-L equations

**Fig. 5** Schematic diagram to show the Fleming's right hand law of the general vorticity equation (6)

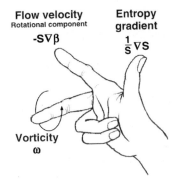

the entropic balance equation (5) and the corresponding vorticity (6), which can be interpreted as follows, when logarithmic entropy gradient, flow velocity are represented by thumb and forefinger respectively, vorticity by middle finger as shown in Fig. 5.

The Eq. (6) is further written using Eq. (5) as

$$\omega = (1/S)\ \nabla S \times (\mathbf{v} - \mathbf{v_D}). \tag{7}$$

where $\mathbf{v_D}$ is the pure divergent component of wind defined as

$$\mathbf{v_D} := -\nabla\alpha \tag{8}$$

For irrotational flow, $\omega = 0$,

$$\mathbf{v} = \mathbf{v_D}, \tag{9}$$

or

$$\mathbf{v} = \mathbf{v_D} + C\ \nabla S \tag{10}$$

where C is an arbitrary constant because $\mathbf{v}$, $\mathbf{v_D}$ and $\nabla S$ are all parallel.

For predominantly rotational flow, $|\mathbf{v}| \gg |\mathbf{v_D}|$, $|\nabla\alpha|$, Eq. (7) is approximated as

$$\omega \approx (1/S)\ \nabla S \times \mathbf{v}. \tag{11}$$

The flow around an entropic source and an entropic sink as given by Eq. (11) is schematically shown in Fig. 6.

The rotational approximation (11) may be used for numerical example to estimate the intensity of vortex which satisfies the entropic balance. We follow Atkins and de Paula (2002) to estimate the entropy generation. The entropy generated and given to the surrounding air from the condensation of water vapor (or evaporation) is $109.1\ \mathrm{J\,K^{-1}\,mol^{-1}}$. Let denote the entropy generated from the phase-change by $\Delta S'_2$. We consider a case for the phase change to occur at the altitude of 2 km, where tornado may develop and the air temperature of the standard

**Fig. 6** The schematic flow diagrams of cyclonic ($\omega > 0$) and anti-cyclonic ($\omega < 0$) flows around the entropy (anomaly) source and sink. The flows are estimated from Eq. (10) which is a simplified form of Eq. (6) for rotational flow and shown by arrowed cyclonic (*blue*) and anti-cyclonic (*red*) curves for the rotational flow velocity **v**

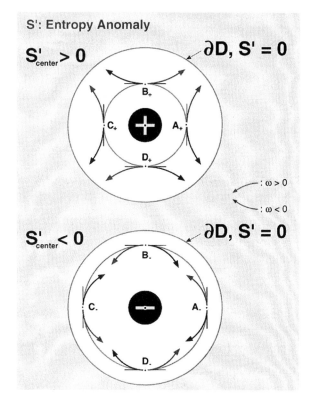

atmosphere at that altitude is 275 K. We need to add to it the entropy increase $\Delta S'_1$ ($= 10.3\,\mathrm{J\,K^{-1}\,mol^{-1}}$) due to the hypothetical temperature change of the water vapor from 275 to 373 K, and $\Delta S'_3$ ($= -23.0\,\mathrm{J\,K^{-1}\,mol^{-1}}$) of water from 373 to 275 K,

$$\Delta S' := \Delta S'_1 + \Delta S'_2 + \Delta S'_3 = 96.4\,\mathrm{J\,K^{-1}\,mol^{-1}}. \qquad (12)$$

For the freezing and melting of occurring at 273 K, we consider similarly, defining the entropy generated by the process as $\Delta S''_2$ ($= 22.0\,\mathrm{J\,K^{-1}\,mol^{-1}}$) with $\Delta S''_1$ ($= -0.04\,\mathrm{J\,K^{-1}\,mol^{-1}}$) of water from 275 to 273 K and $\Delta S'_3$ ($= 0.37\,\mathrm{J\,K^{-1}\,mol^{-1}}$) of ice from 273 to 275 K,

$$\Delta S'' := \Delta S''_1 + \Delta S''_2 + \Delta S''_3 = 22.3\,\mathrm{J\,K^{-1}\,mol^{-1}}. \qquad (13)$$

Consequently, the total entropy change due to phase changes $\Delta S$ in the case of equal molar quantities of water and ice becomes

$$\Delta S := \Delta S' + \Delta S'' = 118.7\,\mathrm{J\,K^{-1}\,mol^{-1}}. \qquad (14)$$

In these calculations, the values of heat capacity of water vapor, liquid water and ice are adopted as 33.58, 75.29 and 37 $\mathrm{J\,K^{-1}\,mol^{-1}}$ respectively from the reference cited above. Here, we assume for simplicity that the values of the heat capacities valid at the temperature 298.15 K and a constant pressure are applicable without serious numerical error to the present case because of our prime interest on the order of magnitudes of tornado.

The magnitudes of vorticity is estimated employing (11) and (14) for appropriate values of the entropy gradient, the standard molar entropy, the distance d between the source and the adjacent sink, and flow velocity. From Eq. (14), the entropy gradient is estimated, because of both of source and sink, as $2 \times 118.7\,\mathrm{J\,K^{-1}\,mol^{-1}}/d$. The standard molar entropy of the air, assuming that it is composed of only $N_2$, $O_2$, is calculated from the value given in the reference cited above as $182.3\,\mathrm{J\,K^{-1}\,mol^{-1}}$, which is used for the value of S in Eq. (14). Employing these values, for some selected values of d and $\mathbf{v}$, the magnitudes of the vorticity $\omega$ become,

$$\omega = 1.30\,10^{-2}\,\mathrm{s^{-1}} \quad \text{for} \quad d = 1\,\mathrm{km}, \; \mathbf{v} = 10\,\mathrm{ms^{-1}}, \tag{15}$$

$$\omega = 1.30\,10^{-1}\,\mathrm{s^{-1}} \quad \text{for} \quad d = 1\,\mathrm{km}, \; \mathbf{v} = 100\,\mathrm{ms^{-1}}, \tag{16}$$

and

$$\omega = 1.30\,\mathrm{s^{-1}} \quad \text{for} \quad d = 100\,\mathrm{m}, \mathbf{v} = 100\,\mathrm{ms^{-1}}. \tag{17}$$

These estimates were made for the case at the altitude of 2 km where the temperature is 2°C. In the case of temperature of $-15$°C, equivalent to the 4.6 km altitude of the standard atmosphere, which is used often for lightning research, the values of vorticity are estimated as about 91% of the estimates given in Eqs. (15)–(17). Therefore, the lower the altitude, the more intense vortex, at the rate of about (5% / km), is expected. This seems also to support the observations of tornadic vortex development in the lower troposphere. Also, if we consider that the entropic gradient near the sink would be much steeper than the estimates made in Eqs. (15)–(17), the entropic balance theory can support the existence of tornadic vortex near the sink adjacent to the source.

# 3 Tornadogenesis

The entropic source and sink under the typical environmental flow conditions of supercell, cited earlier, which may likely produce tornado were schematically shown in Fig. 2. The water vapor brought in by moist lowest level SE wind condenses and arises in supercell and meets with middle level dry SW or westerly wind and then the upper dry westerlies. When the supercell is tornadic, hook echo, mesocyclone, wall cloud and tornado are observed at SW corner area of the supercell as demonstrated schematically in the figure, which shows also two gust fronts on the ground. The one located eastwards is produced evaporation cooled downdraft under the major central portion of supercell, and the one westwards is produced by evaporation of hydrometeors as known as RFD near the west side of the supercell.

Tornado develops at the southwest corner of supercell as schematically shown by the coil-shaped sign.

The warming and cooling occur due to cloud-physical phase changes, such as condensation and evaporation, or freezing and melting, or vapor deposition (or sublimation) and sublimation, and will associate with entropic changes of the air adjacent to the hydrometeors. The airflow around the source and sinks are obtained from the entropic balance equation (5) and the corresponding vorticity (11). In Fig. 7, the three dimensional components of flow rotations, $\omega$, around the major source and sinks are shown by $R_{V1}$, $R_{V2}$, $R_{H1}$, $R_{H2}$, $R_{H3}$, and $R_{H4}$ where the subscripts V and H represent the direction of flow rotation along vertical and horizontal axes respectively. The flow rotation, $\omega$, balances with the entropic gradient and the flow velocity are, using the Fleming's law (Fig. 5) and the flow around symmetric source and sink

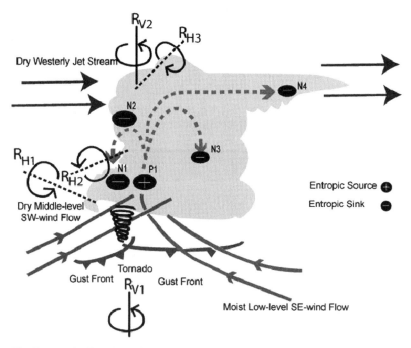

**Fig. 7** Entropically balanced direction of flow rotation for each major sinks and source (Fig. 6), and their between areas. The three dimensional components of flow rotations, $\omega$, around the major source and sinks are shown by $R_{V1}$, $R_{V2}$, $R_{H1}$, $R_{H2}$, $R_{H3}$, and $R_{H4}$ where the subscript V and H represent the direction of flow rotation along vertical and horizontal axes respectively

(Fig. 6), They serve to explain flow patterns around source and sinks, the overshooting mechanism of hydrometeors, cloud-top bump of supercell, and intensification of rear downdraft of supercell and tornado, as will be discussed in the next section.

# 4 Comparison of Entropic Balance Model with Other Tornadogenesis Models

In the last several decades, a vast number of observations in broad variety were made on supercell storms and tornadoes. They were well summarized and represented in a variety of schematic models and theories including computer simulations by many researchers. So, we will compare the entropic balance model with these published schematic models and the theories.

## 4.1 Non-supercell: Updraft Stage

The supercell is characterized by well developed long-lasting and primarily cyclonically rotating storm in the environmental wind field prescribed. Before the supercell is developed, the convective storm is filled with updraft airflow and condensation is predominant. The airflow spreads horizontally at the cloud-top level because of the westerlies and the stability of the layer. Accordingly, an entropic source due to the latent heat produced by condensation exists in the updraft and weaker sinks spread away in surrounding broad areas at large distance from the source (Fig. 8). This is the case of internally irrotational, $\omega \sim 0$, but of strong updraft due to the convergence of $\mathbf{v}$, from Eqs. (8), (9), or (10). It results into stretch and intensification of the vorticity converged externally from the environmental wind flow (e.g., Atkinson 1981; Klemp 1987; Bluestein 1993). It does possibly into initiation of tornadogenesis or tornado-like vortex development, separately from or in advance of tornadogenesis (Lemon 1977; Lemon and Doswell III 1979). See Sect. 5 of this article primarily for the linear model.

## 4.2 Non-supercell (or Supercell Stage I): Initial Downdraft Stage

Figure 9 shows initial downdraft stage of supercell storm, predominated by the major sink development, associated with the major source, which is produced by evaporation of cloud particles and cools the surrounding air causing downdraft. As the downdraft develops, gust front is formed on and near the ground. If the sink locates close to the source, less than a few kilometer, tornado would develop around the sink if the entropy gradient is large enough. It seems to have the properties of non-supercell tornado discussed by Wakimoto and Wilson (1989). Recognizable rotating supercell is characterized by developing downdraft, and may be called "downdraft

## Non-Supercell; Updraft Stage

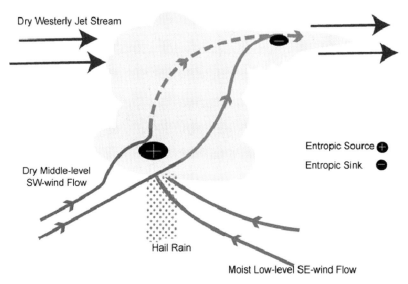

**Fig. 8** Schematic diagram of major flow pattern in non-supercell, updraft stage

## Supercell I: Initial Downdraft Stage

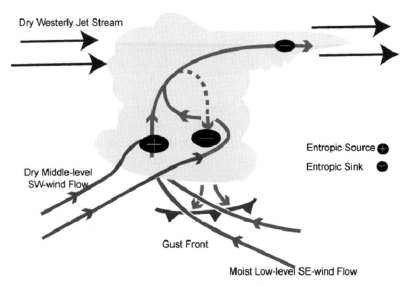

**Fig. 9** Schematic diagram of non-supercell, (or supercell I), initial downdraft stage

stage". Supercell development is divided into three stages. At the stage I of supercell is characterized by continuing updraft with well developed downdraft adjacent to the main updraft and gust front on the ground surface in the downstream side of supercell as seen in the schematic Fig. 9. The broken arrowed curve represents the downward trajectory of hydrometeors which evaporates and cools the adjacent air and then creates an entropic sink. As it get stronger, strong downdraft and a gust front develop at low levels. The flow around the source and sink, anti-cyclonic and cyclonic respectively, are shown also in Fig. 9. The circulation suggested by Chisholm and Renick (1972) [A Fig. 133], is similar in major features to those of Fig. 9.

The updraft due to the convergence of environmental flow intensifies also the environmental vorticity influx. One of the sources of the vorticity is the vertical shear of horizontal wind and generated from upward tilting of the horizontal vortex tube at the low altitudes near the ground surface. The tilting model of the horizontal vortex tube has a difficulty of generating always a pair of plus and minus vortices in the updraft, and one of the pair is expectedly to be eliminated by downdraft. But the downdraft splits the pair into two pairs if the downdraft develops in the updraft, or it is located in the storm away from the major portion of the pair. One of the pair vortices is not easy to be eliminated. However, the tilting model has another problem, fluid dynamically unacceptable too, that is, the end of the horizontal vortex tube is parallel along the ground surface and it is not perpendicular to the ground contrasting against the many visual observations that the end of observed tornadoes, when it touches the ground, is always perpendicular. To avoid the problem, Noda and Niino (2005) interpreted their computer simulation result as if the horizontal tube extends only in the storm outflow front.

On the other hand, by the entropic balance model, the downdraft has two roles; the one is to suppress vorticity, at and near the sink $N_3$, if it locates sufficiently away from the source P1, as seen in Fig. 9, and the other intensifies vorticity if $N_3$ locates sufficiently close to the source P1 and possibly proceed to the Stage II and then Stage III, tornadogenesis.

## 4.3 Supercell Stage II: Intermediate Stage; Overshooting Mechanism of Cloud Particles and Entropically Balanced Rotation in 3-D

The stage II is characterized by, together with one major source and two major sinks, one in upstream side at higher level and the other in downstream side at lower level of supercell, as schematically shown in Fig. 10. The upper level, upstream side sink and the lower level source create negative entropy gradient in the vertical direction. The downward entropy gradient and the eastward westerly jet stream are balanced with the cyclonic rotation along the southward axis. It is the overshooting mechanism of hydrometeors which was generated by condensation and sublimation from the entropic source at lower level. The rotation $R_{H2}$ along the horizontal axis

## Supercell II: Intermediate Stage

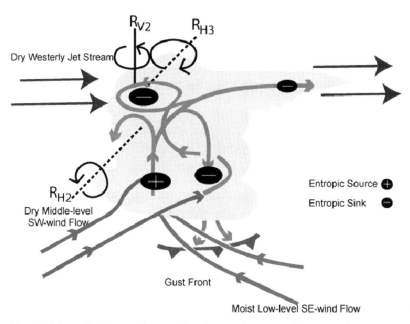

**Fig. 10** Schematic diagram of supercell II, intermediate stage, flows around source and sinks, and entropically balanced rotation in 3-D, for non-tornadic or possibly tornadic storm. The rotation $R_{H2}$ is instrumental to overshoot the hydrometeors against the strong upper westerlies The rotation $R_{H3}$ together with the updraft from the entropic source $P_1$ is working to form derby-hat shape top of the supercell. The rotation $R_{V2}$ is responsible to develop upper mesocyclone

together with the westward momentum carried in the updraft, as seen in Figs. 7 and 10, is instrumental to overshoot the hydrometeors against the strong upper westerlies. The rotation $R_{H3}$ together with the updraft from the entropic source $P_1$ is working to form derby-hat shape top of the supercell.

Also around the upper level sink, the horizontal entropic gradient, symmetric in the vertical axis, is balanced with the westerlies and cyclonic rotation along the vertical axis as shown by $R_{V2}$ in Figs. 7 and 10. This stage seems to be evidenced by the visual sketch and photographic report of Bluestein and Parks (1983), [B Fig. 3.34]. The mammatus clouds in both the downstream and upstream of the upper westerlies seem to show the entropic sinks which spread horizontally due to evaporation-cooled heavier air with remaining un-evaporated cloud forming mammatus cloud. The overshooting of hydrometeors is apparent in the upstream side against the headwind westerly, which evaporates also forming mammatus cloud in a similar manner. The multi-cell structure and the circulation suggested by Chisholm and Renick (1972), [A Figs. 123 and 124], may be understood in the major features from those of Figs. 7 and 11. The overshooting mechanism is important for generation of echo vault and tornadogenesis as will be seen in the several references such

## Supercell III: Downdraft Stage

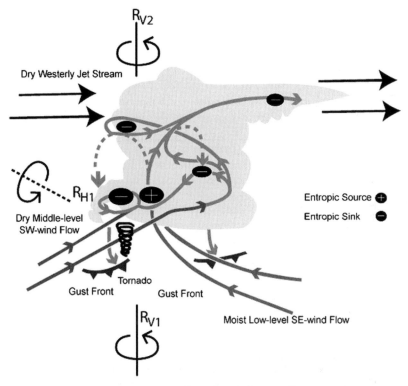

**Fig. 11** Schematic diagram of supercell III, downdraft stage, flows around source and sinks, and entropically balanced rotation $R_{V1}$ along the vertical axis, for tornadic storm. The downdraft created by cooling the surrounding air due to evaporation and sublimation in and around the sink generates another gust front west to the first one as schematically shown in this figure. Furthermore the downdraft brings in the upper-level dry westerlies and the middle-level dry SW or westerly inflow down and evaporate more the hydrometeors, intensifying the downdraft to the ground. The entropic balance rotation $R_{H1}$ along the horizontal axis will intensify the rear down draft of the middle-level SW or westerly dry air inflow

as Browning (1965) [A Fig. 120]. He showed a clear photograph of range height indicator (RHI) vertical section showing the echo vault. The overshooting of hydrometeors is clearly indicated in the schematic vertical section model of a severe multi-cell hailstorm by Browning et al. (1976) [A Fig. 126]. Further references will be found in the next section of supercell stage III: tornadogenesis.

## 4.4 Supercell Stage III: Downdraft Stage; Tornadogenesis

The overshot hydrometeors fall down to the same level of the source and evaporate creating sink. If the distance between the sink and source is small, a strong vorticity along the vertical axis should exist as seen from the entropic balance equations (5), (6) and (11) and the numerical estimates (15)–(17). The downdraft created by cooling the surrounding air due to evaporation and sublimation in and around the sink generates another gust front west to the first one as schematically shown in Fig. 11. Furthermore the downdraft brings in the upper-level dry westerlies and the middle-level dry SW or westerly inflow down and evaporates more the hydrometeors, intensifying more the downdraft to the ground. Indeed, the entropic balance rotation $\mathbf{R}_{H1}$ (Figs. 7 and 11) along the horizontal axis will intensify the rear downdraft of the middle-level SW or westerly dry air inflow. The rear downdraft will be more intensified if SW wind becomes more westerly wind, as seen from Fig. 11. The surface analyses made by Fujita (1958) [A Fig. 134 (a) (b)], from surface barograph traces, wind, radar data, and surface damage survey, can be well understood from the tornadogenesis mechanism of the entropic balance theory. The vertical cross section of radar reflectivity echo pattern across the hock echo made by Lemon (1977) [B Fig. 3.32(b)], which shows overhang echo pattern produced by the overshot hydrometeors evaporated around the sink at the lowest level. The cyclonic flow around the sink is also clearly shown by him [B Fig. 3.32(a)]. The schematic plan view of a tornadic thunderstorm near the surface by Lemon and Doswell III (1979) [B Fig. 3.41], is in a good agreement with the entropic balance theory. The entropic balance theory suggests from Eqs. (6) and (11) that the vortex center locates at the maximum entropy gradient, likely between the sink and the source. This agrees with the analyses of the observations just cited above. Furthermore the classical air flow models of supercell proposed by Browning and Ludlam (1962) [A Fig. 132 (a)], Browning (1964) [A Fig. 132 (c)], Newton (1966), and fine structure found by Wurman and Gill (2000) seem to be explainable as the flow patterns suggested by the entropic balance theory.

The application of the entropic balance theory presented so far suggests development of tornadic vortex, mesocyclone, wall cloud and tornado, at or near the entropic sinks as shown by $\mathbf{R}_{V1}$ and $\mathbf{R}_{V2}$ along the vertical axes in Figs. 7 and 11, if the lower sink locates in close proximity, about few km or less, to the source, in supercell. It requires certain environmental wind field, such as moist air influx of SE wind at low altitudes and dry air influxes of the southwest or more effectively westerly wind at the middle altitudes and the dry westerlies at high altitudes. Also, playing important roles for tornadogenesis are the overshooting of hydrometeors affected by the rotation $\mathbf{R}_{H2}$ along the horizontal axis and the dry rear downdraft produced under the rotational mechanism of $\mathbf{R}_{H1}$ along also the horizontal axis. These dry air influxes are instrumental to develop intense entropic sinks, consequently tornado. We will next examine initial, incipient, stage of tornadogenesis.

# 5 Linear (and Semilinear) Initial Stage

The nonlinear system expressed by Eqs. (5), (6) or (7) and definition (8) is linearized by defining the basic field expressed by $(\bar{\alpha}, \bar{\beta}, \bar{v}, \bar{S})$ by in a simplest form appropriate for the supercell environment discussed in Introduction,

$$\partial_z \overline{S} := S_z \tag{18}$$

$$(\partial_z \overline{u}, \partial_z \overline{v}, 0) := (U_z, V_z, 0). \tag{19}$$

where $S_z$, $U_z$, $V_z$, and in the following equations (20)–(27) U, V, S., $\overline{\alpha}_x$, $\overline{\alpha}_y$, $\overline{\beta}_x$, and $\overline{\beta}_y$ are all constants.

Because of the linear assumption, we have the following expressions;

$$\overline{S} = S_0 + S_z z \tag{20}$$

$$\overline{u} = U + U_z z \tag{21}$$

$$\overline{v} = V + V_z z \tag{22}$$

$$\overline{w} = 0 \tag{23}$$

$$\overline{u}_D = U \tag{24}$$

$$\overline{v}_D = V. \tag{25}$$

$$\overline{\alpha} = -\overline{\alpha}_x x - \overline{\alpha}_y y \tag{26}$$

$$\overline{\beta} = -\overline{\beta}_x x - \overline{\beta}_y y. \tag{27}$$

These Eqs. (20), (21), (22), (23), (24), (25), (26)-and (27) should satisfy Eqs. (5), (6) or (7), as

$$\overline{v} = -\nabla\overline{\alpha} - \overline{S}\nabla\overline{\beta} \tag{28}$$

and the vorticity of the basic field, $\overline{\omega}$, becomes

$$\overline{\omega} = (1/\overline{S})\,\nabla\overline{S} \times (-\overline{S}\nabla\overline{\beta}) \tag{29}$$

$$= (1/\overline{S})\,\nabla\overline{S} \times (\overline{v} - \overline{v}_D) \tag{30}$$

where

$$\overline{v}_D := -\nabla\overline{\alpha}. \tag{31}$$

The vertical vorticity of the basic field $\overline{\omega}_z$ vanishes,

$$\overline{\omega}_z = 0. \tag{32}$$

The corresponding semi linear small perturbation equations of the first order superimposed on the basic fields are

$$v' = v'_D - \overline{S}\nabla\beta' - S'\nabla\overline{\beta} \tag{33}$$

and

$$\omega' = (1/\overline{S})\,\nabla S' \times (\overline{v} - \overline{v}_D) + (1/\overline{S})\,\nabla\overline{S} \times (v' - v'_D). \tag{34}$$

Let $\omega_z'$ to be the perturbation vorticity. Since $\partial_x \overline{S} = 0$ and $\partial_y \overline{S} = 0$, Eq. (34) becomes

$$\omega_z' = (1/\overline{S}) \, \nabla S' \times [S_o/S_z \cdot (U_z, V_z, 0)]. \tag{35}$$

This equation shows extremely important implication on the question why the origin of tornadic vorticity was suspected to exist in the planetary boundary layer. Because of convective mixing and near neutral stability, $S_z$ is considered to vanish in the boundary layer, and consequently $\omega_z'$ increases near the ground, and it became easier to suspect origin of tornadogenesis in the low levels. Also, since in the neutral stratification with small $S_z$, $\partial_z \overline{S} = 0$ and the horizontal perturbation vorticities, $\omega_x'$ and $\omega_y'$, therefore 3-D perturbation vorticity is expressed in the similar way of Eq. (35) as

$$\omega' = (1/\overline{S}) \, \nabla S' \times [S_o/S_z \cdot (U_z, V_z, 0)], \tag{36}$$

and consequently the horizontal rotations of increased magnitudes in the low levels near the ground may well be considered to attribute to RFD and sharp turned updraft of southerly current. It is interesting to note that these results are because that $S_z$ appears as the denominator term of perturbation vorticity equation. It is because of important roles of Lagrange multipliers which have $S_z$ in their denominator. We can see it as follows:

From Eqs. (21), (22), (26), (27) and (28), we have the following equations

$$U_o + U_z z = -\overline{\alpha}_x - (S_o + S_z z) \, \overline{\beta}_x \tag{37}$$

$$V_o + V_z z = -\overline{\alpha}_y - (S_o + S_z z) \, \overline{\beta}_y \tag{38}$$

Since Eqs. (37) and (38) should be valid for any z, $\overline{\alpha}_x, \overline{\beta}_x, \overline{\alpha}_y, \overline{\beta}_y$ are given as

$$\overline{\alpha}_x = U_o + S_o/S_z U_z \tag{39}$$

$$\overline{\beta}_x = -U_z/S_z \tag{40}$$

$$\overline{\alpha}_y = V_o + S_o/S_z V_z \tag{41}$$

$$\overline{\beta}_y = -V_z/S_z. \tag{42}$$

It should be noted that Lagrange multiplier (adjoint) is the coefficient of the equation as the constraint, in Lagrangian, which vanishes so that it makes difficult to see the roles of the multiplier explicitly. One of properties of Lagrange multiplier (adjoint) is known as a sensitivity index (Caccusi 1981; Enrico and Vukicevic 1992; Park and Droegemeier 2000). In this Sect. 5, we demonstrated a new finding that the Lagrange multipliers possess additional hidden property which seems to explain stronger vorticity in the low level atmosphare in case of tornadogenesis.

Figures 12, 13, 14 and 15 show four extreme linear cases for point source and sink for two constant shear cases, in the middle and lower troposphere respectively. The linear cases can be balanced with cyclonic and anticyclonic rotations, which

**Fig. 12** Sign of rotation: Linear approximation with constants $U_z > 0$ and $V_z < 0$ in the neighborhood of an entropic point source. Note that the rotations are primarily of a *dipole* type around the entropic source, different from the rotation shown in Fig. 11, supercell III; Downdraft stage, which is primarily of *monopole* type around an entropic source

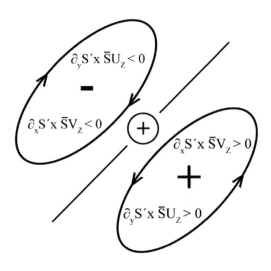

**Fig. 13** Sign of rotation: Linear approximation with constants $U_z > 0$ and $V_z < 0$ in the neighborhood of an entropic point sink. See the same note as Fig. 12 except around an entropic sink

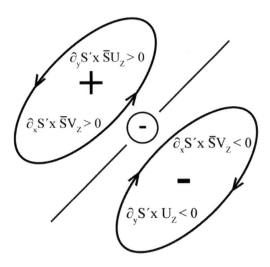

seem to be analogous to and in the category of the earlier mechanisms discussed by several researchers (cited in earlier sections). Figures 12, 13, 14 and 15 demonstrate sign of rotations in linear approximation, for constant $U_z > 0$ and $Vz < 0$, around each of entropic point source and sink seperately, and $U_z > 0$ and $Vz > 0$ for near

**Fig. 14** Sign of rotation: Linear approximation with constants $U_Z > 0$ and $V_Z > 0$ (near the ground) in the neighborhood of an entropic point source. See the same note as Fig. 12

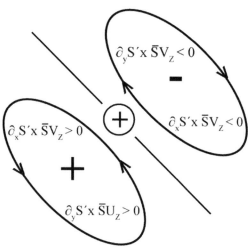

Sign of Rotation : Linear, around Entropic Source
$U_z > 0$ , $V_z > 0$ (near the ground)

**Fig. 15** Sign of rotation: Linear approximation with constants $U_Z > 0$ and $V_Z > 0$ (near the ground) in the neighborhood of an entropic point sink. See the same note as Fig. 12, except around an entropic sink

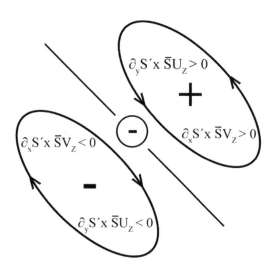

Sign of Rotation : Linear, around Entropic Sink
$U_z > 0$ , $V_z > 0$ (near the ground)

the ground surface, respectively. The rotational centers become as a *dipole* at the early stage of tornadogenesis. Later in the mature tornadogenesis, the velocity and vorticity equations to be used should be nonlinear, Eqs. (5) and (6) respectively, with *monopole* rotational center. It is also important to note that the axis of vortex is tilted, not stretched in the major updraft , but determined by the entropic balance in the environmental wind-shear at the beginning stage of tornadogenesis.

It can be extended for non-supercell type and line sources and sinks (Wakimoto and Wilson 1989), analogous to RFD-gust front. *Consequently, the nonlinear entropic balance equations become important to consider for monopole type supercell, gust front and tornadogenesis, in contrast to its linear approximations for earlier dipole type tornadogenesis.* The results of the mathematical and dynamical analysis in the preceding Sects. 1–4, suggest that the entropic balance model is a candidate to be used in the most efficient and accurate data assimilation for practical and operational analysis and prediction of supercell development and tornadogenesis.

# 6 Remarks on Advanced Detection of Tornadogenesis and Downburst

If we find appropriate model for tornadogenesis mechanism, we may able to use presently available meteorological data and develop appropriate data assimilation for practically accurate prediction of tornadogenesis.

Development of variational data assimilation method with a reduced order or short-cut Kalman filter are interesting to determine from radar reflectivity and Doppler velocity measurements. It may include the most probable hydrometeor trajectory in the practical form of probability density function in variational Kalman filter formulation for tornadogenesis (Sasaki 2003; Rozier et al. 2007). It should be noted here that recently an encouraging research presentation was made at the National Weather Center seminar held on 6 September 2007, by Glen Romine, University of Illinois, Urbana-Champaign. He developed a method of using dual polarimetric radar observation with two-moment microphysical retrieval and Kalman filter based assimilation for retrieval of detailed physical state estimates of storms and the adjacent environment, especially around RFD (Romine 2007).

Important is the recent finding of short temporal scale, 34 s, of microburst (downburst) development and rapid downward spread by the National Weather Radar Testbed Phased Array Radar (NWRT PAR), (Forsyth et al. 2007; Henselman et al. 2006; Henselman 2007). The entropic balance can be extended through Clebsch transformation to downburst analysis and data assimilation, pending future development.

Technology for quick and accurate detection of entropic source and sink with sufficient accuracy and the associated data assimilation can be developed on the basis of the entropy theory, to fulfill present practical needs. Such a development project is being conducted at Weathernews Inc., Makuhari, Japan.

**Acknowledgements** The author would like to express his appreciation to the encouragement and support given to this research by Office of Vice President for Research, the University of Oklahoma, USA, and Weathernews, Inc., Japan. Also the author would like to expresses his sincere appreciation to faculty and scientists of the institutions, especially, Ewha Womans University, Seoul, Korea, National Weather Center of the University of Oklahoma, USA, National Severe Storms Laboratory, and Dissert Research Institute, NOAA, Tokyo University, Kyoto University, Weathernews, Inc., National Institute for Earth Science and Disaster Research, Japan Meteorological

Agency, Japan, for their valuable discussions for the invited presentations given at the conference, seminars and colloquiums. Also, the author thanks Mrs. M. Matsutori of Kyoto University and Mr. Willy Chong of the University of Oklahoma for the technical assistances given to prepare this manuscript.

# References

Alderman EJ et al (1999) A numerical simulation of cyclic mesocyclogenesis. J Atmos Sci 56: 2045–2069

Asai T (1970a) Three dimensional features of thermal convection in a plane Couette flow. J Meteor Soc Japan 48: 18–29

Asai T (1970b) Stability of a plane parallel flow with variable vertical shear and unstable stratification. J Meteor Soc Japan 48:129–139

Atkins P, de Paula J (2002) Atkins' physical chemistry (7th ed.). Oxford University Press, New York

Atkinson BW (1981) Meso-scale atmospheric circulations. Academic Press, London (marked by A with figure number of this book when used for reference in the present article)

Bateman H (1932) Partial differential equations of mathematical physics. Cambridge Univ Press, Cambridge (reprinted by Dover Publ., 1944)

Bluestein HB (1993) Synoptic-dynamic meteorology in midlatitudes, Vol II, Observations and theory of weather systems. Oxford Univ Press, New York (marked by B with figure number of this book when used for reference in the present article)

Bluestein HB, Parks CR (1983) A synoptic and photographic climatology of low-precipitation severe thunderstorms in the southern plains. Mon Wea Rev 111:2034–46

Browning KA (1964) Airflow and precipitation trajectories within severe local storms which travel to the right of the winds. J Atmos Sci 21: 634–639

Browning KA (1965) Some inferences about the updraught within a severe local storm. J Atmos Sci 22: 669–677

Browning KA (1968) The organization of severe local storms. Weather 23:429–434

Browning KA (1977) The structure and mechanism of hailstorms. Hail: A review of hail, science and hail suppression. Meteor Monogr 38 Amer Meteor Soc: 116–122

Browning KA, Ludlam FH (1962) Airflow in convective storms. Quart J Meteor Soc 88: 117–135

Browning KA et al (1976) Structure of an evolving hail storm. V. Synthesis and implications for hail growth and hail suppression. Mon Wea Rev 104: 603–610

Burgess D et al (1982) Mesocyclone evolution statistics. Preprints, 12th conf. on Severe Local Storms, San Antonio, Texas, Amer Meteor Soc: 422–424

Byers HR, Braham RR (1949) The Thunderstorm. Dept. Commerce, USA

Caccusi DG (1981) Sensitivity theory for non-linear system. I: Nonlinear functional analysis approach. J Math Phys 22: 2794–2803

Chandrasekar C (2007) CASA vision of dense low-level network. National Symposium on Multifunction Phased Array Radar: Laveraging technology for a next-generation national radar system. October 10–12, 2007, National Weather Center, Norman Oklahoma

Chisholm A J, Renick JH (1972) The kinematics of multicell and supercell Alberta hail stdies. Research Council of Alberta Hail Studies, Report 72–2: 24–31

Clebsch A (1859) Über ein algemeine transformation der Hydrodynamischen Gleichungen. Crelle J für Math 54:293–312

Davies-Jones R (1984) Streamwise vorticity: The origin of updraft rotation in supercell storms. J Atmos Sci 41: 2991–3006

Davies-Jones R, Brooks HE (1993) Mesocyclogenesis from a theoretical perspective. The Tornado: Its Structure, Dynamics, Prediction, and Hazards. Geophys Monogr 79, Amer Geophys Union: 105–114

Dutton JA (1976) The ceaseless wind. McGraw-Hill, New York

Enrico RM, Vukicevic T (1992) Sensitivity analysis using an adjoint of the PSU-NCAR meso-scale model. Mon Wea Rev 120: 1644–1660

Forsyth DE et al (2007) Update on the National Weather Radar Testbed (phased array). Preprint 23rd Conf. on Interactive Information Processing Systems for Meteorology, Oceanography, and Hydrology, San Antonio, TX, Amer Meteor Soc CD-ROM,7.4

Fujita T (1958) Mesoanalysis of the Illinois tornadoes of April 9, 1953. J Meteor 15: 288–296

Henselman PD (2007) Rapid sampling of severe storms by the National Weather Radar Testbed Phased Array Radar. Presented at National Weather Center Seminar on 02 October 2007

Henselman PD et al (2006) Comparison of storm evolution characteristics: The NWRT and WSR-88D. Preprint, 23rd Conf. on Severe Local Storms, St Louis, MO, Amer Meteo Soc

Honda Y et al (2005) A pre-operational variational data assimilation for nonhydrostatic model at Japan Meteorological Agency: Formulation and preliminary results. Quart J Roy Meteor Soc 131: 3465–3475

Ishikawa Y, Koizumi K (2002) One-month cycle experiments of the JMA mesoscale 4-dimensional variational data assimilation (4D-Var) system. Research Activities in Atmospheric and Oceanic Modelling WNO/TD-No.1105 1: 1.26–1.27

Jeffreys H (1928) Some cases of instability in fluid motion. Proc Roy Soc A 118: 195–208

Kalney E (2003) Atmospheric modeling, data assimiltion and predictability. Cambridge University Press, Cambridge

Kawabata T et al (2007) An assimilation experiment of the Nerima heavy rainfall with a cloud-resolving nonhydrostatic 4-dimensional variational data assimilation system. J Meteor Soc Japan 85: 255–276

Klemp JB (1987) Dynamics of tornadic thunderstorms. Ann Rev Fluid Mech 19: 369–402

Klemp JB, Rotunno R (1983) A study of the tornadic region within a supercell thunderstorm. J Atmos Sci 40: 359–377

Klemp JB, Wilhelmson RB (1978) The simulation of three-dimensional convective storm dynamics. J Atmos Sci 35: 1070–1096

Koizumi K et al (2005) Assimilation of precipitation data to JMA mesoscale model with a four-dimensional variational method and its impact on precipitation forecast. Sola 1: 45–48

Kuo YH et al (1996) Variational assimilation of precipitable water using a non-hydrodynamic mesoscale adjoint model. Part I: Moisture retrieval and sensitivity experiments. Mon Wea Rev 124: 122–147

Lamb H (1932) Hydrodynamics, 2nd ed. Cambridge Univ. Press, Dover Publ, New York

Lemon LR (1977) New severe thunderstorm radar identification techniques and warning criteria: A preliminary report. NOAA Tech. Memo. NWS NSSFC-1, TDU, NSSFC, Kansas City

Lemon LR, Doswell III CA (1979) Severe thunderstorm evolution and mesocyclone structure as related to tornadogenesis. Mon Wea Rev 117: 1184–1197

Leslie LM (1971) The development of concentrated vortex: A numerical study. J Fluid Mech 48: 1–21

Lewis JM et al (2006) Dynamic data assimilation: A least squares approache. Cambridge University Press, Cambridge

Lilly DK (1982) The development and maintenance of rotation in convective storms. In: Bengtsson L and Lighthill J (eds) Intense atmospheric vortices. Springer-Verlag, New York, pp 149–160

Mahfouf JF, Rabier F (2000) The ECMWF operational implementation of four-dimensional variational assimilation-Part II: Experimental results with improved physics. Quart J Roy Meteor Soc 126: 1171–1190

McQuarrie DA, Simon JD (1997) Physical chemistry: A molecular approach. University Science Books, Sansalito, California

Newton CW (1966) Circulations in large sheared cumulonimbus. Tellas 18: 669–712

Noda A (2002) Numerical simulation of supercell tornadogenesis and its structure. (Doc. of Science dissertation in Japanese) Graduate College of Science, Tokyo University, Tokyo

Noda A, Niino H (2005) Genesis and structure of a major tornado in a numerically-simulated supercell storm: Importance of vertical velocity in a gust front. SOLA 1: 5–8

Ogura Y, Phillips NA (1962) Scale analysis of deep and shallow convection in the atmosphere. J Atmos Sci 19: 173–179

Park SK, Droegemeier KK (2000) Sensitivity analysis of a 3D convective storm: Implications for variational data assimilation and forecast error. Mon Wea Rev 128: 184–197

Park SK, Zupanski D (2003) Four-dimensional variational data assimilation For Mesoscale and storm-scale application. Meteor Atmos Phys 82: 173–208

Rasmussen EN et al (1982) Evolutionary characteristics and photogrammetric determination of wind speeds within the Tulia outbreak tornadoes 28 May 1980. Preprints, 12th Conf. on Severe Local Storms. San Antonio, Texas, Amer Meteor Soc: 301–304

Romine G (2007) Assesment and assimilation of polarimetric radar observations: Improving convective storm analyses. Presented at National Weather Center, University of Oklahoma, senimar on 6 September, 2007 (personal communication)

Rotunno R (1981) On the evolution of thunderstorm rotation. Mon Wea Rev 109: 577–586

Rotunno R, Klemp JB (1982) The influence of the shear-induced pressure gradient on thunderstorm rotation. Mon Wea Rev 110: 136–151

Rotunno R, Klemp JB (1985) On the rotation and propagation of numerically simulated supercell thunderstorms. J Atmos Sci 42: 271–292

Rozier D et al (2007) A reduced-order Kalman filter for data assimilation in physical oceanography. SIAM Rev 49: 449–465

Sasaki Y (1955) A fundamental study of the numerical prediction based on the variational principle. J Meteor Soc Japan 33: 262–275

Sasaki Y (1958) An objective analysis based on the variational method. J Meteor Soc Japan 36: 77–88

Sasaki Y (1999) Tornado and hurricane—Needs of accurate prediction and effective dissemination of information. (in Japanese). J Visualization Soc Japan 19, 74: 187–192

Sasaki Y (2003) A theory of variational assimilation with Kalman filter type constraints: Bias and Lagrange multiplier. Mon Wea Rev 131: 2545–2554

Sasaki Y, Baxter TL (1982) The gust front. Thunderstorms: A social, scientific, & technological documentary, Vol. 2, Thunderstorm morphology and dynamics, U. S. Department of Commerce, National Oceanic and Atmospheric Administration, Environmental Research Laboratories: 281–296

Seko H et al (2004) Impacts of GPS-derived water vapor and radial wind measured by Doppler radar on numerical prediction of precipitation. J Meteor Soc Japan 82: 473–489

Seo EK, Biggerstaff MI (2006) Impact of cloud model microphysics on passive microwave retrievals of cloud properties. Part II: Uncertainty in rain, hydrometeor structure, and latent heating retrievals. (personal communication, through the seminar of National Weather Center, University of Oklahoma)

Simpson JE (1972) The effect of the lower boundary on the head of a gravity current. J Fluid Mech 53: 759–768

Takeda T (1971) Numerical simulation of precipitating convective cloud: the formation of a "long-lasting" cloud. J Atmos Sci 28: 350–376

Trapp RJ, Davies-Jones R (1997) Tornadogenesis with and without a dynamical pipe effect. J Atmos Sci 54: 113–133

Tsuboki K (2007) Recent progress of simulation on convective storm and tornadogenesis (in Japanese). Tornado Symposium held on January 13, 2007, organized by Japan Meteorological Agency (personal communication)

Tsuyuki T (1996a) Variational data assimilation in the tropics using precipitation data. Part I: Column model. Meteo Atmos Phys 60: 87–104

Tsuyuki T (1996b) Variational data assimilation in the tropics using precipitation data. Part II: 3D model. Mon Wea Rev 124: 2545–2561

Tsuyuki T (1997) Variational data assimilation in the tropics using precipitation data. Part III: Assimilation of SSM/I precipitation rates. Mon Wea Rev 125: 1447–1464

Tsuyuki T et al (2002) The JMA mesoscale 4D-Var system and assimilation of precipitation and moisture data. Proceedings of the ECMWF/GEMEX Workshop on Humidity Analysis (8–11 July 2002, Reading, UK), ECMWF: 59–67

Tsuyuki T, Miyoshi T (2007) Recent progress of data assimilation methods in Meteorology. J Meteor Soc Japan 65B: 331–361

Wakimoto RM, Wilson JW (1989) Non-supercell tornadoes. Mon Wea Rev 117: 1113–1140

Wilhelmson RB, Klemp JB (1981) A three-dimensional numerical simulation of splitting severe storms on 3 April 1964. J Atmos Sci 38: 1581–1600

Wicker LJ, Wilhelmson RB (1995) Simulation and analysis of tornado development and decay within a three dimensional supercell thunderstorm. J Atmos Sci 52: 2675–2703

Wurman J, Gill S (2000) Finescale radar observations of the Dimmitt, Texas (2 June 1995), tornado. Mon Wea Rev 128: 2135–2164

Xue M et al (1995) ARPS Version 4.0 User's Guide. The center for analysis and prediction of storms, University of Oklahoma

Zou X et al (1993) Variational data assimilation with moist threshold processes using NMC spectral model. Tellas 45A: 370–387

Zrnic D S et al (2007) Agile-beam phased array radar for weather observations. BAMS Nov 2007: 1753–1766

Zupanski D (1993a) The effects of discontinuities in the Betts-Miller cumulas convection scheme on four-dimensional variational data assimilation. Tellas 45A: 511–524

Zupanski D (1997) A general weak constraint applicable to operational 4DVAR data assimilation systems. Mon Wea Rev 125:2275–2292

Zupanski D, Mesinger F (1995) Four dimensional data assimilation of precipitation data. Mon Wea Rev 123: 1112–1127

Zupanski M (1993b) Regional four-dimensional variational data assimilation in a quasi-operational forecasting environment. Mon Wea Rev 121: 2396–2408

# Radar Rainfall Estimates for Hydrologic and Landslide Modeling

Kang-Tsung Chang, Jr-Chuan Huang, Shuh-Ji Kao and Shou-Hao Chiang

**Abstract** The spatial variability of rainfall is an important input to hydrologic and landslide modeling. The scarcity of rain gauges, especially in mountainous watersheds, has been a common problem in incorporating spatially variable rainfall fields into hydrologic and landslide models. This study describes the application of rainfall estimates derived from multi-sensor radar data for two storm events. For both stream discharge simulation and landslide modeling, the results are encouraging. Statistics for comparing the simulated with the observed show that radar rainfall is more accurate than gauged rainfall in simulating stream discharge. A logit model developed from landslides and radar rainfall for one storm event is able to predict over 80% of landslides triggered by another storm event. This study suggests that radar rainfall, although it still has some shortcomings, can be an alternative to gauged rainfall for modeling applications.

## 1 Introduction

As the major driving force to most hydrologic systems, rainfall is a necessary input to hydrologic and landslide modeling. An important issue, when considering rainfall as an input, is the spatiotemporal variability of rainfall. For hydrologic modeling, the distributed (and semi-distributed) approach has evolved for the last two decades to account for the spatial variability of hydrologic variables. However, the approach has been hampered by a lack of rainfall data with appropriate spatial resolution and reliable quality (Kalinga and Gan 2006). For landslide modeling, past studies have used rainfall as an input to physically-based landslide models (e.g. Montgomery and Dietrich 1994; Iverson 2000) and considered rainfall as an explanatory variable for logistic regression analysis (e.g. Dai and Lee 2003). Again, a common problem with these studies has been the unavailability of rainfall data.

Researchers have coped with insufficient local rainfall data by using reference gauges (e.g., Aleotti 2004), Thiessen polygons (e.g., Godt et al. 2006), or spatial

K.-T. Chang (✉)
Kainan University, Taoyuan, Taiwan,
e-mail: chang@uidaho.edu

S.K. Park, L. Xu, *Data Assimilation for Atmospheric, Oceanic and Hydrologic Applications*, DOI 10.1007/978-3-540-71056-1_6,
© Springer-Verlag Berlin Heidelberg 2009

interpolation (e.g., Guzzetti et al. 2004). Nevertheless, the result has been unsatisfactory, leading to an underutilization of a model or a mismatch between a complex model and its driving forces (Boyle et al. 2001; Casadei et al. 2003).

One feasible approach for considering the spatial variability of rainfall is radar technique, which has assisted weather predictions for over 40 years but has only appeared in hydrological applications for a decade or so (Krajewski and Smith 2002). The potential benefit of using radar data is particularly large in mountainous areas, where the rugged terrain and the altitudinal effect impose significant limitations to the real-time operation of a rain gauge network (Hossain et al. 2004). Moreover, some of these mountainous areas have a high frequency of large rainfall accumulations that can lead to catastrophic flood events. Early warning with the use of weather radar for assessing the likelihood of floods and landslides can help mitigate the loss of lives and properties (Dinku and Anagnostou 2002).

Taiwan experiences an average of four to five typhoons (tropical hurricanes) a year. Typhoons bring torrential rains to the island, especially in mountainous catchments, triggering floods and landslides. Although Taiwan has 471 rain gauges, most are located in the lowland areas on the west side of the Central Mountain Range. This is why, for the past several years, government agencies in Taiwan have turned to radar-derived rainfall data for a forecasting and warning program for flash floods and landslides. However, the program's effectiveness has not been established. The main purpose of this study is to evaluate the usefulness of radar data from two typhoon events for stream discharge simulation and landslide modeling in Taiwan.

This paper first describes radar-derived rainfall estimates and the weather radar program in Taiwan. Then it covers the study area, two typhoon events, gauged rainfall and stream discharge data, landslide data, and radar rainfall estimates. Hydrologic modeling, landslide modeling, and their results are followed. The paper concludes with a short conclusion.

# 2 Radar-Derived Rainfall Estimates

In recent years, an increasing number of researchers have used radar-derived rainfall estimates as a data source. There are two main reasons for adopting radar data. First, algorithms have become available for deriving rainfall estimates from weather radar data. For example, the Next Generation Weather Radar (NEXRAD) program in the United States offers rainfall estimates derived from Weather Surveillance Radar-1988 Doppler (WSR-88D) data over the entire country (Fulton et al. 1998; Carpenter and Georgakakos 2004a, b; Reed et al. 2007). Second, in areas such as mountainous catchments, where rain gauges are sparsely distributed, radar data can better capture the spatial variability of rainfall fields than gauged data (Yang et al. 2004; Segond et al. 2007). Radar rainfall estimates from the NEXRAD program have a spatial resolution of 4 km. Other studies have reported finer-resolution estimates at 2 km (Segond et al. 2007) and 1 km (Chang et al. 2007; Norbiato et al. 2007).

Radar rainfall estimates have proven to be useful for analysis and simulation of runoff hydrographs, floods, and extreme rainfall events (Yang et al. 2004; Kalinga and Gan 2006; Haberlandt 2007; Norbiato et al. 2007; Segond et al. 2007). Likewise, rainfall data derived from NEXRAD reflectivity imagery have been used for predicting debris flows (Morrissey et al. 2004; Chen et al. 2007).

The accuracy of rainfall estimates from radar data, however, can be complicated by a number of factors such as miscalibrated radar, ground clutter, beam blockage, beam overshooting, non-uniform vertical profile of reflectivity, and uncertainty in the reflectivity-to-rain rate conversion (Howard et al. 1997; Krajewski and Smith 2002; Steiner and Smith 2002). Several previous studies have shown that, compared to gauged data, radar rainfall estimates tend to be underestimated, especially for stratiform precipitation (Stellman and Fuelberg 2001; Kalinga and Gan 2006; Chiang et al. 2007). Therefore, different algorithms have been proposed to calibrate radar rainfall estimates with gauging station data (Wilson 1970; Xin et al. 1997; Fulton et al. 1998; Hossain et al. 2004; Kalinga and Gan 2006; Haberlandt 2007).

To better understand the spatial distribution of precipitation in mountainous areas, Taiwan's Central Weather Bureau (CWB) has collaborated with the U.S. National Oceanic and Atmospheric Administration's National Severe Storms Laboratory to deploy the QPESUMS (quantitative precipitation estimation and segregation using multiple sensors) system (Gourley et al. 2002). The multi-sensor system uses four Doppler weather radars to cover the island and the adjacent ocean. Using an algorithm developed by Zhang et al. (2005), the system combines data from multiple radars and transforms the data from a spherical coordinate to a Cartesian grid. For rainfall estimates, the system uses base (lowest elevations without blockages) reflectivity data with a spatial resolution of $0.0125°$ ($\sim 1.25\,$km) in both longitude and latitude and a temporal resolution of 10 min.

# 3 Study Area and Data

## 3.1 Study Area

The Shihmen Reservoir watershed, with a drainage area of $760\,$km$^2$ is located in northern Taiwan (Fig. 1). The reservoir is a major reservoir in Taiwan for agricultural irrigation, hydropower generation, public water supply, and flood prevention. Elevations in the watershed range from $135\,$m in the northwest to $3529\,$m in the southeast, with generally rugged topography. Nearly 90% of the study area is forested. Because of high relief and abundant rainfall, the soil type is mainly Lithosols (Entisols). Lithologic formations are mainly composed of sandstone, shale, and slate. The climate is influenced by typhoons in summer and the northeast monsoon in winter. The mean annual temperature is 21°C, with a mean monthly temperature of 27.5°C in July and 14.2°C in January. The annual precipitation

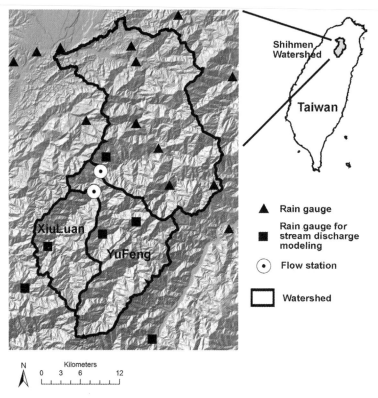

**Fig. 1** The Shihmen reservoir watershed, YuFeng and XiuLuan subwatersheds, rain gauges, and flow stations

averages 2370 mm, with large rainfall events occurring from May to September during the typhoon season.

The Shihmen Reservoir watershed consists of a number of subwatersheds including YuFeng with a drainage area of 335.29 km² and XiuLuan with a drainage area of 115.93 km² (Fig. 1). Because 75% of the annual precipitation occurs from May to September, stream discharges in these two subwatersheds differ significantly between the wet and dry seasons. In YuFeng, the mean daily discharge is 257.1 mm in the wet season, 98.4 in the dry season, and 178.4 annually. In XiuLuan, the mean daily discharge is 89.0 mm in the wet season, 32.5 in the dry season, and 60.9 annually.

Nineteen rain gauge stations are located in and around the Shihmen Reservoir watershed (Fig. 1). They were used for calibrating radar rainfall estimates. Two flow stations are located at the outlet of the YuFeng and XiuLuan subwatersheds (Fig. 1). Observed stream discharge data came from these two stations, thus limiting the study area for stream discharge simulation to the two subwatersheds. On the other hand, the study area for landslide modeling is the entire Shihmen Reservoir watershed.

## 3.2 Typhoons Aere and Haitang

From August 23 to 25, 2004, Typhoon Aere crossed the northern tip of the island in an east-west direction before turning southwestward. The typhoon affected Taiwan, especially the northern half, for three days. During its peak intensity on August 24, Typhoon Aere had a 200-km storm radius and a low pressure reading of 955 hPa, packing winds of $140\,km\,hr^{-1}$ and gusts to $175\,km\,hr^{-1}$. Thirty-four people were killed as a result of the storm, including 15 died as a landslide buried a remote mountain village in the north. The silt accumulation from the upland landslides and stream scouring forced to stop the outlet of drinking water from the Shihmen Reservoir and jammed the water supply pipes for days (Chen et al. 2006). Based on the damage to properties and human lives, Typhoon Aere was the worst typhoon that struck northern Taiwan in recent years.

Typhoon Haitang rotated off the east coast of Taiwan for more than six hours before making landfall on July 18, 2005. The storm then moved west-northwesterly across the center of the island and entered the Taiwan Strait on July 19. During its peak intensity, Typhoon Haitang had a low pressure reading of 915 hPa, packing winds of $195\,km\,hr^{-1}$ and gusts to $260\,km\,hr^{-1}$. From July 17 to 20, 2005, the storm brought torrential rains to the east coast and in mountainous areas, triggering warnings against landslides, mudslides and flash floods. Twelve people were reported dead and three missing from the typhoon event.

## 3.3 Gauged Rainfall and Stream Discharge Data in Subwatersheds

Table 1 shows the means of rainfall data from the six gauge stations (Fig. 1) in and around the subwatersheds for the two typhoon events. Based on the total rainfall and the rainfall intensity, Typhoon Aere is more intense than Haitang by a factor of two to three. Table 2 shows the total discharge, peak discharge, and lag time for the two subwatersheds. During Typhoon Aere, the total discharge nearly equals the total rainfall, and the peak discharge the maximum rainfall intensity, in YuFeng. This kind of correspondence, however, is not found during Typhoon Haitang, a much weaker rainfall event.

It must be noted that stream discharge data are susceptible to errors in measurement from a number of sources. For short-term flow gauges, measurements of high

**Table 1** Averages of rainfall data from six rain gauges in YuFeng and XiuLuan during typhoons Aere and Haitang

| Typhoon | Total rainfall (mm) | Duration (hr) | Avg. R.I. (mm/hr) | Max. R.I. (mm/hr) |
|---------|---------------------|---------------|-------------------|-------------------|
| Aere    | 1089                | 60            | 17.0              | 54.1              |
| Haitang | 431                 | 75            | 5.8               | 28.0              |

R.I.: rainfall intensity.

**Table 2** Stream discharge data in YuFeng and XiuLuan during typhoons Aere and Haitang

| Typhoon | Flow station | Total discharge (cms) | Peak discharge (cms) | Lag time (hr) |
|---------|--------------|-----------------------|----------------------|---------------|
| Aere    | YuFeng       | 102657.5              | 5293.1               | 7             |
|         | XiuLuan      | 25616.6               | 861.4                | 7             |
| Haitang | YuFeng       | 16479.1               | 864.0                | 1             |
|         | XiuLuan      | 8328.1                | 234.8                | 1             |

flows may be affected by extrapolation of the rating curves beyond the limit of calibration, bypassing of the flow gauges, and other problems arising during flood events. Moreover, an extreme flood may change the channel geometry and bias the rating curve.

## 3.4 Landslide Data

Landslides triggered by Typhoon Aere were interpreted and delineated by comparing ortho-rectified aerial photographs taken before and after the typhoon. Compiled by the Aerial Survey Office of Taiwan's Forestry Bureau from the stereo pairs of 1:5000 aerial photographs, these color orthophotographs have a pixel size of 0.35 m and an estimated horizontal accuracy of 0.5 m. Landslides triggered by Typhoon Haitang were interpreted and delineated from 2-m FORMOSAT-2 panchromatic satellite images.

Typhoon Aere triggered 653 new landslides, covering 4.87% of the Shihmen Reservoir watershed. Typhoon Haitang triggered 587 new landslides, covering 2.18% of the study area. Most observed slope failures were shallow landslides on soil mantled slopes with depths less than 2 m. Table 3 summarizes the descriptive statistics of landslides triggered by the two typhoon events.

**Table 3** Landslide data and statistics

| Statistics     | Number | Area  | Maximum | Minimum | Mean | Std |
|----------------|--------|-------|---------|---------|------|-----|
| After Aere     | 653    | 369.7 | 46.7    | <0.01   | 0.57 | 2.1 |
| After Haitang  | 587    | 94.9  | 3.0     | <0.01   | 0.16 | 0.2 |

Area unit: hectare.

## 3.5 Radar Rainfall Estimates

The CWB provided radar base reflectivity data for August 23–25, 2004, corresponding to the event of Typhoon Aere, and for July 17 to 20, 2005, corresponding to the event of Typhoon Haitang, for the Shihmen Reservoir watershed. We also secured ground rainfall measurements for the same time period from 19 automatic rain gauges in and around the watershed.

Developed by the CWB, the method for estimating ground-level area rainfall from radar measurements involves two basic steps. First, radar reflectivity $Z$, measured in $mm^6 m^{-3}$, is converted into rainfall rate $R$, measured in $mm\ hr^{-1}$, by the following $Z$-$R$ power relationship (Marshall et al. 1947; Wilson 1970):

$$Z = 32.5R^{1.65} \tag{1}$$

The parameter values of 32.5 and 1.65 proposed by Xin et al. (1997) are reported to be more accurate than other values for fast moving convective storms.

Rainfall estimates can be improved when rain gauge observations are used to calibrate radar data (Brandes 1975). The calibration method developed by the CWB uses the inverse distance weighted method (IDW) to first create a grid representing the deviations between $R$ and gauged hourly rainfall measurements:

$$dev_0 = \frac{\sum dev_i W_i}{\sum W_i} \tag{2}$$

$W_i = 1/d_i^2$, if $d_i \leq 30$ km; $W_i = 0$, otherwise; where $dev_0$ represents the estimated deviation at cell 0, $dev_i$ the deviation at rain gauge $i$, $W_i$ the weight at rain gauge $i$, and $d_i$ the distance between cell 0 and rain gauge $i$. To complete the calibration, the deviation grid is added to the $R$ grid to obtain the final calibrated rainfall grid.

For this study, we summed the 10-min radar reflectivity data by hour and divided the sum by six for the hourly average. Then we followed Eqs. (1) and (2) to convert the hourly average reflectivity data into hourly rainfall data. The projection of the hourly rainfall grid from geographic to plane coordinates resulted in a 36 by 55 grid with a spatial resolution of 1 km. Using the hourly rainfall data, we derived various rainfall factors associated with the typhoons. Figures 2 and 3 display the maximum 3-hr rainfall intensity, total rainfall, and total duration associated with Typhoon Aere and Typhoon Haitang, respectively.

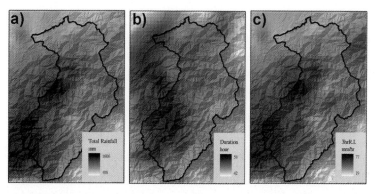

**Fig. 2** Radar-derived maximum 3-hr rainfall intensity, total rainfall, and total duration associated with Typhoon Aere

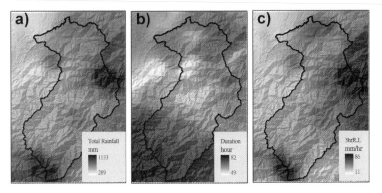

**Fig. 3** Radar-derived maximum 3-hr rainfall intensity, total rainfall, and total duration associated with Typhoon Haitang

# 4 Hydrological Modeling

## 4.1 Stream Discharge Model

The discharge of a stream comprises the surface and subsurface flows. For many mountainous catchments, the variable source area is a major concept to interpret the rainfall-runoff process. The concept is implemented in TOPMODEL (Beven and Kirkby 1979), which measures the subsurface flow in this study. Because the rainfall intensity caused by typhoons can reach as high as 50 mm hr$^{-1}$ and even higher, the surface flow must also be taken into account. To enhance the surface flow representation, this study introduces the flow path unit response function into TOPMODEL.

The unit response function uses the network width function and hydrodynamics to represent the surface flow. The network width function proposed by Kirkby (1976) is a histogram representing the distances from each grid cell to the outlet. Combining Manning's equation and the energy dissipation theory (Molnar and Ramirez 1998), the network width function is convolved to obtain the unit response function (Liu et al. 2003). This method can provide a physically based, yet simple, function for the routing of flows through a river network (Franchini and O'Connell 1996). Therefore, this study used the approximate solution derived from the diffusive transport approach (Liu et al. 2003):

$$U(t) = \frac{1}{\sigma\sqrt{2\pi \cdot t^3/t_0^3}} \exp\left[-\frac{(t-t_0)^2}{2\pi t/t_0}\right] \quad (3)$$

where $U(t)$ [1/T] is the flow path unit response function; $t_0$ [T] is the average travel time of the cell to outlet along flow path and $\sigma$ [T] is the standard deviation of the flow time. The spatially distributed parameters $t_0$ and $\sigma$ are retrieved from digital elevation models (40-m resolution). Accordingly, each flow path has different parameters depending on its length and physical characteristics. Note that Mannings' surface roughness, $n$, is embedded implicitly in Eq. (3). The total surface flow hydrograph at the watershed outlet is obtained by a convolution integral of the flow response from all grid cells:

$$UH(t) = \frac{\int_A RC \cdot U(t) dA}{\int_A RC dA} \tag{4}$$

where $UH(t)$ is the instantaneous unit hydrograph of the watershed, and $RC$ is the runoff coefficient of the grid cell, assumed to be dependent on slope, soil type and land use. The summation of each cell ($dA$) completes the calculation.

The subsurface flow, $Q_i$ [$L^3/T$], is expressed in TOPMODEL by:

$$Q_i = Q_0 \exp(-m \cdot \overline{S}_D)$$
$$Q_0 = A \exp(-\lambda) \tag{5}$$

where $Q_0$ is defined as an outflow parameter related to soil hydraulic properties and topography; $m$ is the soil characteristic parameter. $\overline{S}_D$ is the average of storage deficit for the entire watershed. $Q_0$ [$L^3/T$] is the discharge when storage deficit equals zero. $A$ is the watershed area and $\lambda$ is the averaged topographic index of the entire watershed. The topographic index of each cell is defined by:

$$\lambda_i = \ln \left( \frac{a_i}{T_o \cdot \tan \beta_i} \right) \tag{6}$$

where $a_i$ is the specific contributing area derived from Tarboton's D∞ method (1997), and $\beta_i$ is the local slope gradient calculated by Zevenbergen and Thorne's method (1987). $T_o$ is the soil transmissivity defined as the product of hydraulic conductivity.

Following the steady state assumption in TOPMODEL, the storage deficit for each grid cell is:

$$S_{D,i} = \overline{S}_D + \frac{1}{m} \left[ \lambda - \ln \left( \frac{a_i}{T_o \tan \beta_i} \right) \right] \tag{7}$$

where $S_{D,i}$ is the storage deficit for $i$th cell. If an initial discharge, $Q_{t=0}$ is known, Eq. (4) can be inverted to give a value for $\overline{S}_D$ at $t = 0$ as:

$$\overline{S}_{D,t=0} = -\frac{1}{m} \ln \left( \frac{Q_{t=0}}{Q_0} \right) \tag{8}$$

Therefore, $\overline{S}_D$ is determined at the first time step.

Collectively, only four global variables are essential to route the model: surface roughness ($n$), runoff coefficient ($RC$), soil characteristic parameter ($m$), and transmissivity ($T$). Because this study focused on the effect of the spatial variability of rainfall on stream discharge, all parameters were assumed to be homogeneous. This uniform assumption allowed us to better discriminate differences between areal rainfall patterns. This study also did not calibrate or manipulate the simulations; instead, we used reported values or historical events to determine the parameter values. In accordance with previous studies (Liu et al. 2003, 2005), the Manning's roughness coefficient was set at $0.08 \, \text{m}^{-1/3}$ and the runoff coefficient at 0.35. The transmissivity and soil characteristic parameter were determined to be $0.23 \, \text{m} \, \text{hr}^{-1}$ and 5.72 respectively, using the master recession curve and historical events (Lamb and Beven 1997).

## 4.2 Spatial Interpolation of Gauged Data

To compare with radar rainfall estimates, this study used IDW to interpolate rainfall fields from the hourly records of the six rain gauge stations in and around the two subwatersheds. IDW enforces that the estimated value of a point (or cell) is influenced more by nearby known points than those farther away. An important characteristic of IDW interpolation is that all predicted values are within the range of maximum and minimum values of the known points. Another common method for rainfall estimation is Thiessen polygons. But because only one gauge station is located within the XiuLuan subwatershed, the method was not used in this study.

## 4.3 Comparison of Hyetographs from Radar and Gauged Rainfall

Figures 4 and 5 show the hyetographs from radar and gauged rainfall estimates during the two typhoons in YuFeng and XiuLuan, respectively. The gray zone, serving as the reference, is the mean of rainfall from the six rain gauges. For Typhoon Aere, the total areal rainfall estimates from radar and IDW are 110.4 and 105.3% of the mean in YuFeng, and 121.7 and 118.1% of the mean in XiuLuan. For Typhoon Haitang, the total areal rainfall estimates from radar and IDW are 96.1 and 104.1% of the mean in YuFeng, and 92.6 and 101.9% of the mean in XiuLuan. For both subwatersheds, radar estimates are higher than IDW estimates for Typhoon Aere but are lower than IDW estimates for Typhoon Haitang, a contrast to be investigated in future research.

## 4.4 Stream Discharge Simulation

Figures 6 and 7 show the simulated stream discharges associated with rainfall from the two typhoons in YuFeng and XiuLuan, respectively. The shapes of the simulated hydrographs generally agree with the observed. To better examine the correspondence between the simulated and the observed, this study used the following

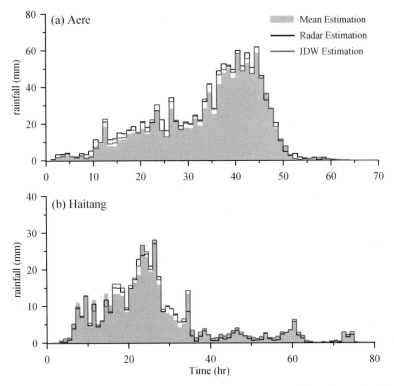

**Fig. 4** The hyetographs derived from radar and IDW areal rainfall estimates in YuFeng: (**a**) Typhoon Aere and (**b**) Typhoon Haitang. The *gray zone* is the mean of the six rain gauges

four statistics: efficient coefficient (EC), error of discharge volume (EQV), error of peak discharge (EQP), and error of time to peak (ET). EC is defined by (Nash and Sutcliffe 1970):

$$EC = 1 - \frac{\sum\limits_{t=1}^{T} \left(Q_0^t - Q_m^t\right)^2}{\sum\limits_{t=1}^{T} \left(Q_0^t - \overline{Q_0}\right)^2} \tag{9}$$

where $Q_0$ is observed discharge, $Q_m$ is modeled discharge, and $Q^t$ is discharge at time $t$. The coefficient presents the perfect match with the value of 1.0. EQV [–] is defined as the ratio of simulated volume to observed discharge, and EQP [–] is defined as the ratio of simulated peak to observed peak. For EQV and EQP, a more accurate simulation will have a value closer to 1.0. ET [hr] is defined as the difference in time to discharge peak, and a smaller ET means a more accurate simulation.

Based on the four statistics, Table 4 compares the performance of radar and IDW rainfall estimates for the two typhoons in YuFeng and XiuLuan. The ET values are the same in all cases. For the other three statistics, radar estimates perform noticeably better than IDW estimates, most likely due to the better spatial resolution of

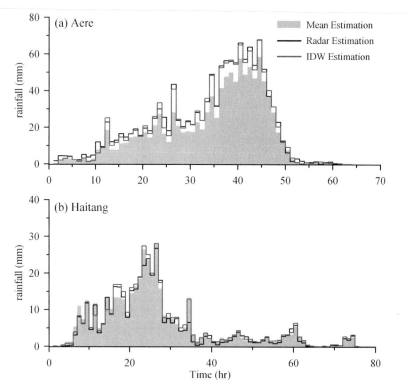

**Fig. 5** The hyetographs derived from radar and IDW areal rainfall estimates in XiuLuan: (**a**) Typhoon Aere and (**b**) Typhoon Haitang. The *gray zone* is the mean of the six rain gauges

radar rainfall. The only exception is Typhoon Aere/YuFeng, in which IDW estimates have a slightly better performance than radar estimates. Table 4 also shows that the same three statistics are generally more accurate in YuFeng than XiuLuan. This may be due to a higher rain gauge density in YuFeng (6 gauges for 335 km$^2$) than XiuLuan (1 gauge for 116 km$^2$).

# 5 Landslide Modeling

## 5.1 Logit Model

In recent years, many researchers have used logistic regression to predict probabilities of landslide occurrence by analyzing the functional relationships between the instability factors and the past distribution of landslides (Guzzetti et al. 1999; Dai and Lee 2003; Ohlmacher and Davis 2003; Ayalew and Yamagishi 2005; Can et al. 2005; Wang et al. 2005; Yesilnacar and Topal 2005). The assumption is that the same factors that caused landslides in the past will trigger landslides in the future.

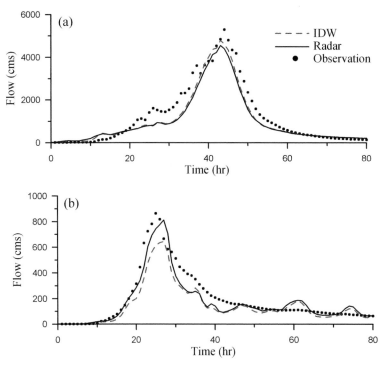

**Fig. 6** The observed (*black dot*) and simulated hydrographs (*black line* from radar and *dashed line* from IDW rainfall estimates) in YuFeng: (**a**) Typhoon Aere and (**b**) Typhoon Haitang

Logistic regression is useful when the dependent variable is categorical (e.g., presence or absence) and the explanatory variables are categorical, numeric, or both (Menard 2002). The logit model has the following form:

$$\text{logit}(y) = a + b_1 x_1 + b_2 x_2 + b_3 x_3 + \cdots + e \tag{10}$$

where the logit of $y$ is the dependent variable, $x_i$ is the ith explanatory variable, $a$ is a constant, $b_i$ is the ith regression coefficient, and $e$ is the error term. The logit of $y$ is the natural logarithm of the odds:

$$\text{logit}(y) = \ln\left(\frac{p}{1-p}\right) \tag{11}$$

where $p$ is the probability of the occurrence of $y$ and $p/(1-p)$ is the odds.

A logit model can be evaluated by the receiver operating characteristic (ROC). The ROC measures the fitness of a model on the basis of true positive (proportion of incidences correctly reported as positive) and false positive (proportion of incidences erroneously reported as positive). Typically, a probability value of 0.5 is used to determine whether the model has made a correct prediction ($> 0.5$) or not ($< 0.5$). Additionally, Cox and Snell $R^2$ and Nagelkerke $R^2$ measure how well the

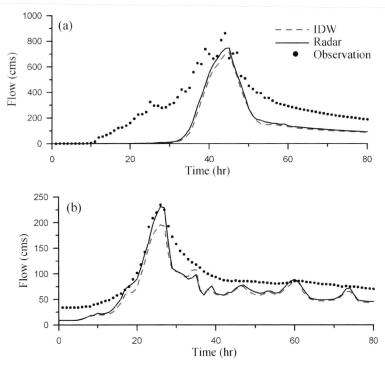

**Fig. 7** The observed (*black dot*) and simulated hydrographs (*black line* from radar and *dashed line* from IDW rainfall estimates) in XiuLuan: (**a**) Typhoon Aere and (**b**) Typhoon Haitang

**Table 4** Statistics compared the observed with the simulated stream discharge in YuFeng and XiuLuan during typhoons Aere and Haitang

|          |     | Aere  |      | Haitang |      |
|----------|-----|-------|------|---------|------|
|          |     | Radar | IDW  | Radar   | IDW  |
| YuFeng   | EC  | 0.82  | 0.84 | 0.86    | 0.77 |
|          | EQV | 0.82  | 0.86 | 0.90    | 0.76 |
|          | EQP | 0.94  | 0.99 | 0.99    | 0.79 |
|          | ET  | −1    | −1   | 2       | 2    |
| XiuLuan  | EC  | 0.63  | 0.56 | 0.77    | 0.70 |
|          | EQV | 0.51  | 0.47 | 0.76    | 0.71 |
|          | EQP | 0.92  | 0.88 | 0.98    | 0.82 |
|          | ET  | 0     | 0    | 1       | 1    |

explanatory variables can predict and explain the dependent variable. Cox and Snell $R^2$ cannot achieve a maximum of 1, whereas Nagelkerke $R^2$ stretches the $R^2$ value to range from 0 to 1.

For this study, the dependent variable ($y$) represented landslide (1) or stable area cell (0) and the explanatory variables were maximum 3-hr rainfall intensity ($x_1$) and total duration ($x_2$). The logit model is significant at the 0.01 level, with ROC = 0.76,

Cox & Snell $R^2 = 0.27$, and Nagelkerke $R^2 = 0.36$. Both explanatory variables are significant at the 1% level, with total duration being slightly more important than maximum 3-hr intensity in explaining landslide occurrence.

## 5.2 Critical Rainfall Conditions

To use rainfall data for predicting landslides, critical rainfall intensity and total duration can be derived from the logit model. Substituting logit ($y$) by Eq. (11) and ignoring the error term, Eq. (10) becomes:

$$\ln\left(\frac{p}{1-p}\right) = a + b_1 x_1 + b_2 x_2 \tag{12}$$

Equation (12) can be rewritten as:

$$x_1 = \frac{1}{b_1}\left[\ln\left(\frac{p}{1-p}\right) - (b_2 x_2 + a)\right] \tag{13}$$

By specifying the $p$ value (e.g., 0.5) and an $x_2$ value in Eq. (13), we can compute a corresponding $x_1$ value. By going through the computation twice, we can get two pairs of $x_1$ and $x_2$ values to plot a straight line representing the specified $p$ value (e.g., 0.5). Figure 8 shows such lines representing the probabilities of landslide occurrence at 0.2, 0.4, 0.5, 0.6, and 0.8. Figure 8 also shows dots for landslide and stable-area cells. The location of a dot or cell corresponds to its maximum 3-hr rainfall intensity and total duration values, and the dot symbol shows whether landslide is present in the cell or not.

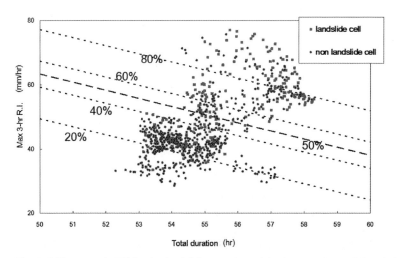

**Fig. 8** Different probabilities for landslide occurrence based on maximum 3-hr rainfall intensity and total duration

## 5.3 *Model Validation*

Landslides triggered by Typhoon Haitang were used for validating the probability model derived from the landslide and rainfall data associated with Typhoon Aere. Using 0.5 as the threshold, the model correctly predicted 81.6% of landslides triggered by Typhoon Haitang. In other words, 81.6% of landslides fall within areas with probabilities greater than 0.5 as predicted by the model (Fig. 9).

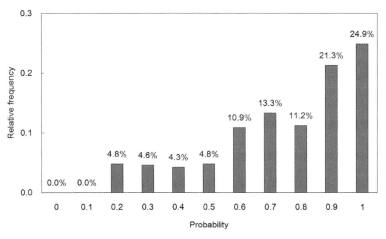

**Fig. 9** A comparison of relative frequencies of landslides triggered by Typhoon Haitang and probabilities calculated from the probability model developed from the landslide and radar rainfall data associated with Typhoon Aere

# 6 Conclusions

This study used radar rainfall estimates from two typhoon events for stream discharge simulation and landslide modeling. The shapes of the simulated hydrographs using radar rainfall estimates generally agreed with the observed. Moreover, the statistics for comparing the simulated with the observed show radar rainfall to be more accurate than gauged rainfall in simulating stream discharge. The results from landslide modeling have been equally encouraging. The logit model relating landslides triggered by Typhoon Aere and rainfall factors derived from radar data is significant at the 0.01 level with a ROC of 0.76. More importantly, the probability model based on the logit model is able to predict 81.6% of landslides triggered by Typhoon Haitang.

Rainfall data with adequate spatial resolution and reliable quality are necessary for developing stream discharge and landslide models. Radar rainfall estimates appear to be a promising data source. 1-km rainfall estimates from radar data are adequate for modeling at the watershed level but may not be sufficient for more detailed analysis. This resolution problem, however, also applies to other types of data such as soils and lithology that are needed for environmental modeling. Previous studies have reported the underestimation problem with radar rainfall estimates. This calls for further research to assess whether typhoons of different rainfall characteristics require different calibration methods for radar rainfall.

# References

Aleotti P (2004) A warning system for rainfall-induced shallow failures. Eng Geol 73:247–265

Ayalew L, Yamagishi H (2005) The application of GIS-based logistic regression for landslide susceptibility in the Kakuda-Yahiko Mountains, Central Japan. Geomorphology 65:15–31

Beven KJ, Kirkby MJ (1979) A physically based, variable contributing area model of basin hydrology. Hydrol Sci B 24:43–69

Boyle DP, Gupta HV, Sorooshian D (2001) Toward improved stream flow forecast: Value of semi-distributed modeling. Water Resour Res 37:2749–2759

Brandes EA (1975) Optimizing rainfall estimates with the aid of radar. J Appl Meteorol 14:1339–1345

Can T, Nefeslioglu HA, Gokceoglu C, Sonmez H, Duman TY (2005) Susceptibility assessments of shallow earthflows triggered by heavy rainfall at three catchments by logistic regression analyses. Geomorphology 72:250–271

Carpenter TM, Georgakakos KP (2004a) Continuous streamflow simulation with the HRCDHM distributed hydrologic model. J Hydrol 298:61–79

Carpenter TM, Georgakakos KP (2004b) Impacts of parametric and radar rainfall uncertainty on the ensemble streamflow simulations of a distributed hydrologic model. J Hydrol 298:202–221

Casadei M, Dietrich WE, Miller NL (2003) Testing a model for predicting the timing and location of shallow landslide initiation in soil-mantled landscapes. Earth Surf Proc 28:925–950

Chang K, Chiang S, Feng L (2007) Analyzing the relationship between typhoon-triggered landslides and critical rainfall conditions. Earth Surf Proc doi:10.1002/esp.1611

Chen C, Lee W, Yu F (2006) Debris flow hazards and emergency response in Taiwan. In Lorenzini G., Brebbia CA, Emmanouloudis DE (eds) Monitoring, simulation, prevention and remediation of dense and debris flows. WIT Press, Southampton, Boston, pp 311–320

Chen C, Lin L, Yu F, Lee C, Tseng C, Wang A, Cheung K (2007) Improving debris flow monitoring in Taiwan by using high-resolution rainfall products from QPESUMS. Nat Hazards 40:447–461

Chiang Y, Chang F, Jou BJ, Lin P (2007) Dynamic ANN for precipitation estimation and forecasting from radar observations. J Hydrol 334:250–261

Dai FC, Lee CF (2003) A spatiotemporal probabilistic modelling of storm-induced shallow landsliding using aerial photographs and logistic regression. Earth Surf Proc 28:527–545

Dinku T, Anagnostou EN (2002) Improving radar-based estimation of rainfall over complex terrain. J Appl Meteorol 41:1163–1178

Franchini M, O'Connell PE (1996) An analysis of the dynamic component of the geomorphologic instantaneous unit hydrograph. J Hydrol 175:407–428

Fulton RA, Breidenbach JP, Seo D, Miller DA (1998) The WSR-88D rainfall algorithm. Weather Forecast 13:377–395

Godt JW, Baum RL, Chleborad AF (2006) Rainfall characteristics for shallow landsliding in Seattle, Washington, USA. Earth Surf Proc 31:97–110

Gourley JJ, Maddox RA, Howard KW, Burgess DW (2002) An exploratory multisensor technique for quantitative estimation of stratiform rainfall. J Hydrometeor 3:166–180

Guzzetti F, Carrara A, Cardinali M, Reichenbach P (1999) Landslide hazard evaluation: a review of current techniques and their application in a multi-scale study, Central Italy. Geomorphology 31:181–216

Guzzetti F, Cardinali M, Reichenbach P, Cipolla F, Sebastiani C, Galli M, Salvati P (2004) Landslides triggered by the 23 November 2000 rainfall event in the Imperia Province, Western Liguria, Italy. Eng Geol 73:229–245

Haberlandt U (2007) Geostatistical interpolation of hourly precipitation from rain gauges and radar for a large-scale extreme rainfall event. J Hydrology 332:144–157

Hossain F, Anagnostou EN, Dinku T, Borga M (2004) Hydrological model sensitivity to parameter and radar rainfall estimation uncertainty. Hydrol Process 18:3277–3291

Howard KW, Gourley JJ, Maddox RA (1997) Uncertainties in WSR-88D measurements and their impacts on monitoring life cycles. Weather Forecast 12:166–174

Iverson RM (2000) Landslide triggering by rain infiltration. Water Resour Res 36:1897–1910.

Kalinga OA, Gan TY (2006) Semi-distributed modelling of basin hydrology with radar and gauged precipitation. Hydrol Process 20:3725–3746

Kirkby MJ (1976) Tests of random network model and its application to basin hydrology. Earth Surf Proc 1:197–212

Krajewski WF, Smith JA (2002) Radar hydrology: rainfall estimation. Adv Water Res 25:1387–1394

Lamb R, Beven K (1997) Using interactive recession curve analysis to specify a general catchment storage model. Hydrol Earth Syst Sci 1:101–113

Liu YB, Gebremeskel S, De Smedt F, Hoffman L, Pfister L (2003) A diffusive approach for flow routing in GIS based flood modeling. J Hydrol 283:91–106

Liu YB, De Smedt F, Hoffmann L, Pfister L (2005) Assessing land use impacts on flood processes in complex terrain by using GIS and modeling approach. Environ Model Assess 9:227–235

Marshall JS, Langille RC, Palmer WM (1947) Measurement of rainfall by radar. J Meteorol 4:186–191

Menard S (2002) Applied logistic regression analysis. 2nd ed. Sage, Thousand Oaks, CA

Molnar P, Ramirez JA (1998) Energy dissipation theories and optimal channel characteristics of river network. Water Resour Res 34:1809–1818

Montgomery DR, Dietrich WE (1994) A physically based model for topographic control on shallow landsliding. Water Resour Res 30:1153–1171

Morrissey MM, Wieczorek GF, Morgan BA (2004) Transient hazard model using radar data for predicting debris flows in Madison County, Virginia. Environ Eng Geosci 10:285–296

Nash JE, Sutcliffe JV (1970) River flow forecasting through conceptual models 1: A discussion of principles. J Hydrol 10:282–290

Norbiato D, Borga M, Sangati M, Zanon F (2007) Regional frequency analysis of extreme precipitation in the eastern Italian Alps and the August 29, 2003 flash flood. J Hydrol 345:149–166

Ohlmacher GC, Davis JC (2003) Using multiple logistic regression and GIS technology to predict landslide hazard in northeast Kansas, USA. Eng Geol 69:331–343

Reed S, Schaake J, Zhang Z (2007) A distributed hydrologic model and threshold frequency-based method for flash flood forecasting at ungauged locations. J Hydrol 337:402–420

Segond M, Wheater HS, Onof C (2007) The significance of spatial rainfall representation for flood runoff estimation: A numerical evaluation based on the Lee catchment, UK. J Hydrol 347:116–131

Steiner M, Smith JA (2002) Use of three-dimensional reflectivity structure for automated detection and removal of nonprecipitating echoes in radar data. J Atmos Oceanic Technol 19:673–686

Stellman KM, Fuelberg HE (2001) An examination of radar and rain gauge-derived mean areal precipitation over Georgia watershed. Weather Forecast 16:133–144

Tarboton DG (1997) A new method for the determination of flow directions and contributing areas in grid digital elevation models. Water Resour Res 33:309–319

Wang H, Liu G, Xu W, Wang G (2005) GIS-based landslide hazard assessment: an overview. Prog Phys Geog 29:548–567

Wilson JW (1970) Integration of radar and raingage data for improved rainfall measurement. J App Meteor 9: 489–497

Xin L, Reuter G, Larochelle B (1997) Reflectivity-rain rate relationships for convective rainshowers in Edmonton. Atmos-Ocean 35:513–521

Yang D, Koike T, Tanizawa H (2004) Application of a distributed hydrological model and weather radar observations for flood management in the upper Tone River of Japan. Hydrol Proc 18:3119–3132

Yesilnacar E, Topal T (2005) Landslide susceptibility mapping: a comparison of logistic regression and neural networks methods in a medium scale study, Hendek region (Turkey). Eng Geol 79:251–266

Zevenbergen LW, Thorne CR (1987) Quantitative analysis of land surface topography. Earth Surf Proc 12:47–56

Zhang J, Howard K, Gourley JJ (2005) Constructing three-dimensional multiple-radar reflectivity mosaics: Examples of convective storms and stratiform rain echoes. J Atmos Oceanic Technol 22:30–46

# High-Resolution QPE System for Taiwan

Jian Zhang, Kenneth Howard, Pao-Liang Chang, Paul Tai-Kuang Chiu,
Chia-Rong Chen, Carrie Langston, Wenwu Xia, Brian Kaney and Pin-Fang Lin

**Abstract** Over the last five years the Central Weather Bureau of Taiwan and the
United States NOAA/National Severe Storms Laboratory have been involved in a research and development initiative to improve the monitoring and prediction of flash
floods, debris flows, and severe storms for the Taiwan environment. The initiative
has produced a system that integrates observations from weather radars, rain gauges,
satellites, and numerical weather prediction model fields to produce high resolution
(1 km to 500 m) and rapid update (10-min) rainfall and severe storm monitoring
products. These prototype products are assessed for potential use by government
agencies and emergency managers for flood, flash flood, and mudslide warnings
and water resource managements. The system also facilitates collaborations with
academic communities for research and development of radar applications including QPE and nowcasting. This paper overviews the system structure and products,
the research activities supporting the system, and the challenges faced in producing
high resolution, accurate QPE for Taiwan.

## 1 Introduction

Floods, flash floods, and mudslides are top causes for weather-related disasters in
Asian countries, causing huge losses of property and human lives each year. Warnings and predictions of floods and mudslides in Taiwan are especially challenging
because of the complex and steep terrain and abundance of rainfall produced by
typhoons. To improve the monitoring and prediction of flash floods, debris flows,
and severe storms for the Taiwan environment, the Central Weather Bureau (CWB)
and the water resource agency (WRA) of Taiwan and the United States National
Oceanic and Atmospheric Administration (NOAA)/National Severe Storms Laboratory (NSSL) have been involved in a research and development (R&D) initiative
over the last five years. The goal of the initiative was to develop a multi-sensor

J. Zhang (✉)
NOAA/OAR National Severe Storms Laboratory and Cooperative Institute for Mesoscale
Meteorological Studies, The University of Oklahoma,
Norman, OK 73072, USA, e-mail: jian.zhang@noaa.gov

S.K. Park, L. Xu, *Data Assimilation for Atmospheric, Oceanic and Hydrologic*     147
*Applications*, DOI 10.1007/978-3-540-71056-1_7,
© Springer-Verlag Berlin Heidelberg 2009

quantitative precipitation estimation (QPE) system that provides very high spatial (1 km to 500 m) and temporal (10 min) resolution rainfall and severe storm products by integrating observations from weather radar, rain gauge, satellite, and numerical weather prediction model analyses. As a result of the collaboration, a system of quantitative precipitation estimation (QPE) and segregation using multiple sensors (QPESUMS) was developed. The QPESUMS system produced experimental QPE products for government agencies and emergency managers for flood, flash flood, and mudslide warnings and water resource managements. The system also facilitates collaborations with academic communities for research and development of radar applications including QPE and nowcasting.

Beginning in year of 2007, the second generation of the QPESUMS system, which is a part of a high-resolution QPE and quantitative precipitation forecast (QPF) (HRQ2) system, was developed. The new high-resolution QPE (HRQ) system was built upon QPESUMS with a new infrastructure that facilitates adaptable data integration from new sources and efficient algorithm updates that are in sync with the equivalent USA system at NSSL. Since late June 2007, the new HRQ system has been undergoing alpha test, generating experimental high-resolution, rapid update rainfall and severe storm products.

This paper overviews the new system structure and products and presents initial results from the alpha test and evaluations. The research activities supporting the system and the challenges in producing high resolution, accurate QPE for Taiwan are also discussed.

## 2 System and Products Overview

An overview flowchart of the new HRQ system is shown in Fig. 1. Multiple data sources are used in four major modules that consist the HRQ system, which are: (1) single radar process; (2) 3-D and 2-D radar mosaic; (3) QPE; and (4) verification. The data sources include the base level data from four S-band radars, satellite infrared imagery data, numerical weather prediction model hourly analyses, rain gauge networks, upper air soundings, and surface observations. A detailed flowchart illustrating all the scientific algorithms and data flows are shown in Fig. 2. Descriptions of the four major modules are provided below.

### 2.1 Single Radar Processes

There are four scientific algorithms in single radar processes: reflectivity quality control (QC), vertical profile of reflectivity (VPR), single radar Cartesian (SRC) grid, and single radar hybrid scan reflectivity. Relationships of the four algorithms with other scientific components in the HRQ system are shown in Fig. 2.

Reflectivity data are quality controlled to remove non-precipitation echoes such as ground/sea clutter, electronic interferences, and anomalous propagations. The reflectivity QC module includes a neural network and several pre- and post-processing

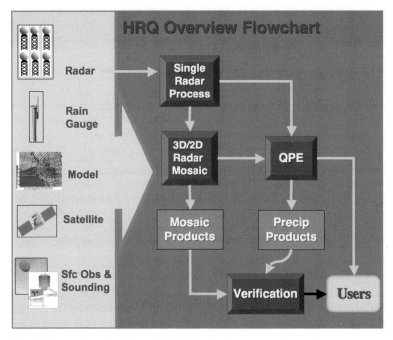

**Fig. 1** An overview flowchart of the HRQ system

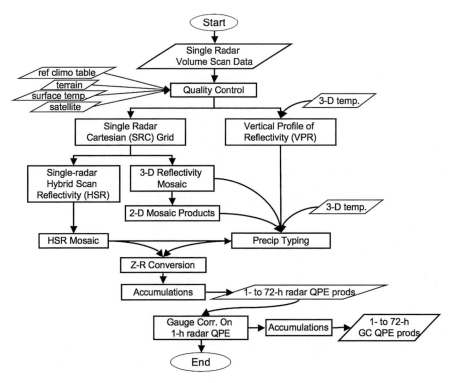

**Fig. 2** Scientific components and associated data flow in the HRQ system

steps. The neural network is based on statistics of horizontal and vertical reflectivity structure (Lakshmanan et al., 2007) such as smoothness, maximum intensity and depth of the echoes. The pre- and post-processing, on the other hand, utilize spatial and temporal image filters and heuristic rules based on terrain, reflectivity climatology, and environmental data. Figure 3 shows an example composite reflectivity field before and after the quality control in the real-time HRQ system. The QC was effective in removing non-precipitation echoes for this case, especially the sea clutter to the north of the RCWF radar (Fig. 3).

Each volume scan of the QC'd reflectivity field is objectively analyzed onto a 3-D Cartesian grid called "single radar Cartesian" (SRC) grid. The SRC grid has a horizontal resolution of ∼1 km × 1 km and 31 levels ranging from 500 m to 18 km above mean sea level. The domain of the SRC is pre-defined for each radar (Fig. 4), and their grid cells are perfectly aligned. The objective analysis scheme includes a nearest neighbor approach on range-azimuth plane and an exponential interpolation in the elevation direction (Lakshmanan et al., 2006). The vertical influence of radar observations is confined between half a beam width above the highest elevation angle and half a beam width below the lowest elevation angle.

Vertical profiles of reflectivity are derived from volume scans of the QC'd reflectivity in its native spherical coordinates. Detailed discussions about the VPRs can be found in Zhang et al. (2008). The VPRs are used to identify bright band layer (Zhang et al., 2008) as well as warm rain processes (Xu et al., 2008).

From the 3-D SRC grid, non-missing reflectivities from the lowest height level are found. These reflectivities represent the precipitation observations closest to the ground that the radar could obtain. They constitute a 2-D field equivalent to the

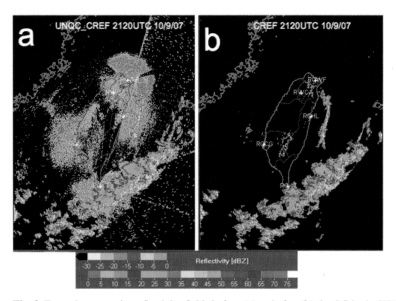

**Fig. 3** Example composite reflectivity fields before (**a**) and after (**b**) the QC in the HRQ system

**Fig. 4** Single radar Cartesian domains (*dashed lines*) for each of the four radars and the HRQ domain (*light-shaded box* in the middle)

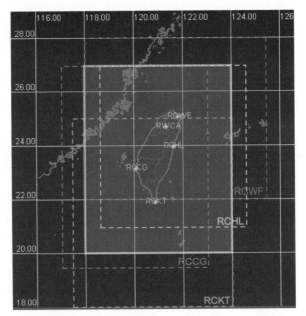

"hybrid scan reflectivity" (HSR) in Fulton et al. (1998) except that the latter was derived from the spherical coordinates. The hybrid scan reflectivity come from different heights at different locations due to the fact that (1) the height of the radar beam increases with range, and (2) terrain clearance and blockages vary spatially.

## 2.2 3-D and 2-D Radar Mosaic

The single radar reflectivity Cartesian grids from the four radars are combined into a regional 3-D reflectivity mosaic grid (re flowchart in Fig. 2). The mosaic domain is shown as the light grey-shaded area in Fig. 4, and it has the same horizontal and vertical resolutions as the SRC grids. The grid is in the cylindrical equidistant map projection and has a resolution of $0.01°$ (longitude) $\times 0.01°$ (latitude), which is $\sim 1\,\text{km}$ in east-west (x-) direction and $\sim 1.112\,\text{km}$ in north–south (y-) direction. Since the four radars have overlapped coverage areas, an exponential distance weighting function is used when multiple radar observations cover a single grid cell (Zhang et al., 2005). The mosaic also allows radar data from different times within a 20-min time window to be combined because observations from the four radars are not fully synchronized. Figure 5 shows example horizontal cross-sections of the 3-D reflectivity mosaic from the real-time HRQ system. The data voids at the lower altitudes (0.5 and 1.0 km, Fig. 5a and b) inside the Taiwan Island were due to blockages of radar observations by mountainous terrains.

A suite of severe storm products, such as probability of severe hail (POSH, Witt et al., 1998), echo top, vertically integrated liquid (VIL, Greene and Clark, 1972),

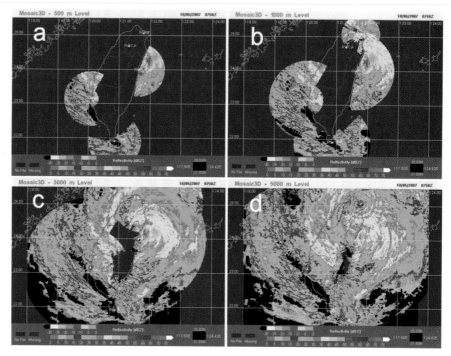

**Fig. 5** Example horizontal cross-sections of the 3-D reflectivity mosaic field at 0.5 (**a**), 1.0 (**b**), 3.0 (**c**), and 5.0 (**d**) km above mean sea level. The images were obtained from the HRQ system and valid at 07:50UTC on 10/6/2007

and VIL density (VILD, Amburn and Wolf, 1997) are calculated form the 3-D reflectivity mosaic grid and the Local Analysis and Prediction System (LAPS, http://laps.fsl.noaa.gov/taiwan/ taiwan_home.html) 3-D temperature analysis. Figure 6 shows examples of aforementioned products for a severe storm event that occurred in southwest Taiwan on 3 August 2007.

## 2.3 Radar QPE

Single radar hybrid scan reflectivity fields are mosaiced to produce a regional hybrid scan reflectivity field on the HRQ domain. The HSR mosaic scheme and associated weighting functions are defined below:

$$HSR = \frac{\sum_i w_L^i \bullet w_H^i \bullet SHSR^i}{\sum_i w_L^i \bullet w_H^i} \tag{1}$$

$$w_L = \exp\left(-d^2/L^2\right) \tag{2}$$

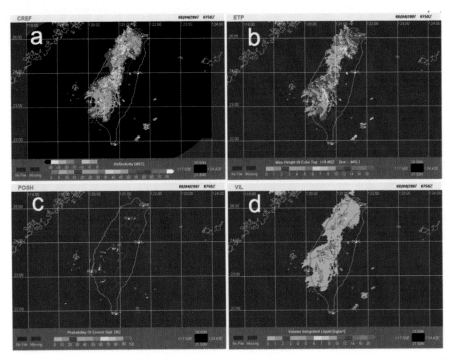

**Fig. 6** Example severe storm products, CREF (**a**), ETOP (**b**), SHI (**c**), and VIL (**d**), from the HRQ system valid at 07:50UTC on 8/4/2007

$$w_H = \exp\left(-h^2/H^2\right) \tag{3}$$

Here *HSR* represents the mosaiced hybrid scan reflectivity, $i$ is the radar index, *SHSR* is the single radar hybrid scan reflectivity field. There are two parts in the weighting function, one in the horizontal direction ($W_L$) and another in the vertical direction ($W_H$). Variable $d$ represents the distance between the analysis point and the radar, and $h$ represents the height (above ground level) of the single radar HSR bin. Parameters $L$ and $H$ are adaptable shape factors of the two weighting functions and their default values in the current HRQ system are 50 and 3 km, respectively.

Precipitation typing consists of a series of physically based heuristic rules as shown in Fig. 7. Each grid point is assigned a precipitation type based on 3-D reflectivity structure and the LAPS 3-D temperature field. The first step in precipitation typing is the identification of hail. A grid cell is considered to contain hail if the VILD value exceeds 2.0 g/m³ (adaptable parameter). A warm rain identification and delineation scheme (Xu et al., 2008) is then applied to the non-hail pixels. Hourly averaged, volume scan-mean VPRs from each radar are examined. If the slope of a VPR below the freezing level is negative, then the radar from which the VPR is derived is identified as a "warm-rain radar". All the echoes above 30 dBZ within a pre-defined radius of the "warm-rain radar" are labeled as warm rain. And any

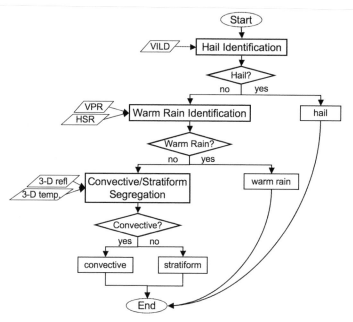

**Fig. 7** A detailed flowchart of the precipitation typing algorithm in the HRQ system

echoes 30 dBZ or higher that is contiguous to the warm rain region is defined as warm rain as well.

For all the pixels that are not warm-rain or hail, a convective/stratiform segregation is applied. A pixel is identified as convective if (a) a reflectivity at any height in the column is greater than 50 dBZ or (b) a reflectivity is greater than 30 dBZ at −10°C height or above. All the grid columns that are not identified as convective are classified as stratiform.

Figure 8 shows an example precipitation type and the associated HSR fields during Typhoon Krosa that affected Taiwan during 5–7 October 2007. Most of heavy precipitation bands were identified as warm rain process (green-colored areas in Fig. 8b) because VPRs from RCWF and RCCG showed large negative slopes below the freezing level (Fig. 9a and c). Even though a bright band signature existed in both VPRs, the rapid increase of mean reflectivity with decreasing height clearly indicated significant hydrometeor growth (both in number and in size) in the lower atmosphere that is very common in a warm rain process. Therefore all the echoes higher than 30 dBZ (in HSR, Fig. 8b) within 150 km (adaptable parameter) radius of the two radars were defined as warm rain. In addition, all the areas with 30 dBZ or higher reflectivity and contiguous to any region already defined as warm rain were also delineated as warm rain (Fig. 8a).

The VPR from RCHL showed a nearly constant reflectivity below the freezing level. For RCKT, the mean reflectivity appeared to be decreasing with decreasing height. Both VPRs were not classified as warm-rain, and the heavy precipitation bands over the southeast ocean were delineated as convective (red-color in Fig. 8a).

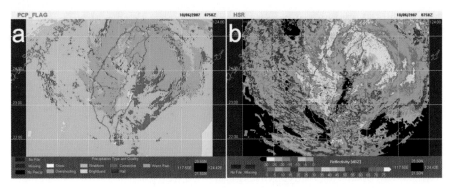

**Fig. 8** An example precipitation type (**a**) and the associated hybrid scan reflectivity (**b**) fields from Typhoon Krosa valid at 07:50UTC on 10/6/2007. The colors in *a* represent stratiform (*cyan*), convective (*red*), warm rain (*green*), and potential bright band contamination (*yellow*). More detail can be found in the text

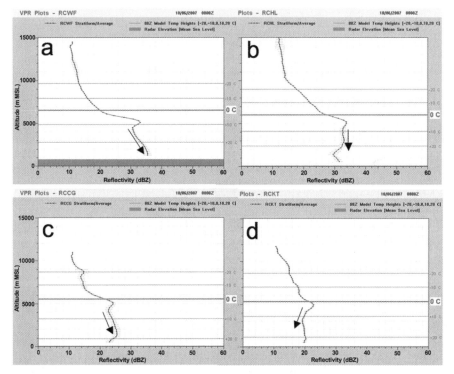

**Fig. 9** Hourly averaged volume scan-mean vertical profiles of reflectivity from RCWF (**a**), RCHL (**b**), RCCG (**c**), and RCKT (**d**) during Typhoon Krosa valid at 08:00UTC on 10/6/2007. The *brown lines* represent, from bottom to top, heights of 20, 10, 0, −10, and −20°C. temperatures

The remaining large areas of light precipitation were classified as stratiform (cyan-color in Fig. 8a). Areas where the HSRH is above the freezing level were specially flagged (yellow-color in Fig. 8a) to indicate that the radar QPE in these regions were less reliable.

These precipitation typing results are still preliminary. Further evaluations using ancillary data such as 3-D temperature and moisture fields will be performed in the next HRQ development cycle.

The instantaneous radar precipitation rate is obtained by applying Z-R relationships to the mosaiced HSR according to the precipitation type at each grid cell. Three Z-R relationships (Fig. 10) are used in association with the precipitation type field. For hail pixels, the convective Z-R is applied with a cap at 49 dBZ (adaptable).

The radar precipitation rate field is calculated every 10 min in correspondence to radar volume scan schedule. And then the one and three hourly accumulations are updated using the latest rate field. Longer-term accumulations (i.e., 6, 12, 24, 48, and 72 hr QPEs) are calculated every hour on the top of the hour, by aggregating hourly accumulations.

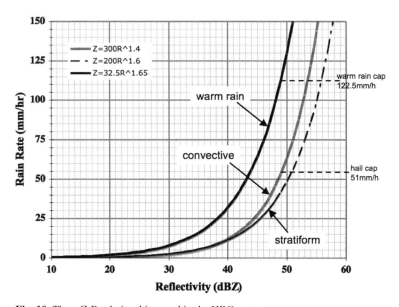

**Fig. 10** Three Z-R relationships used in the HRQ system

## *2.4 Local Gauge Correction of the Radar QPE*

The CWB has a dense and well-maintained rain gauge network that consists of over 500 stations. The gauge data are available in real-time every 5 min. These ground-based in situ rainfall observations provide very good data sets for correcting the radar QPE for various errors, such as those due to calibration offset, unpresentative Z-R relationships, vertical profile of reflectivity, and so on.

The local gauge correction (LGC) of radar QPE involves the following steps:

1. Computing gauge-radar biases at each gauge stations using hourly rainfalls from gauge observations and from radar estimates at co-located grid cells. That is:

$$dR^n = G_{1h}^n - R_{1h}^n \tag{4}$$

Here $n$ is the gauge index; $G_{1h}^n$, $R_{1h}^n$ and $dR^n$ are the gauge observed 1-h rainfall, radar estimated 1-h rainfall, and the gauge-radar bias at the $n$th gauge station, respectively.

2. Interpolating the gauge-radar biases from gauge stations onto the HRQ analysis grid, i.e.,

$$dR_{i,j} = \frac{\sum\limits_{n=1}^{M} \frac{1}{r_n^2} \bullet dR^n}{\sum\limits_{n=1}^{M} \frac{1}{r_n^2}} \tag{5}$$

Here $i$ and $j$ are 2-D grid cell indices, $dR_{i,j}$ the interpolated gauge-radar bias at the $(i, j)$th grid cell, and $r$ is the distance between the grid cell $(i, j)$ and the gauge station $n$. $M$ is the total number of gauges in an influence region of 30 km radius from the grid cell. If $M$ is greater than 6, then only the nearest 6 gauges are used in the interpolation. This constraint is imposed to preserve fine gradients in precipitation distributions.

To assure the quality of the interpolated bias field, a sanity check based on the consistency between radar and gauge rainfalls is implemented. All gauge-radar bias pairs are excluded from the interpolation if the following criteria are met:

$$\frac{R_{1h}^n}{G_{1h}^n} \geq 5; \quad for \quad G_{1h}{}^n \geq 1mm; \tag{6}$$

$$R_{1h}^n - G_{1h}^n \geq 20mm; \quad for \quad G_{1h}^n < 1mm; \tag{7}$$

3. Applying the gridded gauge-radar biases to the radar-based QPE, i.e.,

$$R_{i,j}^{GC} = R_{i,j} + dR_{i,j} \tag{8}$$

Here $R_{i,j}$ and $R_{i,j}^{GC}$ are the radar-based QPE and local gauge corrected radar QPE at the grid cell, respectively.

4. Replacing the final LGC QPE at all grid cells that co-locate with a gauge station with the gauge observed values;

5. Applying a land mask to the LGC QPE field and cutting off the data over the ocean because no gauge correction is available in the ocean area.

Figure 11 shows initial results of radar QPEs before and after the local gauge correction for the Typhoon Krosa. The radar network provides a much larger coverage than does the gauge network and serves as an excellent surveillance tool for weather

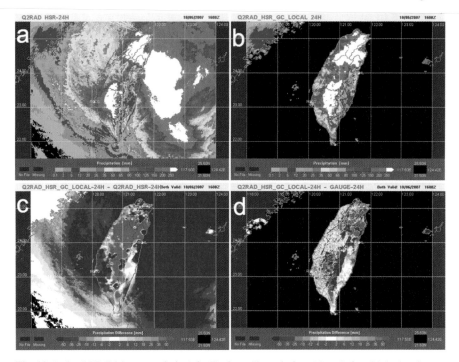

**Fig. 11** Radar QPE (24-h accumulation) for Typhoon Krosa before (**a**) and after (**b**) the local gauge correction. The difference field (after–before) between the two is shown in *c* and the difference between the local gauge corrected radar QPE and a gauge analysis is shown in *d*

systems over the entire island and the surrounding ocean. The difference between the two (Fig. 11c) indicates that the radar QPE was underestimating in the north and south central Taiwan, and overestimating in the southwest Taiwan. The causes of these biases are still under investigations. The difference between the LGC radar QPE and the gauge analysis (Fig. 11d) revealed many fine structures that were observed by radar, but not by the gauge network. These fine-resolution precipitation gradients may be important for mudslide warnings and predictions in complex terrain regions.

## 2.5 Verification

The current HRQ system contains a real-time product visualization and verification component where all products can be viewed via a web page. The real-time rain gauge observations are compared to the HRQ precipitation products, and mean bias (ratio), correlation coefficient, and root-mean-square error (RMSE) between the HRQ products and gauge observations are calculated for all the accumulation periods (1-h to 72-h). This component provides developers as well as users a useful tool for evaluations of real-time products on a 7/24 basis. The evaluations facilitate new research efforts and result in improvements and enhancements to the real-time HRQ products. Several such research and enhancements are presented in the next section.

# 3 Supporting Researches and Challenges

The experimental real-time HRQ products have been assessed continuously since the implementation of the system in late June of 2007 at the CWB. After the initial evaluations, several issues were identified for the radar based QPE as shown in Fig. 12. The issues include:

(a) Residual ground clutter (area #1 in Fig. 12);
(b) Data holes near the east coast (area #2 in Fig. 12); and
(c) Discontinuities along the azimuth direction (e.g., areas 3–6 in Fig. 12) of the radars.

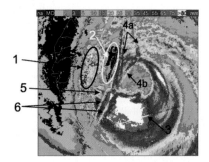

**Fig. 12** An example 3-h precipitation accumulation product from the experimental HRQ system valid at 20:00UTC on 8/17/2007

## 3.1 Residual Ground Clutter

Examinations of raw and quality controlled reflectivity data indicated that the ground clutter along the south central mountain range (area #1 in Fig. 12) were from the 3rd tilt of RCCG. According to beam propagations under standard atmospheric refraction conditions, the main beam of the 3rd tilt should clear the ground. Therefore the ground clutter in the QPE are either anomalous propagations or due to that side lobes of the 3rd tilt are hitting mountain tops and generating significant returns. To mitigate these ground clutter, a manual hybrid scan map was created based on reflectivity climatologies for RCCG and for each of the other radar (Fig. 13). The climate hybrid scan maps take into account non-standard ground clutter as well as blockages shown in reflectivity climatologies that are not represented in the standard beam blockage maps. All bins below the climate hybrid scan tables are removed in

**Fig. 13** Climate hybrid scan tables for RCWF (**a**), RCHL (**b**), RCCG (**c**) and RCKT (**d**). The colors represent the lowest tilts that are free of ground clutter and significant blockage

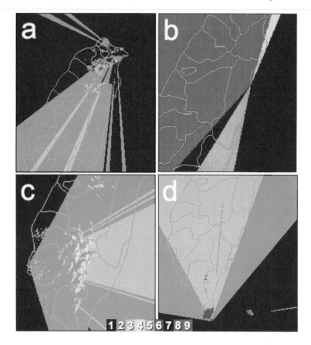

**Fig. 14** Same as in Fig. 12 but after the following changes in the HRQ system: (1) applied climate hybrid scan tables and (2) applied cross radial interpolation for small gaps in azimuth direction. See text for detail

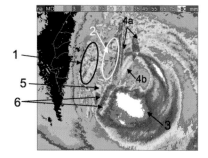

a QC post-processing. The abnormally high amounts on the south central mountain-tops were greatly reduced after the climate hybrid scan tables were applied (area #1 in Fig. 14). The climate hybrid scan table from RCHL also helped mitigating the holes in area #2 (Figs. 14 vs. 12) because the initial hybrid scan table assumed 360° scan while RCHL actually operates in sector scans (Fig. 13b).

## 3.2 Cross-Radial Gap Filling

The radial gap in area #3 in Fig. 12 was due to a blockage on the first tilt in RCKT (Fig. 13d). In order to mitigate this type of small radial gaps, a cross-radial interpolation is developed and implemented. The interpolation uses data from the non-blocked radials surrounding the blocked radial, and linearly interpolate across the azimuth direction to fill in the gap. The linear interpolation only applies to gaps of $\leq 5°$ for ranges 0–50 km, $\leq 4°$ for 51–100 km, $\leq 3°$ for 150–200 km, $\leq 2°$ for 201–300 km, and $1°$ for ranges 301 km and beyond. In addition, a linear interpolation is applied to range gaps of 5 km or less. These gap-filling steps successfully mitigated the discontinuities in area #3 (Fig. 14).

## 3.3 HSR Mosaic Weighting Function

The discontinuities indicated by "4a" in Fig. 12 were due to the discontinuities in the RCWF hybrid scan (Fig. 13a) and a large $L$ parameter (200 km) used earlier in the HSR mosaic scheme (re Sect. 2.3). The large $L$ resulted in a flatter weighting function $W_L$ (re Eq. [2] in Sect. 2.3) and larger influence radius of single radar HSR fields. This allowed discontinuities in the RCWF HSR to have a large impact on the mosaiced HSR field in area "4a" even though the area is much closer to RCHL than to RCWF. To correct for this, the $L$ value was reduced to 50 km and the discontinuities were eliminated (Fig. 15) after the new $L$ was applied. The new $L$ value also largely mitigated the discontinuities in area "4b" (reduced impact of RCKT with respect to RCHL, see Fig. 13d) and in area #5 (reduced impact of RCCG with respect to RCKT, see Fig. 13c).

**Fig. 15** Same as in Fig. 14 except that the value of $L$ was changed from 200 to 50 km

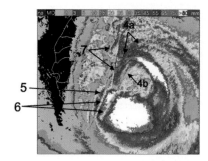

## 3.4 Remaining Issues and Challenges

After the aforementioned improvements and enhancements, there are still issues remaining. The discontinuities in areas #6 and #7 on Fig. 15 are among the several remaining issues. These were associated with hybrid scan reflectivity changes from upper to lower tilts or vice versa. The area #6 corresponds to the hybrid scan of RCKT (Fig. 13d) and #7 to RCHL (Fig. 13b). A potential solution is to apply a linear ramping in the hybrid scan reflectivity field across boundaries between the upper and lower tilts. The ramping code is underdevelopment. Other challenges include sea clutter (not shown) contamination and uncertainties in Z-R relationships.

# 4 Summary

The second generation of the QPESUMS system was developed and started to run in real-time experimentally at the Central Weather Bureau of Taiwan since June 2007. The final objective of the system is to support government agencies and emergency managers for operational flood, flash flood, and mudslide warnings and water resource managements. The system integrates data from weather radar, rain gauges, model, and satellite and produce severe weather and QPE products at 1 km resolution and 10-min update cycle.

The experimental products have been under extensive assessments and preliminary results are presented in this paper. The assessments facilitated improvements and enhancements to various components in the current system, such as the generation of new climate hybrid scan tables, optimization of radar mosaic weighting function, and radar data gap filling. Also as a result of the assessments, new algorithms such as sea clutter removal and cross-tilt reflectivity discontinuity mitigation are underdevelopment.

The new high-resolution QPE (HRQ) system is designed to be adaptive to different radar networks. Initial efforts have been undergoing to integrate a dual-pol radar (RCCU) data from the National Central University of Taiwan into the HRQ 3-D mosaic. The future work includes evaluations of QPE accuracies before and after integration of the dual-pol radar data and further refinements of the radar QPE.

**Acknowledgements** The funding for this research was provided by Central Weather Bureau, Water Resource Agency, and Soil and Water Conservation Bureau of Taiwan.

# References

Amburn, SA, Wolf PL (1997) VIL density as a hail indicator. Wea Forecast 12, pp 473–478

Fulton R, Breidenbach J, Seo D-J, Miller D, O'Bannon T (1998) The WSR-88D rainfall algorithm. Wea Forecast 13, pp 377–395

Greene DR, Clark RA (1972) Vertically integrated liquid water—a new analysis tool. Mon Wea Rev 100, pp 548–552.

Lakshmanan V, Fritz A, Smith T, Hondl K, Stumpf GJ (2007) An automated technique to quality control radar reflectivity data. J Appl Meteor 46, 288–305

Lakshmanan V, Smith T, Hondl K, Stumpf GJ, Witt A (2006) A real-time, three dimensional, rapidly updating, heterogeneous radar merger technique for reflectivity, velocity and derived products. Wea Forecast 21, 802–823

Witt A, Eilts MD, Stumpf GJ, Johnson JT, Mitchell ED, Thomas KW (1998) An enhanced hail detection algorithm for the WSR-88D. Wea Forecast 13, pp 286–303

Xu X, Howard K, Zhang J (2008) An automated radar technique for the identification of tropical precipitation. J Hydrometiorol, accepted, doi: 10.1175/2007 JHM954.1.

Zhang J, Langston C, Howard K (2008) Bright band identification based on vertical profiles of reflectivity from the WSR-88D. J Atmos Ocean Tech 25, 1859–1872.

Zhang J, Howard K, Gourley JJ (2005) Constructing three-dimensional multiple radar reflectivity mosaics: Examples of convective storms and stratiform rain echoes. J Atmos Ocean Tech 22, 30–42

# Assimilation of Satellite Data in Improving Numerical Simulation of Tropical Cyclones: Progress, Challenge and Development

Zhaoxia Pu

**Abstract** Tropical cyclones originate over the ocean where conventional data are very sparse. Satellites provide a very useful source of data for studying tropical cyclones. This paper gives a brief review on the recent progress in satellite data assimilation for improving the numerical simulations and forecasts of tropical cyclones. The general problems and challenges in tropical cyclone forecasting and satellite data assimilation and potential improvements in the future are also addressed in light of the most recent research.

## 1 Introduction

As witnessed in recent years, the social, ecological and economic impacts of tropical cyclones can be devastating. Tropical cyclones, especially cyclones with hurricane intensity, are one of the nature's most intense phenomena and one of the coastal resident's greatest fears. They threaten the maritime industry, devastate coastal regions and cause floods and erosion inland through torrential rainfall, high winds and severe storm surges. Owing to the great social and economic impact of tropical cyclones, it is of great importance that the track and intensity of the tropical cyclones be accurately predicted many hours in advance. Therefore, accurate forecasts of tropical cyclones and their associated precipitation have been listed as a high-priority research area in the U.S. (e.g., Elsberry and Marks 1999) and other regions around the world (e.g., Meng et al. 2002; Wu and Kuo 1999).

Tropical cyclones originate over the ocean, where conventional meteorological observations tend to be sparse. Due to lack of the data, deficiencies in model initial condition can lead to inaccurate tropical cyclone forecasts. Our knowledge of the physical processes that control tropical cyclone evolution is also limited. Thus, accurately forecasting tropical cyclone track and intensity has been a challenging problem for numerical weather prediction. It has been recognized that the additional observations over the ocean has a positive impact on tropical cyclone

Z. Pu (✉)

Department of Meteorology, University of Utah, 135 S 1460 E, Rm. 819, Salt Lake City, UT, USA, e-mail: Zhaoxia.Pu@utah.edu

S.K. Park, L. Xu, *Data Assimilation for Atmospheric, Oceanic and Hydrologic Applications*, DOI 10.1007/978-3-540-71056-1_8,
© Springer-Verlag Berlin Heidelberg 2009

forecasts. For instance, the U.S. National Oceanic and Atmospheric Administration (NOAA) synoptic surveillance missions are very useful in improving individual tropical cyclone track forecasts (e.g., Franklin and DeMaria 1992; Franklin et al. 1993; Burpee et al. 1996; Aberson 2002; Aberson and Sampson 2003; Aberson and Etherton 2006). Other research programs such as the "Dropwindsonde Observations for Typhoon Surveillance near the Taiwan Region" (DOTSTAR) has also demonstrated the positive impact of the targeted observations on the track forecasts of the typhoons over the North Pacific Ocean (Wu et al. 2007).

Along with advancements in remote sensing techniques, the amount of usable satellite data has increased rapidly in the last two decades. Many satellite derived data products become useful sources for tropical cyclone analyses and forecasting (e.g., Velden et al. 1992; Leslie et al. 1998; Velden et al. 1998; Pu et al. 2002; Zhu et al. 2002; Pu and Tao 2004; Hou et al. 2004, Chen 2007; Zhang et al. 2007). Owing to the improvement of large-scale forecast models, data assimilation techniques, and the use of satellite and aircraft reconnaissance observations, tropical cyclone track forecasts have improved significantly during the last two decades (e.g., in U.S., the official error trends are documented online at www.nhc.noaa.gov/verification). To date, many types of satellite data are available on a daily basis. Satellite data assimilation has become an active research field for tropical cyclone studies. The purpose of this paper is to give a brief review on the recent progress in satellite data assimilation for improving tropical cyclone forecasts. The general problems and challenges in tropical cyclone forecasts and data assimilation and the future development of satellite data assimilation in improving tropical cyclone forecasts will also be discussed.

The paper is organized as follows. Sample studies on the impact of satellite data on tropical cyclone simulations and forecasts are reviewed in Sect. 2. The challenges in tropical cyclone forecasts and current problems in satellite data assimilation are discussed in Sect. 3. Recent development and potential improvement are addressed in Sect. 4. Concluding remarks are made in Sect. 5.

## 2 Impact of Satellite Data in Improving Numerical Simulation of Tropical Cyclones

Many previous studies demonstrated that satellite data have a positive impact on the forecasts of tropical cyclones. This section gives a brief literature review on recent progress in assimilating a few types of sample satellite data in hurricane simulations and forecasts.

### 2.1 Satellite Derived Rainfall Rates

Development and evolution of tropical cyclones usually are accompanied by convection and precipitation processes. Satellite observations, such as these acquired from the US Department of Defense's Special Sensor Microwave Imager (SSM/I) and the National Aeronautics and Space Administration's (NASA) Tropical Rainfall Measuring Mission (TRMM), are able to measure the rainfall features associated with the tropical cyclones. Depending upon the surface rain rates derived from the SSM/I, Krishnamurti et al. (1991) developed a method to physically initialize the Florida State University global cumulus parameterization spectral model. A comparison study was then conducted by Tibbetts and Krishnamurti (2000) to evaluate the performance of four different rain-rate algorithms in hurricane track forecast using the physical initialization method, and found that SSM/I rain-rate products developed by NOAA produced the best prediction results. Hou et al. (2000) assimilated the SSM/I and TRMM Microwave Image (TMI) derived surface rainfall and total precipitatble water into the NASA Goddard Earth Observing System (GEOS) global analysis with a one-dimensional variational data assimilation (1DVAR) minimization procedure. The applications demonstrated that the assimilation of the TMI and SSM/I satellite rainfall rates results in improvements in tropical cyclone track forecasts in the GEOS global model (Hou et al. 2004). With the fifth-generation Pennsylvania State University-National Center for Atmospheric Research Mesoscale Model (MM5), Pu et al. (2002) found that the TMI rain rate had a significant impact on the mesoscale numerical simulation of Supertyphoon Paka (1997). In addition, the impact of TMI surface rainfall rate in the tropical cyclone forecast was further investigated by Pu and Tao (2004) using the MM5 four-dimensional variational data assimilation (4DVAR) scheme. Results show that inclusion of TMI data is beneficial, as the data assimilation helps the model produce realistic rain-bands and the asymmetric structure of Hurricane Bonnie (1998). At the same time, assimilation of rainfall data improves the quantitative precipitation forecasting during the storm evolution. In addition, they also found that assimilation of the TMI rainfall data is sensitive to the error characteristics that were specified in the data assimilation system.

## 2.2 Satellite Derived Wind Information

It has been well recognized that the environmental vertical wind shear is an important factor that influences the tropical cyclone structure and development. Rogers et al. (2003) suggested that tropical cyclone convection and precipitation structures are closely related to the magnitude of environmental vertical wind shear. Merrill (1988) and Frank and Ritchie (1999) concluded that weak environmental vertical wind shear is a necessary condition for tropical cyclone deepening. Paterson et al. (2005) suggested that the favored wind shear for the rapid intensification of a tropical storm is $2 \sim 4\,\mathrm{m\,s^{-1}}$, while Black et al. (2002) found that Hurricane Olivia (1994) intensified remarkably with the vertical wind shear of $8\,\mathrm{m\,s^{-1}}$. Although there is controversy about the impact of environmental wind shear in tropical cyclone development and intensification, assimilation of the wind information into the numerical models generally show a positive impact on tropical cyclone forecasts.

Early studies by Velden et al. (1992) and Leslie et al. (1998) indicated that satellite wind data has a significant impact on hurricane track forecasting. Soden et al. (2001) performed a series of experimental forecasts to evaluate the impact of Geostationary Operational Environmental Satellite-8 (GOES-8) derived cloud track and water vapor winds on hurricane track predictions from the Geophysical Fluid Dynamics Laboratory (GFDL) hurricane prediction system. Over one hundred cases were examined from eleven different storms covering three seasons (1996–1998). Results demonstrated that the assimilation of the GOES winds leads to statistically significant improvements for all forecast periods, with the relative reductions in track error ranging from $\sim 5\%$ at 12 h to $\sim 12\%$ at 36 h. In a recent study, Zhang et al. (2007) also indicated that the assimilation of GOES-8/9 data resulted in an improved forecast for Hurricane Lili (2002) in its weakening stage.

Despite of the positive impact of GOES wind data, Chen (2007) examined the impact of the SSM/I satellite-derived winds and NASA Quick Scatterometer (QuickSCAT) ocean surface vectors on numerical simulations of hurricane Isidore (2002). Results show that the assimilation of either SSM/I or QuickSCAT satellite winds strengthened the cyclonic circulation in the analysis. However, the increment from the QuikSCAT wind analysis is more complicated than that from the SSM/I analysis due to the correction of the storm location. Assimilation of the QuickSCAT wind data enhances the air–sea interaction processes and improves the simulated intensity for Isidore. The storm structure was better simulated and the storm track forecast is clearly improved, in particularly during the later period of the simulation. However, a lack of information about the wind direction from SSM/I data prevented it from having much of an effect.

In a recent study, Pu et al. (2008) assimilated the dropwindsonde, GOES-11 rapid-scan atmospheric motion vectors and NASA QuikSCAT near-surface wind data, collected during NASA Tropical Cloud Systems and Processes (TCSP) field experiment in July 2005, into an advanced research version of the Weather Research and Forecasting (WRF) model using its three-dimensional variational (3DVAR) data assimilation system. The impacts of the mesoscale data assimilation on WRF numerical simulations of tropical storms Cindy and Gert (2005) near landfall are examined. Results show that the assimilation of dropwindsonde and satellite wind data into the WRF model improves the forecasts of the two tropical storms up to the landfall time. Specifically, the QuikSCAT wind information is very important for improving the storm track forecast, whereas the dropwindsonde and GOES-11 wind data are also necessary for improved forecasts of intensity and precipitation. The assimilation of satellite wind and dropwindsonde data improves the quantitative precipitation forecasts near landfall of the tropical storms.

## 2.3 Satellite Derived Temperature and Moisture Profiles

Tropical cyclone environmental thermodynamic conditions are the one of the most important factors that influence tropical cyclone development (Emanuel 1986;

Emanuel et al. 2004). The assimilation of atmospheric temperature and moisture observations has proven beneficial to tropical cyclone forecasts. One of the relevant studies on assimilation of the retrieved temperature profiles from the Advanced Microwave Sounding Unit (AMSU-A) showed a positive impact on the construction of a more realistic vortex in the hurricane initialization (Zhu et al. 2002). In addition, Wu et al. (2006) incorporated the retrieved temperature and humidity profiles from Atmospheric Infrared Sounder (AIRS, aboard the NASA Aqua satellite) into the MM5 model with a nudging technique to study the influences of the Saharan Air Layer (SAL) on the formation of Hurricane Isabel (2003). They found that incorporating the AIRS data resulted in a better simulation of the large-scale flow pattern and the timing and location of the formation of Hurricane Isabel (2003) and its subsequent track.

In addition to the AIRS retrieved temperature and moisture profiles, Zhang et al. (2007) found that assimilating the sounding data (temperature and dewpoint temperature profiles) acquired from the Moderate Resolution Imaging Spectroradiometer (MODIS) aboard the NASA Terra and Aqua spacecrafts also improves the outer-core thermodynamic features of the Hurricane Lili (2002) from MM5 model simulations. Although the data has a slight impact on the track forecast, improvements in the model's intensity prediction have clearly been seen as the simulations with data assimilation result in the lessening of the incorrect prediction of intensification.

## 2.4 Satellite Radiances

All of the aforementioned studies assimilated satellite derived (retrieved) data products. In reality, satellite direct measurements are mostly in the form of the radiances or brightness temperatures. In order to eliminate the additional data errors introduced by the retrieval process, it has been suggested to directly assimilate the satellite radiances instead of the retrieved data products in operational numerical weather prediction (e.g., Derber and Wu 1998).

Studies have been conducted to determine the effectiveness of the direct assimilation of satellite radiances into the mesoscale model for tropical cyclone studies. Zou et al. (2001) investigated the influence of clear-sky GOES brightness temperature data on the numerical simulations of Hurricane Felix (1995). Results show that the assimilation of GOES satellite data modified not only the environmental flow but also the structure of the initial vortex, which is located over a region devoid of satellite data. This modification resulted in a reduction of the 12-h forecast errors verified by radiosonde data. Differences in the prediction of Hurricane Felix with and without satellite data were also found in the prediction of the upper-level jet, the cold temperature trough ahead of the hurricane, the size of the hurricane eye, and the location of the maximum hydrometeor. Zhang et al. (2004) assimilated clear-sky ATOVS microwave radiances to typhoon track prediction. Results showed significant improvement in the typhoon track forecasts.

# 3 Challenges in Tropical Cyclone Prediction and Satellite Data Assimilation

## 3.1 Challenges in Tropical Cyclone Prediction

Although in the past decade hurricane forecasts have been improved significantly due to advances in satellite remote sensing techniques and data assimilation, compared with the improvement in tropical cyclone track forecasts, intensity forecast errors have improved only slightly over the same period (Elsberry and Marks 1999; McAdie and Lawrence 2000; Houze et al. 2006; Rogers et al. 2006). Rogers et al. (2006) noted that the official 48-h track forecast errors have decreased by nearly 45% while the official tropical cyclone intensity forecast errors have decreased by only 17%. They also commented that lack of improvement in numerical forecasts of tropical cyclone intensity can be attributed to three factors: (1) inaccurate initial structure of the storm vortex and environment in numerical models; (2) limitations in numerical models themselves; (3) inadequate understanding of the physics of tropical cyclones and their development. Among all other factors, one of the major reasons for the slow pace of improvement in intensity forecasts may be deficiencies in the collection and assimilation of real-time inner core data into numerical weather prediction models; most of the previous satellite data assimilation efforts contributed to the improvement in hurricane environmental conditions. Due to cloud and rain effect, satellite retrieved products usually have large uncertainties under cloudy and precipitating areas, thus most of the cloudy and rain-effected data are rejected during the quality control procedures in data assimilation. In addition, most of recent data assimilation systems only assimilate the clear-sky radiances. *Therefore, few observations in tropical cyclone inner core regions are assimilated into the model, thus the model may not resolve the real inner core structure and intensity of hurricanes.*

## 3.2 Vortex Initialization

Uncertainties in tropical cyclone initialization are one of the fundamental reasons for the current lack of skill in tropical cyclone forecasts. Commonly, forecasts of track and intensity changes for tropical cyclones, especially the mature hurricanes require accurate representation of their vortices in model initial conditions. However, vortices contained in large-scale analyses from operational centers are often too weak and sometimes misplaced. The conventional observations in the vicinity of the hurricane are usually sparse. Due to the aforementioned problem, few satellite observations are assimilated in the tropical cyclone inner core regions. Therefore, the use of so-called "bogus vortices" is often adopted in numerical simulations and forecasts (Kurihara et al. 1993; Leslie and Holland 1995; Pu and Braun 2001; Braun et al. 2006).

A bogus vortex is an artificial vortex specified according to the estimated size of the cyclone (the radius of maximum winds), its position, and its intensity (the maximum velocity or minimum sea-level pressure). Traditionally, such bogus vortices have been defined by some empirical methods and are directly implanted into the larger-scale environment. Although some successful simulations, including the prediction of hurricane movement and structure, have been conducted using bogus vortices (e.g., Kurihara et al. 1993; Trinh and Krishnamuti 1992), a weakness in such an approach is the inconsistency of the vortices with the properties of the prediction model (Iwasaki et al. 1987; Mathur 1991). A more advanced scheme has been proposed by Kurihara et al. (1993) at GFDL to overcome such defects. The method shows a substantial improvement in track prediction for operational hurricane forecasts (Bender et al. 1993). The success of the GFDL technique indicates the importance of having a dynamically and thermodynamically consistent initial vortex that is compatible with the resolution and physics of the hurricane prediction model. As a natural extension of GFDL's method, Zou and Xiao (2000) have proposed a so-called "variational bogus vortex scheme" to improve the initial vortex by using a 4DVAR technique. They show very encouraging results for Hurricane Felix (1995). On the other hand, Zhu et al. (2002) used NOAA AMSU-A derived three-dimensional temperature to generate the initial vortex by solving the nonlinear balance relationships and then integrating the hydrostatic equation. They obtained fairly good results in the simulation of Hurricane Bonnie (1998).

Although previous studies have made significant improvements in hurricane initialization and forecasts, many problems remain. First, although the GFDL scheme offers a smooth start for integration of numerical prediction model, the vortex itself is still *artificial*. In the same way, the variational bogus vortex scheme can ensure dynamic consistency between the initial vortex and numerical models, but cannot guarantee the completely *realistic* vortex structure. Therefore, in some cases, the vortices still spin-up at the beginning of forecasts (Pu and Braun 2001). Also, due to both the lack of conventional observations and the difficulties in the use of satellite observed inner core data, especially these rain and cloud-affected data, *initialization of hurricane inner core structures with realistic dynamic and thermodynamic observations have not yet been addressed in many previous studies.*

## 3.3 Sensitivity of Hurricane Intensity Forecasts to Physical Processes

In addition to the importance of tropical cyclone initialization to data assimilation, a lack of understanding of the environmental, physical, and thermodynamic conditions that control the hurricane intensity change has also been an urgent problem. A better understanding of the hurricane intensity change would eventually contribute to an increased ability to make accurate forecasts. Many studies have been dedicated to this topic. The majority of the research literature on hurricane intensity focuses on the prestorm thermodynamic environment (e.g., Emanuel 1986 and 1988, Emanuel et al. 2004; Bister and Emanuel 1998), certain properties of the atmospheric environment such as the vertical shear of the horizontal wind (e.g., Jones 2000; Frank and Ritchie 2001), and dynamical features such as disturbances in the upper troposphere (e.g., Molinari and Vollaro 1989, 1990, 1995). In addition, recent studies also made progress in understanding the physical and microphysical processes that have a large impact on hurricane intensity forecasts. In one of most recent studies, Zhu and Zhang (2006) examined the effects of various cloud microphysical processes on hurricane intensity, precipitation and inner-core structure using a series of 5-day explicit simulations of Hurricane Bonnie (1998) using the MM5 model. The results indicated that varying cloud microphysical processes produced little cahnge in hurricane track but resulted in pronounced departures in hurricane intensity and inner-core structures. McFarquhar et al. (2006) investigated the effects of three different microphysical schemes in the MM5 model on a high-resolution simulation (2 km) of Hurricane Erin (2001). They found that the different microphysical schemes caused only marginal differences in the hurricane track forecast but reproduced notable differences in the intensity forecast, although at the same time they also found that the representation of boundary layer processes is crucial in determining the strength of the simulated intensity. Cecil and Zipser (1999) indicated that ice water content and liquid water content have implications with regards to hurricane intensity. High correlations were found between future hurricane intensity and satellite-based 85-GHz ice-scattering signature within one degree radius of the cyclone center.

Results of aforementioned previous sensitivity studies imply that *there is strong association between the hurricane intensification and cloud microphysical and other physical processes. However, so far, assimilation of cloudy satellite data and cloud properties into the NWP still remains a challenging problem* (Errico et al. 2007; Weng 2007). Although satellite observations in the visible, infrared, and microwave provide a great deal of information on clouds and precipitation, many of these available data have not yet been assimilated into the numerical models. Therefore, *in addition to the research efforts to improve the physical parameterization schemes in the numerical model, assimilation of tropical cyclone inner core data and cloud data might be a way to enhance the forecast accuracy of tropical cyclone intensity and structure.*

# 4 Recent Development

Recent developments in advanced radiative transfer models (e.g., Kummerow et al. 2001; Liu and Weng 2002; Weng 2007) have brought to light the assimilation of cloud and rain-affected satellite radiances into the numerical weather forecast models. These radiative transfer models take the cloud scattering and emission from surface materials into account thus they are able to simulate the cloudy satellite radiances from numerical model outputs. As an example, an advanced radiative transfer model (e.g., Kummerow 1993, Kummerow et al. 2001) was used in simulating the satellite observed brightness temperature by taking the inputs as the temperature and hydrometeor profiles from MM5 model simulated Super Typhoon Paka (1997) over the eastern Pacific Ocean at 5-km horizontal grid resolution. The TMI observed brightness temperature and precipitation were then compared with the corresponding simulated quantities. During the numerical simulation period (up to 72 h simulation from 0000 UTC 12 December 1997), the simulated Typhoon intensity (represented by maximum surface wind and minimum sea-level pressure) was *very close to* the observed intensity (figure not shown here, details can be found in Pu et al. 2002). The simulated track errors are also in a relatively reasonable range (typically 10~60 km).

Figures 1 and 2 illustrated the comparison of simulated fields and corresponding observations at 0500UTC 13 December 1997 and 1500 UTC 13 December 1997, respectively. The numerical simulation results were either directly compared with satellite measurements [e.g., TRMM precipitation radar (PR) reflectivity and rain fall rates retrieval] or were input into an advanced radiative transfer model, namely, the Kummerow's radiative transfer model (see detail in Kummerow 1993, Kummerow et al. 2001) to produce similar quantities as in the observations (TMI microwave brightness temperature). As seen from both figures, simulated properties have a similar horizontal distribution pattern to that of the observations, including the close pattern between observed near surface rainfall rate (instantaneous) and the simulated surface rainfall rate (calculated from hourly accumulated rainfall). However, both simulated radar reflectivity and 10 GHz BT are much higher than that

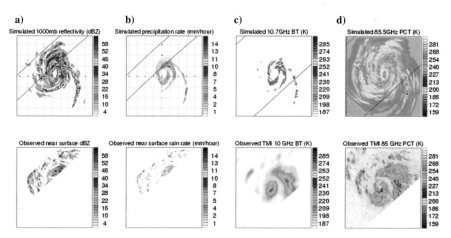

**Fig. 1** Comparison of the model simulated and satellite observed quantities of super Typhoon Paka at 0500UTC 13 December 1997. (**a**) simulated (*top*) and TRMM PR observed (*bottom*) reflectivity (dBZ); (**b**) Simulated hourly accumulated rainfall (*top*) and retrieved instantaneous rainfall rate from TRMM PR reflectivity (*bottom*); (**c**) Simulated 10 GHz horizontal polarized brightness temperature (*top*) and corresponded TRMM Microwave Imager (TMI) measurements (*bottom*); (**d**) Simulated 85 GHz Polarization Corrected Temperature (PCT, *top*) and TMI measured PCT (*bottom*). All figures covered the same areas between 4°N − 11°N and 156°E − 166°E

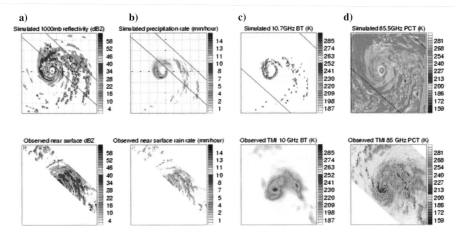

**Fig. 2** Same as Fig. 1, except for the comparison at 1500 UTC 13 December 1997 for the domain between 6°N–13°N and 155°E–165°E

of the corresponding observations, indicating the simulated rain water content is greater than that of the observation. At the same time, a smaller simulated 85 GHz Polarization Corrected Temperature (PCT, Spencer et al. 1989) suggests that there are more ice scattering signatures from simulated ice water contents. Except for the possible bias from the radiative transfer calculation, the results indicate *that the simulated cloud (microphysics) properties in the storm inner core may be wrong.* The results clearly indicate the following: in order to obtaining the reasonable hurricane inner core structures in the numerical simulations, the assimilation of cloudy-satellite radiances is certainly necessary.

Fortunately, there has been great progress in the atmospheric sciences community in developing fast and accurate radiative transfer models able to better handle the scaterring from atmospheric constitutes (e.g., cloud, precipitation and aerosols) and surface materials (Weng 2007). This progress leads to further development in the uses of the satellite data in cloudy and precipitating areas. In a most recent study, Weng et al. (2007) has developed a hybrid variational scheme (HVAR) to produce the vortex analysis associated with tropical storms. The scheme uses a radiative transfer model including scattering and emission for radiance simulation and allows for direct assimilation of rain-affected radiances from satellite microwave instruments. Through a 4DVAR scheme, the atmospheric temperature from AMSU and surface parameters from the Advanced Microwave Scanning Radiometer (AMSR-E) are assimilated into global forecast model outputs to improve analysis. The HVAR was applied for two hurricane cases in 2005 hurricane season, resulting in the improved analyses of three-dimensional structures of temperature and wind fields as compared with operational model analysis fields. Results also show that HVAR reproduces detailed structures for the hurricane warm core at the upper troposphere. Both lower-level wind speed and upper-level divergence are enhanced with reasonable asymmetric structure.

# 5 Concluding Remarks

Tropical cyclones originate over the ocean, where the conventional data are very sparse. Satellite observations provide a very useful data source for studying tropical cyclones. In many previous studies, various types of satellite data were assimilated into the numerical models and have been proven to be useful for improving the tropical cyclone forecasts. However, despite the great impact of satellite and other data sets on tropical cyclone track forecasts, tropical cyclone intensity and structure forecasts remain a great challenge in practice. In addition to our limited understanding in the physical mechanism and processes that control the tropical cyclone intensity change, deficiencies in the model initial conditions contribute to the limited forecast skill in hurricane intensity and track forecast. Specifically, initialization of hurricane inner core structures with realistic dynamic, thermodynamic and hydrometeor observations has not yet been addressed in many studies. At the same time, many data assimilation systems only assimilate the clear-sky radiances. Cloud and rain-affected retrieved data are rejected by the data assimilation system during the data quality control process. Recent developments in advanced radiative transfer models brings hope to be able to fully use the available satellite data in the near future. It is anticipated that the assimilation of the cloud and rain-effected satellite information will result in better a representation of the tropical cyclone's environmental conditions, along with inner core and outer rainband structures, which would lead to better forecasting and a clear understanding of the tropical cyclone evolution and intensity change.

**Acknowledgements**  This study is supported by NASA EOS program Grant No. NNX08AD32G.

# References

Aberson SD (2002) Two years of operational hurricane synoptic surveillance. Wea Forecast 17:1101–1110

Aberson SD, Sampson CR (2003) On the predictability of tropical cyclone tracks in the Northwest Pacific Basin. Mon Wea Rev 131:1491–1497

Aberson SD, Etherton BJ (2006) Targeting and data assimilation studies during Hurricane Humberto (2001). J Atmos Sci 63:175–186

Bender MA, Ross RJ, Tuleya RE, Kurihara Y (1993) Improvements in tropic cyclone track and intensity forecasts using the GFDL initialization system. Mon Wea Rev 121:2046–2061

Bister M, Emanuel KA (1998) Dissipative heating and hurricane intensity. Meteor Atmos Phys 50:233–240

Black ML, Gamache JF, Marks FD, Samsury CE, Willoughby HE (2002) Eastern Pacific Hurricanes Jimena of 1991 and Olivia of 1994: The effect of vertical shear on structure and intensity. Mon Wea Rev 130:2291–2312

Braun SA, Montgomery MT, Pu Z (2006) High-resolution simulation of Hurricane Bonnie (1998). Part I: The organization of eyewall vertical motion. J Atmos Sci 63: 19–42

Burpee RW, Aberson SD, Franklin JL, Lord SJ, Tuleya RE (1996) The impact of omega dropwindsondes on operational hurricane track forecast models. Bull Amer Meteor Soc 77:925–933

Cecil DJ, Zipser EJ (1999) Relationship between tropical cyclone intensity and satellite-based indicators of inner core convection: 85-GHz ice scattering signature and lightning. Mon Wea Rev 127: 103–123

Chen SH (2007) The Impact of assimilating SSM/I and QuikSCAT satellite winds on Hurricane Isidore simulation. Mon Wea Rev 135:549–566

Derber, JC, Wu WS (1998) The use of TOVS cloud-cleared radiances in the NCEP SSI analysis system. Mon Wea Rev 126:2287–2299

Elsberry RL, Marks FD (1999) The hurricane landfall workshop summary. Bull Amer Meteor Soc 80:683–685

Emanuel KA (1986) An air-sea interaction theory for tropical cyclones. Part I: Steady-state maintenance. J Atmos Sci 43:585–604

Emanuel KA (1988) The maximum intensity of hurricanes. J Atmos Sci 45:1143–1155

Emanuel KA, DesAutels C, Holloway C, Korty R (2004) Environmental control of tropical cyclone intensity. J Atmos Sci 61:843–858

Errico RM, Bauer P, Mahfouf JF (2007) Issues regarding the assimilation of cloud and precipitation data. J Atmos Sci 64:3785–3798

Frank WM, Ritchie EA (1999) Effects of environmental flow upon tropical cyclone structure. Mon Wea Rev 127:2044–2061

Frank WM, Ritchie EA (2001) Effects of vertical wind shear on the intensity and structure of numerically simulated hurricanes. Mon Wea Rev 129:2249–2269

Franklin JL, DeMaria M (1992) The impact of omega dropwindsonde observations on barotropic hurricane track forecasts. Mon Wea Rev 120:381–391

Franklin JL, Lord SJ, Feuer SE, Marks FD (1993) The kinematic structure of hurricane Gloria (1985) determined from nested analyses of dropwindsonde and Doppler radar data. Mon Wea Rev 121:2433–2451

Hou AY, Ledvina DV, Da Silva AM, Zhang SQ, Joiner J, Atlas RM (2000) Assimilation of SSM/I-derived surface rainfall and total precipitatble water for improving the GEOS analysis and climate studies. Mon Wea Rev 128: 509–537

Hou AY, Zhang SQ, Reale O (2004) Variational continuous assimilation of TMI and SSM/I rain rates: Impact on GEOS-3 hurricane analyses and forecasts. Mon Wea Rev 132: 2094–2109

Houze RA, Chen SS, Lee WC, Rogers RF, Moore JA, Stossmeister GJ, Bell MM, Ceterone J, Zhao W, Brodzik SR (2006) The Hurricane Rainband and intensity change experiment: observations and modeling of Hurricanes Katrina, Ophelia, and Rita. Bull Amer Meteor Soc 87:1503–1521

Iwasaki T, Nakano H, Sugi M (1987) The performance of a typhoon track prediction model with cumulus parameterization. J Meteor Soc Japan 65:555–570

Jones SC (2000) The evolution of vortices in vertical shear. Part III: Baroclinic vortices. Quart J Roy Meteor Soc 126: 3161–3185

Krishnamurti TN, Xue J, Bedi HS, Ingles K, Oosterhof D (1991) Physical initialization for numerical weather prediction over the tropics. Tellus 43:53–81

Kurihara Y, Bender MA, Ross RJ (1993) An initialization scheme of hurricane models by vortex specification. Mon Wea Rev 121: 2030–2045

Kummerrow C (1993) On the accuracy of the Eddington approximation for radiative transfer in the microwave frequencies. J Geophys Res 98: 2757–2765

Kummerrow C, Hong Y, Olson WS, Yang S, Adler RF, McCollum J, Ferraro R, Petty G, Shin DB, Wilheit TT (2001) The evolution of the Goddard profiling algorithm (GPROF) for rainfall estimation from passive microwave sensors. J App Meteor 40:1801–1821

Leslie LM, Holland GJ (1995) On the bogussing of tropical cyclones in numerical models: A comparison of vortex profiles. Meteor Atmos Phys 56:101–110

Leslie LM, LeMarshall JF, Morison RP, Spinoso C, Purser RJ, Pescod N, Seecamp R (1998) Improved hurricane track forecasting from the continuous assimilation of high quality satellite wind data. Mon Wea Rev 126:1248–1258

Liu Q, Weng F (2002) A microwave polarimetric two-stream radiative transfer model. J Atmos Sci 59: 2396–2402

Mathur MB (1991) The National Meteorological Center's quasi- lagrangian model for hurricane prediction. Mon Wea Rev 119:1419–1447

McAdie CJ, Lawrence MB (2000) Improvement in tropical cyclone track forecasting in the Atlantic basin, 1970–98. Bull Amer Meteor Soc 81:989–997.

McFarquhar GM, Zhang H, Heymsfield G, Hood R, Dudhia J, Halverson JB, Marks F (2006) Factors affecting the evolution of Hurricane Erin (2001) and the distributions of hydrometeors: Role of microphysical processes. J Atmos Sci 63:127–150

Meng Z, Chen L, Xu X (2002) Recent progress on tropical cyclone research in China. Adv Atmos Sci 19: 103–110

Merrill RT (1988) Environmental influences on hurricane intensification. J Atmos Sci 45:1678–1687

Molinari J, Vollaro D (1989) External influences on hurricane intensity. Part I: Outflow layer eddy angular momentum fluxes. J Atmos Sci 46:1093–1105

Molinari J, Vollaro D (1990) External influences on hurricane intensity. Part II: Vertical structure and response of the hurricane vortex. J Atmos Sci 47:1902–1918

Molinari J, Vollaro D (1995) External influences on hurricane intensity. Part III: Potential vorticity structure. J Atmos Sci 52:3593–3606

Paterson LA, Hanstrum BN, Davidson NE, Weber HC (2005) Influence of environmental vertical wind shear on the intensity of hurricane-strength tropical cyclones in the Australian region. Mon Wea Rev 133:3644–3660

Pu Z, Braun SA (2001) Evaluation of bogus vortex techniques with four-dimensional variational data assimilation. Mon Wea Rev 129:2023–2039

Pu Z, Tao WK, Braun SA, Simpson J, Jia Y, Halverson J, Hou A, Olson W (2002) The impact of TRMM data on mesoscale numerical simulation of supertyphoon Paka. Mon Wea Rev 130:2248–2258

Pu Z, Tao WK (2004) Mesoscale assimilation of TMI data with 4DVAR: Sensitivity study. J Meteor Soc Japan 82:1389–1397

Pu Z, Li X, Velden C, Aberson S, Liu WT (2008) Impact of aircraft dropsonde and satellite wind data on the numerical simulation of two landfalling tropical storms during TCSP. Wea Forecast 23:62–79

Rogers R, Chen S, Tenerelli J, Willoughby H (2003) A numerical study of the impact of vertical shear on the distribution of rainfall in Hurricane Bonnie (1998). Mon Wea Rev 131:1577–1599

Rogers R, Aberson S, Black M, Black P, Cione J, Dodge P, Dunion J, Gamache J, Kaplan J, Powell M, Shay N, Surgi N, Uhlhorn E (2006) The intensity forecasting experiment: A NOAA multiyear field program for improving tropical cyclone intensity forecasts. Bull Amer Meteor Soc 87:1523–1537

Soden BJ, Velden CS, Tuleya RE (2001) The impact of satellite winds on experimental GFDL hurricane model forecasts. Mon Wea Rev 129:835–852

Spencer RW, Goodman HM, Hood RE (1989) Precipitation retrieval over Land and Ocean with the SSM/I: Identification and characteristics of the scattering signal. J Atmos Ocean Tech 6:254–273

Tibbetts RT, Krishnamurti TN (2000) An intercomparison of hurricane forecasts using SSM/I and TRMM rain rate algorithm(s). Meteo Atmos Phys 74:37–49

Trinh VT, Krishnamuti TN (1992) Vortex initialization for typhoon track prediction. Meteor Atmos Phys 47:117–126

Velden CS, Hayden CM, Franklin JL, Lynch JS (1992) The impact of satellite-derived winds on numerical hurricane track forecasting. Wea Forecast 7:107–118

Velden CS, Olander TL, Wanzong S (1998) The impact of multispectral GOES-8 wind information on Atlantic tropical cyclone track forecasts in 1995. Part 1: Dataset methodology, description and case analysis. Mon Wea Rev 126: 1202–1218

Weng F (2007) Advances in radiative transfer modeling in support of satellite data assimilation. J Atmos Sci 64: 3799–3807

Weng F, Zhu T, Yan B (2007) Satellite data assimilation in numerical weather prediction models. Part II: Uses of rain-affected radiances from microwave observations for hurricane vortex analysis. J Atmos Sci 64: 3910–3925

Wu CC, Kuo YH (1999) Typhoons affecting Taiwan: Current understanding and future challenges. Bull Amer Meteor Soc 80: 67–80

Wu CC, Chou KH, Lin PH, Aberson S, Peng, MS, Nakazawa T (2007) The impact of dropwindsonde data on typhoon track forecasts in DOTSTAR. Wea Forecast 22:1157–1176

Wu L, Braun SA, Qu JJ, Hao X (2006) Simulating the formation of Hurricane Isabel (2003) with AIRS data. Geophys Res Lett 33:L04804

Zhang H, Xue J, Zhu G, Zhuang S, Wu X, Zhang F (2004) Application of direct assimilation of ATOVS microwave radiances to typhoon track prediction. Adv Atmos Sci 21:283–290

Zhang X, Xiao Q, Fitzpatrick PJ (2007) The impact of multisatellite data on the initialization and simulation of Hurricane Lili's (2002) rapid weakening phase. Mon Wea Rev 135: 526–548

Zhu T, Zhang DL, Weng F (2002) Impact of the advanced microwave sounding unit measurements on hurricane prediction. Mon Wea Rev 130:2416–2432

Zhu T, Zhang DL (2006) Numerical simulation of Hurricane Bonnie (1998). Part II: Sensitivity to varying cloud microphysical processes. J Atmos Sci 63: 109–126

Zou X, Xiao Q (2000) Studies on the Initialization and simulation of a mature hurricane using a variational bogus data assimilation scheme. J Atmos Sci 57: 836–860

Zou X, Xiao Q, Lipton AE, Modica GD (2001) A numerical study of the effect of GOES sounder cloud-cleared brightness temperatures on the prediction on Hurricane Felix. J Appl Meteor 40: 34–55

# Diagnostics for Evaluating the Impact of Satellite Observations

Nancy L. Baker and Rolf H. Langland

**Abstract** The adjoints of the numerical weather prediction (NWP) model and data assimilation system may be used together to objectively determine the observation impact – or whether a given observation platform or observing system improves or degrades the subsequent NWP forecast.

The observation impact is a very specific measure of forecast impact, as it depends upon the choice of forecast metric, the suite of observations assimilated, the data assimilation system, and the NWP forecast model. This chapter presents an overview of data assimilation adjoint theory, the observation impact calculation, and the appropriate choices for the forecast metric. Several applications of the observation adjoint technique are presented to illustrate its usefulness to help identify systematic problems with the observing network, to quantify the value of different observing platforms, to monitor the quality of the observing network, and for channel selection for satellite radiometers.

## 1 Introduction

This research was originally motivated more than ten years ago by preparations for the then upcoming FASTEX (The Fronts and Atlantic Storm-Track Experiment, Joly et al. 1997). The adaptive observation targeting methods for FASTEX included the singular vector approach (Palmer et al. 1998; Gelaro et al. 1999; Bergot 1999; Buizza and Montani 1999) and the gradient sensitivity technique (Langland and Rohaly 1996; Bergot 1999). These methods sought to identify the sensitive regions where small errors could amplify and dominate the short- to medium-range forecast error (Rabier et al. 1996). However, neither one of these targeting methods was able to take into account how the observations would be used by the data assimilation system, nor the presence of other observations in the region. Partly as a result of this, assimilation of the special targeted observations from FASTEX led to both degraded and improved forecasts (Doerenbecher and Bergot 2001).

N.L. Baker (✉)

Marine Meteorology Division, Naval Research Laboratory, Monterey, CA 93943–5502, USA

e-mail: nancy.baker@nrlmry.navy.mil

S.K. Park, L. Xu, *Data Assimilation for Atmospheric, Oceanic and Hydrologic Applications*, DOI 10.1007/978-3-540-71056-1_9,

© Springer-Verlag Berlin Heidelberg 2009

In 1996, Roger Daley[1] derived the equations for this missing piece – that of the adjoint of the data assimilation system. Baker (2000) and Baker and Daley (2000) explored observation adjoint sensitivity in an idealized context. By 2000, Roger Daley had constructed the adjoint of the NRL[2] Atmospheric Variational Data Assimilation System (NAVDAS). As it turned out, the observation-space formulation of NAVDAS was ideal for developing the adjoint, as it simply required applying the analysis operators in a reverse sense. Since that time, this area of research has expanded rapidly, and several operational NWP centers routinely use the adjoint of their data assimilation systems to monitor the quality of the observing network, diagnose assimilation system deficiencies, and make decisions on which observations to assimilate.

An overview of the data assimilation adjoint theory will be presented in Sect. 2, with idealized examples presented in Sect. 3. A discussion of the choice of an appropriate forecast metric is in Sect. 4, along with the definition of the observation impact, while Sect. 5 presents several applications of the observation adjoint technique for data monitoring, channel selection for satellite radiometers, and identifying systematic errors in the observations or the way the observations are used by the data assimilation system. Section 6 presents conclusions and discussion of ideas for future use of observation impact.

## 2 Derivation of Observation and Background Sensitivity

Following Baker and Daley (2000), the derivation of the observation and background sensitivity begins with the three-dimensional analysis problem and the analysis equation (Daley 1991),

$$\mathbf{x_a} = \mathbf{x_b} + \mathbf{K}(\mathbf{y} - H\{\mathbf{x_b}\}) \tag{1}$$

Following the notation given by Ide et al. (1997), the vector of observations is given by $\mathbf{y}$, the background vector is given by $\mathbf{x_b}$, and the analysis vector is given by $\mathbf{x_a}$. In general, the application of the observation or forward operator $H$ represents any necessary spatial and temporal interpolations from the forecast model background to the observation location and time. If the observed quantity is not directly related to the model state variables, then $H$ also represents the transformation from the forecast values to the observed quantity. For satellite radiances (or brightness temperatures), $H\{x_b\}$ represents the forward radiative transfer model applied to $\mathbf{x_b}$ and computes forecast or background radiances. The differences between the observation and the background $(\mathbf{y} - H\{\mathbf{x_b}\})$ are the components of the *innovation vector*, while $\mathbf{x_a} - \mathbf{x_b} = \mathbf{K}(\mathbf{y} - H\{\mathbf{x_b}\})$ is the *correction vector*. If one considers linear analysis problems only, then

---

[1] Roger Daley (1943–2001) was a senior visiting scientist at the Naval Research Laboratory from 1995 to 2001.

[2] Naval Research Laboratory.

$$\mathbf{x_a} = \mathbf{x_b} + \mathbf{K}(\mathbf{y} - \mathbf{Hx_b}). \tag{2}$$

The matrix $\mathbf{H}$ is the Jacobian matrix corresponding to the forward operator $H\{x_b\}$ linearized about the background state vector. This approximation is valid for the radiative transfer relationship between temperatures and radiances, but is not valid for moisture retrievals. The Kalman gain (or weight) matrix, $\mathbf{K}$, is given by

$$\mathbf{K} = \mathbf{BH}^T(\mathbf{HBH}^T + \mathbf{R})^{-1}, \tag{3}$$

where $\mathbf{B}$ is the background error covariance matrix and $\mathbf{R}$ is the observation error covariance matrix.

Equation (2) may be rewritten as

$$\mathbf{x_a} = \mathbf{x_b} - \mathbf{KHx_b} + \mathbf{Ky} = (\mathbf{I} - \mathbf{KH})\mathbf{x_b} + \mathbf{Ky}, \tag{4}$$

where $\mathbf{I}$ is the identity matrix. The sensitivity of the analysis to the observations $\partial\mathbf{x_a}/\partial\mathbf{y}$ and the sensitivity of the analysis to the background $\partial\mathbf{x_a}/\partial\mathbf{x_b}$ is derived first. Following Gelb (1974), the vector gradient of a vector is a matrix and is given by $\partial\mathbf{x}^T/\partial\mathbf{y} = \mathbf{A}$, or $\partial x_i/\partial y_k = a_{ki}$. Using this relationship, the following equations may be derived:

$$\partial\mathbf{x_a}/\partial\mathbf{y} = \mathbf{K}^T, \tag{5a}$$

$$\partial\mathbf{x_a}/\partial\mathbf{x_b} = (\mathbf{I} - \mathbf{KH})^T = \mathbf{I} - \mathbf{H}^T\mathbf{K}^T, \tag{5b}$$

where

$$\mathbf{K}^T = (\mathbf{HBH}^T + \mathbf{R})^{-1}\mathbf{HB}. \tag{6}$$

The cost function $J$ is a scalar measure of some aspect of the forecast of interest (see Sect. 4.1 for a specific example). The gradient or sensitivity of $J$ with respect to the analysis or initial conditions for the forecast is given by $\partial J/\partial\mathbf{x_a}$, and is determined by the adjoint of the tangent forward propagator based on the nonlinear trajectory. In this paper, this quantity will be referred to as the analysis sensitivity vector.

The sensitivity of the forecast aspect to the observations $(\partial J/\partial\mathbf{y})$, is referred to as the *observation sensitivity vector*, while the sensitivity of the forecast aspect to the background field $(\partial J/\partial\mathbf{x_b})$, is referred to as the *background sensitivity vector*. Using the chain rule and (5), the observation and background sensitivity vectors may be written as

$$\partial J/\partial\mathbf{y} = \frac{\partial\mathbf{x_a}}{\partial\mathbf{y}}\frac{\partial J}{\partial\mathbf{x_a}} = \mathbf{K}^T\partial J/\partial\mathbf{x_a}, \tag{7a}$$

$$\partial J/\partial\mathbf{x_b} = \frac{\partial\mathbf{x_a}}{\partial\mathbf{x_b}}\frac{\partial J}{\partial\mathbf{x_a}} = (\mathbf{I} - \mathbf{H}^T\mathbf{K}^T)\partial J/\partial\mathbf{x_a} \tag{7b}$$

Expanding the terms in the transposed Kalman gain matrix $(\mathbf{K}^T)$ in (6) using (3) gives the following expressions for the observation and background sensitivity vectors,

$$\partial J/\partial \mathbf{y} = (\mathbf{HBH}^T + \mathbf{R})^{-1}\mathbf{HB}\partial J/\partial \mathbf{x}_a, \tag{8a}$$

$$\partial J/\partial \mathbf{x}_b = [\mathbf{I} - \mathbf{H}^T(\mathbf{HBH}^T + \mathbf{R})^{-1}\mathbf{HB}]\partial J/\partial \mathbf{x}_a. \tag{8b}$$

In the analysis problem (1), $\mathbf{K}$ is the matrix of weights given to the *innovation* or difference between the observation and the background value (in observation space). In an analogous sense, the matrix $\mathbf{K}^T$ is the matrix of weights given to the sensitivity of $J$ with respect to the analysis for the observation sensitivity problem (7a).

It is also convenient for display purposes to define the *analysis space projection of the observation sensitivity vector*,

$$\mathbf{H}^T\partial J/\partial \mathbf{y} = \partial J/\partial \mathbf{x}_a - \partial J/\partial \mathbf{x}_b. \tag{9}$$

Thus, (8a) and (8b) give a method of calculating the gradient (or sensitivity) of the forecast aspect with respect to the observations $\partial J/\partial \mathbf{y}$ and the gradient of the forecast aspect with respect to the background field $\partial J/\partial \mathbf{x}_b$.

## 3 Idealized Examples of Observation Sensitivity

In this section, two idealized examples from Baker (2000) are presented to illustrate some of the basic principles of observation and background sensitivity. For the first example, consider a two-dimensional horizontal univariate (e.g., geopotential height) case. The gradient of $J$ with respect to the initial conditions $\partial J/\partial \mathbf{x}_a$ is specified on the two-dimensional domain using simple trigonometric and exponential functions and is shown in Fig. 1, where Fig. 1a has small spatial scale features and Fig. 1e has large spatial scale features. A hypothetical situation in which $\partial J/\partial \mathbf{x}_a$ straddles a coastline is simulated in Fig. 1, with a data-dense continent on the left-hand side of the figure and a data-void ocean is on the right-hand side of the figure.

The observation locations are shown by the "+" signs in Fig. 1b, c and 1f, g. An observation is assumed to be available at every gridpoint, and the observation errors are assumed to be uncorrelated. Over the continent, the observation error variance $\varepsilon_r^2 = 1.0$, while over the ocean, the observation error variance is set to $1.0 \times 10^{6.}$ As the background error variance equals 1.0, this allows the observation sensitivity (which is in observation space) to be readily contoured. The background error covariance $\mathbf{B}$ is specified using the special Second Order Autoregressive Function (SOAR) where $\rho_b$ is the background error correlation between two analysis grid points given by

$$\rho_b(x_i, x_j, y_i, y_j) = (1 + r/L_b)\exp(-r/L_b), \tag{10}$$

$L_b$ is the background error correlation length scale, and r is the distance between grid points. The background error correlation length is specified as $L_b = 2.42\Delta x$.

The observation and background sensitivity vectors corresponding to the small-scale analysis sensitivity vector are shown in Fig. 1b, c, respectively, while the

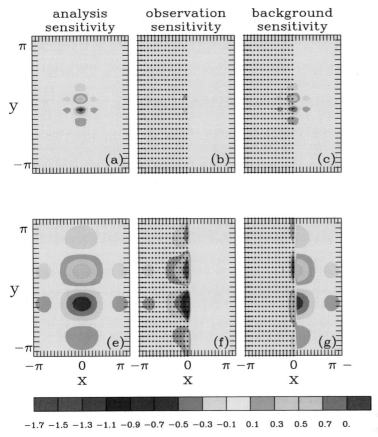

**Fig. 1** Simulated sensitivity vectors for small- (**a, b, c**) and large-scale (**e, f, g**) analysis sensitivity vectors. The imposed analysis sensitivity vectors are shown in (**a**) and (**e**), the observation sensitivity vectors in (**b**) and (**f**), and the background sensitivity vectors in (**c**) and (**g**). The observation locations are given by the "+". The color scale is indicated along the bottom. (Panels d and h are not shown). From Baker and Daley (2000)

observation and background sensitivity vectors corresponding to the large-scale analysis sensitivity vector are shown in Fig. 1f, g.

We first consider the well-observed continental interior portion far from the coastline. For the small-scale case (Fig. 1a, b, c), even though the analysis sensitivity gradient is well sampled in the continental interior, there is no sensitivity to the observations, only to the background. For the large-scale case of Fig. 1e, f, the observation sensitivity is the same as the analysis sensitivity in the continental interior and there is no sensitivity to the background. This apparent discrepancy occurs because the background error correlation length scale is relatively large, implying that the background errors are primarily large-scale, and the analysis will use the observations primarily to reduce the large-scale errors (Daley, 1991). Because the small-scale background errors are implicitly assumed to be relatively small, the observations have very little effect on these small spatial scales. Thus, for this example, the small-scale features of the analysis are derived primarily from the background and the large-scale features are derived primarily from the observations.

In an adjoint context, this means that the background sensitivity will be derived primarily from the small scales of the analysis sensitivity and the observation sensitivity will be derived primarily from the large scales.

The region along the boundary between the well sampled and the unsampled areas is considered next. For the small-scale analysis sensitivity gradient, the sensitivity to coastal observations is only slightly larger than in the well-sampled continental interior (Fig. 1b). For the large-scale case (Fig. 1e, f, g), the situation is completely different. In a narrow region along the coastal boundary, both the sensitivity to the observations (Fig. 1f) and the sensitivity to the background (Fig. 1g) are greater in magnitude than the analysis sensitivity (Fig. 1e) at the same gridpoint. The background sensitivity is of opposite sign to the observation and analysis sensitivities in this coastal region, which is consistent with (9). This phenomenon was defined by Baker and Daley (2000) as observation and background super-sensitivity, respectively, and can be explained with the following example.

For this experiment, the analysis sensitivity gradient (Fig. 2a) is composed of a combination of the large- and small-scale sensitivity patterns from Fig. 1. The background error correlation is modeled using the SOAR function for heights from (1) with $L_b = 3.6\Delta x$; this value was chosen as it maximizes the observation sensitivity for multiple observations. The background error standard deviation is set to 1.0 and the observation error standard deviation is set to 0.1. Twenty observations are placed at gridpoints in a "Z" shape across the centers of the analysis sensitivity gradient pattern (Fig. 2a).

The resulting analysis space projection of the observation sensitivity vector ($\mathbf{K}^T \partial J / \partial \mathbf{y}$) and the background sensitivity vector are displayed in Fig. 2b, c respectively. The largest observation sensitivity occurs for the circled observation at the lower right end of the "Z" in Fig. 2b. The effect of the observation density gradient is more pronounced when the density change occurs in a region of significant analysis sensitivity gradient amplitude (e.g., for the circled observation). When the observations extend all of the way across the analysis sensitivity gradient, the amplitude of the observation sensitivity is less (for example, for the observation indicated by the arrow). Thus, the largest observation sensitivity does not necessarily occur where the analysis sensitivity gradient is a maximum, but where there is a large change in observation density and the analysis sensitivity gradient is both large scale and sufficiently large in magnitude. This result is analogous to the coastal example in Fig. 1.

Assuming that the background field is zero and the observations equal one, the resulting two-dimensional analysis from (2) is given in Fig. 2d. The homogenous, isotropic nature of the correlation function, shown for the circled location in Fig. 2e, is evident and only a hint of the "Z" configuration of the observations can be seen.

The dependence of the univariate observation sensitivity on the observation density can be understood by graphically examining the various terms in the observation sensitivity equation, e.g.,

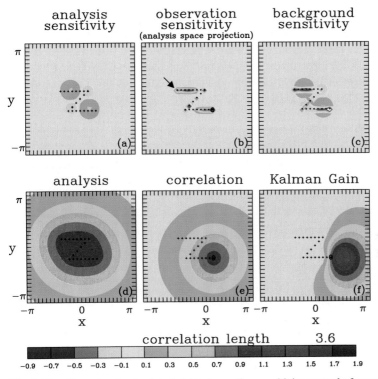

**Fig. 2** Two-dimensional univariate height observation sensitivity example for multiple observations. The largest sensitivity occurs for the height observation identified by the circle; the other height observations are given by the "+". In this plot, $L_b = 3.6\Delta x$, and $\varepsilon_r/\varepsilon_b = 0.1$. (**a**) The imposed analysis sensitivity gradient, (**b**) the analysis space projection of the observation sensitivity vector, (**c**) the background sensitivity, (**d**) the corresponding analysis, assuming an innovation of 1.0 m at each observation location, (**e**) the background error correlation function corresponding to the circled observation, and (**f**) the row of $\mathbf{K}^T$ for the circled observation. The color scale is at the bottom. From Baker (2000)

$$\partial J/\partial \mathbf{y} = \mathbf{K}^T \partial J/\partial \mathbf{x}_a = (\mathbf{HBH}^T + \mathbf{R})^{-1}\mathbf{HB}\partial J/\partial \mathbf{x}_a.$$

The term $\mathbf{HB}$ is the background error correlation between the observation locations and every gridpoint. The row of this matrix corresponding to the circled observation is shown in Fig. 2e. It is symmetric in appearance and is essentially the same for every observation location (given the constraints of a finite domain). The inverse error covariance matrix $(\mathbf{HBH}^T + \mathbf{R})^{-1}$ is in observation space and is not plotted. The term $\mathbf{HBH}^T$ is the background error correlation between observation locations. Since the observation errors are assumed to be spatially uncorrelated, the matrix $\mathbf{R}$ is simply the diagonal matrix of the observation error variances $\varepsilon_r^2$. The row of the transpose of the Kalman gain matrix ($\mathbf{K}^T$) corresponding to the circled observation is plotted in Fig. 2f. The resulting pattern is not symmetric, but has large values adjacent to the circled observation.

Plots of the appropriate row of $\mathbf{K}^T$ for all of the 20 observations are shown in Fig. 3. The most striking features are the very large lobes (in both size and magnitude) that occur for observations that are relatively isolated from their neighbors. Observations that are located near the center of the pattern have much smaller maxima and minima of the Kalman gain (in both size and magnitude). These variations in $\mathbf{K}^T$ are due to the matrix $(\mathbf{HBH}^T + \mathbf{R})^{-1}$. Observations that are farther from the

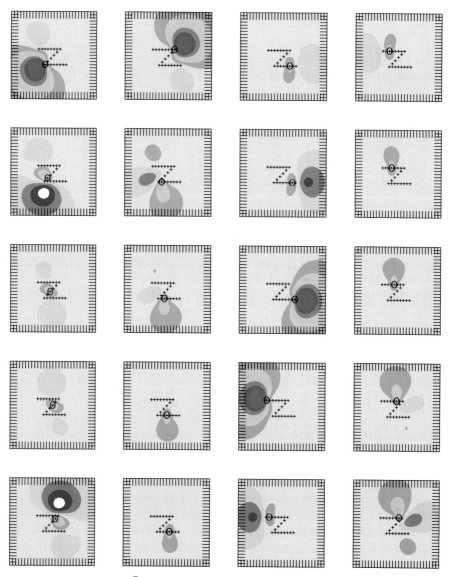

**Fig. 3** Plots of the row of $\mathbf{K}^T$ for the 20 observations shown in Fig. 2. The row of $\mathbf{K}^T$ corresponds to the circled observation in each panel. The grid domain and color scale corresponds to Fig. 2. From Baker (2000)

other observations are less correlated with them and this leads to the large asymmetry in the Kalman gain. For the forward analysis problem, this implies that isolated observations contain more independent information than observations with close neighbors and are thus given greater weight in the analysis. In the adjoint sense, this indicates that isolated observations have larger "adjoint weights" ($\mathbf{K}^T$) and potentially greater observation sensitivity.

The resulting observation sensitivity vector (8a) is the result of the matrix-vector multiplication between $\mathbf{K}^T$ and $\partial J/\partial \mathbf{x}_a$. This can be visualized for any given observation by mentally summing, gridpoint by gridpoint over the domain, the product of $\partial J/\partial \mathbf{x}_a$ (Fig. 2a) and the row of the $\mathbf{K}^T$ corresponding to that observation (Figs. 2e or 3). This exercise can be used to explain the larger observation sensitivity value for the observation at the lower-right end of the "Z" (indicated by the circle) when compared to the observation at the upper-left end of the "Z" (indicated by an arrow). The magnitude and shape of the row of $\mathbf{K}^T$ is the same for both observations. However, the circled observation is located in a region with larger values of the analysis sensitivity gradient, so that the non-zero portions of $\mathbf{K}^T$ overlap larger values of the analysis sensitivity gradient. Hence, the circled observation has the larger observation sensitivity.

It is evident that the sensitivity for a given observation depends upon the overlap between the amplitude and spatial extent (length scale) of the appropriate row of $\mathbf{K}^T$, and the amplitude and spatial extent (length scale) of the analysis sensitivity gradient $\partial J/\partial \mathbf{x}_a$. The observation sensitivity is maximized when the maxima or minima of the analysis sensitivity gradient coincide with the maxima or minima of the row of $\mathbf{K}^T$ such that the overall observation sensitivity contributions from the projection of $\mathbf{K}^T$ onto $\partial J/\partial \mathbf{x}_a$ are of the same sign.

The Kalman gain or weight matrix is arguably the most important term for both the linear analysis and observation adjoint sensitivity problems. The properties (amplitude, sign, and length scale) of this matrix are a function of several factors, with the background error correlation length scale being the dominant factor. The scale of $\mathbf{K}^T$ decreases as $L_b$ decreases, and the influence of the analysis sensitivity gradient in the immediate vicinity of the observation becomes more marked (with correspondingly less influence from the adjacent sub-structures of the analysis sensitivity gradient). In addition, neighboring observations will not contribute to the observation sensitivity at a particular location if the correlation length scale is too small. If $L_b$ is too long, then the maximum amplitude of $\mathbf{K}^T$ may not coincide with either the observations or analysis sensitivity gradient extrema (particularly for the more isolated observations).

# 4 Application to Real Problems

Observation sensitivity, while an interesting quantity in its own right in that it lends insight into how observations are used by the assimilation system, is not capable of quantifying whether a given observation/observing system improves or degrades the forecast. An observation may have large observation sensitivity, but if the innovation is zero, its impact on the resulting analysis and forecast will be minimal, other than through its effects on $\mathbf{K}^T$. Similarly, an observation may have small observation sensitivity, but that does not imply that the resulting changes to the analysis will be small, only that they will not be in the direction needed to effectively change for forecast aspect $J$. To estimate the impact of the observations on the forecast, some other measure of observation impact is required.

## 4.1 Forecast Metrics and the Cost Function J

The first step is to define the cost function $J$ or aspect of the forecast of interest. In principle, it could be defined as any first-order differentiable function of the model forecast. The tangent linear approximation generally limits the forecast length to three days or less. One frequently-used (Rabier et al. 1996, Morneau et al. 2006, Gelaro et al. 2007) forecast aspect is the quadratic measure of forecast error given by

$$e_f = \langle (\mathbf{x}_f - \mathbf{x}_t), \mathbf{C}(\mathbf{x}_f - \mathbf{x}_t) \rangle \tag{11}$$

where $\mathbf{x}_f$ is the forecast and $\mathbf{x}_t$ is the verification state, given by the analysis valid at the forecast time. The matrix $\mathbf{C}$ is symmetric matrix of weights, typically for either a dry or moist static total energy, for all model levels.

## 4.2 Observation Impact Calculations

Doerenbecher and Bergot (2001) used a linear combination of the innovation and the observation sensitivity observation impact to assess the impact of the FASTEX special observations. Subsequently, Langland and Baker (2004a) developed the following definition for observation impact.

We begin by considering two 24 hr forecast trajectories, one starting from the analysis background (nominally a 6 hr forecast from the previous update cycle), and one starting from the current analysis (t = 0 hrs) as shown in Fig. 4. The forecast error for the background trajectory is given by $e_g$, while the forecast error in the analysis trajectory is given by $e_f$. The cost function $J$ is defined to be the difference between the forecast errors ($\Delta e_f^g = e_f - e_g$), and is due solely to the assimilation of the observations at the analysis time (t = 0 hrs). In this case, $e_f$ and $e_g$ represent forecast errors of terrain pressure and temperature, wind, and humidity between the surface and the upper troposphere, about 250 hPa. Following Langland and Baker (2004a), $\Delta e_f^g$ can be estimated using the observation sensitivity gradients as

$$\delta e_f^g = \left\langle (\mathbf{y} - \mathbf{H}\mathbf{x}_b), \mathbf{K}^T \left\{ \frac{\partial e_f}{\partial \mathbf{x}_a} + \frac{\partial e_g}{\partial \mathbf{x}_b} \right\} \right\rangle, \tag{12}$$

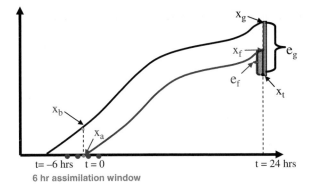

**Fig. 4** Illustrations of the two trajectories used for the calculation of the observation impact. The trajectory ending at $\mathbf{x_g}$, with error $e_g$, begins from the background $\mathbf{x_b}$, while the trajectory ending at $\mathbf{x_f}$, with error ef, begins from the analysis $\mathbf{x_a}$. Both trajectories are verified against the analysis $\mathbf{x_t}$ at t = 24 hrs. The assimilation of the observations at t = 0 hrs moves the forecast trajectory from the dark green line to the blue line. Adapted from Langland and Baker (2004a)

Equation (12) provides a single measure of the global observation impact for all observations. Negative values indicate a reduction in forecast error, while positive values indicate an increase in forecast error. The global observation impact can be easily partitioned into observation subsets, for example, for a given observing platform (e.g. radiosondes), or by satellite channel. It is important to remember that the computation of the observation impact in (12) always involves the entire set of observations, and changing the properties of even one observation changes the scalar measure for all other observations.

The accuracy of the observation impact results depends upon the accuracy of the adjoint-based calculations in (12). One way to assess this accuracy is to compare the nonlinear forecast error differences $e_f - e_g$ to those computed using adjoint techniques. In Fig. 5, the red line indicates the full nonlinear NOGAPS[3] model moist static energy error in $\mathrm{J\,kg^{-1}}$ (ordinate) for the forecasts starting from $\mathbf{x_b}$ for the period from 01 Jan 2008 to 02 Mar 2008 (abscissa). The yellow line indicates the same error metric, but for the forecast trajectory starting from $\mathbf{x_a}$. The difference in these two errors $(e_f - e_g)$ is given by the blue line, while the green line is the adjoint estimate of the error difference, as computed by (12) using the adjoint of NAVDAS[4]. The adjoint-based estimate recovers, on average, approximately 80% of the nonlinear forecast error differences. The systematic underestimation of the actual nonlinear forecast error differences $\Delta e_f^g$ by observation adjoint impact $\delta e_f^g$ is likely due to neglected moist physics in the TLM (i.e. convection), and nonlinearities in the dry dynamics. It is interesting to note how closely the adjoint-based estimates track the day-to-day variations in the nonlinear forecast error differences.

---

[3] Navy Operational Global Atmospheric Prediction System. NOGAPS is the operational spectral forecast model and is described in Hogan et al. (1999). The adjoint is described in Rosmond (1997).

[4] NAVDAS is an observation-space 3D-Var system, and is described in Daley and Barker (2001). The NAVDAS adjoint development is discussed in Langland and Baker (2004b).

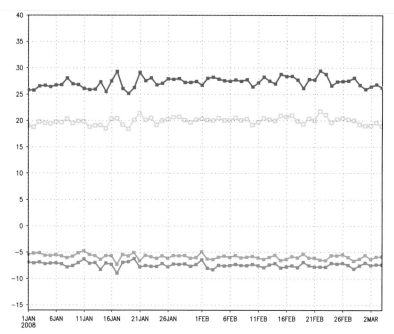

**Fig. 5** Comparison of the nonlinear forecast error for the 30 hr forecast starting from $\mathbf{x}_b$ (*red line*) and the 24 hr forecast starting from $\mathbf{x}_a$ (*yellow line*). The difference in these errors is given by the blue line, while the green line denotes the adjoint-based impact computed from (12)

## *4.3 Nonlinearity Considerations*

Errico (2007) and Gelaro et al. (2007) have examined the effects on nonlinearity on the interpretation of the partial sums used to bin the observation impact by type (platform, station, channel, etc.) Their results indicate that the observation impact measure proposed in Langland and Baker (2004a, b) is essentially equivalent to the 3rd order approximation of the change in a quadratic measure of forecast error (see above) due to the observations. Gelaro et al. (2007) demonstrated that a higher than first order approximations is required to accurately assess the observation impact. The dominant nonlinearity in (12) is apparently due to the quadratic nature of the cost function (11), although other nonlinearities arise because the second and third order terms have a dependence upon the innovations and the forecast starting from the analysis $\mathbf{x}_a$. Despite these nonlinearities, the authors concluded that observation impact measure proposed by Langland and Baker (2004a) yields reasonable estimates. However, they noted that their assessment was made using large numbers of observations, and cautioned that these conclusions may not hold for small subsets of observations, or even large subsets with comparable impact magnitudes.

# 5 Application of Observation Adjoint Techniques

The observation impact has been successfully used by scientists at the NRL and Fleet Numerical Meteorology and Oceanography Center (FNMOC) to diagnose a number of data quality issues in operational observation suite used by NAVDAS. The observation impact is routinely computed at 0000 UTC for the operational real-time NAVDAS/NOGAPS forecast run. The archived observation impact statistics have been used to identify errors in the observation meta data, such as incorrect station elevations in the Master Station List. Monitoring of these statistics also permitted the correct identification of a temporary wind direction bias of around 30° with the Lihue (91165) radiosonde observations during December 2004.

One obvious question is whether removing observations flagged as non-beneficial translates to an increase in forecast skill. The following example illustrates that this is indeed the case. During January and February 2006, the MTSAT atmospheric motion vectors (AMVs) produced by the Cooperative Institute for Meteorological Satellite Studies at University of Wisconsin were systematically providing non-beneficial impact (as indicated by the red shading) in the southern hemisphere (see Fig. 6a). The corresponding mean zonal wind innovations were also unusually large and negative (the blue shading in Fig. 6b). After reviewing the observation impact statistics, all MTSAT AMVs more that 39° from the satellite sub-point were removed from the operational assimilation run. An Observation System Experiment (OSE) was run, removing this data from the assimilation system for the period of 16 February to 27 March 2006. The 500 hPa height anomaly correlations indicated around 3 hrs forecast improvement in both hemispheres at 5 days. The data providers at CIMSSS were simultaneously contacted, and problem was quickly traced to a processing issue related to a lack of a unique time stamp for each scan-line, with a resulting in a wind speed bias that increased to the south (as a function of true scan time, Chris Velden, personal communication). CIMSS scientists rapidly imple-

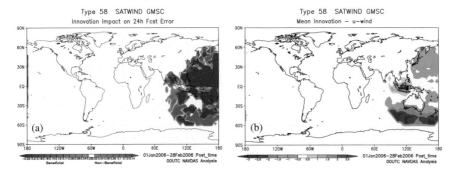

**Fig. 6** Total observation impact (*left panel*; J kg$^{-1}$) for MTSAT winds for the period from 01 January 2006 to 28 February 2006. The green shading indicates beneficial impact, while the red shading indicates non-beneficial impact. The mean zonal wind innovation (m s$^{-1}$) for the same period is shown in the *right panel*. The slow wind speed biases due to the observation processing problem are shown by the blue shading in the southern hemisphere

mented a fix to their wind processing system, and the observation impact and mean innovation values decreased to within their normal ranges (not shown).

## 5.1 Channel Selection for AIRS

The observation impact has also been used with the NASA AQUA AIRS (Advanced Infrared Sounder) radiance observations as a method to identify beneficial and non-beneficial channels for assimilation (Ruston et al. 2006). With the observation impact approach, the observation impact is generated from a several-week assimilation run using the subset of channels under consideration. The observation impact statistics are generated, and the non-beneficial observations are eliminated from the next forecast trial. This approach differs from other channel selection methods, such as entropy-reduction, adjoint sensitivity and Kalman filter sensitivity. The entropy-reduction method iteratively selects the satellite channel that most effectively reduces the background error (Rodgers, 1996; Fourrié and Rabier, 2004). The adjoint sensitivity method (Doerenbecher and Bergot, 2001) is also an iterative approach, and preferentially selects the channel with the maximum observation sensitivity. The Kalman filter sensitivity (Bergot and Doerenbecher, 2002) iteratively chooses the channel that gives the maximum decrease in the forecast error variance. Fourrié and Rabier (2004) showed that the last 3 methods produced similar results (in terms of the computed degrees of freedom for signal), even though the selected channels are not the same. In particular, the adjoint-based methods tend to favor channels with information in the sensitive regions of the lower troposphere, while the entropy-reduction methods tend to include more information from the upper troposphere.

An example of the AIRS and AMSU observation impact for the NAVDAS/ NOGAPS system for an assimilation run from 19 to 25 August 2006 is shown in Fig. 7. For this example, the initial channel set was selected based on channel usage at other operational NWP centers. The channels are grouped by frequency and labeled by channel number. Positive numbers along the abscissa indicate an increase in forecast error (in J kg$^{-1}$) due to that channel, while negative numbers indicating a reduction in forecast error. The forecast metric is the moist static energy error norm, as described earlier.

Some of the non-beneficial impact is easily explained – in the operational version of NAVDAS, only linear observation operators are allowed, and only those with sensitivity to the temperature state vector. Thus, it is not surprising that the water vapor channels increase the forecast error. The responses for AMSU channels 8 and 10 are not as readily explained, nor are the varied responses for the short wave $CO_2$ channels.

The channels that contributed to the forecast degradation were removed, along with the ozone and water vapor channels, and the assimilation run was repeated. The results are presented in Fig. 8.

Overall, the majority of the channels now act to reduce the forecast error. Interestingly, some of the channels that were previously beneficial are now non-beneficial.

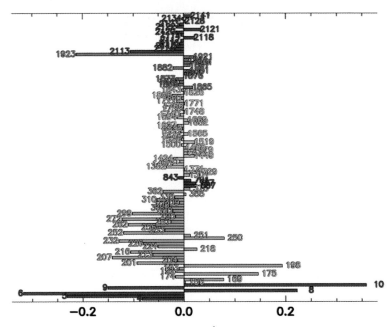

**Fig. 7** Observation impact (abscissa, in J kg$^{-1}$) computed using moist static energy-weighted error norm for the NAVDAS/NOGAPS system for August 19–25, 2006. The impact for the AIRS and AMSU channels is plotted along the ordinate, where the numbers beside the bars indicate the channel number, with color-coding as follows: short-wave $CO_2$ (*blue and gray*); water vapor (*light blue*); long wave window (*brown*), long wave $CO_2$ (*green*), and the AMSU microwave (*red*). Figure courtesy of Ben Ruston

The reason why this occurs is not known, however examining the problem in a 1D-Var context may lend some insight into its nature. It's also possible that this is an indication of the nonlinearity aspects of the problem discussed in Sect. 4.3.

## 5.2 AMSU-A Radiance Assimilation Example

The AMSU-A observation impact for NAVDAS/NOGAPS is shown in Fig. 9. The strongest forecast reductions are for the mid-tropospheric channels (AMSU Channels 5, 6, 7, 8). The most striking feature is the non-beneficial impact for AMSU channels 8 and 9 for the NRL system. This last result is contradictory with observation impact statistics from the NASA/GMAO GEOS-5 (Gelaro, personal communication), Environment Canada 4DPSAS (Pellerin, personal communication) and ECMWF 4D-Var (Cardinali, personal communication). These results suggest a problem with the assimilation of these channels in NAVDAS/NOGAPS, possibly due to the relatively low model top (effectively around 4 hPa), or insufficient model and analysis resolution near the tropopause.

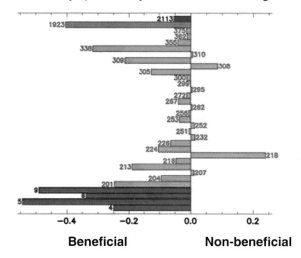

**Assessment of AQUA sensors**
AMSU/A, AIRS longwave 14-13µm,
AIRS shortwave 4.474µm, **AIRS shortwave 4.180µm**

AQUA sensitivity specified by channel number: Aug 15-26, 2006

**Beneficial**                    **Non-beneficial**

**Fig. 8** Same as Fig. 7, but for the observation impact after the non-beneficial channels were removed. Figure courtesy of Ben Ruston

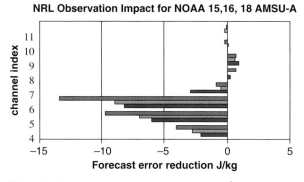

**NRL Observation Impact for NOAA 15,16, 18 AMSU-A**

**Fig. 9** Total observation impact (abscissa in J kg$^{-1}$) versus channel index (ordinate) for NOAA-15 (*blue*), 16 (*gray*) and 18 (*green*) AMSU-A observations assimilated by NAVDAS/NOGAPS for 1 January to 28 February 2006 at 00 UTC

## 5.3 Observation Impact for the Global Observing Network

The observation impact for all observations assimilated by the NAVDAS/NOGAPS system daily at 00 UTC is shown in Fig. 10 for the 30 days ending 25 September 2007. In the left panel, the observation impact is binned according to observing platform and expressed as the total impact, or reduction in forecast error, over the

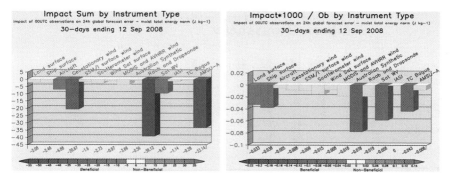

**Fig. 10** Total (*left panel*) and per ob (*right panel*) 30 day 00 UTC observation impact (J kg$^{-1}$) for the major observing platforms. Negative values (*green*) indicate a reduction in forecast error. White bars indicate essentially neutral impact. Orange bars (not present) would indicate an increase in forecast error

30 days. In the right panel, the total impact is first scaled (by 1000) and then divided by the number of observations in each bin.

In terms of total impact, for the NRL system, radiosondes are still the dominant source of information in the 00 UTC analyses, followed by the AMVs and AMSU-A radiances. The satellite total precipitable water observations rank as fourth. These results are analogous to those found with OSEs, although it should be noted that the two are not equivalent. The adjoint-based methods quantify the reduction in forecast error due to an observing system, in the context of all observations assimilated. The OSE approach gives the impact relative to a control run when an observing system is removed from the assimilation.

On a per observation basis, the two observing platforms with the most beneficial impact are the Australian surface pressure pseudo-observations (PAOBs) and the tropical cyclones pseudo-observations generated from the tropical cyclone warning messages. The total observation counts for each type are small, so it is possible that some of the nonlinear aspects discussed in 4.3 are relevant; however, these relative rankings have remained consistent over several years of monitoring.

## 5.4 Observation Impact as a Function of Observation Time

The final example (Fig. 11) is from the Environment Canada, and shows the mean Southern Hemisphere observation impact per observation for the 4D-PSAS system as a function of the time of the observation within the assimilation window. A clear trend is apparent, indicating that the observations at the end of the assimilation window have the largest impact. Preliminary results from NRL's four-dimensional data assimilation system, NAVDAS-AR (NAVDAS-Accelerated Representer, Xu and Rosmond 2005; Rosmond and Xu 2006) show a similar trend. These results are particularly relevant for satellite-data providers, as they show that reduced data latency is of critical importance.

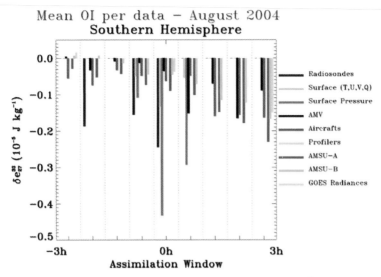

**Fig. 11** Mean observation impact per observation (ordinate in J kg$^{-1}$) for the EC 4D-PSAS system, as a function of observation time relative to the assimilation window. The observing platforms are color-coded and given in the legend. © Her Majesty The Queen in Right of Canada, Environment Canada, [2008] Reproduced with the permission of the Minister of Public Works and Government Services Canada

## 6 Concluding Remarks

The adjoint of the data assimilation system has proven to be a useful tool to help identify systematic problems with the observing network, to quantify the value of different observing platforms, and as a way to compare different data assimilation systems. One must keep in mind that the observation impact, as calculated via (12) or (14) is a very specific measure of forecast impact, as it depends upon the choice of $J$, the suite of observations assimilated, the data assimilation system, and the NWP forecast model. Despite the similar results obtained using different NWP systems, the observation impact is, strictly speaking, valid only for the data assimilation system and NWP forecast system used to compute the statistics[5]. Ultimately, that is its greatest strength, and potentially, its greatest weakness.

The ongoing intercomparisons between different operational NWP centers will help clarify the extent that the observation impact informs us about issues with the global observing network and data assimilation systems in general.

**Acknowledgements** The authors gratefully acknowledge support from the Naval Research Laboratory, under program element 062435N, project number BE-435-046, and the Joint Center for Satellite Data Assimilation. We thank our Monterey colleagues Tim Hogan (NRL), Randy Pauley (FNMOC), Ben Ruston (NRL) for their contributions. We also thank Simon Pellerin, Stéphane Laroche, Josée Morneau, and Monique Tanguay (EC) for their contributions to this paper.

---

[5] The same is true for OSEs and OSSEs (Observing System Simulation Experiments).

# References

Baker NL (2000) Observation adjoint sensitivity and the adaptive observation-targeting problem. Ph.D. dissertation, Naval Postgraduate School, 265 pp. Available from the Naval Research Laboratory, Monterey, CA 93943.

Baker NL, Daley R (2000) Observation and background adjoint sensitivity in the adaptive observation-targeting problem. Q J R Meteorol Soc 126: 1431–1454.

Bergot T (1999) Adaptive observations during FASTEX: A systematic survey of upstream flights. Q J R Meteorol Soc 125: 3271–3298.

Bergot T, Doerenbecher A (2002) A study of the optimization of the deployment of targeted observations using adjoint-based Methods. Q J R Meteorol Soc 128: 1689–1712.

Buizza R, Montani A (1999) Targeting observations using singular vectors. J Atmos Sci 56: 2965–2985.

Daley R (1991) Atmospheric data assimilation, Cambridge University Press, 457 pp.

Daley R, Barker E (2001) NAVDAS: Formulation and Diagnostics. Mon Wea Rev 129: 869–883

Doerenbecher A, Bergot T (2001) Sensitivity to observations applied to FASTEX cases. Nonlinear Proc Geophys 8: 467–481.

Errico R (2007) Interpretation of ad adjoint-derived observational impact measure. Tellus 59A: 273–276.

Fourrié NA, Rabier F (2004) Cloud characteristics and channel selection for IASI radiances in meteorologically sensitive areas. Q J R Meteorol Soc 130: 1839–1856.

Gelaro R, Langland RH, Rohaly GD, Rosmond TE (1999) An assessment of the singular vector approach to targeted observing using the FASTEX data set. Q J R Meteorol Soc 125: 3299–3327.

Gelaro R, Zhu Y, Errico R (2007) Examination of various-order adjoint-based approximations of observation impact. Meteorologische Zeitschrift, 16: 685–692.

Gelb A (1974) Applied optimal estimation. MIT Press, 374 pp.

Hogan TF, Rosmond TE, Pauley RL (1999) The navy operational global atmospheric prediction system: Recent changes and testing of gravity wave and cumulus parameterizations. Preprints, 13th Conf Numerical Weather Prediction, Denver, CO, Amer Meteorol Soc, pp 60–65.

Ide K, Courtier P, Ghil M, Lorenc AC (1997) Unified notation for data assimilation: operational, sequential and variational. J Meteorol Soc Japan, 75 1 B: 181–189.

Joly A, Jorgensen D, Shapiro MA, Thorpe A, Bessemoulin P, Browning KA, Chalon J-P, Clough SA, Emanuel KA, Eymard L, Gall R, Hildebrand PH, Langland RH, Lemaitre Y, Lynch P, Moore JA, Persson POG, Snyder C, Wakimoto R (1997) The Fronts and Atlantic Storm-track Experiment (FASTEX): Scientific objectives and experimental design. Bull Amer Meteorol Soc, 78: 1917–1940.

Langland RH, Rohaly GD (1996) Adjoint-based targeting of observations for FASTEX cyclones. Preprints, 7th Conf Mesoscale Processes, Reading, UK, Amer Meteorol Soc, pp 369–371.

Langland RH, Baker NL (2004a) Estimation of observation impact using the NRL atmospheric variational data assimilation adjoint system. Tellus, 56A: 189–201.

Langland, RH, Baker NL (2004b) A technical description of the NAVDAS adjoint system. NRL/MR/7530-04-8746. Available from the Naval Research Laboratory, Monterey, CA, 93943, 62 pp.

Morneau J, Pellerin S, Laroche S, Tanquay M (2006) Estimation of the adjoint sensitivity gradients in observation space using the dual (PSAS) formulation of the Environment Canada operational 4D-Var. Proceedings, 2nd THORPEX Intl Science Symp, Landshut, Germany, 4–8 December 2006, pp 162–163.

Palmer TN, Gelaro R, Barkmeijer J, Buizza R (1998) Singular vectors, metrics and adaptive observations. J Atmos Sci 55: 633–653.

Rabier F, Klinker E, Courtier P, Hollingsworth A (1996) Sensitivity of forecast errors to initial conditions. Q J R Meteorol Soc, 122: 121–150.

Rodgers, C. D. (1996) Information content and optimization of high spectral resolution measurements. Optical Spectroscopic Techniques and Instrumentation for Atmospheric and Space

Research II, SPIE Vol. 2830, 136–147. Published by the International Society for Optical Engineering, PO Box 10, Bellingham, Washington, 98227-0010, USA, 2830.

Rosmond TE (1997) A technical description of the NRL adjoint modeling system, NRL/MR/7532/97/7230 Available from the Naval Research Laboratory, Monterey, CA 93943, 55 pp.

Rosmond T, Xu L (2006) Development of NAVDAS-AR: non-linear formulation and outer loop tests. Tellus, 58A, 45–58.

Ruston B, Blankenship C, Campbell W, Langland R, Baker N (2006) Assimilation of AIRS data at NRL. 15th Intl TOVS Study Conf, Maratea, Italy, 4–10 October 2006.

Xu L, Rosmond R (2005) Development of NAVDAS-AR: formulation and initial tests of the linear problem. Tellus, 57A, 546–559.

# Impact of the CHAMP Occultation Data on the Rainfall Forecast

Hiromu Seko, Yoshinori Shoji, Masaru Kunii and Yuichi Aoyama

**Abstract** Impacts of the occultation data observed by CHAMP were investigated using the mesoscale four-dimensional data assimilation system of the Japan Meteorological Agency. We assimilated refractivity data provided from GFZ by using the method of Chen et al., in which the vertical correlation of the observation error was considered. This method was extended to apply the refractivity average along the path from a GPS satellite to a LEO satellite. When this extended method was applied to predict a rainfall system that developed in northern Japan, the rainfall system, which could not be developed without the assimilation of the occultation data, was reproduced. This result demonstrated that occultation data is useful for improving heavy rainfall forecasts.

## 1 Introduction

In general, low-level humid airflows are supplied into the rainfall systems when heavy rainfalls occur. Because the rainfall amount and intensity are largely influenced by water vapor distribution (i.e., horizontal extent of low-level humid region and vertical profile of water vapor), the three-dimensional distribution of water vapor is essential for improving rainfall forecasts. Due to recent developments of satellite observation techniques (e.g., TRMM Microwave Imager (TMI) and Global Positioning System (GPS)), the horizontal distribution of precipitable water vapor (PWV), which is defined as the water vapor amount in the vertical column, can be obtained. Improvement of rainfall forecast by assimilation of PWV data has been reported thus far. Nakamura et al. (2004) found that GPS-derived PWV data improved the rainfall forecast in some cases when they were assimilated into the initial fields of the Mesoscale Model[1] (MSM) (JMA 2002) of the Japan Meteorological Agency (JMA). Koizumi and Sato (2004) also confirmed that GPS-derived PWV data on

H. Seko (✉)
Meteorological Research Institute, 1-1 Nagamine, Tsukuba, Ibaraki 305-0052, Japan

[1] MSM was changed from a hydrostatic version to non-hydrostatic version in September 2004. The hydrostatic version of MSM was used in this study.

S.K. Park, L. Xu, *Data Assimilation for Atmospheric, Oceanic and Hydrologic Applications*, DOI 10.1007/978-3-540-71056-1_10,
© Springer-Verlag Berlin Heidelberg 2009

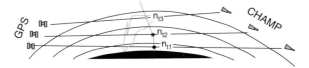

**Fig. 1** Schematic illustration of occultation. Dots denote tangent points on the paths from the GPS satellites and the CHAMP satellite

land and TMI-derived PWV data on the sea complementarily improved the rainfall forecast of the model. Besides PWV, vertical profiles of refractivity that are obtained by occultation observation are also expected to improve the rainfall forecast because refractivity is a function of temperature and water vapor.

In occultation observation, a low Earth orbit (LEO) satellite receives signals transmitted from GPS satellites. The altitude through which the signal passes depends on the positions of the GPS and LEO satellites (Fig. 1). For example, the signal path slices the lower atmosphere as the GPS satellite viewed from the LEO satellite declines to the horizon. Because signals pass various altitudes of the atmosphere, they are exploited for estimating the refractivity profile. In contrast, PWV data observed by TMI and ground-based GPS do not provide information about the vertical distribution of water vapor. Providing vertical water vapor profiles is one of the merits of occultation data. It has been reported that the upper dry atmosphere enhances convections even if the temperature is warmer than surrounding areas (e.g., Kato 2006); therefore, the occultation data, which provides the vertical water vapor distribution, is expected to be useful for improving rainfall forecasts.

Radio signals transmitted from GPS satellites are delayed and bent as they pass through the atmosphere. The general analysis procedure of occultation data uses bending angles of paths that are obtained from the temporal variation of signal delays observed by LEO satellites. A refractivity profile is calculated from the bending angles of the paths with the assumption of uniform distribution of refractivity. Accuracy of the vertical refractivity profile has been reported thus far. Wickert et al. (2004) and Kuo et al. (2004) compared the observed refractivity profiles with the analyses from European Center of Medium-Range Forecasts (ECMWFs) or the data from the global radio sonde network and found a negative bias in the lower troposphere, especially in the tropical region. Chen et al. (2006) also investigated the error of refractivity data caused by the assumption of the horizontally uniform distribution of refractivity. Regarding the assimilation experiments using occultation data, Ozawa et al. (2005) demonstrated that the numerical forecast was improved when the bending angles of the paths were assimilated into the Global Spectrum Model (GSM) of JMA. Based on their results, occultation data of the Challenging Mini-satellite Payload (CHAMP) began to be assimilated operationally into GSM in March 2007 at JMA.

However, few assimilation experiments using the meso-scale model have been performed thus far. Huanga et al. (2005) investigated the influences of GPS occultation refractivity data on the simulations of typhoons, which passed Taiwan (Typhoon Nari in September 2001 and Typhoon Nakri in July 2002). They found that forecasts of the typhoon track and accumulated rainfall improved when the CHAMP data were assimilated by the 3d-Var data assimilation system. Besides improving the forecast of typhoons, water vapor fields modified by the assimilation of occultation data are expected to improve the forecast of mesoscale phenomena. Thus, the data assimilation experiments of other phenomena (e.g., heavy rainfall associated with the Baiu front) by using occultation data are desired.

In this study, the impact of the refractivity profile on the rainfall system developed along the Baiu front was investigated by the Meso-scale 4-dimensional Data Assimilation System (hereinafter abbreviated as Meso 4D-DAS) of JMA (Koizumi et al. 2005), which has been developed for the numerical prediction of severe weather phenomena (e.g., heavy rainfalls). The vertical profile of the refractive index, which is defined as (refractivity-1)$\times 10^6$, provided by the Information Systems and Data Center (ISDC) of GeoForschungsZentrum Potsdam (GFZ) was used as the assimilation data. This refractive index data is referred to as CHAMP/GFZ data. The vertical resolution of CHAMP/GFZ data is as high as 200 m. In general, the observation error of high-resolution data is correlated. Thus, high-resolution data should be thinned out, or its correlation should be considered in the assimilation. In addition to the vertical correlation of observation error, it should be noted that the refractive index data provided by ISDC is reported with locations of the ray path tangent point, of which positions are closest to the Earth in the path of a GPS satellite and CHAMP. CHAMP observes the integrated signal delay along the path from GPS satellites to CHAMP, not the tangent point value (Fig. 1). The refractive index at a tangent point is obtained from the integrated signal delay along the path by assuming that the refractivity is distributed uniformly in the horizontal direction. This study will demonstrate the validity of this assumption, in addition to the influence of thinning on the assimilation. Furthermore, the assimilation methods by which these refractivity characteristics were considered are introduced, and the impact of occultation data is investigated.

Section 2 presents the statistical values of refractive index data (e.g., bias and RMS). Section 3 discusses the rainfall event to which the assimilation experiment was applied. The impact of the conventional data is also presented. Section 4 explains the assimilation methods by which the vertical correlation of observation error and the horizontally uneven distribution of refractivity were considered. Section 5 presents the assimilation results using these methods. Section 6 is a summary.

## 2 Statistical Features of Occultation Refractive Index Data

First, the bias and RMS of CHAMP/GFZ data were examined. Figure 2a presents the distribution of tangent points of the lowest paths (see Fig. 1) observed from 1 to 31 July 2004. Because this period was only one month, the tangent points of CHAMP/GFZ data were not distributed uniformly. However, the occultation data were expected to be useful assimilation data since many tangent points exist over the ocean, where observation data are much fewer than over the land. The bias and

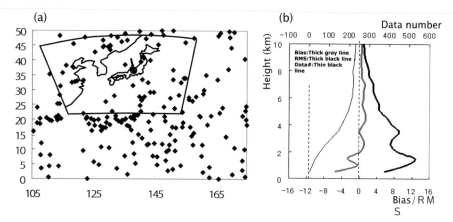

**Fig. 2** (**a**) Distribution of refractive index data provided by CHAMP/ISDC of GFZ. (**b**) Vertical profiles of bias and RMS of the D-value. The thin line indicates the total number of occultation data points

RMS of the refractive index were estimated from the difference between the observed refractive index and first-guess data (D-value). In this study, the first-guess data was produced from the output of the MSM of JMA. The horizontal grid interval of MSM was 10 km. The simplified Arakawa-Schubert scheme and large-scale condensation were used as convective parameter schemes in MSM. First-guess values of the refractive index at the tangent point were obtained from temperature, water vapor, and pressure and produced by spatial interpolation of the grid point values of MSM.

Bias and RMS profiles of D-value in the fan-shaped area in Fig. 2a, which is a forecast domain of MSM, are presented in Fig. 2b. The number of tangent data points became fewer as altitude decreased because the signal from the GPS satellite became weaker as the path became lower. The amount of data under a height of 2 km became less than 50 profiles. Besides the insufficient amount of data, the D-value had a large negative bias under the height of 2 km, as previous studies pointed out (e.g., Wickert et al. 2004, Kuo et al. 2005). For these reasons, tangent data points below 2 km were not used as assimilation data in this study. The RMS of D-values became larger as the height of the tangent point became lower, fluctuating especially below a height of 4 km.

Next, the bias and RMS of CHAMP/GFZ were compared with other occultation data. Figure 3 reveals the bias and RMS of CHAMP and Satelite de Aplicaciones Cientificas-C (SAC-C) refractive index data provided by the Global Environmental & Earth Science Information System (GENESIS) of the Jet Propulsion Laboratory (JPL). These data are hereinafter referred to as CHAMP/JPL and SAC-C/JPL. The areas and periods of these data were the same as those of CHAMP/GFZ. The biases of CHAMP/JPL and SAC-C/JPL were larger than that of CHAMP/GFZ, and the negative bias existed at a height below 5 km for CHAMP/JPL and below 7 km for SAC-C/JPL. It has been reported that the negative biases were caused by

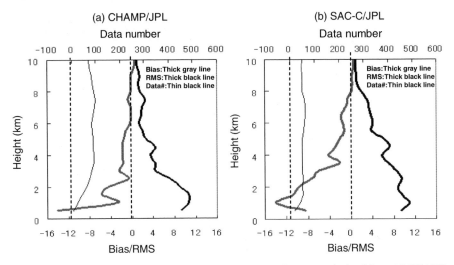

**Fig. 3** Vertical profiles of bias and RMS of the D-value. D-values were obtained from (**a**) CHAMP data and (**b**) SAC-C data provided by JPL. The thin line indicates the total number of occultation data points

multi-path phenomena (e.g., Kuo et al. 2005). In the multi-path phenomena, signals transmitted from GPS satellites earlier reached LEO satellites with regular signals, after passing through a super-refractivity region or reflecting from the sea surface (Sokolovskiy 2003). These signals contaminated the information of the regular signals and caused the negative bias. Large bias existed in CHAMP/JPL and SAC-C/JPL data of July 2004 because refractivity data were estimated geometrically, while the Full Spectrum Inversion (FSI) method was used in GFZ to remove the influence of the multi-path signals by using the received signal intensity. Based on these comparisons, this study used the CHAMP/GFZ as assimilation data.

# 3 Rainfall Event of Assimilation Experiment and Assimilation Results of Conventional Data

## 3.1 Rainfall Event on 16 July 2004

The occultation data were assimilated to the rainfall event in the assimilation experiments. On 16 July 2004, the Baiu front, which is a stationary rainy front extending from China to Japan in early summer, crossed northern Japan (Fig. 4a). Along the Baiu front, a well-developed cloud region was clearly seen in northern Japan in an IR image of 1500 JST (Japan Standard Time; 0000 UTC corresponds to 0900 JST) (Fig. 4b). Figure 5 indicates the three-hour rainfall amount estimated from the reflectivity intensity data of conventional radar and the precipitation amount from rain

(a)                                        (b)

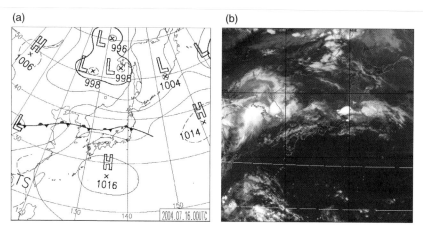

**Fig. 4** (a) Surface weather chart at 0900 JST on 16 July 2004 and (b) IR images observed by GMS at 1500 JST on 16 July 2004

gauge network data. The intense rainfall region extended from northwest to south-east under the well-developed cloud region. At Ohyu, located in the intense rainfall region (cross in Fig. 5), daily rainfall of 143 mm was recorded by the Automated Meteorological Data Acquisition System (AMeDAS). Besides this intense rainfall, weak rainfall regions were generated along the western side of northern Japan. Be-cause the weak rainfall regions were in the mountain regions, it was deduced that these regions were produced by the topography. After 2100 JST, the intense rainfall region moved to the eastern side of Japan, and then the rainfall intensity in northern Japan became weaker.

When the intense rainfall in northern Japan was maintained, the tangent point of CHAMP data passed on the southern side of the intense rainfall region at 1206 JST. Figure 6 indicates the observed refractive index profile of CHAMP/GFZ and

**Fig. 5** Three-hour rainfall distribution at 1800 JST. Rainfall was estimated from the conventional radar reflec-tivity by calibrating with rain gauge data. The circle denotes the position of the tangent point, and the cross denotes the position of Ohyu

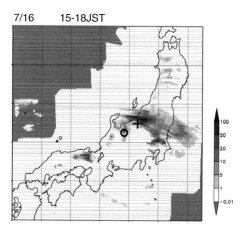

7/16      15-18JST

**Fig. 6** Vertical profile of refractive index at the tangent point, which was located in northern Japan (denoted by a circle in Fig. 5). The black line indicates the profile observed by CHAMP at 1206 JST on 16 July 2006. The gray line is the first-guess profile obtained from the forecast of MSM

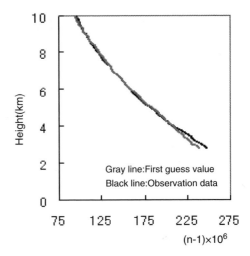

the corresponding first-guess profile. Above a height of 4 km, the first-guess refractive index, which was produced from the output of MSM, was similar to the observed one. However, the first-guess index below 4 km was smaller than that of the observed value. Because the refractive index in the lower atmosphere was more influenced by water vapor than by temperature, this difference indicated that the water vapor of MSM was drier than that of the actual one. Thus, it was expected that assimilation of this refractive index profile would increase the rainfall amount near the tangent point.

## 3.2 Assimilation Results of Conventional Data

Meso 4D DAS of JMA was used as the data assimilation system of this study. Because the microphysical processes of Meso 4D DAS were the same as those of MSM, the assimilation results that were not influenced by the difference of physical processes (e.g., microphysical process, radiation schemes) directly revealed the impact of the occultation data.

Before describing the assimilation method of occultation data, the rainfall regions predicted from the analyzed fields that were obtained by assimilation of conventional data are explained. In this study, surface meteorological data including ships and buoys, upper sounding data, Aircraft Communication and Recording System (ACARS), and Advanced TIROS Operational Vertical Sounder (ATOVS) data were used as conventional data. Because the occultation of Fig. 6 occurred at 1206 JST, the conventional data from 0900 JST to 1500 JST were assimilated.

When the conventional data were assimilated, the rainfall regions in the mountain area along the western side of northern Japan were reproduced (Fig. 7). However,

7/16        15-18JST           (FT=3)

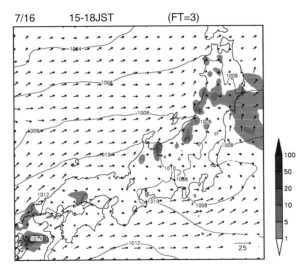

**Fig. 7** Three-hour rainfall distribution at 18000 JST (Forecast time=3). Initial fields were obtained by assimilation of the conventional data. Contour lines denote sea-level pressure, and vectors indicate horizontal wind at the lowest layers of MSM

the intense rainfall region crossing northern Japan was not reproduced. This failure indicated that the conventional data could not adequately reproduce the intense rainfall region in northern Japan. Hereinafter, this result is referred to as CNTL.

# 4 Assimilation Methods of Refractivity Data

## *4.1 Brief Description of Assimilation Methods*

The refractive index assimilation experiments in this study are listed in Table 1. To demonstrate the usefulness of occultation data, the refractive index data was added to conventional data in the assimilation. The impact of the occultation data was evaluated by comparing the rainfall regions predicted from the analyzed fields with the observed rainfall distribution. Namely, when the rainfall distribution became close to the observed distribution, it was deduced that the occultation data had a positive impact on the rainfall forecast.

As mentioned in the introduction, a few aspects of the occultation data assimilation method need improvement. One concern is the vertical correlation of the observation error. The resolution of the tangent point data provided by ISDC/GFZ is as high as 200 m. Therefore, it stands to reason that the observation error should have a vertical correlation that cannot be ignored. In general, the high-resolution data is thinned out to reduce the correlation of the observation error. However, the thinning might remove the tangent point data that improve the rainfall forecast.

**Table 1** List of assimilation experiments

| Experiment name | Assimilation data | Refractivity index data | Assimilation method |
|---|---|---|---|
| CNTL | Conventional | – | – |
| HT | Conventional data refractivity index | Tangent point | High resolution |
| LT | Conventional data refractivity index | Tangent point | Thinning |
| VCT | Conventional data refractivity index | Tangent point | High resolution vertical correlation |
| HP | Conventional data refractivity index | Path | High resolution |
| VCP | Conventional data refractivity index | Path | High resolution vertical correlation |

For example, if the low-altitude data whose refractivity was much larger than the first-guess data in Fig. 6 is removed, the impact of the refractive index data is likely to become smaller. To simulate the tangent point data without omitting any refractive index data, the observation error covariance, which expresses the relationship between other layers' errors, was estimated; the tangent point data were then assimilated using this covariance of observation error. Assimilation experiments using the thinned tangent point data or the error covariance were performed to determine the influence of the vertical correlation of the tangent point data. These experiments are referred to as HT (High-resolution Tangent point data), LT (Low-resolution Tangent point data), and VCT (Vertical-Correlation-considered Tangent point data).

Another concern is the uneven distribution of refractivity. The general method for estimating refractivity profiles at tangent points is explained below. In the estimation of tangent point data, bending angles of the path from a GPS satellite to a LEO satellite are used as the input data. A bending angle is a sum of the local bending angles caused by vertical gradients of refractivity along the ray-path. To obtain the local bending of the tangent point, bending angles at other points must be subtracted from the bending angle of the path. In this step of the estimation, refractivity is assumed to be uniformly distributed. Namely, the refractivity gradients along the path of which the altitudes are higher than its tangent point are assumed to be equal to the gradients at the same altitude's tangent points estimated from the upper path data, when they are subtracted from the bending angle of the path.

Because of the influence of weather disturbances (e.g., rainfall systems) or latitudinal variation of temperature or water vapor, this assumption is not always satisfied. This expectation can be confirmed by using first-guess data because the first-guess fields express weather disturbances and latitudinal variations. Figure 8a depicts the vertical position of the lowest path, the refractivity profile of which is plotted in Fig. 6. The height of the tangent point, which was the lowest point of the path, was 2800 m. The altitude of the path increased as the point in the path became farther from the tangent point. The path length below a height of 10 km was 600 km (range of ir was from −30 to +30). The azimuth direction of the lowest path was

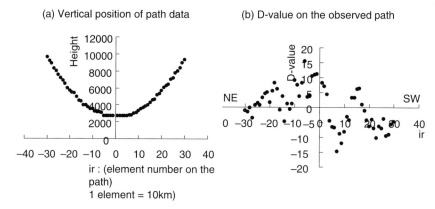

**Fig. 8** (**a**) Vertical positions of the lowest paths observed at 1206 JST on 16 July 2006. (**b**) Difference of refractivity between the observed tangent point profiles and the first-guess values along the lowest path

213 degrees. Thus, the northeastern part of the path penetrated the intense rainfall region in northern Japan. The difference between the observed tangent point refractivity profiles and the first-guess refractivity is depicted in Fig. 8b. The difference became smaller as the point in the path receded from the tangent point, because the absolute values of both refractive indexes became smaller at the upper atmosphere. The difference near the tangent point was asymmetrical and fluctuated greatly because the northeastern part of the path passed the rainfall region. This fluctuating difference indicated that the assumption of uniform refractivity distribution was not satisfied in this event. Therefore, the path data, which allowed the uneven distribution of refractivity, was reproduced from the tangent point data, and was then assimilated. The procedures for reproducing the path data are described in the next section.

When path data, of which paths passed through each tangent point, were reproduced, vertical resolution of the path data was set to 200 m. Thus, the vertical correlation of the path data must also be considered. In addition to the experiments using tangent point data, two experiments, HP (high-resolution path data) and VCP (vertical-correlation-considered path data), were performed.

## 4.2 Detailed Description of Individual Methods

*HT Case:* The refractive index profile at the tangent points with a vertical interval of 200 m, provided by GFZ, was assimilated. The observation error covariance of the HT case was a diagonal matrix because the vertical correlation of the observation error was not considered. Diagonal elements were produced from the simplified vertical profile of the RMS of the D-value (Fig. 1). Because the RMS of the D-value consisted of the observation error and forecast error, the observation error of each level was estimated by assuming that the observation error was equal to the forecast error. This observation error is called the Tangent Point Observation Error (TPOE). The experiments in this study used only one occultation profile; therefore, the cost function of occultation data was much less than that of conventional data. Because one purpose of this study was to demonstrate the potential of occultation data to improve the rainfall forecast, TPOEs that were multiplied by 0.1 were used as the diagonal elements of the observation error covariance. The observation operator, which is the relationship between the grid points of MSM and the observation data, was produced by following two steps. First, the temperature and mixing ratio of water vapor at the tangent points were obtained by spatial interpolation. Because the occultation occurred at 1206 UTC, the grid point data of 1200 UTC were used in the interpolation. Second, the refractive index ($N$) at the tangent points was produced from the interpolated values by using

$$N = 77.6 \frac{P}{T} + 3.73 \times 10^5 \frac{e}{T^2} \tag{1}$$

where $T$ is temperature (K), $e$ is partial pressure of water vapor (hPa), and $P$ is pressure (hPa).

*LT Case:* Thinning was performed so that the vertical resolution became 1 km. In general, the water vapor of the lower layer influences the rainfall amount more sensitively. Thus, the refractive index at the lowest level was not removed when the data was thinned out. Because it was expected that the impact of the assimilation would become weaker when the tangent point data were thinned out, TPOE multiplied by 0.01 was used as the diagonal element of the observation error covariance of the LT case. The observation operator was the same as that of the HT case , except for the vertical resolution of the tangent point data.

*VCT Case:* The vertical correlation of the observation error was estimated by following Chen et al. (2005). In their method, the difference between the observed refractive index and the corresponding first guess ($DN_{obs(t)-mdl(t)}$) consists of two deviations, the observation error ($DN_{obs\_err}$) and forecast error ($DN_{mdl(t')-mdl(t'-12)}$). The deviation originating from the forecast error was obtained from the outputs of the numerical model with the NMC method (Parish and Deber 1992). The difference between the two numerical forecasts of which the initial times lagged by 12 hours was assumed to be produced by the forecast error, and the deviation of the forecast time $t$ was estimated by the temporal interpolation of the difference. Deviation due to observation error was obtained by subtracting this forecast deviation from $DN_{obs(t)-mdl(t)}$

$$x_i = DN^i_{obs\_err} = DN^i_{obs(t)-mdl(t)} - \frac{t}{12} DN^i_{mdl(t')-mdl(t'-12)} \tag{2}$$

where $i$ is the vertical level of the tangent point profiles. The thickness of each vertical level in this study was 200 m, which was the same as the vertical resolution of CHAMP/GFZ data.

Vertical correlations of the observation errors were produced using

$$r_{k_1 k_2} = \frac{\overline{x_{k_1} x_{k_2}} - \overline{x_{k_1}}\,\overline{x_{k_2}}}{\sqrt{\overline{x_{k_1}^2} - \left(\overline{x_{k_1}}\right)^2}\sqrt{\overline{x_{k_2}^2} - \left(\overline{x_{k_2}}\right)^2}} \tag{3}$$

where $r$ is the correlation between levels $k_1$ and $k_2$. The bars over the values indicate the average of all tangent profiles from 1 to 31 July 2004, including the rainfall event of this study. Figure 9a illustrates the vertical correlation during this period within the domain of MSM. Horizontal and vertical axes are the relative number of vertical levels from the lowest levels of tangent point profiles. The top levels in both axes correspond to an altitude of 10 km. The correlation of lower levels (circles in Fig. 9a) was abnormally high because of insufficient data. The abnormally large data were removed, and the vertical correlation was simplified by averaging the correlation with the coordinate of the difference of levels (Fig. 9b). The observation error covariance was produced from the simplified correlation by multiplying the observation errors of two levels (Fig. 9c);

$$\mathbf{R} = \begin{pmatrix} r_{11}\sigma_1^2 & r_{21}\sigma_1\sigma_2 & \cdots & r_{n1}\sigma_1\sigma_n \\ r_{12}\sigma_1\sigma_2 & r_{22}\sigma_2^2 & \cdots & \vdots \\ \vdots & \vdots & \ddots & \vdots \\ r_{1n}\sigma_1\sigma_n & \cdots & \cdots & r_{nn}\sigma_n^2 \end{pmatrix} \tag{4}$$

where $\sigma_k$ is the observation error of level $k$. In this case, TPOE multiplied by 0.1 was used as $\sigma_k$.

Because the observation error covariance matrix was smaller than $50 \times 50$, the inverse of covariance matrix, which was used in the cost function, could be solved directly. The relationship between the tangent point and model grid points was the same as in the HT case. Thus, the observation operator of the HT case was used.

*HP Case:* The refractive index averaged along the path was used as the observation data in the HP case. Path data allowed uneven distribution of the refractive index along the path and was produced from the tangent point refractivity data by

(a) Correlation of observation error      (b) Simplified correlation of obs. error      (c) Simplified covariance of obs. error

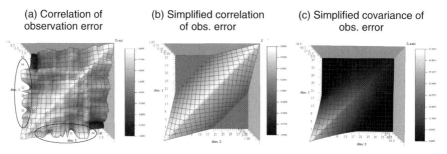

**Fig. 9** (**a**) Vertical correlation of observation error of tangent point data, (**b**) simplified vertical correlation of observation error, and (**c**) simplified covariance of tangent point data. Correlation and covariance were produced from the observation deviations from 1 to 31 July 2004. Horizontal and vertical axes are the relative number of vertical levels from the lowest levels of tangent point profiles. Top levels in the axes correspond to an altitude of 10 km

$$n_{path} = \left. \sum_{tp=lowest\_level\_of\_data}^{10\,km\_level\_of\_data} l_{tp}n_{tp} \middle/ \sum_{tp=lowest\_level\_of\_data}^{10\,km\_level\_of\_data} l_{tp} \right. \tag{5}$$

where $n_{tp}$ is the refractive index and $l_{tp}$ is the path length within the depth of each level. Because water vapor that influenced the refractive index was very small at an altitude of 10 km, the top altitude of the path in the estimation of the average refractive index was set to 10 km. Because the bending angle of the path was very small, the positions of paths were assumed to be straight lines that penetrated the tangent points, although the bending angle is generally used as the input data in estimating tangent point data.

The observation operator was also replaced with that of the path data. The path data of the refractive index (i.e., the refractive index averaged along the path) was produced by the following procedures. First, refractive indexes at grid points of MSM were produced from the outputs of MSM (i.e., temperature, mixing ratio of water vapor and pressure) using Eq. (1). The path from a tangent point to both ends of the path, of which the altitudes were 10 km, was divided into elements with lengths of 10 km, so that the tangent point was located at the center of the lowest elements. The refractive index at the center of the elements was obtained from the grid point refractive index by spatial interpolation and then averaged along the path.

Figure 10 depicts the vertical profiles of the observed path data and first-guess path data. The absolute value of the refractivity became smaller than that of the tangent point because the small refractive index at the upper tangent point was included in the path data. As for the difference between path data, the observation data below a height of 5 km was larger than first-guess data, as it was in tangent point data; thus, assimilating this path data was expected to increase rainfall. The observation error for the path data was determined in the same manner as for the path refractive index data. Namely, the observation error of path data was estimated by the path-length

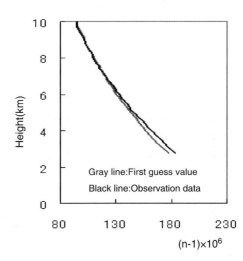

**Fig. 10** Vertical profile of the path data. Path data was the refractive index average along the path. Vertical axis indicates the height of tangent point of the paths

weighted average of the observation error of the tangent point. The observation error obtained by this method is called as the Path Observation Error (POE). In this case, the vertical correlation of path data was not considered. Thus, POE multiplied by 0.1 was used as the diagonal element of the observation error covariance.

*VCP Case:* Path data indicating which paths passed through the tangent points were produced by the same procedure as in the HP case. Because the vertical resolution of the path data was as high as 200 m, the vertical correlation of observation error was considered. The vertical correlation of the observation error of path data was determined in the same manner as in the VCT case (Fig. 11). The correlation decreased as the difference of levels increased. Compared with the correlation in the VCT case, the vertical correlation was larger and smoother. In the path data, the refractive index data of the upper atmosphere, where refractivity was distributed more uniformly, was included in the path data. This caused differences in the vertical correlation. POE multiplied by 0.1 was used as the observation error in Eq. (4). The observation operator was the same as that in the HP case.

(a) Correlation of observation error          (b) Simplified covariance of obs. error

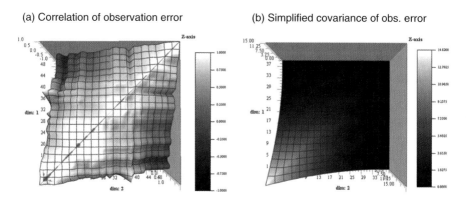

**Fig. 11** Same as Fig. 9a and c except for the path data

# 5 Results of the Assimilation Experiments

## 5.1 Rainfall Distribution Predicted from the Assimilated Fields

This section explains the results of the aforementioned assimilation methods by presenting rainfall distributions that were predicted from the assimilated fields. As explained in Sect. 3.2, conventional data could not adequately reproduce the intense rainfall region crossing the northern part of Japan. If the rainfall distribution predicted from the assimilated fields were similar to the observed distribution, it could be concluded that the occultation data was useful for improving the rainfall forecast and that the assimilation method was appropriate for occultation data.

*HT Case*: As mentioned in Sect. 4.2, the impact of occultation data in this study was expected to be small because only one data profile was assimilated. Thus, TPOE, which was produced from the simplified D-value, was multiplied by 0.1 and then used as the diagonal element of observation error covariance. When the tangent point data were assimilated with this observation error, the rainfall region that crossed the northern part of Japan was increased slightly, although the rainfall region was still smaller than the observed ones (Fig. 12). This result indicated that occultation refractivity data had the potential to reproduce the rainfall.

*LT Case:* The rainfall region was slightly improved in the HT case. Its impact was expected to become smaller when the tangent point data was thinned. Thus, TPOE multiplied by 0.01 was used as the diagonal element of observation error covariance to determine the influence of thinning more clearly, although this small observation error was not desirable because it makes the observation data seem incredibly accurate. When all tangent data were assimilated using this small observation error, the rainfall regions became larger than that of the HT case (Fig. 13a). When the tangent point data was thinned out so that its vertical resolution became 1 km, the rainfall regions became smaller (Fig. 13b). This result indicated that the thinning, which is generally performed to reduce the correlation of the observation error, was not suitable for the occultation data.

7/16 15-18JST

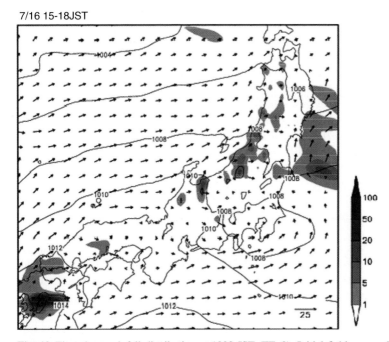

**Fig. 12** Three-hour rainfall distribution at 1800 JST (FT=3). Initial fields were obtained by assimilation of the conventional data and tangent point data. TPOE multiplied by 0.1 was used as the diagonal element of observation error covariance. Contour lines denote sea-level pressure, and vectors denote horizontal wind at the lowest layers of MSM

(a) 7/16 15-18JST (200 m)            (b) 7/16 15-18JST (1 km)

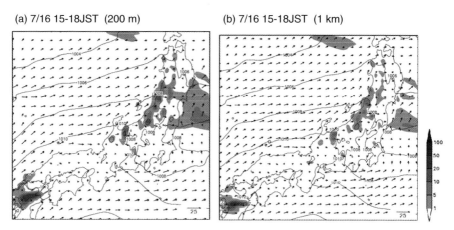

**Fig. 13** Same as Fig. 12 except for the observation error and the vertical resolution of refractive index data. TPOE multiplied by 0.01 was used as the diagonal element of the observation error covariance. Tangent point data of (**b**) were thinned so that the vertical resolution became 1 km

*VCT Case:* In this case, the vertical correlation of the observation error was considered by introducing the off-diagonal elements in the observation error covariance matrix, the production of which was explained in Sect. 4.2. TPOE multiplied by 0.1 was used as the observation error in Eq. (4). When the vertical correlation was introduced, the intense precipitation region in northern Japan was adequately reproduced (Fig. 14). This large impact indicated that introducing the observation error covariance that expressed the vertical correlation was useful for reproducing the intense rainfall region in this case study.

*HP Case:* In the HP case, the path data was assimilated without considering the vertical correlation of the observation error. As explained in Sect. 4.2, POE was produced by averaging TPOE with the weight of the path length in each level. This POE was also multiplied by 0.1 and then used as the diagonal element of observation error covariance. When the path data was assimilated, the impact became weaker than that of the HT case (Fig. 15). It was deduced that the information of low-level atmosphere that moistened the air was diluted. Although the path data seemed to be suitable for the occultation data, this method did not improve the rainfall forecast of this case study.

*VCP Case:* POE multiplied by 0.1 was used as the observation error in Eq. (4) for estimating the observation error covariance. When the path data was assimilated, the intense rainfall region in northern Japan was reproduced where it was observed. Although the simulated rainfall intensity in northern Japan was slightly greater than the observed intensity, the simulated rainfall regions became similar to the observed one (Fig. 16). Because the characteristics of occultation data (i.e., large vertical correlation of observation error and the integrated data of refractivity) were considered, this method was the most appropriate in this study for the occultation data.

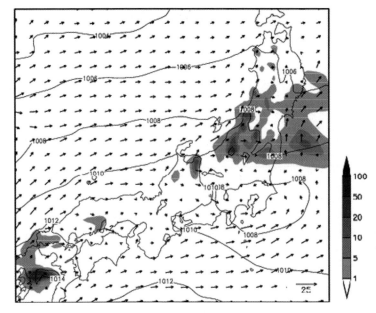

**Fig. 14** Same as Fig. 12 except for consideration of the vertical correlation of observation error

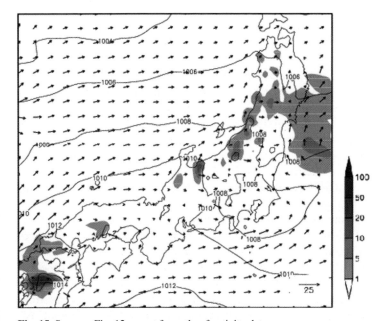

**Fig. 15** Same as Fig. 12 except for path refractivity data

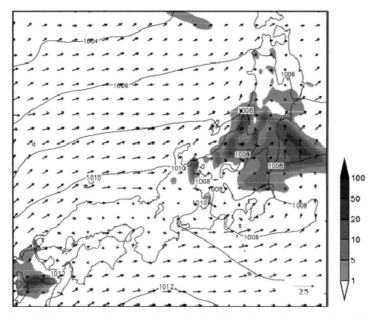

**Fig. 16** Same as Fig. 12 except for path refractivity data and consideration of the vertical correlation of observation error

## 5.2 Difference of Assimilated Fields of Water Vapor and Temperature from the CNTL

The aforementioned rainfall regions are closely related to the initial fields produced by assimilation of the occultation data. How were the initial water vapor and temperature fields modified by the assimilation of occultation data?

Before showing how the initial fields differ from CNTL, the rainfall distribution predicted from the initial fields produced by assimilation of GPS-derived PWV data is explained. To monitor the movement of the tectonic plates, the Geographical Survey Institute (GSI) deployed more than 1000 GPS receivers over Japan. When PWV data observed by this GPS network are assimilated, the rainfall forecast is expected to be improved. The difference of water vapor and temperature in this experiment can be used as a reference when the rainfall forecast is improved by assimilating GPS-PWV data. The method of assimilating GPS-derived PWV in this study was the same as that of Seko et al. 2004, in which the rainfall forecast was improved by assimilating GPS-PVW data. In addition to the conventional data, PWV observed by more than 210 GPS receivers, whose altitudes were within 50 m from the model topography, were assimilated. The data assimilation window was three hours, from 1200 JST to 1500 JST, on 16 July 2004. This experiment is called GPS-PWV.

Figure 17 depicts the rainfall regions for the GPS-PWV case. When GPS-PWV data were assimilated, the intense rainfall region in northern Japan was well reproduced. The northern part of the predicted intense rainfall disappeared, and the

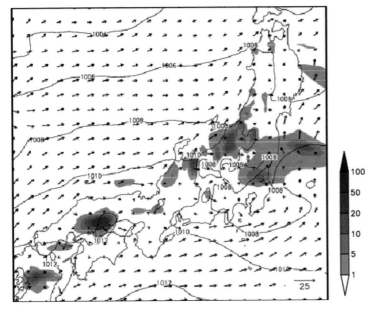

**Fig. 17** Three-hour rainfall distribution at 1800 JST (FT=3). Initial fields were obtained by assimilation of the GPS-derived PWV data

simulated rainfall region in northern Japan became more similar to the observed one. Thus, the GPS-PWV case was considered a good reference for the experiments of the occultation data assimilation.

Figure 18 presents the difference fields of water vapor and temperature at the lowest level of MSM. Compared with the initial fields of CNTL, water vapor increased and temperature decreased near the tangent point with all methods. These changes

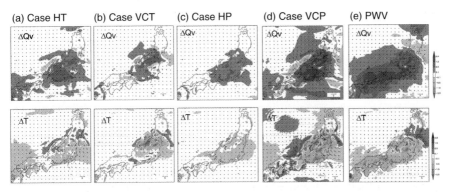

**Fig. 18** Difference of initial fields from those of the CNTL case. Upper and lower panels indicate the difference of water vapor and temperature at the lowest level of MSM. Arrows in each panel indicate differences of horizontal wind at the lowest level of MSM

were favorable for generating and enhancing the intense rainfall. When vertical correlation of the observation error at the tangent point was considered, the difference became larger and the modified region was more localized near the tangent point (Fig. 18b). When the path data was assimilated, the difference became smaller, and the modified region was slightly smaller than that of the HT case (Fig. 18c). These differences supported the dilution of the influence of the low-level tangent point data. When the vertical correlation of the path data was considered, the large difference region extended to central Japan (Fig. 18d). The difference distribution near the tangent point was most similar to that of the GPS-PWV case. This similarity indicated that the method of the VCP case was the most proper for the occultation data in this case.

# 6 Summary

The results of this study are summarized as follows.

1. A few kinds of occultation refractivity data were investigated statistically; the most accurate data, provided from GFZ, were then assimilated into a numerical model.
2. The impacts of several assimilation methods were investigated. When the path data was assimilated in consideration of the vertical correlation of observation error, an intense rainfall region, which was not reproduced by the conventional data, was reproduced. This method was concluded to be a useful assimilation method for occultation data, because of the consideration of the occultation data characteristics as well as the large improvement of the rainfall forecast.

This study demonstrated the improvement of the rainfall forecast by occultation data through the results of an assimilation experiment. However, a few issues remain. One important issue is the observation errors of the occultation data. The observation errors (i.e., TPOE and POE) multiplied by 0.1 were used to determine the impact of one occultation profile data more clearly. However, this observation error seemed to be too small to use in operational data assimilation, in which many profiles are assimilated simultaneously. More investigation of the observation error, especially for operational assimilation, is needed. Another issue is the insufficient number of assimilation experiments. Assimilation of refractive index profiles data improves water vapor distribution, and then affects the rainfall distribution, the development of typhoon, and so on. Investigations of impacts on various weather phenomena are also desired.

**Acknowledgements** CHAMP/GFZ data was provided by Dr. Wickert of GFZ. The data of CHAMP/JPL and SAC-C/JPL were provided by GENESIS of JPL. GPS data observed by GEONET was provided by GSI. We used the Meso 4D DAS developed by NPD of JMA. Initial and boundary data of first guesses were also provided by NPD of JMA. The authors express their thanks to Dr. Wickert and the staff of ICDS/GFZ, GENESIS/JPL, GSI, and NPD of JMA; Prof. T. Tsuda of Kyoto University; Mr. E. Ozawa of NPD of JMA; and members of the project "Application of GPS radio occultation data to the studies of temperature and humidity variations in the tropical troposphere." Dr. Y. H. Kuo, Dr. S. Sokolovskiy, and Dr. T. Iwabuchi provided useful comments on the assimilation method for the occultation data. The authors extend their appreciation to Prof. T. Tsuda, Mr. E. Ozawa, Dr. Y. H. Kuo, Dr. S. Sokolovskiy, Dr. T. Iwabuchi, and members of the project "Application of GPS radio occultation data to the studies of temperature and humidity variations in the tropical troposphere." This study was part of the above-mentioned, project which was supported by the Ministry of Education, Culture, Spots, Science and Technology as a part of the Japan Earth Observation System Promotion Program.

# References

Chen S-H, Vandenberghe F, Huang C-Y (2006) Error Characteristics of GPS retrieved refractivity using a simulation study. J Meteor Soc Japan 84:477–496

Chen S-H, Kuo Y-H, Sokolovskiy S, Huang C-Y (2005) Estimation of observational errors of GPS radio occultation soundings, Second GPS RO Data Users Workshop, 22–24 August 2005, Lansdowne, USA

Haung C-H, Kuo H-Y, Chen S-H, Vandenberghe F (2005) Improvement in typhoon forecasting with assimilated GPS occultation refractivity. Wea Forecast 20:931–953

JMA (2002) Outline of the operational numerical weather prediction at a Japan mesoscale model. 185pp

Kato T (2005) Formation factors of 2004 Niigata-Fukushima and Fukui heavy rainfall and problems in the precipitations using a Cloud-resolving model. SOLA 1:1–4

Koizumi K, Ishikawa Y, Tsuyuki T (2005) Assimilation of precipitation data to the JMA Mesoscale Model with a four-dimensional variational method and its impact on precipitation forecasts. SOLA 1:45–48

Koizumi K, Sato Y (2004) Impact of GPS and TMI precipitable water data on mesoscale numerical weather prediction model forecasts. J Meteor Soc Japan 82b:453–457

Kuo Y-H, Wee T-K, Sokolovskiy S, Rocken C, Schreiner W, Hunt D, Anthes RA (2005) Inversion and error estimation of GPS radio occultation data. J Meteor Soc Japan 82b:507–531

Nakamura H, Koizumi K, Mannoji H (2004) Data assimilation of GPS precipitation water vapor into the JMA mesoscale numerical weather prediction model and its impact on rainfall forecasts. J Meteor Soc Japan 82b:441–451

Ozawa E, Tada H, Aoyama Y (2005) Development of GSM 3Dvar assimilation system for occultation data. Report of outlays for promoting science and technology 'Application of precise satellite positioning for monitoring the Earth's environment', in Japanese

Parish DF, Deber JC (1992) The national meteorological center's spectral statistical interpolation analysis system. Mon Wea Rev 120:1747–1763

Seko H, Kawabata T, Tsuyuki T, Nakamura H, Koizumi K, Iwabuchi T (2004) Impacts of GPS-derived water vapor and radial wind measured by Doppler Radar on numerical prediction of precipitation. J Meter Soc Japan 82:473–489

Sokolovskiy S (2003) Effect of superrefraction on inversion of radio occultation signals on the lower troposphere. Radio Sci 38:1058, doi:10.1029/2002RS002728

Wickert J, Schmidt T, Beyerle G, König R, Reigber C, Jakowski N (2004) The radio occultation about CHAMP: Operational data analysis and validation of vertical atmospheric Profiles. J Meteor Soc Japan 82b:381–395

# Parameter Estimation Using the Genetic Algorithm in Terms of Quantitative Precipitation Forecast

Yong Hee Lee, Seon Ki Park, Dong-Eon Chang, Jong-Chul Ha and Hee-Sang Lee

**Abstract** In this study, the optimal parameter estimation is performed for both physical and computational parameters in a mesoscale meteorological model, and its impact on the quantitative precipitation forecasting (QPF) is assessed for a heavy rainfall case occurred at the Korean peninsula in June 2005. Experiments are carried out using the PSU/NCAR MM5 model and the genetic algorithm (GA) for two parameters: the reduction rate of the convective available potential energy in the Kain-Fritsch (KF) scheme for cumulus parameterization, and the Asselin filter parameter for numerical stability. The fitness function is defined based on a QPF skill score. It turns out that each optimized parameter significantly improves the QPF skill. Such improvement is maximized when two optimized parameters are used simultaneously. Our results indicate that optimizations of computational parameters as well as physical parameters and their adequate applications are essential in improving model performance.

## 1 Introduction

Numerical weather/climate prediction models contain numerous parameterizations for physical processes and numerical stability. Parameterizations are based on physical laws but typically contain parameters whose values are not known precisely. The values of the parameters directly or indirectly affect the performance of model, and thus uncertainties in parameter values may lead to sensitive results from some models, especially with high resolution and sophisticated microphysics (e.g., Park and Droegemeier, 1999). Accordingly, optimal estimation of parameters is one of several essential factors in improving the accuracy of numerical forecasts.

Recently, efforts have been made to obtain better estimation of parameters for numerical forecast models using various methods such as the variational technique using a full-physics adjoint model (Zhu and Navon, 1999), the Bayesian stochastic

S.K. Park (✉)
Department of Environmental Science and Engineering, Ewha Womans University,
Seoul 120–750, Republic of Korea
e-mail: spark@ewha.ac.kr

S.K. Park, L. Xu, *Data Assimilation for Atmospheric, Oceanic and Hydrologic Applications*, DOI 10.1007/978-3-540-71056-1_11,
© Springer-Verlag Berlin Heidelberg 2009

inversion (BSI) based on multiple very fast simulated annealing (VFSA) (Jackson et al., 2004), the downhill simplex method (Severijns and Hazeleger, 2005), and the ensemble Kalman filter (EnKF) (Aksoy et al., 2006).

The genetic algorithm (GA), which is a global optimization technique, has also been applied to some optimization problems including parameter estimation, especially for simple models. Compared to traditional optimization methods based on the gradient of a function, the GA is more appropriate when the function includes some complexities and/or discontinuities (Barth, 1992). Major advantages of the GA include that: (1) derivatives of fit function with respect to model parameters are not required; and (2) nonlinearity between the model and its parameters can be handled (Holland, 1975; Goldburg, 1989; Charbonneau, 2002).

Based on the optimization of model parameters such as the biharmonic horizontal diffusion coefficient, the ratio of the transfer coefficient of moisture to the transfer coefficient of sensible heat, and the Asselin filter coefficient, Zhu and Navon (1999) demonstrated that the positive impact of the optimally-estimated parameter values persists for longer than that of the optimized initial conditions. The parameter estimation problems have been explored to a wide scope including the land surface parameters (Jackson et al., 2004), the radiation and cloud parameters (Severijns and Hazeleger, 2005), the vertical eddy mixing coefficient (Aksoy et al., 2006), and even for the purpose of experiment design (Barth, 1992, Hernandez et al., 1995). However, attempt has seldom been made on the parameter estimation problem associated with the quantitative precipitation forecasting (QPF).

This study focuses on optimal parameter estimation to improve the QPF skill in a mesoscale meteorological model using the GA. Section 2 describes characteristics of the model and parameters, and Sect. 3 explains the details of GA for parameter estimation. Discussions on results appear in Sect. 4, and conclusions are provided in Sect. 5.

## 2 Case, Model and Experiments

A heavy rainfall case in Korea is selected for experiments. It occurred in the west-central part of the Korean peninsula, associated with a summer monsoon front, with a local maximum 6-hr accumulated rainfall of 100 mm in Seoul from 1200 UTC to 1800 UTC 26 June 2005.

In this study, the 5th-generation Pennsylvania State University-National Center for Atmospheric Research Mesoscale Model (MM5) version 3.6.3 (Grell et al., 1994) is employed. The computational domain consists of $218 \times 181$ grids in the horizontal, with a resolution $\Delta x = \Delta y = 18$ km, and 35 layers in the vertical. The model domain is shown in Fig. 1a along with terrain. The MM5 is integrated up to 12 hrs starting from 0600 UTC 26 June 2005, with a time step $\Delta t = 45$ s. Schemes for physical processes include: the MRF PBL (Hong and Pan, 1996), the Kain-Fritsch (KF) cumulus parameterization (Kain, 2003), the Dudhia radiation (Dudhia, 1989)

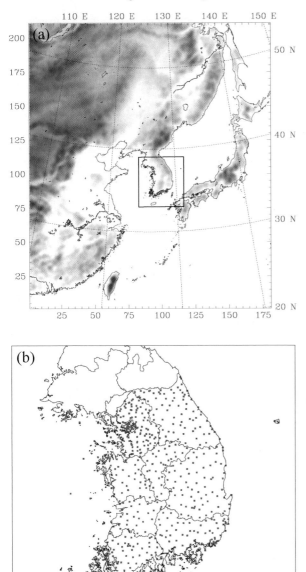

**Fig. 1** (**a**) Domain configuration of MM5 along with topography. The inner box indicates the verification area. (**b**) The location of the 592 rain gauge stations in Korea used for the QPF verification. The average distance of rain gauge stations is about 18 km

and RRTM, the Schultz microphysics (Schultz, 1995), and the five-layer soil scheme (Dudhia, 1996).

For the parameter estimation study, we focus on the closure assumption of the KF parameterization and the Asselin filter coefficient in MM5. The "closure" in the KF parameterization relates the intensity of convective activity to the resolved-scale properties in a model, and the assumes that convection consumes at least 90% of the environmental convective available potential energy (CAPE) over an advective time period, bounded by a maximum of 1 hr and a minimum of 30 min (Kain, 2003). However, Saito et al. (2006) found that this setting tended to over stabilize the model atmosphere, making strong rainfall decrease with time in the forecast period of 18 hours. To prevent this undesirable excessive stabilization of the model atmosphere, the reduction rate of CAPE in the column, for a single application of the KF scheme, is diminished from the default value to 85%. In this study, an optimal parameter estimation experiment will be carried out to obtain the optimal value of the "closure assumption" in the KF scheme.

The temporal differencing in MM5 consists of leapfrog steps with an Asselin filter (Asselin, 1972). Splitting of the solution often associated with the leapfrog scheme is avoided by using this filter. It is applied to all variables $\alpha$ as

$$\hat{\alpha}^t = (1 - 2\nu)\alpha^t + \nu(\alpha^{t+1} + \hat{\alpha}^{t-1}), \tag{1}$$

where $\hat{\alpha}$ is the filtered variables, and $\nu \in [0, 1]$ is the Asselin filter parameter. The value of $\nu$ is set to 0.1 in MM5 for all variables (Grell et al., 1994). However, Bryan and Fritsch (2000) found that the Asselin filter parameter used in MM5 is a source of the unphysical thermodynamic structures. Another parameter estimation experiment in this study will be focused to obtain the optimal value of the Asselin coefficient. The lower bound of the Asselin coefficient is set to 0.01, while its upper bound is set to 0.3.

## 3 Methodology of Parameter Estimation

This study aims at performing optimal estimations of two parameters in MM5 using the GA. The GA is a global optimization approach based on the Darwinian principles of natural selection. This method, developed from the concept of Holland (1975), aims to efficiently seek the extrema of complex function – see Goldberg (1989) for a detailed description. Deb (2000) discussed an efficient constraint handling for the GA.

A key concept in the GA is the chromosome. A chromosome contains a group of numbers that completely specifies a candidate during the optimization process. Typically, the GA uses crossover, mutation, and reproduction to provide structure to a random search. The GA restricted to mere reproduction and mutation is a version of stochastic random search. The incorporation of the crossover operator, which mates two chromosomes, provides a qualitatively different search, one that has no counterpart in stochastic grammars. Crossover works by finding, rewarding and re-combining "good" segments of chromosomes, and the more faithfully the segments of the chromosomes represent the better we can expect genetic algorithms to per-form. The GA also uses randomization heavily in choosing a chromosome that will propagate to future generations. In general, the average fitness of individuals in-creases with each generation, through the process of natural selection. In each suc-cessive generation, individuals with bad genes are weeded out while those with good genes propagate their genetic code. The genetic code that determines the fitness of an individual is termed, logically enough, the chromosome of that individual. Given a chromosome, the GA should be able to ascertain its fitness. The performance and search time depend on the number of bits, the size of a population, the mutation and crossover rates, choice of features and mapping from chromosomes to the parameter itself, the inherent difficulty of the problem and possibly parameters associated with other heuristics.

For the parameter estimation experiments in this study, a GA package called the PIKAIA (Charbonneau, 2002) is employed. Each generation has 20 chromosomes. The crossover probability is set to 0.85, implying that 85% of the chromosomes in a generation are allowed to crossover in an average sense. The maximum and minimum mutation probability is set to 0.05 and 0.005, respectively.

Internally, the PIKAIA seeks to maximize a function $f(X)$ in a bounded $n$-dimensional space,

$$X \equiv (x_1, x_2, \cdots, x_n), \qquad x_k \in [0.0, 1.0] \, \forall k \tag{2}$$

In our problem, there exist two adjustable parameters, i.e., $n = 2$. Then we may associate the reduction assumption of the KF scheme $\varepsilon$ with $x_1$ and the Asselin filter parameter $v$ with $x_2$. The ranges of parameters are $0 \le \varepsilon(x_1) \le 0.95$ and $0.01 \le v(x_2) \le 0.3$, respectively.

The function to be optimized (i.e., Fitness) is defined by using a QPF skill score, the equitable treat score (ETS) (Schaefer, 1990),

$$\text{Fitness} = \sum_i ETS_i, \quad i = 1, 2, \cdots, 100 \tag{3}$$

where $i$ is the precipitation threshold in mm. Here, the ETS is defined as:

$$ETS = \frac{H - R}{F + O - H - R}, \tag{4}$$

where $H$ is the number of hits, and $F$ and $O$ are the numbers of samples in which the precipitation amounts are greater than the specified threshold in forecast and observation, respectively, and $R$ is the expected number of hits in a random forecast $-R = FO/N$, where $N$ is total number of points being verified.

Each generation includes 20 individual MM5 runs as a function of $\varepsilon$ and $v$. Every individual run with the two parameters is encoded by chromosomes and returns the accumulated rainfall to determine the fitness; thus the fitness function is dynamically coupled to the MM5 model. In each generation, the two parameters make independent search for the optimal solution concurrently; hence there exists no feedback between the two parameters.

Clearly, the relationship between the fitness function and the model parameters is strongly nonlinear. Therefore any other robust estimator can be substituted with little or no changes to the overall procedure of GA. Under the right conditions the GA has shown to converge to good solutions remarkably quickly and has the advantage that the rate of convergence varies in accordance with the complexity of the search space (Goldberg, 1989; Holland, 1975).

## 4 Results

In this study, two parameters in MM5 are optimized (i.e., the closure assumption of KF scheme, $\varepsilon$, and the Asselin filter coefficient, $v$) to improve the QPF skill using the GA. Figure 2 depicts performance of chromosomes in terms of generations. In the first few generations, the performance of chromosomes improves significantly as the GA discovers and populates the best regions in the search space. The zeroth generation consisted of 20 chromosomes chosen randomly. The spread of fitness at the zeroth generation is quite large. This implies that both parameters (i.e., $\varepsilon$ and $v$) exert sensitive impact on the precipitation forecasts because the fitness function is defined in terms of a QPF skill score. The performance of average chromosomes

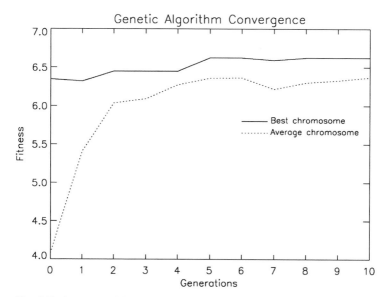

**Fig. 2** Performance of the best chromosome in each generation and of an average chromosome in a generation. A typical GA does its tuning in stages called generations. The final solution has $(\varepsilon, v) = (0.01, 0.25)$

improved exponentially, up to the second generation (i.e., 60 runs MM5), as the GA discovers and populates the best regions in the search space. This implies that evolution for only a few generations is sufficient to obtain optimal estimation parameters. The final solutions of the two parameters are $\varepsilon = 0.0111781 \approx 0.01$ and $v = 0.2498580 \approx 0.25$, through global optimization in the fitness space which has multiple minima (not shown).

Figure 3 compares the ETSs computed for the 6-hr accumulated rainfall at forecast period of 6–12 hr from the following four experiments using: (1) the default parameter (CNTL; $\varepsilon = 0.9$, $v = 0.1$); (2) no convective parameterization scheme (NC; $\varepsilon =$none, $v = 0.1$); (3) the revised KF parameter (KF; $\varepsilon = 0.01$, $v = 0.1$); (4) the revised Asselin filter parameter (AF; $\varepsilon = 0.9$, $v = 0.25$); and (5) the revised parameters for both the KF scheme and the Asselin filter (KF-AF; $\varepsilon = 0.01$, $v = 0.25$). The ETSs of the default run dropped rapidly with increasing threshold values reaching lower than 0.1 at thresholds larger than 30 mm. In general, it can be noticed that the GA-estimated parameters give positive effect on increasing the QPF skill, either independently or together.

In the original KF scheme, $\varepsilon$ is set to 0.9; that is, the convection consumes the pre-existing CAPE by 90%. However, the GA-estimated value (i.e., $\varepsilon = 0.01$) is quite different from the original. This implies that the convective rainfall in the selected case requires almost no consumption of the pre-existing CAPE. It is notified that, compared with convective systems in the North America, those in the East

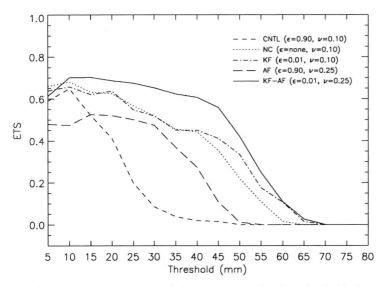

**Fig. 3** The ETSs of precipitation forecasts in terms of various thresholds (in mm). Curves denote scores for the 6–12-hr accumulated rainfall forecasts for experiments CNTL (control run with default values), NC (no convective parameterization), KF (using optimized parameter in the KF scheme), AF (using optimized parameter in the Asselin filter), and KF-AF (using optimized parameters from both the KF scheme and the Asselin filter)

Asia and nearly saturated up to the mid-troposphere; thus resulting in a smaller amount of CAPE, especially prior to and during heavy rainfall (see Lee et al., 1998; Hong, 2004). Therefore, in applying the KF scheme to convective rainfalls in the East Asia, it might be essential to assume slow or almost no consumption of the pre-existing CAPE (e.g., Saito et al., 2006); however, it does not necessarily mean that the KF scheme is not applicable to the QPF study in this region.

Compared with the no-convective parameterization experiment (i.e., NC), the KF scheme revised with the GA-estimation (i.e., KF) shows much higher ETSs at thresholds larger than 40 mm (see Fig. 3). It suggests that the KF scheme is still useful but with an optimized value of $\varepsilon$ in accordance with the environment that consumes the CAPE slowly for a heavy rainfall event in the East Asia.

The revised Asselin filter (i.e., AF; $v = 0.25$) also brings about improvement in the ETSs for thresholds of 15–50 mm. Generally, the Asselin filter with $v = 0.25$ removes $2\Delta t$ waves and reduces the amplitude of $4\Delta t$ waves by one-half, but with little effect on longer-period waves; that is, it acts as a low-pass filter in time. In contrast, the Asselin filter with default value ($v = 0.1$) serves as a high-pass filter so that some short-period waves, including gravity waves, are not filtered out. Although the Asselin filter is used for the purpose of numerical stability, the result indicates that its impacts on the QPF are considerable; thus it should be treated with care.

The experiment using both parameters estimated through the GA (i.e., KF-AF) produced the highest ETSs for almost all thresholds, exceeding 0.6 at thresholds lower than 45 mm. It is notable that the QPF skill increases prominently when two revised parameters are used together in the model. This suggests that simultaneous optimization and use of all uncertain parameters, both physical and computational, are essential in improving model performances.

Figure 4 represents a 6-hr accumulated rainfall for the forecast time from 6 to 12 hr for two experiments: (1) with the default values (i.e., $\varepsilon = 0.9$, $v = 0.1$; Fig. 4b) and (2) with the GA-estimated values (i.e., $\varepsilon = 0.01$, $v = 0.25$; Fig. 4c). During the 6-hr period between 1200 UTC and 1800 UTC 26 June 2005, a heavy rainfall occurred in the west-central part of the Korean peninsula with a local maximum of 100 mm in Seoul (Fig. 4a). The default experiment failed to simulate the amount of rainfall – only 25 mm at the region where more than 90 mm is observed (cf. Fig. 4a and b). Meanwhile, the experiment with the GA-estimated parameters simulated the localized heavy rainfall quite well with 70 mm peak rainfall (cf. Fig. 4a and c).

Overall, it is notable that optimization of parameters improves the QPF significantly, in both location and amount of rainfall, especially when two optimized parameters are used simultaneously. In the selected heavy rainfall case, the closure assumption of the KF scheme is reduced from 90% (default) to 1% while the Asselin filter coefficient is adjusted from 0.1 to 0.25 after optimal estimation.

The uniqueness problem in parameter estimation is ultimately related to the issue of parameter identifiability (Navon, 1997). Since the GA is basically a random search algorithm, the parameter indentifiability can be assessed by repeating the GA run, each composed of 200 MM5 runs (i.e., 10 generations $\times$ 20 chromosomes), until the best chromosomes start to repeat with some regularity.

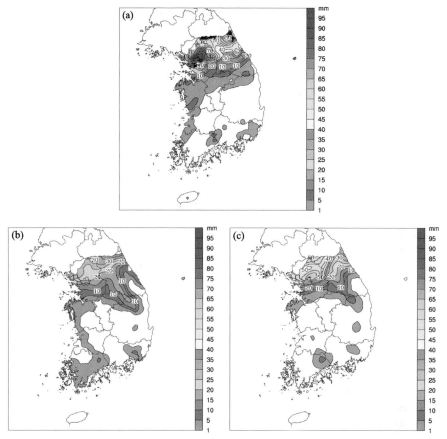

**Fig. 4** A 6-hr accumulated precipitation (in mm) of (**a**) observation, and experiments with (**b**) the default parameters and (**c**) the GA-estimated parameters, ending at 1800 UTC 26 June 2005. The model output is interpolated to the rain gauge locations for verification

## 5 Conclusions

In this study, optimal estimation of parameters in a mesoscale meteorological model (MM5) is performed in the purpose of improving the QPF skills for a heavy rainfall case in the Korean peninsula, employing a global optimization technique called the genetic algorithm (GA). The GA is applied to find out optimal parameters directly using the QPF skill score as a fitness (cost) function.

The GA is robust to complexity and nonlinearity in the model and thus providing more flexible and direct way of solving in parameter estimation. Therefore, nonlinear relations between the fitness function and the model parameters are well treated in the GA. However, the evolution in GA must accommodate physical constraints associated with development and growth so that all possible paths would not be searched in the genetic parameter space (see Deb, 2000).

The model parameters selected for optimal estimation are the reduction rate of the convective available potential energy (CAPE) in the Kain-Fritsch (KF) scheme, $\varepsilon$, for convective parameterization (i.e., physical parameter) and the Asselin filter coefficient for numerical stability, $\nu$, (i.e., computational parameter). The optimized solutions are $(\varepsilon, \nu) = (0.01, 0.25)$. The GA discovered and populated the best regions in the search space only in a few generations.

Each optimized parameter similarly exerted a favorable influence on the heavy rainfall forecast by improving the QPF skill. Further significant improvement in the QPF skill was achieved when two optimized parameters were used simultaneously in the model. This implies that an interaction between optimized physical and computational parameters works favorably to bring about potentially best performance of a numerical model. Therefore, optimizations of computational parameters as well as physical parameters and adequate use of optimized parameters are essential in improving model performance.

It is noteworthy that the optimally-estimated reduction rate of the CAPE in the KF scheme is much smaller than the default value, which is consistent with previous studies depicting slow consumption of pre-existing CAPE in the heavy rainfall cases in the East Asia (Lee et al., 1998; Hong, 2004); thus representing well the characteristic of the selected rainfall environment. Such tendency in the KF parameter will be further investigated with more heavy rainfall cases at least near the Korean peninsula.

**Acknowledgements** This research was performed for the project of "A study on improving short-range precipitation forecast skill" funded by the Korea Meteorological Administration (KMA). The second author is partly supported by the Korea Research Foundation Grant funded by the Korean Government (MOEHRD) (R14-2002-031-01000-0).

# References

Aksoy A, Zhang F, Nielsen-Gammon JW (2006), Ensemble-based simulations state and parameter estimation with MM5. Geophys Res Lett 33:L12801, doi:10.1029/2006GL026186

Asselin R (1972) Frequency filter for time integrations. Mon Wea Rev 100:487–490

Barth NH (1992) Oceanographic experiment design II: Genetic algorithms. J Atmos Oceanic Technol 9: 434–443

Bryan GH, Fritsch JM (2000) Unphysical thermodynamic structures in explicitly simulated thunderstorms. 10th PSU/NCAR Mesoscale Model User's Workshop, Boulder, CO, NCAR, Available from http://www.mmm.ucar.edu/mm5

Charbonneau P (2002) An introduction to genetic algorithms for numerical optimization. NCAR Tech Note TN-450+IA, 74pp

Deb K (2000) An efficient constraint handling method for genetic algorithm. Comput Methods Appl Mech Eng 186:311–338

Dudhia J (1989) Numerical study of convection observed during the winter monsoon experiment using a mesoscale two dimensional model. J Atmos Sci 46:3077–3108

Dudhia J (1996) A multi-layer soil temperature model for MM5. Preprints, Sixth PSU/NCAR Mesoscale Model User's Workshop, Boulder, CO, NCAR, 49–50

Goldburg D (1989) Genetic algorithms in search, optimization and machine learning. Addison-Wesley, Reading, MA, 432pp

Grell GA, Dudhia J, Stauffer DR (1994) A description of the fifth-generation Penn State/NCAR Mesoscale Model (MM5). NCAR Tech. Note TN-398+STR, 138pp

Hernandez F, Traon R-Y, Barth NH (1995) Optimizing a drifter cast strategy with genetic algorithm. J Atmos Oceanic Technol 12:330–345

Holland J (1975) Adaptation in natural and artificial systems. University of Michigan Press, Ann Arbor, 228pp

Hong S-Y (2004) Comparison of heavy rainfall mechanisms in Korea and the Central US. J Meteor Soc Japan 5:1469–1479

Hong S-Y, Pan HL (1996) Nonlocal boundary layer vertical diffusion in a medium-range forecast model. Mon Wea Rev 124:2322–2339

Jackson C, Sen MK, Stoffa PL (2004) An efficient stochastic Bayesian approach to optimal parameter and uncertainty estimation for climate model predictions. J Climate 17:2828–2841

Kain J (2003) The Kain-Fritsch convective parameterization: An update. J Appl Meteoro 43:170–181

Lee D-K, Kim H-R, Hong S-Y (1998) Heavy rainfall over Korea during 1980–1990. Korean J Atmos Sci: 1:32–50

Navon IM (1997) Practical and theoretical aspects of adjoint parameter estimation and indetifiability in meteorology and oceanography. Dyn Atmos Ocean 27:55–79

Park SK, Droegemeier KK (1999) Sensitivity analysis of a moist 1D Eulerian cloud model using automatic differentiation. Mon Wea Rev 127:2180–2196

Saito K, Fujita T, Yamada U, Ishida J-I, Kumagai Y, Aranami K, Ohmori S, Nagasawa R, Kumagai S (2006) The operational JMA nonhydrostatic mesoscale model. Mon Wea Rev 134:1266–1298

Schaefer JT (1990) The critical success index as indicator of warning skill. Wea Forecast 5:570–575

Schultz P (1995) An explicit cloud physics parameterization for operational numerical weather prediction. Mon Wea Rev 123:3331–3343

Severijns CA, Hazeleger W (2005) Optimizing parameters in an atmospheric general circulation model. J Climate 18:3527–3535

Zhu Y, Navon IM (1999) Impact of parameter estimation on the performance of the FSU global spectral model using its full-physics adjoint. Mon Wea Rev 127:1497–1517

# Applications of Conditional Nonlinear Optimal Perturbations to Ensemble Prediction and Adaptive Observation

Zhina Jiang, Hongli Wang, Feifan Zhou and Mu Mu

**Abstract** Conditional nonlinear optimal perturbation (CNOP), which is a natural extension of the singular vector into the nonlinear regime, is applied to ensemble prediction study and the determination of sensitive area in adaptive observations. The purpose of this paper is to summarize the recent progresses of the authors in these fields.

For the ensemble prediction part, a quasi-geostrophic model is used under the perfect model assumption. A series of singular vectors (SVs) and CNOPs have been utilized to generate the initial ensemble perturbations. The results are compared for forecast lengths of up to 14 days. It is found that the forecast skill of samples, in which the first singular vector (FSV) is replaced by CNOP, is comparatively higher than that of samples composed of only SVs in the medium range (day 6 ∼ day 14). This conclusion is valid under the condition that analysis error is a kind of fast-growing ones regardless of its magnitude, whose nonlinear growth is faster than that of FSV in the later part of the forecast.

The potential application of CNOP to identify the data-sensitive region in targeting strategy is explored by using the 5th generation Pen-State University/National Center for Atmosphere Research mesoscale Model (MM5) and its adjoint system. The differences between FSVs and CNOPs and their evolutions are studied in two precipitation cases in July 2003 and in August 1996 respectively. It is found that the structures of CNOPs differ much from those of FSVs as well as the developments of their total energies. The results of sensitivity experiments indicate that the forecast results are more sensitive to the CNOP-type initial errors than to FSV-type in terms of total energy. These results suggest that it is feasible to use CNOP to identify the sensitive region in adaptive observations.

**Keywords:** Singular vector · Conditional nonlinear optimal perturbation · Ensemble prediction · Adaptive observation · Weather

M. Mu (✉)
State Key Laboratory of Numerical Modelling for Atmospheric Sciences and Geophysical Fluid Dynamics (LASG), Institute of Atmospheric Physics, Chinese Academy of Sciences, Beijing 100029, China
e-mail: mumu@lasg.iap.ac.cn

S.K. Park, L. Xu, *Data Assimilation for Atmospheric, Oceanic and Hydrologic Applications*, DOI 10.1007/978-3-540-71056-1_12,
© Springer-Verlag Berlin Heidelberg 2009

# 1 Introduction

Singular vector (SV) analysis has proved to be helpful in understanding the linear instability properties of atmospheric and oceanic flows (Buizza et al. 1993; Buizza and Palmer 1995; Farrell and Moore 1992; Ehrendorfer 2000). SVs are the linear perturbations that are determined from the constraint that they produce the maximum possible linear growth in a given norm over a finite interval of time. Lorenz (1965) first considered SVs in a meteorological context in a predictability study with a low-dimensional barotropic model. At present, the SV approach is widely used in the study of predictability, ensemble prediction and targeting observations (Molteni and Palmer 1993; Molteni et al. 1996; Ehrendorfer and Tribbia 1997; Palmer et al. 1998; Frederiksen 2000; Buizza et al. 2005).

At the European Centre for Medium-Range Weather Forecasts (ECMWF), SV approach proved to be successful to generate initial condition perturbations for ensemble prediction (Mureau et al. 1993). Ehrendorfer and Tribbia (1997) demonstrated that forecast-error covariance can be predicted most efficiently using an ensemble constructed in the subspace of the leading analysis-error covariance SVs under assumptions of linearity of error growth and normality of errors, which presented the theoretical justification for the use of SVs in ensemble prediction systems. For simplicity, the total-energy SVs are probably a reasonable substitute for analysis-error covariance SVs (Molteni et al. 1996). Gelaro et al. (2002) computed the SVs with an analysis-error variance metric, and the results showed that the leading SV are consistent with the expected distribution of analysis errors. Hamill et al. (2003) calculated the flow-dependent analysis-error covariance SVs and further suggested that operational ensemble forecasts based on total energy SVs could be improved by changing the type of singular vector used to generate initial perturbations. Leutbecher (2005) studied the impact of using SVs computed from 12-h forecasts instead of analyses on the ECMWF ensemble prediction system, which showed that SVs calculated from forecasts could be used to disseminate the ensemble forecasts earlier or to allocate more resources to the nonlinear forecasts. At present, refining use of SVs has become one of the trends for development of ensemble predictions.

SV is utilized to identify the data-sensitive region in targeting strategy as well. The concept of adaptive or targeted observations is based on the idea of adding supplementary observation information in data-sensitive regions to the conventional observing network in order to improve numerical forecast skill in the region we concerned (Emanuel et al. 1995; Snyder 1996). This observational network is adaptive in the sense that the location of these measurements varies from day-to day. Several objective methods have been utilized to identify "target" regions for the exampling of supplementary observations. These methods are mainly based on three different types of techniques, including adjoint technique, ensemble technique and inverse integral technique, for example, SVs, Ensemble Transform Kalman Filter (ETKF, Bishop et al. 2001). In general, SV targeting emphasizes the rapid growth of forecast errors from small initial condition uncertainties (Buizza et al. 2007). It has been implemented during field experiments such as FASTEX (Joly et al. 1999), NOR-PEX (Langland et al. 1999a), and NATReC (Petersen et al. 2006). The results from these field experiments show that, in average, the forecast skill can be improved by assimilation of targeted observations (Gelaro et al. 1999, Langland et al. 1999b, Langland 2005; Buizza et al. 2007).

Note that SV is established on the fact the initial perturbation is sufficiently small and the time period is not too long, so that the evolution of the perturbation can be governed by the tangent linear model (TLM). However, there is no consensus about the forecast period for which the linearity assumption is valid (Errico et al. 1993; Lacarra and Talagrand 1988; Mu et al. 2000). Keeping in mind that the motions of the atmosphere and the oceans are both dominated by complicated nonlinear systems and the linear assumption may have some limitations, Barkmeijer (1996) tried a modification technique to the linearly fastest-growing perturbation and attempted to construct fast-growing perturbations for the nonlinear regime. Nevertheless, he also recognized that his technique may not necessarily result in the nonlinearly fastest-growing perturbation. The similar procedure was described in Oortwijn and Barkmeijer (1995) and Oortwijn (1998). Then, Mu and Duan (2003) proposed a new method called conditional nonlinear optimal perturbation (CNOP), which can acquire the maximum value of the objective function according to physical problems. This method has been used in many research fields (Duan et al. 2004; Mu et al. 2004; Duan and Mu 2006; Duan et al. 2007; Mu and Zhang 2006; Mu et al. 2007a).

As early as 1997, Anderson pointed out that using singular vector decomposition (SVD) to determine the directions in which the evolution would be most sensitive in only relevant in a linear regime and failed to give information about the likelihood of the extreme ensemble perturbations. Gilmour and Smith (1997) also presented that there are limits of linearity assumptions in the construction of ensemble perturbations. Zhang et al. (2002) found that error growth rates at small scales depend on the difference amplitude, suggesting that nonlinearity is important. Errico et al. (2002) declared that the evolution of the perturbations with structures consistent in size and shape with initial condition uncertainty can be substantially nonlinear, even after only a single day's growth. Realizing the limitations of linear singular vectors in ensemble forecast and adaptive observation, the authors have investigated the applicability of CNOP to these fields. The present paper gives a summary of the progresses on these two issues. One is to compare the use of CNOPs with SVs to the application of ensemble prediction by using a quasi-geostrophic model, which is presented in Mu and Jiang (2007). The other one is to explore the potential application of CNOP to adaptive observations strategy. Emphasis is placed on the identification of the data-sensitive region by using The 5th generation Pen-State University/National Center for Atmosphere Research mesoscale Model (MM5) (Dudhia 1993) and its adjoint system (Zou et al. 1997). The main contents are taken from Mu et al. (2007b), which is in Chinese.

## 2 The Method: CNOP

In this section, we briefly introduce the method of conditional nonlinear optimal perturbation. Assume the initial condition $U|_{t=0} = U_0$, the propagator $M$ is well-defined; $U(T) = M_T(U_0)$ is the solution of the nonlinear model at time $T$. Perturbations $u_0$ to the initial condition $U_0$ result in deviations from the original trajectory, so that the system follows a new trajectory $\tilde{U}(T) = M_T(U_0 + u_0) = U(T) + u(T)$. The nonlinear evolution of $u_0$ is defined as $u(T) = M_T(U_0 + u_0) - M_T(U_0)$. We also assume that for sufficiently small perturbations, the perturbations $u(T)$ can be determined by integrating the tangent linear model for time $T$: $u(T) \approx M_T(U_0)u_0$, where M represents the tangent linear propagator.

For chosen norms $\| \cdot \|_1$ and $\| \cdot \|_2$ (the subscript represents the different norms), CNOP is the initial perturbation $u_{0\sigma}^*$ which makes the objective function $J(u_0)$ acquire the maximum under the initial constraint condition $\|u_0\|_1 \leq \sigma$,

$$J(u_{0\sigma}^*) = \max_{\|u_0\|_1 \leq \sigma} J(u_0) \tag{1}$$

where

$$J(u_0) = \|u(T)\|_2 = \|M_T(U_0 + u_0) - M_T(U_0)\|_2 \tag{2}$$

$U_0$ and $u_0$ are the initial basic state and perturbation respectively, and $\sigma$ is a presumed positive constant representing the magnitude of the initial uncertainty. In the above objective function, the perturbations $u(T)$ is determined by the nonlinear model. Note that, the norms used in the objective function and the initial constraint condition may be same, which depends on the physical problems.

The optimization algorithm employed is the spectral projected gradient 2 (SPG2) (Birgin et al. 2000), which calculates the least value of a function of several variables subject to box or ball constraints. The great strength of the SPG2 method is its ability to solve problems with higher dimensions.

To capture the maximum of $J(u_0)$ with the constraint $\|u_0\|_1 \leq \sigma$, we may calculate the minimum of a new objective function with the same constraint $\|u_0\|_1 \leq \sigma$. The new objective function is defined as

$$J_1(u_0) = -[J(u_0)]^2 = -\|M_T(U_0 + u_0) - M_T(U_0)\|_2^2 \tag{3}$$

Through variation method and adjoint technique, the gradient of the objective function $J_1$ with respect to initial perturbation $u_0$ can be acquired, which should be supplied when using the SPG2 method. According to the definition of CNOP, CNOP is the global maximum of the cost function. However, there exists the possibility that the cost function attains its local maximum in a small neighborhood of a point in the phase space. Such an initial state is called local CNOP. CNOP and local CNOP possess clear physical meanings. For example, Duan et al. (2004) revealed that CNOP (local CNOP) acquired on the climatological background state is most likely to evolve into El Nino (La Nina) event and acts as the optimal precursors for El Nino (La Nina).

To compare with the corresponding CNOP, FSV is also generated using the SPG2 method with a very small constraint condition. Here, the objective function for FSV is a modified version of CNOP, which is obtained by replacing the nonlinear evolution of the initial perturbation with integration of the tangent linear model. That is

$$J(u_0) = \|u(T)\|_2 \approx \|M_T(U_0)u_0\|_2 \tag{4}$$

Due to the linear characteristics of SV, we yield another SV by multiplying the original SV by a constant, so as to make its norm also equal to $\sigma$. The other singular vectors are obtained by a general iterative power method (Farrell and Moore, 1992; Golub and Van Loan, 1996).

# 3 Ensemble Prediction Experiments

## 3.1 The Model and Experimental Design

In this part a two-dimensional nondimensional barotropic quasi-geostrophic (QG) model is adopted, which is as follows,

$$\begin{cases} \dfrac{\partial q}{\partial t} + J(\psi, q) = 0 \\ q = \nabla^2 \psi - F\psi + f_0 + f_0 h_s \qquad \text{in} \quad \Omega \times [0, T] \\ \psi|_{t=0} = \psi_0 \end{cases} \tag{5}$$

where the dependent variable is stream function $\psi(x, y, t)$, $q$ is the potential vorticity, $F = \frac{f^2 L^2}{gH}$ the Planetary Froude number, $f_0$ nondimensional Coriolis parameter, $H$ the typical characteristic height with the value of $H = 10^4$ m in this paper, $h_s$ nondimensional topography. The domain is $\Omega = [0, 2X] \times [0, 2Y]$, with double periodical boundary conditions. Typical characteristic length and velocity are $L = 10^6$ m, and $U = 10 \text{ms}^{-1}$, $F = 0.102$, $f_0 = 10.0$.

In our numerical approach, the five-point difference scheme and Arakawa finite difference scheme are employed to discretize the Laplacian operator and the Jacobian operator respectively. The temporal discretization is carried out by using Adamas-Bashforth scheme. $\Omega = [0, 6.4] \times [0, 3.2]$. The grid spacing $\Delta x = 0.2$ and the time step $\Delta t = 0.006$ corresponding to 200 km and 10 min respectively.

For simplicity, the topography is defined as a function of y-coordinate, similar to the one in experiment 2 or 3 in Mu and Zhang (2006). Here, it is given by

$$h_s(y) = h_0 \times \left( \sin\left(\frac{4\pi y}{2Y}\right) + 1 \right), \quad \text{where } h_0 = 0.112 \tag{6}$$

The energy norm is defined as

$$\|\psi\|_E{}^2 = \int_\Omega \left( |\nabla\psi|^2 + F|\psi|^2 \right) d\Omega \tag{7}$$

which is used in the objective function, $\nabla = (\partial/\partial x, \partial/\partial y)$.

And the $L_2$ norm is

$$\|\psi\|_{L_2}{}^2 = \int_\Omega \psi^2 d\Omega \tag{8}$$

which is used in the initial constraint condition, where $\psi$ is the stream function.

This study is carried out under a perfect model scenario. First, the nonlinear model is used to produce a reference atmospheric state starting from a specified, presumed exact initial condition. This state is regarded as a "true" state with which all predictions will be compared. The state of control forecast is then produced with the same model starting from the analysis field, which is generate by a simple four-dimensional variational (4-Dvar) data assimilation. Both SVs and CNOPs are calculated by using the state of control forecast as basic state, with optimization time of two days. The magnitudes of SVs and CNOPs, measured in $L_2$ norm, are equal to that of analysis errors. The perturbations are added to or subtracted from the analysis field, which is the initial state of the control forecast, to generate the perturbed initial ensemble states. Forecasts from perturbed initial ensemble states and one control forecast constitute one sample. The ensemble perturbations for sample 1 (denoted as S1) are composed of SVs. Sample 2 (denoted as S2) is obtained by replacing the FSV by CNOP in S1, with the other members remaining the same. Monte Carlo method (MC) is based on its interpretation as a random sample of a probability distribution (Leith, 1974), which is also used to be compared with S1 and S2. For S1 and S2, 7 members are generated respectively and for Monte Carlo, 23 members are generated.

## 3.2 Numerical Results

Four experiments are performed under different magnitudes of analysis errors. In the first three experiments, the true initial state $\psi_0^t$ is given by

$$\psi_0^t = 0.5\sin\left(\frac{\pi x}{X}\right) + 1.0\sin\left(\frac{\pi y}{Y}\right) + 0.5 \tag{9}$$

Its energy norm is $\|\psi_0^t\|_E = 6.582$ and $L_2$ norm is $\|\psi_0^t\|_{L_2} = 4.234$. The $L_2$ norm of analysis error in Exp. 1 is 0.0664, which account for 1.57% of that of the true state respectively. For experiment 2 and experiment 3, the $L_2$ norms of analysis error are 0.0340 and 0.0158 respectively. In the fourth experiment, the true initial state $\psi_0^t$ is given by

$$\psi_0^t = 1.0\sin\left(\frac{\pi x}{X}\right) + 1.0\sin\left(\frac{\pi y}{Y}\right) + 5.0 \tag{10}$$

Its energy norm is $\left\| \psi_0^t \right\|_E = 10.156$ and its $L_2$ norm $\left\| \psi_0^t \right\|_{L_2} = 23.073$. The $L_2$ norm of analysis error is 0.0476, accounting for 0.206% of that of the true state.

If the nonlinear growth of SV, measured in root mean squared errors, is greater (smaller) than that of analysis error in the early (later) part of the forecast, and the nonlinear growth of CNOP is the greatest all along, the type of analysis errors is regarded as fast-growing errors (shown in Fig. 1a). If the nonlinear growth of analysis error, measured in root mean squared errors, is smaller than the growths of SV and CNOP during the whole forecast time, the type of analysis errors is regarded as slow- growing errors (Fig. 1b).

Ensemble mean forecast is adopted as a principle tool for assessing forecast skills. The root mean squared (rms) distance in phase space with the forecast lead time between the ensemble mean of forecasts and the true state is illustrated in Fig. 2. The results show that for the first three experiments, the analysis errors belong to fast growing errors. Under this condition, in the early part of the forecast, there is no distinct difference between the mean forecasts of S1 and S2. In the later part of the forecast, ensemble S2 provides the best mean forecasts. With the reduction of the magnitudes of analysis error, S2 makes the mean forecast approach that of ensemble S1 gradually. We also investigate the case of slowly growing analysis errors, which is illustrated in experiment 4. The results show that the control run can provide a good forecast. S1 and S2 even make the forecast worse. Besides, in this case even though MC has more ensemble members than SV/CNOP method, it almost makes no improvement of the forecast skill during the whole forecast period in the above four experiments. This suggests that when analysis errors belong to fast-growing type, compared with SV method, CNOP method can represent the evolution of initial analysis error better, and consequently makes the ensemble forecast better. While analysis errors belong to slow-growing type, both SV and CNOP methods cannot represent the evolution of initial analysis error well, and yield the failure of the ensemble forecast.

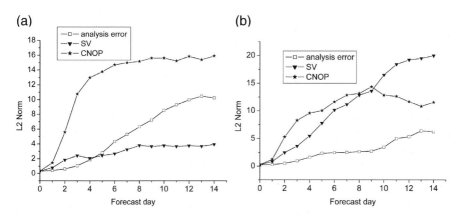

**Fig. 1** Nonlinear growths of analysis error, 1st SV and CNOP with the forecast time in terms of $L_2$ norm. (**a**) Exp. 1. (**b**) Exp. 4

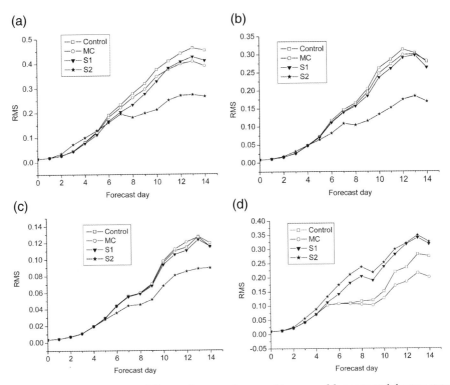

**Fig. 2** The root mean squared distance between the ensemble mean of forecasts and the true state with the forecast time. (**a**) Exp. 1 (**b**) Exp. 2 (**c**) Exp. 3 (**d**) Exp. 4

It is well known that ensemble prediction is an approximate implementation of probability forecasting. Under a perfect model assumption, the ensemble forecast skill mainly depends on whether the initial ensemble perturbations can represent the initial analysis error properly. In the above experiments, if analysis errors belong to fast-growing ones, both SV method and ours can represent the initial analysis errors well, so they make good forecasts. If analysis errors are slow-growing type, the initial analysis errors can be properly represented by neither SV method nor our approach, so they make bad forecasts. The above analysis suggests that understanding the information about analysis error, and then adopting suitable initial ensemble perturbation method are very important. One of the main functions of ensemble prediction is to improve the forecast skill of extreme weather events, of which the initial analysis errors almost all belong to fast-growing ones (otherwise, under a perfect model assumption, slow-growing errors will not yield a failure of the forecast). Consequently, it is reasonable to expect that our new method may be effective to the cases of extreme weather events. Of course, this needs further studies and implementation of experiments.

# 4 Potential Application to Targeted Observation

## 4.1 The Model and Synoptic Situations

MM5 and its adjoint system are used to obtain CNOP and FSV in this part. The moist physical parameterizations included in this study are dry convective adjustment, grid-resolve large-scale precipitation, and a Kuo type cumulus parameterization schemes.

The dry total energy (TE) norm is used in both the objective function and initial constraint condition, which is expressed in a continuous formulation as follows:

$$\|\|_E^2 = \frac{1}{D} \int_D \int_0^1 \left[ u'^2 + v'^2 + \frac{c_p}{T_r} T'^2 + R_a T_r \left( \frac{p'_s}{p_r} \right)^2 \right] d\sigma\, dD \qquad (11)$$

Where $c_p$ and $R_a$ are specific heat at constant pressure and gas constant of dry air, respectively (with numerical values of $1005.7 \, \mathrm{J\,kg^{-1}\,K^{-1}}$ and $287.04 \, \mathrm{J\,kg^{-1}\,K^{-1}}$, respectively), and the integration extends over the full horizontal domain $D$ and vertical direction $\sigma$. Reference temperature $T_r = 270 \, \mathrm{K}$ and surface pressure $p_r = 1000 \, \mathrm{hPa}$. Terms $u'$, $v'$, $T'$ and $p'_s$ are the perturbed zonal and meridional horizontal wind components, temperature and surface pressure, respectively.

The case A studied here is a MeiYu front rainfall process over the Huaihe River basin during July 2003. The simulation was initialized at 0000 UTC 04 July 2003 with NCEP analysis reanalyzed with conventional observations. The case B is a cyclonic rainfall over Huabei plain during August 1996. The simulation was initialized at 0000 UTC 05 August 1996 with the ECMWF reanalysis data. For two cases, the model forecast domain employs 51(61) grid points in the north–south (east–west) direction with 10 vertical layers. 120 km grid resolution is employed with the model center at (32°N, 110°E). The model top pressure is 100 hPa. The initial temperature and wind vector at sigma level 0.45 (about 550 hPa) and the 24 h simulation rain field are showed in Fig. 3a for case A and Fig. 3b for case B. The simulated 24 h accumulation precipitation pattern is well consistent with the observation, but for case A rainfall is not as larger as the observation. This deficiency may partly due to the low horizontal and vertical resolution and the use of simple moist physical process. The optimization period is 24 h. The local projection operator $\mathbf{P}$ is employed in objective function, the value of $\mathbf{P}$ within (without) the verification region (quadrate domain, see Fig. 3) is 1 (0).

## 4.2 Numerical Results

### 4.2.1 Case A

CNOP and local CNOP have been found in this case. We find that CNOP, local CNOP and FSV exhibit baroclinic structures as can be seen in Fig. 4 showing

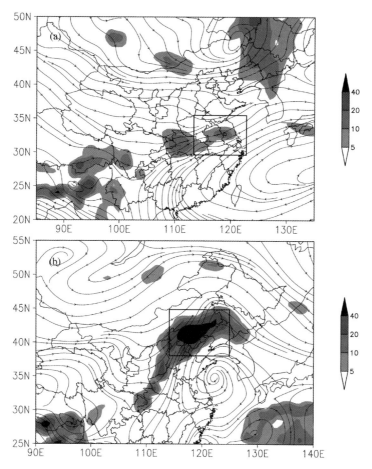

**Fig. 3** The streamline (contour) on 0.45 σ level and the 24-hour simulated accumulation precipitation (mm; *shaded*) at the initial time, (**a**) Case A, (**b**) Case B. Rectangular area is the verification area

temperature in 3-dimensional space. There is a pronounced westward tilt with height in the temperature field.

The horizontal structures of CNOP, local CNOP and FSV exhibit properties of localness, however, their structures are less similar. As an illustration, the temperature and wind perturbations of CNOP, FSV and local CNOP at sigma level 0.45 (about 550 hPa) are shown in Fig. 5. These patterns are quite localized near the trough over the northeast of Tibetan Plateau. At other levels CNOP and local CNOP have also the local characteristic (Figures not shown). Closer inspection of Fig. 5 indicates that CNOP, FSV and local CNOP differ by a slight phase shift in spatial pattern: the pattern of FSV shift right CNOP by about one quarter phase and the pattern of local CNOP shift left CNOP by about one quarter phase.

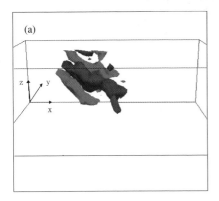
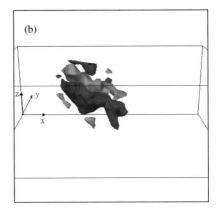

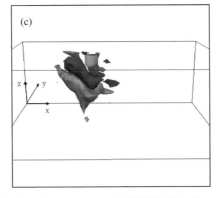

**Fig. 4** 3-dimensional distribution of temperature at 0000UTC 4 July for case A, black for 0.2°C, gray for −0.2°C: (**a**) CNOP; (**b**) FSV; (**c**) local CNOP

It is evident that there is two strong warm centers at about (32°N, 101°E) and (35°N, 109°E) and a moderate cold centre at about (37°N, 111°E) respectively for CNOP (Fig. 5a). From Fig. 5b it is easily seen that there are a strong cold centre at about (33°N, 97°E) and two moderate warm centers lying to the right of the cold center for the FSV. At this level the maximums (negative minimums) of T u and v are 0.8 K (−0.6 K), 0.4 m/s (−1.4 m/s), 0.5 m/s (−0.3 m/s) for CNOP and 0.6 K (−1.0 K), 0.6 m/s (−0.6 m/s), 0.9 m/s (−0.3 m/s) for FSV respectively. From the wind vector of Fig. 5a and b, it can be seen that the location of the center of cyclic perturbation circulation is different.

The structures of local CNOP and CNOP seem to closely resemble each other at this level, but the location center in temperature field has a little different. In addition, the main cold temperature center for CNOP is at (37°N, 111°E) (Fig. 5a) but for that of local CNOP is at (33°N, 92°E) (Fig. 5c).

It is of interest to investigate how the growth of CNOP is during the optimization time interval of 24 h. Figure 6 shows the nonlinear evolutions of CNOP and FSV

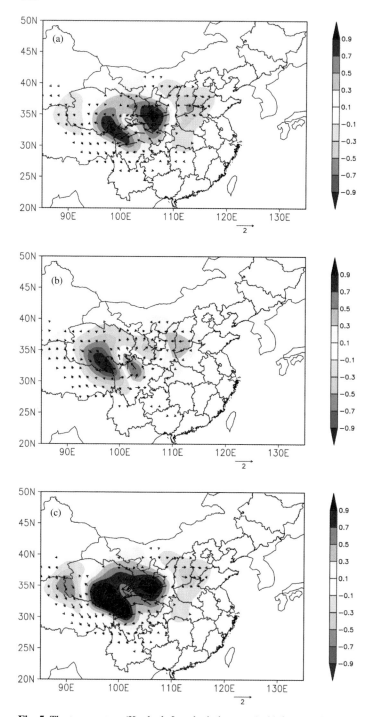

**Fig. 5** The temperature (K; *shaded*) and wind vector (m/s) for case A on 0.45σ level at 0000UTC 4 July: (**a**) CNOP; (**b**) FSV; (**c**) local CNOP

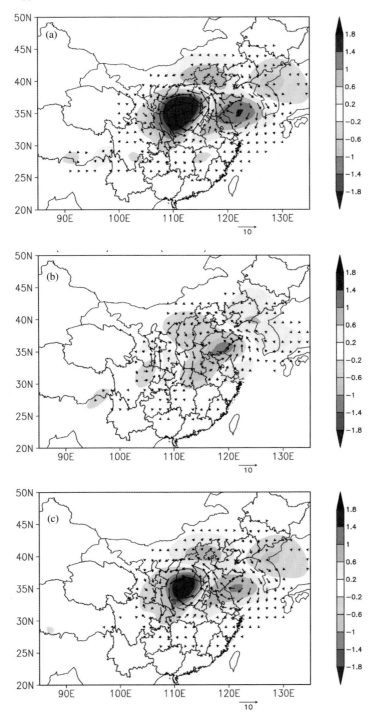

**Fig. 6** The same as Fig. 5, but at 0000UTC 5 July (**a**) CNOP; (**b**) FSV; (**c**) local CNOP

and local CNOP at 0500 UTC. It is clear that the CNOP has moved along the flow on 550 hPa while maintaining its local structures and the horizontal scale becomes larger at the same time. They arrive at the verification domain where the rainfall process happened.

From the temperature and wind perturbations of CNOP, FSV and local CNOP at sigma level 0.45 shown in Fig. 6, it can be seen that there exhibits an apparent wave train pattern in temperature filed. This characteristic of phase shift in spatial patterns is much clearer than that at the initial time.

The maximums (negative minimums) of T, u and v are 0.6 K (−1.5 K), 4.0 m/s (−4.0 m/s), 7.0 m/s (−6.0 m/s) respectively for CNOP at level $\sigma = 0.45$. The non-linear evolution of FSV (Fig. 6b) at optimization finial time is 0.6 K (−1.2 K), 3.5 m/s (−2.5 m/s), 6.0 m/s (−5.0 m/s). These results show that the maximum possible nonlinear perturbation evolution of CNOP is larger than that of FSV. This characteristic is even more obvious at 0.25 sigma level (Figures not shown). The nonlinear evolution of -FSV is also investigated, which is weak comparing with that of FSV (Figures not shown).

The time evolution of TE is shown in Fig. 7. The TE has been scaled by a factor which value is equal to the initial total perturbation energy. For the CNOP and local CNOP, finial energy J increase by a factor of 20.7 and 20.4 over 24 h respectively, and for the FSV (-FSV), finial total energy J is amplified by a factor 14.1 (11.5). It is clear that finial total energy J of evolution of CNOP is about 1/3 larger than that of nonlinear evolution of FSV (-FSV). It becomes apparent that for CNOP much of growth occurs during the last half of the optimization interval and for local CNOP occurs during the last third of the optimization interval. The amplification factor of CNOP and local CNOP is larger than that of FSV (-FSV) all the time.

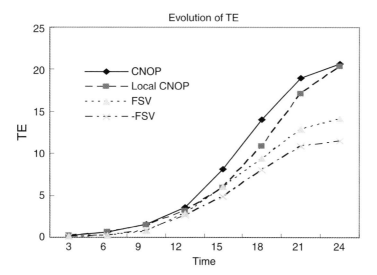

**Fig. 7** The development of total energy in the verification area during the optimal period for case A, CNOP (*solid*), local CNOP (*long dashed*), FSV (*short dashed*) and –FSV(*dash dotted*). The y-axis values are the results of TE divided by the counterpart of initial time

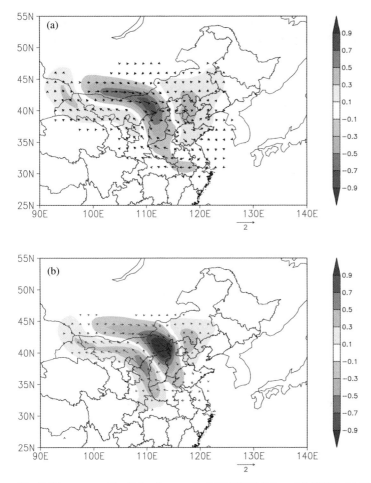

**Fig. 8** The same as Fig. 5, but for case B at 0000UTC 5 Aug. (**a**) CNOP, (**b**) FSV

### 4.2.2  Case B

CNOP and FSV are also investigated for the second synoptic case. Local CNOP is not found in this case. From Fig. 8, it is shown that both CNOP and FSV have the properties of localness, similar to case A, the vertical temperature fields of CNOP and FSV show baroclinic structures (figures not shown).

The main difference from case A is that the CNOP and FSV are located near and in the verification domain. This is because that the cyclone led to this rainfall maintain and move slowly, this property is not like that west flow in the upstream of the verification area in case A.

The temperature and wind perturbations of the nonlinear evolutions of CNOP and FSV at 0000 UTC 6 Aug at sigma level 0.45 (about 550 hPa) are shown in Fig. 9. It is clear that the CNOP has moved along the flow on 550 hPa while the anticyclonic circulation becomes stronger at the same time. They arrive at the verification domain where the rainfall process happened.

The maximums (negative minimums) of T, u and v are 0.6 K (−1.1 K), 3.0 m/s (−4.0 m/s), 2.0 m/s (−5.0 m/s) respectively at this level for CNOP. The nonlinear evolution of FSV at optimization finial time is 0.8 K (−1.2 K), 3.5 m/s (−2.0 m/s), 2.0 m/s (−5.0 m/s). These results show that the maximum possible nonlinear perturbation evolution of CNOP is almost equivalent to that of FSV at this level.

Although the evolution of CNOP at 0.45 sigma level seems almost equivalent to that of FSV, the evolution of TE for CNOP and FSV is much different (Fig. 10). For the CNOP, finial energy increase by a factor of 13.0 over 24 h, and for the FSV and -FSV, finial total energies are amplified by a factor 8.56 and 8.63 respectively. It is clear that finial total energy of evolution of CNOP is about 1/2 larger than that of nonlinear evolution of FSV (-FSV). The amplification factor of CNOP is larger than that of FSV (-FSV) all the time.

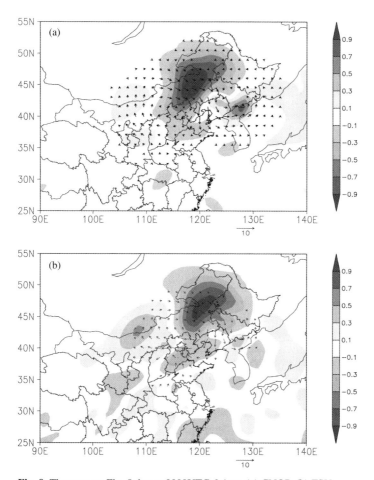

**Fig. 9** The same as Fig. 8, but at 0000UTC 6 Aug. (**a**) CNOP, (**b**) FSV

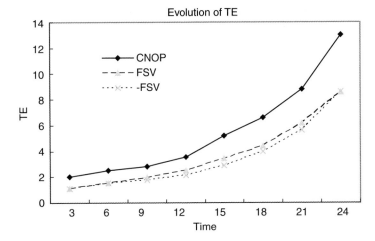

**Fig. 10** The same as Fig. 7, but for case B

SI is also calculated to compare the difference of the evolutions of CNOP and FSV at 0000 UTC 6 Aug. SI between evolutions of CNOP and FSV (-FSV) is 0.60 (−0.36). It indicates that the structures of evolutions of CNOP are different from those of FSV.

## 4.3 Sensitivity Experiments

The above numerical results have shown that the CNOP has the largest impact on the forecast in the verification area. Considering the initial constraint condition stands for the magnitude of analysis error, the question followed is that whether the forecast skill in the verification domain can be improved when the CNOP/FSV type error in initial state is reduced. Several ideal sensitivity experiments are carried out to investigate this problem. The amplitudes of CNOP and FSV are reduced by factors of 0.75, 0.50 and 0.25 respectively, and then experiments are named as C0.75, C0.50, C0.25, F0.75, F0.50 and F0.25 correspondingly. The amplitudes of CNOP and FSV which are not rescaled are named as C1.00 and F1.00 respectively.

### 4.3.1 Results of Case A

Figure 11 shows the evolutions of TE in these experiments for case A. Comparing the results of experiments C1.00, C0.75, C0.50 with those of experiments F1.00, F0.75, F0.50, it is found that, much gain is obtained when the amplitude of CNOP type error is reduced. And, when the amplitude becomes smaller(see C0.50, C0.25, F0.50, F0.25), the nonlinear evolution of TE is not sensitive to perturbation pattern

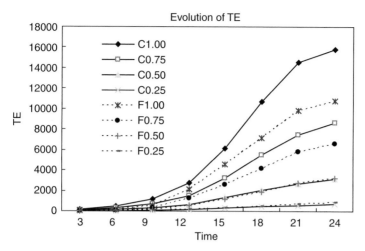

**Fig. 11** The development of total energy for sensitivity trials C1.00, C0.75, C0.50, C0.25, F1.00, F0.75, F0.50, F0.25 for case A, CNOP (*solid lines*), FSV (*short dashed lines*)

under the same amplitude, or expressed as when the amplitude becomes big, nonlinear evolution of TE is sensitive to perturbation structure under the same amplitude.

### 4.3.2 Results of Case B

Figure 12 shows the evolution of TE in these experiments for case B. Clearly, the differences between the evolutions of TE show that, much gain is obtained when the amplitude of CNOP type error is reduced if the factor is larger than 0.5. Comparing these results with those of experiments F1.00, F0.75, F0.50, F0.25, it is found

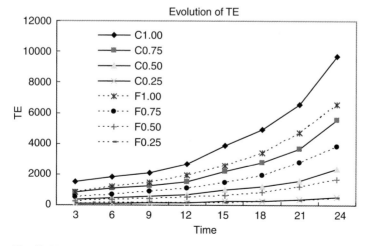

**Fig. 12** The same as Fig. 11, but for case B

that when the amplitude becomes smaller (see C0.50, F0.50, C0.25 and F0.25), the nonlinear evolution of TE is not sensitive to perturbation pattern under the same amplitude.

# 5 Conclusions and Discussions

The above numerical experiments are preliminary exploration of the new method CNOP in two parts: ensemble prediction and adaptive observations.

In the first part, a quasi-geostrophic model has been adopted under a perfect model assumption, so all prediction errors are caused by uncertainties in the initial state. The results show that under the condition that the analysis errors grow rapidly, in the early part of the forecast, there is no distinct difference between the mean forecasts of S1 and S2. In the later part of the forecast, ensemble S2 provides the best mean forecasts. With the reduction of the magnitudes of analysis error, S2 makes the mean forecast approach that of ensemble S1 gradually. We also investigate the case of slowly growing analysis errors. The results show that the control run can provide a good forecast. S1 and S2 even make the forecast worse. That suggests that when analysis errors belong to fast-growing type, compared with SV method, CNOP method can represent the evolution of initial analysis error better, and consequently makes the ensemble forecast better. While analysis errors belong to slow-growing type, both SV and CNOP methods cannot represent the evolution of initial analysis error well, and yield the failure of the ensemble forecast.

Only one CNOP is used as ensemble perturbation in S2, and the ensemble mean forecast is adopted as the evaluation method. Our analyses reveal that the forecast skill of different generation of initial ensemble perturbation depends on the type of the analysis error, which can be understood from the fact that ensemble prediction is an approximate implementation to probability forecast. To investigate the application of CNOP to ensemble forecasts further, more complex models and a large number of experiments and other evaluating methods are needed to confirm the above numerical results. CNOPs of different initial constraint conditions and different optimization time attempt to be applied in the ensemble forecasts, and especially the local CNOP of some basic states may need to be explored. Work is under way to verify if such ensembles can improve the forecast quality better.

Admittedly, our experiments are based on a perfect model scenario, whereas, the assumption is not a realistic analog for actual numerical weather prediction, where model error may be significant or even dominant. So, the application of CNOP to operational ensemble prediction system still has a long way to go which needs to be explored. But it is not the limitation of CNOP itself, other approach, such as LSV , also has such problem. Besides, our results show that when analysis errors belong to fast-growing type, compared with SV method, CNOP method can represent the evolution of initial analysis error better, and consequently makes the ensemble forecast better; On the contrary, the introduction of CNOP may make the forecast bad. But for real atmospheric forecasting, no one can know in a prior whether the analysis errors belong to fast-growing type, so it is difficult for us to choose the method for generation of ensemble perturbations. What is more, if the ensemble prediction using our method forecast some weather events which often do not happen in the real atmosphere, it is not acceptable to use our method to make the ensemble prediction. Considering that the ensemble prediction using the SV method may not forecast the weather events which do happen in the real atmosphere, as a more immediate goal, it is necessary to distinguish the two contrary situations to seek how to combine the various methods to make the ensemble perturbations. It is expected that further study on applications of nonlinear optimization technique to ensemble prediction may offer guidelines for the future development of ensemble forecasting technique.

In the second part, the potential application of CNOP to identify the data-sensitive region in targeting strategy is explored. As the first step, only the difference between CNOP and FSV including their evolutions is discussed. CNOPs have been computed by using a mesoscale model with moist physical process both for a Mei Yu front rainfall case and for a cyclonic rainfall case over an optimization time interval of 24 h.

The results indicate that the structures of CNOP (local CNOP) and FSV exhibit localness and baroclinic properties. The structures of CNOP (local CNOP) as well as their evolution are less similar to those of FSV in terms of similarity index. Besides, it is found that the total energy of CNOP (local CNOP) at optimization time is larger than that of FSV (-FSV). The results of sensitivity experiments indicate that the forecasts are more sensitive to the CNOP-type initial errors than to FSV-type. This suggests that the forecast skill will be improved much from the reduction of CNOP-type error in initial condition than FSV-type. These results suggest that CNOP are feasible to identify the data-sensitive region in adaptive observation. When the CNOPs are available, the procedure SVs had utilized can be employed to specify the targeting area (Buizza and Montani 1999). The targeting area can be defined by the grid point where its total energy is larger than a certain value, for example, 0.5 times its maximum energy. The impact of norm specification on CNOPs needs to be considered also. The analysis error variance metric may be a substitute to energy norm at the initial time. Other cost functions with definite physical means at finial time can be chosen. In addition, we have to solve a high dimensional constrained optimization question to obtain CNOPs. The efficiency of the optimization algorithm is a challenge. These issues will be further studied in the context of an observation system simulation experiment. Based on the operational 4-Dvar data assimilation system, it is not difficult to build a prototype to solve constrained optimization problem in procedure. CNOP is expected to be tested together with other method in real operational targeted observation.

# References

Anderson JL (1997) The impact of dynamical constraints on the selection of initial conditions for ensemble predictions: low-order perfect model results. Mon Wea Rev 125(11):2969–2983

Barkmeijer J (1996) Constructing fast-growing perturbations for the nonlinear regime. J Atmos Sci 53(19): 2838–2851

Birgin EG et al (2000) Nonmonotone spectral projected gradient methods for convex sets. SIAM Journal on Optimization 10(4): 1196–1211

Bishop CH et al (2001) Adaptive sampling with the ensemble transform Kalman filter. Part I: Theoretical aspects. Mon Wea Rev 129:430–436

Buizza R et al (1993) Computation of optimal unstable structures for a numerical weather prediction model. Tellus 45A: 388–407

Buizza R, Palmer TN (1995) The singular-vector structure of the atmospheric global circulation. J Atmos Sci 52(9):1434–1456

Buizza R, Montani A (1999) Targeting observations using sigular vectors. J Atmos Sci 56(17): 2965–2985

Buizza R et al (2005) A comparison of the ECMWF, MSC, and NCEP global ensemble prediction systems. Mon Wea Rev 133(5):1076–1097

Buizza R et al (2007) The value of observations. II: The value of observations located in singular-vector-based target areas. Q J R Meteorol Soc 133:1817–1832

Duan WS et al (2004) Conditional nonlinear optimal perturbations as the optimal precursors for ENSO events. J Geophys Res 109: D23105, doi:10.1029/2004JD004756

Duan WS, Mu M (2006) Investigating decadal variability of El Nino-Southern Oscillation asymmetry by conditional nonlinear optimal perturbation. J Geophys Res 111, Co7015, doi:10.1029/2005JC003458

Duan WS et al (2007) The decisive role of nonlinear temperature advection in El Nino and La Nina amplitude asymmetry. J Geophys Res doi:10.1029/2006JC003974

Dudhia J (1993) A nonhydrostatic version of the Penn State/NCAR mesoscale model: Validation tests and simulation of an Atlantic cyclone and cold front. Mon Wea Rev 121:1493–1513

Emanuel KA et al (1995) Report of the first prospectus development team of the U.S. Weather Research Program to NOAA and the NSF. Bull Amer Meteor Soc 76: 1194–1208

Errico RM et al (1993) Examination of the accuracy of a tangent linear model. Tellus 45A:462–477

Errico RM et al (2002) The workshop in atmospheric predictability. Bull Amer Meteor Soc 9:1341–1343

Ehrendorfer M, Tribbia J (1997) Optimal prediction of forecast error covariances through singular vectors. J Atmos Sci 54(2):286–313

Ehrendorfer M (2000) The total energy norm in a quasigeostrophic model. J Atmos Sci 57: 3443–3451

Farrell BF, Moore AM (1992) An adjoint method for obtaining the most rapidly growing perturbation to oceanic flows. J Phys Oceanogr 22:338–349

Frederiksen JS (2000) Singular vector, finite-time normal modes, and error growth during blocking. J Atmos Sci 57: 312–333

Gelaro R et al (1999) An assessment of the singular vector Approach to targeted observations using the FASTEX data set. Q J R Meteorol Soc 125: 3299–3328

Gelaro R et al (2002) Singular vector calculations with an analysis error variance metric. Mon Wea Rev 130(5): 1166–1186

Gilmour I, Smith LA (1997) Enlightenment in Shadows. In: Kadtke JB, Bulsara A (eds) Applied nonlinear dynamics and stochastic systems near the millennium. AIP, New York, pp 335–340

Golub GH, Van Loan CF (1996) Matrix computation. Johns Hopkins University Press, Baltimore, MD, 694pp

Hamill TM et al (2003) Ensemble forecasts and the properties of flow-dependent analysis-error covariance singular vectors. Mon Wea Rev 131(8):1741–1758

Joly A et al (1999) Overview of the field phase of the fronts and Atlantic storm-track experiment (FASTEX). Q J R Meteorol Soc 125: 3131–3163

Lacarra J, Talagrand O (1988) Short-range evolution of small perturbations in a barotropic model. Tellus 40A:81–95

Langland RH et al (1999a) The North Pacific Experiment (NOPREX298): Targeted observations for improved North American weather forecasts. Bull Amer Meteor Soc 80:1363–1384

Langland RH et al (1999b) Targeted observations in FASTEX: Adjoint based targeting procedures and data impact experiments in IOP/ 7 and IOP/ 8. Q J R Meteor Soc 125:3241–3270

Langland RH (2005) Issues in targeted observing. Q J Roy Meteor Soc 131: 3409–3425

Leith CE (1974) Theoretical skill of Monte Carlo forecasts. Mon Wea Rev 102:409–418

Leutbecher M (2005) On ensemble prediction using singular vectors started from forecasts. Mon Wea Rev 133(10):3038–3046

Lorenz EN (1965) A study of the predictability of a 28-variable model. Tellus 17:321–333

Molteni F, Palmer TN (1993) Predictability and finite-time instability of the northern winter circulation. Q J R Meteorol Soc 119:269–198

Molteni F et al (1996) The new ECMWF ensemble prediction system: methodology and validation. Q J R Meteorol Soc 122:73–119

Mureau R et al (1993) Ensemble prediction using dynamically conditional perturbations. Q J R Meteorol Soc 119:299–323

Mu M et al (2000) The impact of nonlinear stability and instability on the validity of the tangent linear model. Adv Atmos Sci 17:375–390

Mu M, Duan WS (2003) A new approach to studying ENSO predictability: Conditional nonlinear optimal perturbation. Chinese Sci Bull 48:747–749

Mu M et al (2004) The sensitivity and stability of the ocean's thermohaline circulation to finite amplitude perturbations. J Phys Oceanogr 34:2305–2315

Mu M, Zhang ZY (2006) Conditional nonlinear optimal perturbations of a barotropic model. J Atmos Sci 63:1587–1604

Mu M, Jiang ZN (2007) A new approach to the generation of initial perturbations for ensemble prediction: Conditional nonlinear optimal perturbation. Chinese Science Bulletin 52:1457–1462 (in Chinese)

Mu M et al (2007a) A kind of initial errors related to 'spring predictability barrier' for El Nino events in Zebiak-Cane model. Geophy Res Lell 34, L03709, doi:10.1029/2006GL027412

Mu M et al (2007b) A Preliminary application of conditional nonlinear optimal perturbation to adaptive observation. Chinese J Atmos Sci 31(6):1102–1112 (in Chinese)

Oortwijn J, Barkmeijer J (1995) Perturbations that optimally trigger weather regimes. J Atmos Sci 52:3932–3944

Oortwijn J (1998) Predictability of the onset of blocking and strong zonal flow regimes. J Atmos Sci 55:973–994

Palmer TN et al (1998) Singular vectors, metrics, and adaptive observations. J Atmos Sci 55:633–653

Petersen GN et al (2006) Impact of NATReC observations on the UK Global model forecasts. Geophys Res Abs 8:08999

Snyder C (1996) Summary of an informal workshop on adaptive observations and FASTEX. Bull Amer Meteor Soc 77:953–961

Zhang FQ et al (2002) Mesoscale predictability of the "surprise" snowstorm of 24–25 January 2000. Mon Wea Rev 130:1617–1632

Zou XL et al (1997) Introduction to adjoint techniques and the MM5 adjoint modeling system. NCAR Technical Note, NCAR/ TN2435 + STR. 117pp

# Study on Adjoint-Based Targeted Observation of Mesoscale Low on Meiyu Front

Peiming Dong, Ke Zhong and Sixiong Zhao

**Abstract** Targeted observation is such an idea that to reduce the small errors in the initial analysis that induce great errors in the numerical weather forecast by adding observation in the targeted area and so obtain the improvement of the numerical prediction accuracy of a given weather system. Many works associated with this topic have been carrying out, especially by European and USA scientists. The cyclone forming and developing in Pacific and Atlantic Ocean is mainly focused on in their research. It has significant meaning to address the key issues associated with targeted observation of Asian weather systems, together with the feasibility to improve the accuracy of their numerical forecast by using this technology. These weather systems have individual characteristics comparing with that of European and USA.

Study on adjoint-based targeted observation of mesoscale low on Meiyu Front is implemented in this paper. The mesoscale model MM5 and its corresponding tangent linear and adjoint model are used. The scheme of sensitivity analysis based on adjoint method is designed firstly. The linear assumption is verified to remain valid for two days, thus it is proper to use the adjoint-based method for the targeted observation issue of two days numerical forecast of mesoscale low on Meiyu Front. The distributions of sensitivity gradient of different physical fields tend to be consistent. So the targeted observation region decided by this kind of sensitivity analysis will fulfill the requirement of targeted observation. At the same time, it is found that the sensitivity gradient is dominated by the leading vectors. It denotes that it is also proper to decide the targeted observation region by using the leading singular vectors. Following the discussions of these key issues associated with adjoint-based targeted observation of mesoscale low on Meiyu Front, the possibility is verified whether the accuracy of numerical forecast could be improved by this method.

**Keywords:** Targeted observation · adjoint · sensitivity analysis · mesoscale low · meiyu front

P. Dong (✉)
Institute of Atmospheric Physics, Chinese Academy of Sciences, Beijing, 100029, China,
e-mail: dongpm@cams.cma.gov.cn

S.K. Park, L. Xu, *Data Assimilation for Atmospheric, Oceanic and Hydrologic Applications*, DOI 10.1007/978-3-540-71056-1_13,
© Springer-Verlag Berlin Heidelberg 2009

# 1 Introduction

Recent study shows that the improvement of the quality of initial conditions is crucial to improve the accuracy of numerical weather forecast. Some small errors in the initial analysis can induce great errors in the following forecast. By adding some extra observations in the region on where the errors concrete in the initial conditions, it is expected that the errors in the initial analysis could be reduced and the improvement of numerical forecast could be obtained. This method is called targeted observation. It is also named adaptive observation in some references.

Many works about targeted observation have already been carrying out, especially in European and USA (Emanuel et al. 1995; Langland and Rohaly 1996; Joly et al. 1997; Palmer et al. 1998; Langland and Gelaro et al. 1999; Bergot et al. 1999; Berliner et al. 1999; Bishop and Toth 1999; Buizza and Montani 1999; Bishop et al. 2001; Majumdar et al. 2002). With the investigation of the method to identify the sensitivity area of targeted observation, the possibility of improving the numerical forecast is illustrated. A serial of field experiments are also carried out to match the requirement of research simultaneously. Such as, The Fronts and Atlantic Storm-Track Experiment (FASTEX), The North Pacific Experiment (NORPEX) and The Winter Storms Reconaissance Program (WSRP) et al. The targeted observation method was already put into operational use in some numerical weather centers including ECMWF and NCEP.

The cyclone forming and developing in Pacific and Atlantic Ocean is mainly focused on in the research mentioned above. It is a scientific matter for Asian scientist to investigate the potential to improve the numerical forecast of Asian weather system by using this targeted observation method. Especially, Asian weather systems have individual characteristics comparing with that of European and USA. There is definitely a crucial need to discuss the key issues associated with targeted observation of Asian weather systems.

The mesoscale low on Meiyu Front is taken as the research object in our attempt to address this topic. Meiyu Front, which is called Baiu in Japan and Changma in Korea, is the main synoptic system that brings rainfall to these countries during the summer monsoon season every year. Mesoscale low on Meiyu Front is one of the most important mesoscale systems associated with heavy rainfall. It always moves from west to east along the Meiyu Front and brings heavy rainfall with a large coverage along its trace. It affects mainly the middle and lower reaches of the Yangtze River in China. Then it also affects the downstream areas including Korea and Japan during its later stage after it moves over the sea.

With the idea to improve the numerical forecast of mesoscale low on Meiyu Front in mind, a series of study on adjoint-based targeted observation of mesoscale low on Meiyu Front is carried out. Some results are presented in this paper. It is organized in the following way: a brief description of adjoint-based sensitivity analysis is illustrated in Sect. 2. Then the mesoscale low case on Meiyu Front is introduced in Sect. 3. The verification of the ability of the tangent linear and adjoint model to describe small perturbation in the nonlinear model associated with mesoscale low on Meiyu Front is given in Sect. 4. The result of sensitivity analysis and the possibility of improving numerical forecast are presented in Sects. 5 and 6, respectively. Section 7 is a brief discussion and concluding remark on this research topic.

## 2 The Adjoint-Based Sensitivity Analysis

The adjoint-based sensitivity analysis is used to decide the sensitivity area in the initial time. The local errors will be decreased by adding extra observations in this targeted area. The accuracy of numerical forecast is expected for the errors in the sensitivity area are crucial to the growing forecast errors in the integral process of numerical model. The adjoint-based sensitivity analysis is an analysis of sensitivity gradient. The sensitivity gradient

$$\partial J / \partial \vec{x}_{0,t} = L^T P^T E P (\vec{x}_{k,f} - \vec{x}_{k,t})$$

is taken as the gradient of the scalar cost function

$$J = 1/2 \langle P(\vec{x}_{k,f} - \vec{x}_{k,t}), P(\vec{x}_{k,f} - \vec{x}_{k,t}) \rangle_E$$

where $\langle , \rangle_E$ demonstrates the inner product defined through a positively defined Hamilton matrix E, $L^T$ is the adjoint model of the nonlinear forecast model $F$, $P$ is the local projection matrix. It is a diagonal matrix. The element is one for the grid in the sensitivity analysis object domain, while zero for other point. $\vec{x}_{k,t}$ and $\vec{x}_{k,f}$ are the analysis and forecast, respectively. The subscript t and f denote the analysis and forecast state. k indicates that the state is at k time.

The dry energy metric is adopted as E in our study. It is illustrated as following:

$$\langle (\vec{x}_{k,f} - \vec{x}_{k,t}), (\vec{x}_{k,f} - \vec{x}_{k,t}) \rangle_E$$

$$= \iiint \bar{\rho} \left( \frac{u'^2}{2} + \frac{v'^2}{2} + \frac{w'^2}{2} + \frac{1}{2} \frac{g^2}{\bar{N}^2} \frac{\theta'^2}{\bar{\theta}^2} + \frac{1}{2\bar{\rho}^2} \frac{p'^2}{c_s^2} \right) \frac{\partial \bar{p}}{\partial \sigma} dx dy d\sigma$$

the first two items in left is horizontal kinetic energy. The third item is vertical kinetic energy. The fourth and firth items are available potential energy and elastic energy, respectively. Where, $u',v',w',\theta',p'$ are the differences between $\vec{x}_{k,f}$ and $\vec{x}_{k,t}$ of the west-east wind velocity, north-south wind velocity, vertical velocity, potential temperature and perturbed pressure, respectively. $\bar{\theta}$ is the potential temperature of base state. In practical calculation, $\theta'$ and $\bar{\theta}$ are substituted with the difference of temperature and the base state temperature approximately. $\bar{\rho}$ is the density of base state. $\bar{N}$ is Brunt-Väisälä frequency. $\bar{p}$ is reference pressure, and $c_s$ is sound velocity of base state. g is gravitational acceleration.

## 3 The Research Case

The mesoscale low on Meifu Front in this paper is the case that occurred during the period 26–28 June 1999. Figure 1 a-c is the NCEP reanalysis geopotential height at 850 hPa from 0000 UTC 26 to 0000 UTC 28 June 1999, with an interval of 24 hours. The mesoscale low located near (29°N, 107°E) with a value of 1440 gpm. It moved from west to east along the Meiyu Front and developed gradually. At 0000 UTC

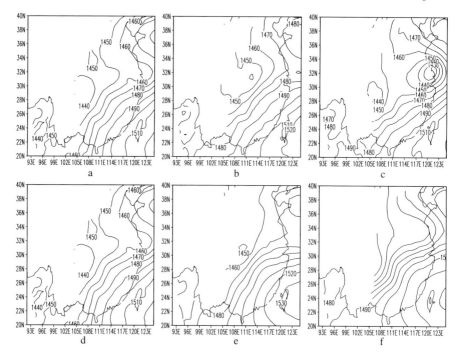

**Fig. 1** Geopotential height of the NCEP reanalysis and simulation of the mesoscale low from 26 to 28 June 1999 at 850 hPa. The interval of contour is 10 gpm. (**a**)–(**c**) are the NCEP reanalysis from 0000 UTC 26 to 0000 UTC 28 June 1999, with an interval of 24 hours; (**d**)–(**f**) are the integral of the original initial analysis

27 June 1999, the low got to the middle reach of the Yangtze River in China. The center was 1439 gpm. At 0000 UTC 28 June, the low arrived at the East China Sea region and deepened further. The center has deepened to 1419 gpm. It moved over the sea and kept on developing and also brought heavy rainfall to the downstream areas including Korea and Japan. The stage of the mesoscale low in China region is mainly focused on in our study.

The NCEP reanalysis data is used as the initial condition to the regional numerical forecast model MM5. The model takes the integral domain with 45 and 43 grids in west–east and north–south direction respectively, and 15 vertical levels. The horizontal grid distance is 60 km. The center of the integral domain locates at (30°N, 112°E). The model domain is illustrated in Fig. 2 as D01. MM5 provides the user with abundant physical schemes, such as the cumulus parameterization scheme, the planetary boundary layer scheme, radiation scheme and explicit moisture scheme. More details about the model could refer to the user guide (Grell et al. 1994). The Grell cumulus parameterization and Blackadar high resolution planetary boundary scheme with no explicit moisture scheme are used here. The tangent linear and adjoint models corresponding to the model MM5, developed by Zou et al. (1997; 1998) are utilized in the followed adjoint-based research. The tangent linear model and the adjoint model take the same physical configuration as the nonlinear model.

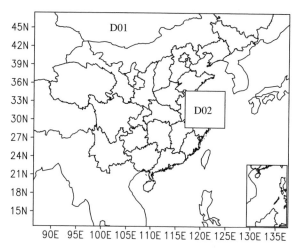

**Fig. 2** The integral domain (D01) and the sensitivity analysis object domain (D02)

It could be seen form Fig. 1 d–f that the numerical forecast of mesoscale low from the original initial condition is a failure. The initial field is the same as the NCEP reanalysis for it comes from the reanalysis (Fig. 1d, a). After the first 24 hours, it presents obvious difference between simulation and reanalysis. The trough originating from the mesoscale low extends to the coast of East China in reanalysis. In the original integral, the trough only extends less (Fig. 1e, b). At 48 hours, the mesoscale low is forecasted to die out, while it developed obviously in true state (Fig. 1f, c).

## 4 The Verification of Linearization Assumption

Before the discussion of how to improve the forecast is carried out, one of the crucial key issues associated with adjoint-based method must be taken into account firstly. That is the validation of linear assumption on which the tangent linear and adjoint approximation are all based. In other words, there is a need to find out whether the tangent linear and adjoint model has ability to describe small perturbation in the nonlinear model and how long the description ability could exists.

In many studies based on the adjoint method, it is assumed that the tangent linear and adjoint model has the description ability in a certain time range. Lorenz and Emannel (1998) conducted experiments to verify the utility of adaptive observation but without expected result. Hansen and Smith (2000) analyzed their experiment and determined that the failure was mainly due to the invalidation of the linearization assumption of small perturbation in the nonlinear model. The validation of the existence of this description ability is especially emphasized by Errico (1997). It is pointed out that the ability is closely associated with the weather system.

Mesoscale low on Meiyu Front has individual thermal and dynamical structure. Therefore, the verification of linearization assumption is a stepping stone for any further work.

A small perturbation $\delta \vec{x}_0$ at initial time $t_0$ is added to form a perturbed initial analysis:

$$\vec{x}_{0,t,+} = \vec{x}_{0,t} + \delta \vec{x}_0$$

The nonlinear forecast model $F$ is integrated from $\vec{x}_{0,t,+}$, a forecast is obtained at $t_k$:

$$\vec{x}_{k,f,+} = F \vec{x}_{0,t,+}$$

The difference between the forecast from the perturbed initial analysis and that of unperturbed initial analysis is regarded as the small perturbation developed in the nonlinear model. The small perturbation developed in linear model could be gotten by the integral of tangent linear model.

The small perturbation of west-east wind velocity $u$ in the tangent linear model and nonlinear models on $\sigma = 0.81$ during the integral process is listed in Fig. 3 once every 24 hours. At the start (Fig. 3a, d), the perturbations are all the same in the two models. At 24 hours (Fig. 3b, e), the central position, intensity and distribution of two kinds of perturbations are still very similar. At 48 hours (Fig. 3c, f), the perturbations are similar to a certain extent, however there are some differences in the central position, and the velocity of the center of tangent linear model is $5 \text{ m s}^{-1}$ greater than that of the nonlinear model.

The distributions of the perturbations of north-south wind velocity $v$, temperature $t$, vapor mixed ratio $q$ are not shown here. They are similar to that of west-east wind velocity $u$. In order to make the results have the general significance, the average of the correlation coefficient ($C_k$, subscript k indicates the value is at k time) of the perturbations $u$, $v$, $t$ and $q$ in 15 $\sigma$ levels is calculated. At 24 hours, $C_{24} = 0.93$. At 48 hours, $C_{48} = 0.78$. It clarifies that two kinds of perturbations tend to be inconsistent as the integral proceeds. Even at 48 hours, they are consistent with each other to certain extent. It shows that the tangent linear model has certain ability to describe the small perturbation in the nonlinear model. It is proper to use the adjoint-based method for the targeted observation issue of two days forecast of mesoscale low on Meiyu Front.

To further verify the sufficiency of the description ability, a practical method following the idea of Klinder et al. (1998) is applied to test whether the adjoint model could calculate the gradient of the J defined above with respect to the initial analysis accurately. That is, to calculate the minimization direction with the minimization algorithm and getting the new, better $\vec{x}_{0,t}$, after several iterations. If the adjoint model could calculate $\partial J / \partial \vec{x}_{0,t}$ correctly, $\vec{x}_{0,t}$ should become optimal, and $J$ should decrease gradually. It means the description ability of the adjoint model could meet the requirement. Here, the dry error energy in the sensitivity analysis object domain D02 (shown in Fig. 2) is calculated for the simulated error of the mesoscale low is mainly concerned. D02 is just the position of the mesoscale low at 0000 UTC 28 June, 1999.

The total dry error energy at 48 hours after iterating 10 times is listed in Table 1. As the iteration proceeds, the total error energy and its all components

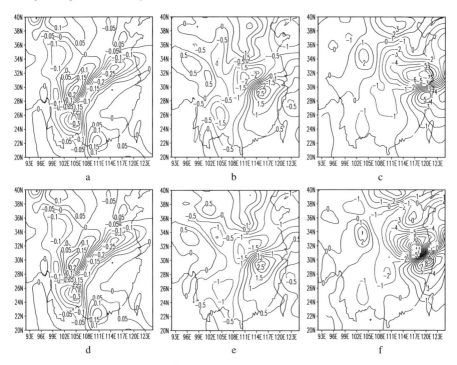

**Fig. 3** The perturbations of u (m s$^{-1}$) in the tangent linear model and nonlinear model from 0000 UTC 26 to 0000 UTC 28 June, 1999 on $\sigma = 0.81$ with an interval of 24 hours. The contour interval in (**a**) and (**d**) is 0.05 m s$^{-1}$, in (**b**) and (**e**) 0.5 m s$^{-1}$, and in (**c**) and (**f**) 1 m s$^{-1}$. (**a**), (**b**), and (**c**) are the perturbations in nonlinear model at 0, 24, and 48 h, respectively. (**d**), (**e**), (**f**) are the perturbations in the tangent linear model at 0, 24, and 48 h, respectively

**Table 1** The total dry error energy and each components at 48 hours after iterating 10 times

| Iteration | Total error energy | Horizontal error kinetic energy | Vertical error kinetic energy | Available potential error energy | Error elastic energy |
|---|---|---|---|---|---|
| 1 | $4.6 \times 10^9$ | $3.3 \times 10^9$ | $1.1 \times 10^5$ | $1.2 \times 10^9$ | $3.1 \times 10^7$ |
| 2 | $4.3 \times 10^9$ | $3.1 \times 10^9$ | $1.1 \times 10^5$ | $1.2 \times 10^9$ | $2.8 \times 10^7$ |
| 3 | $3.7 \times 10^9$ | $2.6 \times 10^9$ | $1.4 \times 10^5$ | $1.0 \times 10^9$ | $2.0 \times 10^7$ |
| 4 | $3.5 \times 10^9$ | $2.5 \times 10^9$ | $1.3 \times 10^5$ | $9.7 \times 10^8$ | $1.8 \times 10^7$ |
| 5 | $3.0 \times 10^9$ | $2.2 \times 10^9$ | $1.0 \times 10^5$ | $8.2 \times 10^8$ | $1.1 \times 10^7$ |
| 6 | $3.2 \times 10^9$ | $2.4 \times 10^9$ | $8.1 \times 10^4$ | $7.9 \times 10^8$ | $9.2 \times 10^6$ |
| 7 | $2.9 \times 10^9$ | $2.1 \times 10^9$ | $9.0 \times 10^4$ | $7.7 \times 10^8$ | $8.6 \times 10^6$ |
| 8 | $2.7 \times 10^9$ | $1.9 \times 10^9$ | $7.6 \times 10^4$ | $7.5 \times 10^8$ | $7.6 \times 10^6$ |
| 9 | $2.5 \times 10^9$ | $1.9 \times 10^9$ | $6.4 \times 10^4$ | $6.9 \times 10^8$ | $6.6 \times 10^6$ |
| 10 | $2.4 \times 10^9$ | $1.8 \times 10^9$ | $9.0 \times 10^4$ | $6.1 \times 10^8$ | $5.6 \times 10^6$ |

tend to decrease. Apart from at iteration six, there is a small reversion. It may be the phenomenon brought by the minimization algorithm. It verifies that the gradient calculation of the adjoint model is correct to certain extent and could fulfill the research requirement.

# 5 Sensitivity Analysis

## 5.1 Intensity of the Sensitivity Gradient

The sensitivity gradient with respect to different physical fields is not uniform in intensity. Table 2 is the integral of $(\partial J/\partial \vec{x}_{0,t})_{var}^2$ in entire atmosphere. The elements of $(\partial J/\partial \vec{x}_{0,t})_{var}^2$ is the square of the elements of $(\partial J/\partial \vec{x}_{0,t})_{var}$. The subscript var denotes the physical fields in the initial analysis including vapor mixed ratio $q_0$, temperature $t_0$, east-west wind velocity $u_0$ and vertical wind $w_0$, perturbing pressure $p_0$. It could be seen in Table 2 that the sensitivity with respect to $q_0$ is most intensive, followed by $t_0$, $u_0$, $w_0$, $p_0$, respectively.

**Table 2** Integral of $(\partial J/\partial \vec{x}_{0,t})_{var}^2$ in entire model atmosphere

| Var | $q_0(J^2(Kg\ Kg^{-1})^{-2})$ | $t_0(J^2(K\ K^{-1})^{-2})$ | $u_0(J^2(ms^{-1})^{-2})$ | $w_0(J^2(m\ s^{-1})^{-2})$ | $p_0(J^2\ Pa^{-2})$ |
|---|---|---|---|---|---|
| $(\partial J/\partial \vec{x}_{0,t})_{var}^2$ | $5 \times 10^{13} \sim 7 \times 10^{14}$ | $2 \times 10^{10} \sim 2 \times 10^{11}$ | $2 \times 10^9 \sim 2 \times 10^{10}$ | $3 \times 10^8 \sim 3 \times 10^9$ | $3 \times 10^4 \sim 3 \times 10^5$ |

However, it could not be decided to which physical field the simulation error is most sensitive in terms of the intensity of sensitivity gradient because the comparison among sensitivity gradient to different variables is in vain. Whether the simulation error is sensitive to one physical field far more than others just could be found by iterating the minimization algorithm.

## 5.2 Spatial Characteristics of the Sensitivity Gradient

The contours of the vertical integral of $(\partial J/\partial \vec{x}_{0,t})_{var}^2$ of $q_0$, $t_0$, $u_0$, $w_0$, $p_0$ in the entire model atmosphere are plotted in Fig. 4 to analyze the spatial characteristics of the sensitivity gradient. It could be found that the sensitivity gradient with respect to $w_0$ distributes in scattered form, while the sensitivity gradient with respect to other physical fields converges together and form isolated sensitive regions. Furthermore, the main sensitive regions of $q_0$, $t_0$, $u_0$ and $p_0$ almost superpose each other.

Introducing the equation:

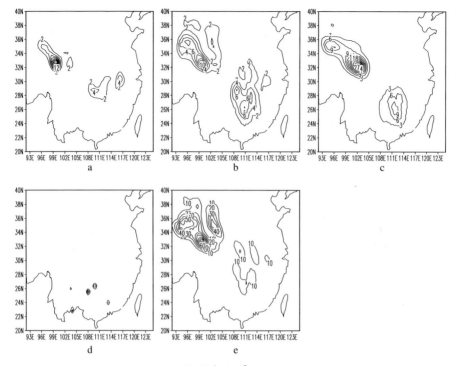

**Fig. 4** The vertical integral of $(\partial J / \partial \vec{x}_{0,t})_{var}^2$ of (**a**) the vapor mixed ratio $q_0(1 \times 10^{11}\,\mathrm{J^2\,Kg^2\,Kg^{-2}})$, (**b**) temperature $t_0(1 \times 10^6\,\mathrm{J^2\,K^2\,K^{-2}})$, (**c**) east-west wind velocity $u_0(1 \times 10^5\,\mathrm{J^2\,m^{-2}\,s^2})$, (**d**) vertical wind $w_0(1 \times 10^4\,\mathrm{J^2\,m^{-2}\,s^2})$ and (**e**) perturbed pressure $p_0(\mathrm{J^2\,Pa^{-2}})$ in the entire model atmosphere. The contour interval of (**a**) is $2 \times 10^{11}\,\mathrm{J^2\,Kg^2\,Kg^{-2}}$, (**b**) is $2 \times 10^6\,\mathrm{J^2\,K^2\,K^{-2}}$, (**c**) is $3 \times 10^5\,\mathrm{J^2\,m^{-2}\,s^2}$, (**d**) is $2 \times 10^4\,\mathrm{J^2\,m^{-2}\,s^2}$ and (**e**) is $10\mathrm{J^2\,Pa^{-2}}$

$$\partial J / \partial \vec{x}_{0,t} \approx \boldsymbol{E}^{1/2} \sum_{i=1}^{N} d_i \lambda_i (\vec{z}_0)_i$$

(Rabier et al. 1996; Gelaro et al. 1998, 1999), where $(\vec{z}_0)_i$ is singular vector at initial time, $\lambda_i$ is singular value responding to $(\vec{z}_0)_i$, which is the growing ratio in the simulation interval and $\lambda_1 \geq \lambda_2 \geq \cdots \geq \lambda_{N-1} \geq \lambda_N$, $d_i$ is the coefficient of simulation error projecting to the normalizing form of a singular vector at the terminal time, $N$ is the total number of singular vectors, and $\boldsymbol{E}$ is the dry energy metric matrix used to calculate the sensitivity gradient and singular vectors. More detail about the definition and calculation of singular vector could be found in Buizza and Palmer (1995).

It is thought that the sensitivity gradient tended to be dominated by the quickly growing perturbations. 30 singular vectors are calculated in this paper. Taking

$$a = \langle \vec{y}, \vec{y} \rangle_{\boldsymbol{E}},$$

the inner product of

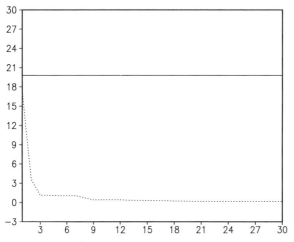

**Fig. 5** $a = \langle \vec{y}, \vec{y} \rangle_E$ (*dot line*) $(1 \times 10^{10}\,\text{J})$ and $b = \langle E^{-1}\partial J/\partial \vec{x}_{0,t}, E^{-1}\partial J/\partial \vec{x}_{0,t} \rangle_E$ (*solid line*) $(1 \times 10^{10}\,\text{J})$ as a function of $N$. The abscissa is the number of singular vector, and the y-coordinate denotes the values of $a$ and $b$

$$\vec{y} = E^{-1}\partial J/\partial \vec{x}_{0,t} - E^{-1/2}\sum_{i=1}^{N} d_i \lambda_i(\vec{z}_0)_i$$

about E is calculated. In Fig. 5, the x-axis is $N$, the dot line is a curve of $a$ changing with $N$, the solid line indicates

$$b = \langle E^{-1}\partial J/\partial \vec{x}_{0,t}, E^{-1}\partial J/\partial \vec{x}_{0,t} \rangle_E$$

Comparing the two lines, at $N = 30$, $a/b = 0.79\%$. It is implies that the first 30 singular vectors are already enough to spell out the sensitivity gradient. Taking $N = 10$, the ratio of $a$ to $b$ is 2.09%, The first several quickly growing singular vectors could make up of the major part of the sensitivity gradient. In another words, the sensitivity gradient mainly reflects the quickly growing perturbation.

### 5.3 Physical Characteristics of the Sensitivity Gradient

The perturbation could be derived from the sensitivity gradient to connect the sensitivity gradient with the physical question. Importing the equation:

$$\delta\vec{x}_0 = -\sum_{i=1}^{5} A_{var_i}(\partial J/\partial \vec{x}_{0,t})_{var_i} / \left\|(\partial J/\partial \vec{x}_{0,t})_{var_i}\right\|_A$$

$(\partial J/\partial \vec{x}_{0,t})_{var_i}$'s elements concerned var is equal to those of $\partial J/\partial \vec{x}_{0,t}$, while other elements are zero. $\|\ \|_A$ defines a metric, which is the absolution of elements of

**Table 3** Error energy at 48 hours with $\vec{x}'_{0,t}$ as the initial analysis corresponding to different $A_{var_i}$

|  | $A_{q0}$ $(KgKg^{-1})$ | $A_{t0}(K)$ | $A_{u0}$ $(ms^{-1})$ | $A_{w0}$ $(ms^{-1})$ | $A_{p0}(Pa)$ | Total error energy $(J)$ |
|---|---|---|---|---|---|---|
| Configuration 1 | 0 | 0 | 0 | 0 | 0 | $4.6 \times 10^9$ |
| Configuration 2 | 0.002 | 0 | 0 | 0 | 0 | $3.8 \times 10^9$ |
| Configuration 3 | 0 | 4 | 0 | 0 | 0 | $3.0 \times 10^9$ |
| Configuration 4 | 0 | 0 | 5 | 0 | 0 | $4.4 \times 10^9$ |
| Configuration 5 | 0 | 0 | 0 | 0.02 | 0 | $4.5 \times 10^9$ |
| Configuration 6 | 0 | 0 | 0 | 0 | 260 | $4.3 \times 10^9$ |
| Configuration 7 | 0.002 | 2 | 5 | 0.02 | 260 | $2.9 \times 10^9$ |

$(\partial J/\partial \vec{x}_{0,t})_{var_i}$ with the maximum absolution. $A_{var_i}$ is the perturbation amplitude introduced artificially. Seven configurations are designed in Table 3. The initial perturbation corresponding to configuration 7 in Table 3 at 500 hPa is shown in Fig. 6.

It could be found that the perturbations show obvious mesoscale characterstics and are mainly near the mesoscale low. It indicates that those small initial errors leading to great forecast error associated with the mesoscale low on Meiyu Front are

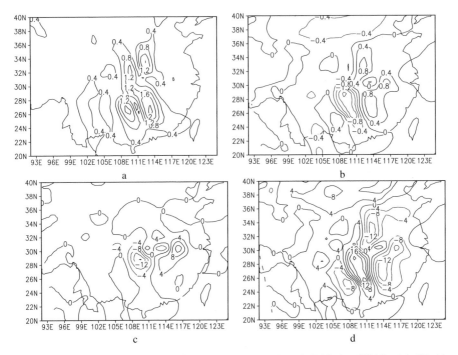

**Fig. 6** The initial perturbation corresponding to configuration 7 in Table 3 at 500 hPa. (**a**), (**b**), (**c**), and (**d**) correspond to $\sqrt{u^2 + v^2}$ (m s$^{-1}$)(horizontal wind velocity perturbation, interval of contour is 0.4 m s$^{-1}$), $t$(K) (temperature perturbation, interval of contour is 0.4 K), $q(1 \times 10^{-4}$ Kg Kg$^{-1}$) (vapor mixed ratio perturbation, interval of contour is $4 \times 10^{-4}$ Kg Kg$^{-1}$), and $h$(gpm) (geopotential height perturbation, interval of contour is 4 gpm), respectively

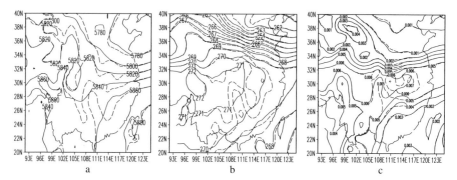

**Fig. 7** Original initial analysis (*solid line*) and the perturbed initial analysis (*dashed line*) corresponding to configuration 7 in Table 3 at 500 hPa. (**a**), (**b**), and (**c**) correspond to the geopotential (interval of contour is 20 gpm), temperature (the interval of contour is 1 K), and vapor mixed ratio (interval of contour is $1 \times 10^{-3}$ Kg Kg$^{-1}$), respectively

also mesoscale. Much mesoscale information in the initial condition is important to improve the accuracy of the numerical forecast.

Further investigation shows that the sensitivity gradient of different physical fields are correlated dynamically and consistently distributed. Putting the perturbation $\delta\vec{x}_0$ in the original initial analysis $\vec{x}_{0,t}$, a perturbed initial analysis $\vec{x}'_{0,t} = \vec{x}_{0,t} + \delta\vec{x}_0$ could be obtained. Figure 7 shows these two analysis fields at 500 hPa. Combining Figs. 6 and 7, the special phase relation between different physical variable could be revealed.

In the case of the temperature and moist fields, in the region of $(24°N, 100°E)$ to $(33°N, 120°E)$, the low and high pressure regions are controlled by high temperature and high moisture, low temperature and low moisture respectively (Fig. 7b, c). In the perturbed analysis, just as the geopotential field does, the dominated temperature and moisture fields also tend to be intensive.

This kind of dynamical correlation also exists between the geopotential height and the horizontal stream in region of $(24°N, 100°E)$ to $(33°N, 120°E)$. Figure 8 gives the geopotential height (the solid line) and horizontal wind vector (the vector) of the difference between the perturbed and the original analysis. There is an obvious balance between the geopotential height and the horizontal wind vector.

The dynamical correlation among the initial perturbation tends to make the distribution of the initial perturbations consistent. In Fig. 6, the initial scalar perturbations are similar to each other. The central position of different scalar perturbations almost superposes each other. Although there are some differences between the initial perturbations of horizontal wind and scalar physical fields, the regions with obvious perturbations of horizontal stream and other physical fields are close to each other, and the central positions also have good spatial corresponding relations.

The dynamical relation and consistent distribution characteristics denote that all kinds of physical fields (including vector and scalars) should be amended at the same time when improving the initial analysis, instead of just only changing some special physical fields.

**Fig. 8** Geopotential (*solid line*) and horizontal wind (*vector*) of the initial perturbation in the region of (24°N, 100°E) to (33°N, 120°E) at 500 hPa

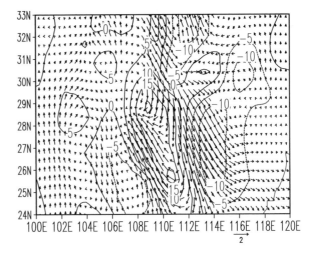

## 6 The Improvement of Numerical Forecast

A new numerical simulation could be obtained by using the perturbed initial analysis. Figure 9 is the integral of the perturbed initial analysis, defined in terms of configuration 7 in Table 3. The time is the same as that of Fig. 1. It could be seen that though there is small difference between the perturbed initial analysis and the original initial analysis. The mesoscale low is simulated correctly at 48 hours forecast. The position is much close to the reanalysis. Just the intensity of low seems a little weak. Taking account of the forecast without development of mesoscale low in the original integral, this result is encouraging. It implies that the improvement of numerical forecast could really be obtained by the correction of the small errors in the initial field.

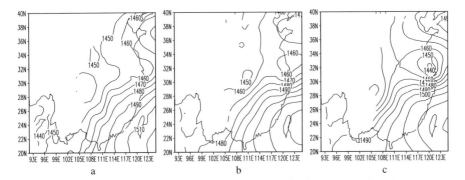

**Fig. 9** Geopotential height of the integral of perturbed initial analysis defined in terms of configuration 7 in Table 3 from 0000 UTC 26 to 0000 UTC28 June 1999 at 850 hPa, with an interval of 24 hours. The interval of the contour is 10 gpm

Other configurations in Table 3 are designed to test the effect of $A_{var_i}$. $A_{var_i}$ is chosen randomly under consideration of rationality. $\vec{x}'_{0,t}$ corresponding to configuration 1 is actually equal to $\vec{x}_{0,t}$, and $\vec{x}'_{0,t}$ corresponding to configurations 2–6 are a new perturbed initial analysis through perturbing the vapor mixed ratio q0, temperature t0, east-west wind velocity u0, vertical wind w0, disturbing pressure p0 separately.

From the simulation error energy at 48 hours with $\vec{x}'_{0,t}$ as the initial condition corresponding to different $A_{var_i}$ (Table 3), it could be found that all $\vec{x}'_{0,t}$ make the error energy smaller than that of original analysis. The result shows that the error correction of each physical variable in the initial analysis all contributes to the improvement of numerical forecast.

# 7 Conclusion and Discussion

Study on adjoint-based targeted observation of mesoscale low on Meiyu Front is implemented in our work. Targeted observation is such an idea that to reduce the small errors in the initial analysis that induce great errors in the numerical weather forecast by adding observation in the targeted area and so obtain the improvement of the accuracy of numerical prediction of a given weather system. European and USA scientists have led the field in this topic. However, there is a need for Asian scientist to investigate this issue associated with Asian weather system. This is what our attempt is made to meet in this paper.

The mesoscale low on Meiyu Front is taken as the research object. Firstly, the linearization assumption is verified to remain valid for two days, and thus it is proper to use the adjoint method for a two day forecast of mesoscale low on Meiyu Front. This is a crucial key issue because whether the targeted observation region is decided by sensitivity gradient analysis or the singular vectors, two popular methods adopted now, they are all adjoint-based.

One requirement must be satisfied when putting targeted observation into practice: the extra observation should be limited to the local region, so the limited number observation instrument such as plane and so on could make observation efficient. The investigation above has shown that the sensitivity gradient is localized, and the sensitivity gradient distributions of different physical fields tend to be consistent. The targeted observation region decided by the sensitivity gradient will fulfill the requirement of targeted observation.

As for the targeted observation based on singular vector, it is important to decide how many and which singular vector should be used to decide the targeted observation region. It is found in our study that the sensitivity gradient is dominated by the leading singular vectors. It implies that it is proper and also enough to use the several leading singular vectors to decide the extra observation network in targeted observation.

It seems that there is dynamical correlation among the sensitivity gradient of different physical variables. It means that a more accurate forecast will possibly be obtained if the initial analysis of all fields are improved simultaneously. What is more, the moisture handles an important role in the process of mesoscale low on Meiyu Front. It is critical to include more accurate moisture processes in the numerical model and enhance the vapor observation.

Perturbing the initial analysis with perturbations derived from the sensitivity gradient improves the simulation of mesoscale low. It means that for the forecast accurate of the mesoscale low on Meiyu Front, there is still a lot to gain by decreasing the small errors in the initial analysis.

Our attempt revealed that targeted observation is really a feasible way to improve the numerical forecast associated with mesoscale low on Meiyu Front. With the research going on, more issues including what extent the observation error is similar to the sensitivity gradient, the observation system simulation experiment of targeted observation et al., are carried out to find out the major outlines of practical operation associated with targeted observation. Some interesting results are obtained (Dong Peiming and Zhang X 2004; Dong Peiming and Zhong K et al. 2006; Zhong Ke et al. 2007a, b). It is expected that these investigation could push the use of this technology method and enhance improvement of the accuracy of numerical forecast associated with Asian weather system that have been demonstrated in European and USA.

**Acknowledgements**  This work was support by the National Natural Science Foundation of China under Grants No. 40405020 and 40775027.

# References

Bergot T, Hello G, Joly A, Malardel S (1999) Adaptive observation: A feasibility study. Mon Wea Rev 127:743–765

Berliner LM, Lu ZQ, Snyder C (1999) Statistical design for adaptive weather observations. J Atmos Sci 56:2536–2552

Bishop CH, Toth Z (1999) Ensemble transformation and adaptive observations. J Atmos Sci 56:1748–1765

Bishop CH, Etherton BJ, Majumdar SJ (2001) Adaptive sampling with the ensemble transform kalman filter. Part I: theoretical aspects. Mon Wea Rev 129:420–436

Buizza R, Palmer TN (1995) The singular-vector structure of the atmospheric general circulation. J Atmos Sci 52:1434–1456

Buizza R, Montani A (1999) Targeting observations using singular vector. J Atmos Sci 56:2965–2985

Dong P, Zhang X (2004) Targeted observations and adjoint sensitivity analysis. Meteorol Sci Technol (in Chinese). 32:1–5

Dong P, Zhong K, Zhao S (2006) Impact of regional uncertainties of the initial state upon numerical forecast of mesoscale low on Meiyu Front. Climatic Environ Res (in Chinese), 9:617–633

Emanuel K, et al. (1995) Report of the first prospectus development team of the U.S. weather research program to NOAA and the NSF. Bull Amer Meteor Soc 76:1194–1208

Errico RM (1997) What is an adjoint model. Bull Amer Meteor Soc 78:2577–2591

Gelaro R, Buizza R, Palmer TN, Klinker E (1998) Sensitivity analysis of forecast errors and the construction of optimal perturbations using singular vectors. J Atmos Sci 55:1012–1037

Gelaro R, Langland R, Rohaly GD, Rosmond TE (1999) An assessment of the singular vector approach to targeted observations using the FASTEX data set. Q J R Meteor Soc 125:3299–3327

Grell GA, Dudhia J, Stauffer DR (1994) A description of the fifth generation Penn State/NCAR mesoscale model (MM5). NCAR/TN-398+STR

Hansen JA, Smith LA (2000) The role of operational constraints in selecting supplementary observations. J Atmos Sci 57:2859–2871

Joly A, et al. (1997) The fronts and atlantic storm-track experiment (FASTEX): Scientific objectives and experimental design. Bull Amer Meteor Soc 78:1917–1940

Klinder E, Rabier F, Gelaro R (1998) Estimation of key analysis errors using the adjoint technique. Q J R Meteor Soc 124:1909–1933

Langland R, Rohaly G (1996) Adjoint-based targeting of observation for FASTEX cyclones. Preprints, Seventh Conf. on Mesoscale Process, Reading, United Kingdom, Amer Meteor Soc 369–371

Langland R, Gelaro R, Rohaly GD, Shapiro MA (1999) Targeted observations in FASTEX: Adjoint based targeting procedures and data impact experiments in IOP/8 and IOP/8. Q J R Meteor Soc 125:3241–3270

Lorenz EN, Emannel K (1998) Optimal sites for supplementary weather observations: Simulation with a small model. J Atoms Sci 55:399–414

Majumdar SJ, Bishop CH, Etherton BJ (2002) Adaptive sampling with the ensemble transform Kalman filter. Part II: field program implementation. Mon Wea Rev 130:1356–1369

Palmer TN, Gelaro R, Barkmeijer J, Buizza R (1998) Singular vectors, metrics, and adaptive observations. J Atmos Sci 55:633–653

Rabier F, Klinker E, Courtier P, Hollingsworth A (1996) Sensitivity of forecast errors to initial conditions. Q J R Meteorol Soc 122:121–150

Zhong K, Wang Y, Dong P, Zhao S (2007a) Adjoint-based sensitivity analysis of a mesoscale low on the Meiyu front and its implications for adaptive observation. Adv Atmos Sci 24:435–448.

Zhong K, Dong P, Zhao S (2007b) Analysis of initial errors of mesoscale low on Meiyu Front by the pseudo-inverse perturbation. Climatic Environ Res (in Chinese), 12:647–658

Zou X, Vandenberghe F, Pondeca M, Kuo YH (1997) Introduction to adjoint techniques and the MM5 adjoint modeling system. NCAR/TN-435-STR

Zou X, Huang W, Xiao Q (1998) A user's guide to the MM5 adjoint modeling system. NCAR/TN-437+IA

# Ocean Data Assimilation: A Coastal Application

Xiaodong Hong, James A. Cummings, Paul J. Martin and James D. Doyle

**Abstract** The Navy Coupled Ocean Data Assimilation (NCODA) system is applied to a period during the Autonomous Ocean Sampling Network II (AOSN II) field campaign conducted in the Monterey Bay area in August 2003. The multivariate analysis of NCODA is cycled with the Navy Coastal Ocean Model (NCOM) in a sequential, incremental, update cycle. In addition to the operational data obtained from the Global Ocean Data Assimilation Experiment (GODAE) server, which included satellite observations of sea-surface temperature (SST) and sea-surface height and insitu surface and sub-surface observations of temperature and salinity, high-density data from aircraft SST observations and high-frequency data from buoys used for the AOSN II field experiment are also assimilated. The results from data assimilative and non-assimilative runs are compared with and verified against observations. Bias and root-mean-square errors of temperature indicate that forecast skill from the data assimilative run exceeds errors from the persistence and the non-assimilative runs. The seasonal thermocline is better represented and the warm bias for both upwelling and relaxation periods is significant reduced.

## 1 Introduction

Coastal ocean forecast uncertainty stems from uncertainty in the ocean model physics and numerics, air-sea fluxes, lateral boundary conditions, initial conditions, and bathymetry. To reduce the uncertainty of an ocean forecast, an ocean data assimilation system is needed to better estimate the ocean state and model parameters by combining observed ocean data with the governing physics and dynamics of the ocean model. There are a number of ocean data assimilation schemes based on different theories and methods to optimally or sub-optimally provide state estimation in space and time. Compared to the use of data assimilation in meteorology, ocean data assimilation has a much shorter history. Blue water models are very well constrained at the surface using altimeter sea surface height (SSH). However, it is difficult to constrain models in coastal areas due to the inherent short time/space scales and

X. Hong (✉)
Marine Meteorology Division, Naval Research Laboratory, Monterey, CA 93943, USA

S.K. Park, L. Xu, *Data Assimilation for Atmospheric, Oceanic and Hydrologic Applications*, DOI 10.1007/978-3-540-71056-1_14,
© Springer-Verlag Berlin Heidelberg 2009

lack of high frequency ocean observations (other than moorings). In recent years, unprecedented in situ observational tools, such as sophisticated autonomous robotic vehicles, aircraft, CODAR (Coastal Ocean Dynamics Applications Radar), drifters, etc., have been increasingly used to collect data qualitatively and quantitatively in various coastal ocean field experiments. These data provide opportunities for testing and validating data assimilative forecast schemes. These evaluations, in turn, can help to identify model deficiencies and lead to improved forecast models.

The ocean data assimilation system used here is NCODA (Cummings 2005). NCODA is currently being applied in real time at Fleet Numerical Meteorology and Oceanography Center (FNMOC) and at the Naval Oceanographic Office (NAV-OCEANO). It can be executed as a stand-alone analysis or cycled with an ocean forecast model in a sequential, incremental, update cycle. The ocean forecast models that have been cycled with NCODA include the Hybrid Coordinate Ocean Model (HYCOM), the Navy Coastal Ocean Model (NCOM), the parallel Ocean Prediction Ocean model (POP), and the Shallow Water Analysis Forecast System (SWAFS), which is based on the Princeton Ocean Model. In this study, the NCOM forecast model (Martin 2000) is cycled with the NCODA system.

In this application, NCOM uses initial and lateral boundary conditions from the Navy's operational global NCOM forecast system. The surface atmospheric forcing for NCOM is provided from real-time forecasts by the Navy's Coupled Ocean/Atmosphere Mesoscale Prediction System (COAMPS®)[1] (Hodur 1997).

The execution of NCODA cycled with NCOM during a coastal field campaign is performed and evaluated in this study. The field campaign is the Autonomous Ocean Sampling Network (AOSN) II conducted in Monterey Bay during August 2003. The objective of AOSN II is to develop an adaptive, coupled, observational/modeling prediction system capable of providing an accurate 3- to 5- day forecast of marine-biology events. The regional circulation near the Monterey Bay can be described in two distinct hydrographic states: upwelling state and relaxed state. The upwelling is driven by the prevailing north/northwesterly wind (i.e., directed towards the south/southeast). Two upwelling centers are formed at headlands to the north of the bay at Point Ano Nuevo and south at Point Sur. During the upwelling periods, there is a cyclonic circulation (or eddy, Tseng and Breaker 2007) in the bay and an anticyclonic California Current meander, also sometimes referred to as the Monterey Bay Eddy (Ramp et al. 2005), offshore of the bay. When the wind relaxes, the upwelling reduces, and the offshore eddy moves into the bay and interacts with the flow over the shelf.

The mesoscale circulations in the vicinity of Monterey Bay are highly complex and variable. It is difficult to model these features correctly due to the errors resulted from model dynamics, model resolution, bathymetry, atmospheric forcing and lateral boundary conditions. Fortunately, we have a data assimilation system in place so that data collected from the field experiment can be used for data assimilative simulation to compensate the drawback in numerical simulation. The purposes of this work are to use observational data obtained from the AOSN II field campaign to

---

[1] COAMPS®, COAMPS is a registered trademark of the Naval Research Laboratory

perform data assimilative simulation of upwelling/relaxation processes in the vicinity of Monterey Bay and to assess the skill of the data assimilative simulation.

Section 2 briefly describes the components of the ocean data-assimilation system. The configuration of the data-assimilation system is described in Sect. 3. Information about the observations that are assimilated is provided in Sect. 4. Results from the assimilative simulation are discussed and compared with the observation from the AOSN II field experiment in Sect. 5. Verification and evaluation of the ocean data assimilation system are shown in Sect. 6. Summary and conclusions are presented in Sect. 7.

# 2 Brief Description of Each Component

## 2.1 NCODA

NCODA is a fully three-dimensional, multivariate, optimum-interpolation (MVOI) (Daley 1991) ocean-data assimilation system that produces simultaneous analyses of temperature, salinity, geopotential (dynamic height), and vector velocity. A complete description of NCODA can be found in Cummings (2005). The formulation is as:

$$\mathbf{x}_a = \mathbf{x}_b + \mathbf{P}_b \mathbf{H}^T (\mathbf{H} \mathbf{P}_b \mathbf{H}^T + \mathbf{R})^{-1} \{\mathbf{y} - \mathbf{H}(\mathbf{x}_b)\} \tag{1}$$

where $\mathbf{x}_a$ is the analysis vector, $\mathbf{x}_b$ is the background vector, $\mathbf{P}_b$ is the background-error covariance matrix, $\mathbf{H}$ is the forward operator, $\mathbf{R}$ is the observation error covariance matrix, and $\mathbf{y}$ is the observation vector.

The observation vector $\mathbf{y}$ contains all of the synoptic temperature, salinity, and velocity observations that are within the geographic and time domains of the forecast model grid and update cycle. The forward operator $\mathbf{H}$ is a spatial interpolation of the forecast model grid to the observation locations performed in three dimensions. Thus, $\mathbf{H} \mathbf{P}_b \mathbf{H}^T$ is approximated directly by the background-error covariance between the observation locations, and $\mathbf{P}_b \mathbf{H}^T$ directly by the error covariance between the observation and grid locations. The quantity $\{\mathbf{y} - \mathbf{H}(\mathbf{x}_b)\}$ is referred to as the innovation vector, $\{\mathbf{y} - \mathbf{H}(\mathbf{x}_a)\}$ is the residual vector, and $\mathbf{x}_a - \mathbf{x}_b$ is the increment (or correction) vector.

The background-error covariances are separated into a background-error variance and a correlation. The correlation is further separated into a horizontal and a vertical component. NCODA uses flow-dependence covariances in the analysis by scaling the horizontal and vertical correlations with a correlation computed from the geopotential height difference between two locations. The horizontal correlation length-scales are specified as the first baroclinic Rossby radius of deformation computed from the historical profile archive (Chelton et al. 1998). The vertical correlation length-scales can be either constant (used in this study), monotonically increasing or decreasing with depth, or varying with background density vertical gradients.

All analysis variables use the same background-error second-order autoregressive correlation model for calculating the horizontal correlations.

Background-error variances vary with location, depth, and analysis variable. They are related to the analysis increments and expectations based on the age of the data on the grid. The background-error variances are allowed to increase with time in the long-term absence of observations until the errors asymptote at the limit of the expected variance, specified as either climate variability or model error. The climate variability is specified in this study due to the lack of availability of model error from the global NCOM system.

The observation errors and the background errors are assumed to be uncorrelated, and errors associated with observations made at different locations and at different times are also assumed to be uncorrelated. Observation errors are computed as the sum of a measurement error and a representation error. Most measurement-error variances are specified as input parameters based on fairly well-known ocean observing errors. One exception is for the geopotential observations. Geopotential errors are computed from the observation errors of the potential temperature and salinity, using the partial derivatives of the equation of state. Representation errors are a function of the resolutions of the model and of the observing network.

Altimeter sea surface height (SSH) is assimilated from synthetic temperature profiles computed using the Modular Ocean Data Assimilation System (MODAS) database. MODAS provides the time-averaged co-variability of SSH and temperature at depth at a fixed location (Fox et al. 2002).

All ocean observations are subject to data quality-control (QC) procedures prior to assimilation. The need for quality control is fundamental in the analysis system; erroneous data can cause an incorrect analysis, while rejecting extreme data can miss important events. The primary purpose of the QC system is to identify observations that are obviously in error, as well as the more difficult process of identifying measurements that fall within valid and reasonable ranges, but are erroneous. A secondary use of the QC system is the creation and maintenance of an analysis-forecast increment database for use in the a posteriori computation of the optimum interpolation statistical parameters. A detailed description of the real-time QC system can be found in Cummings (2006).

## 2.2 NCOM

NCOM is a three-dimensional, primitive equation, free-surface model using the hydrostatic, Boussinesq, and incompressible approximations. Details of the model description can be found in Martin (2000). NCOM is designed to offer the user a range of numerical choices in terms of parameterizations, numerical differencing, and vertical grid structure. NCOM uses a hybrid vertical coordinate system, which allows for the use of all sigma-layers, or all z-levels, or a combination of the sigma-layers for the upper ocean and z-levels below. The model equations are solved on a staggered, Arakawa C-grid. Temporal differencing is leap-frog with an Asselin filter to suppress time splitting. Spatial averages and finite differences are mainly second order with an option for higher-order formulations for advection. The propagation of surface waves and vertical diffusion is treated implicitly. The Mellor-Yamada Level 2.5 turbulence scheme is used for vertical mixing. NCOM forcing includes surface air-sea fluxes, lateral open boundary conditions, tides, and river and runoff discharges.

NCOM has been applied to many locations, including the Adriatic Sea (Pullen et al. 2003) and Monterey Bay (Shulman et al. 2007) to study fine-scale oceanic features under atmospheric forcing with different resolution, and the Gulf of Lion to study the effects of time variation of the surface buoyancy flux on the formation of deep-water convection during the winter season (Hong et al. 2008). A recent application is the development of NCOM ensemble forecasting (Hong and Bishop 2005) using the ensemble transform technique (Bishop and Toth 1999) and adaptive sampling for coastal observations (Hong and Bishop 2006) using the ensemble transform Kalmar filter method (Bishop et al. 2001).

## 2.3 Atmospheric forcing

The atmospheric forcing for NCOM consists of the surface air pressure, wind stress, heat flux, and effective surface salt flux for the momentum, temperature, and salinity equations. For this study, these are provided from COAMPS atmospheric forecasts in hourly frequency. COAMPS is a fully compressible, nonhydrostatic, primitive equation model based on a staggered, C grid and solved using a time-splitting technique with a semi-implicit formulation for the vertical acoustic modes (Hodur 1997; Hodur et al. 2002; Doyle et al. 2008). A Robert time filter is used to damp the computational mode. All derivatives are computed to second-order accuracy and options are provided for forth-order accurate horizontal advection and diffusion. COAMPS uses parameterization schemes for subgrid-scale convection, shortwave and longwave radiation processes, and mixed-phase cloud microphysics. A three-dimensional, multivariate, optimum-interpolation (MVOI), analysis technique is used to map the observations to the model grid and generate the initial conditions for the forecast model for each data assimilation cycle. Quality-controlled data used in the analysis are radiosonde, aircraft, satellite, and surface observations. Additional information about the atmospheric model set-up and forecast skill during AOSN II is discussed in Doyle et al. (2008).

To allow some interactive feedback from the ocean model, the surface latent and sensible heat fluxes for NCOM are computed from the COAMPS wind speed and air temperature and humidity and the NCOM-predicted sea-surface temperature (SST) using the drag coefficient from the standard bulk formulas of Kondo (1975) (Martin and Hodur 2003). The surface salt flux for NCOM is calculated from the computed latent heat flux and the COAMPS precipitation. The extinction of solar radiation in seawater as classified by Jerlov (1976) according to turbidity was used in the model with six optical types to define the subsurface penetration of the COAMPS solar radiation.

# 3 Configuration

The computational domain, grid projection, and horizontal resolution for NCODA and NCOM are the same to minimize the errors from the horizontal interpolation of fields between the two systems. The ocean-model domain covers both the Monterey and San Francisco Bay areas (Fig. 1a) and is within the innermost grid of COAMPS (Fig. 1b) (e.g. Doyle et al. 2008). The grid projection is Lambert conformal with a horizontal spacing of 3 km. In the vertical, NCOM uses a total of 40 layers with 15 sigma-layers in the upper ocean and 25 z-levels in the deeper water. The vertical resolution ranges from 1 to 500 m. NCODA uses 30 standard depth levels with a maximum depth of 3000 m. The horizontal grid size for both NCODA and NCOM is 96 × 168. The 1/8° global NCOM real-time nowcast for 1 August 2003 is used to start NCOM before data assimilation cycle takes place. The lateral boundary conditions for the regional NCOM are also supplied from the global NCOM and updated at a 3 h frequency. The lateral boundary conditions are important for providing large-scale forcing, such as California current, through the open boundaries of the regional model.

The COAMPS forecasts are produced twice daily out to 72 h using a 12 h, incremental, data-assimilation cycle on a quadruple-nested grid system with horizontal resolutions of 81, 27, 9, and 3 km (Fig. 1b). The innermost grid has dimensions 199 × 199 × 30. The COAMPS surface fields from the innermost nest with high-resolution are output on an hourly basis to force NCOM. The high-resolution COAMPS surface winds provide good representation of the narrow

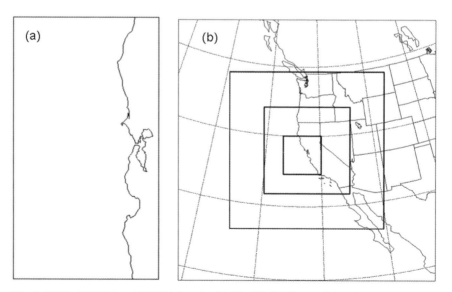

**Fig. 1** (**a**) The NCODA and NCOM domain; (**b**) The COAMPS nested domain

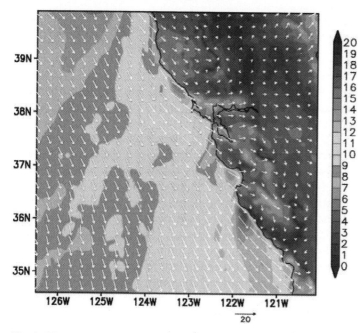

**Fig. 2** COAMPS 10 m wind speed (ms$^{-1}$) valid at 23Z August 11, 2003 from the innermost nest with horizontal resolution of 3 km

bands (about 10 by 50 km) of strong wind stress and wind stress curl (Fig. 2), which are very important to the generation of upwelling along the California coast (Pickett and Paduan 2003). These narrow bands are parallel to the coast and adjacent to major California coastal promontories, causing upwelling through Ekman transport. Observations of SST from satellites have shown that cold-water plumes off northern California are frequently anchored to coastal topography (Kelly 1985).

The assimilative system is performed in a sequential incremental update cycle with an update interval of 24 h. NCODA uses NCOM 24 h forecast fields of temperature, salinity and velocity as first guess fields and assimilates all available observations within the update-cycle time window so that it allows the use of background information closest to the observation time. The analysis fields are used to initialize the NCOM forecast at each analysis time.

In order to quantify the improvement of the forecast by data assimilation, two experiments are conducted for entire month of August 2003. The first experiment is run with data assimilation and produces 72 h forecast at each analysis update time based on a 24 h update cycle. The second experiment is a case with NCOM being integrated forward from August 1 to 31, 2003 without any data assimilation. The first experiment is referred to as the "assimilative run", while the second one is referred to as the "non-assimilative run".

# 4 Observations

Observations used in the ocean analysis include all sources of operational ocean observations. They contain remotely-sensed SST from AVHRR GAC infrared satellite, sea surface height from satellite altimeters, in situ surface and sub-surface observations of temperature and salinity from a variety of sources, such as ships, buoys, expendable bathythermographs, and conductivity-temperature-depth sensors. A description of the operational data sources can be found in Table 1 in Cummings (2005). We display here the types, paths, locations of observation data for one particular analysis time as an example in Fig. 3. The number of observations for each analysis cycle is also provided in the validation section. These data have been quality controlled and archived in the Global Ocean Data Assimilation Experiment (GODAE) server hosted by the Fleet Numerical Meteorology and Oceanography Center (FNMOC).

In addition to the operational observations, aircraft SST data collected during the AOSN II field campaign in August 2003 by the Naval Postgraduate School (NPS) (Ramp 2003) and continuous time series of data from the MBARI buoys m1 (Chavez 2003a) and m2 (Chavez 2003b) are also assimilated. The airborne measurements were made using a twin-engine, eight-seat, Piper Navajo owned and operated by Gibbs Flite Center (Ramp et al. 2005). The plane typically flew below the

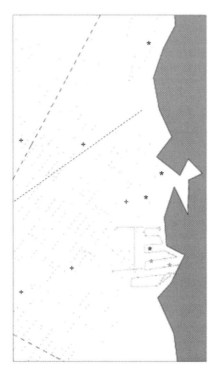

|   | day AVHRR GAC infrared satellite SST |
|---|---|
| . | night AVHRR GAC infrared satellite SST |
| + | ship SST |
| * | fixed buoy SST |
| * | moorings m1 and m2 |
| — | Jason, GFO and Envisat satellite altimeter |
| ... | NPS aircraft |

**Fig. 3** Observation types and locations used in the assimilation for 13 August 2003

quasi-permanent, summertime, stratus deck and provided a spatial context for the relative sparse in situ observations. The airborne data were collected at 1 Hz resolution and included pressure, temperature, dew point, relative humidity, and SST. A Heitronics KT-19 infrared radiation pyrometer was used to measure SST with 0.1°C precision and 0.5°C absolute accuracy. Sample aircraft paths for 13 August 2003 are shown using red-dashed lines in Fig. 3. There are a total of 12 daily flights during the August 2003 AOSN II field experiment, from which the measured SSTs are quality controlled and assimilated in this study.

The m1 and m2 MBARI buoys are located at (36.75°N, 122.03°W) and (36.7°N, 122.39°W) (green-asterisks in Fig. 3), respectively. Both of the moorings provide data in real time from the surface to 300 m at 10-min intervals. These high-frequency temperature and salinity profiles from m1 and m2 are quality controlled and assimilated in this study.

## 5 Comparisons of Assimilative Results with Observations

The upwelling/relaxation features during the AOSN II experiment in August 2003 are explored using the results from the data assimilative simulation. At this time, the winds over Monterey Bay can be described as periods that are upwelling favorable with north/northwesterly or upwelling unfavorable with south/southwesterly (see Doyle et al. 2008). The corresponding wind stresses are shown in Fig. 4a.

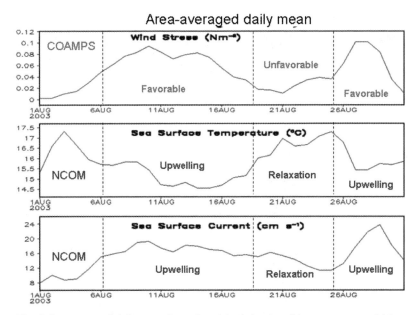

**Fig. 4** Area-averaged daily mean for surface (**a**) wind stress, (**b**) temperature, and (**c**) current. The area is within 36.2–37.2°N and 123–121.6°W

As a result, the regional circulation corresponds to an upwelling state and a re-laxation state. The winds in early August are not favorable for upwelling. From August 7 to 19, the prevailing north/northwesterly winds (i.e., directed towards the south/southeast) are re-established and induce upwelling. Warmer surface waters are forced offshore, allowing cold waters to rise to the surface near the coast. The surface temperature averaged over the Monterey Bay area is significantly reduced and the averaged surface current speed increases (Fig. 4b, c). From August 20 to 24, the winds are light with a south or southwest direction, resulting in relaxation con-ditions. During this period, upwelling along the coast diminished and the warm off-shore water moved shoreward. The area-averaged surface temperature increases and surface current speed decreases. During the latter portion of August, there is a short period of upwelling when the northwesterly wind is onset again around the bay.

Significant diurnal fluctuations in upwelling occur during the data assimilative simulation associated with diurnal fluctuations in the surface atmospheric conditions (see Fig. 5). These resemble a classic sea-breeze circulation pattern forced by large surface heating differences between the coastal marine atmosphere and the Central Valley (Banta et al. 1993). In Fig. 5, at the north upwelling center (Point Ano Neuvo) the simulated SST decreases and the sea-surface salinity (SSS) increases during the upwelling period and vice versa during the relaxation period. The diurnal fluctua-tions for wind stress, SST, and SSS are superimposed on the longer-period changes

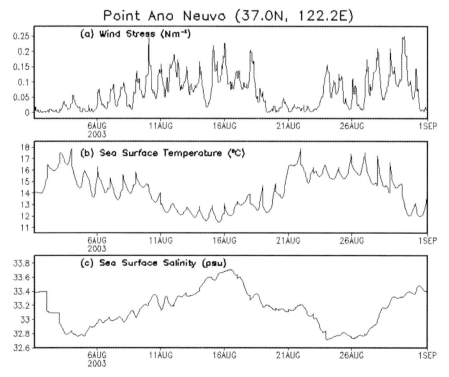

**Fig. 5** Hourly (**a**) surface wind stress, (**b**) temperature, and (**c**) salinity at Point Ano Neuvo

associated with the upwelling and relaxation events. The peak upwelling takes place on August 16 as indicated by the lowest surface temperature and salinity.

## 5.1 Upwelling Period

Two upwelling centers develop off Point Ano Nuevo and Point Sur during the up-welling period from 7 to 19 August. Figure 6a shows the SST forecast for 18 h (valid at 18Z August 15, 2003) using the assimilation of observed data as shown in Fig. 3. The assimilation realistically depicts the signature of the upwelling since it has been proceeding for several days. The coldest upwelled water in the upwelling center off Point Ano Nuevo reaches 11.5 °C at this time. Large horizontal SST gradients oc-cur between the upwelled cold water and the offshore warm water. A cold tongue of upwelled water off Point Ano Nuevo is advected southward across the mouth of Monterey Bay. The plume of upwelled cold water extends southward and joins with the upwelled cold water from Point Sur, resulting in a large, cold-water region

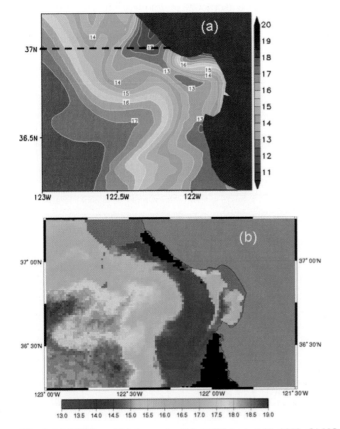

**Fig. 6** (a) SST from 18 h forecast valid at 18Z August 15, 2003. (b) NOAA POES AVHRR HRPT SST at 1858Z August 15, 2003 (NOAA POES AVHRR, Courtesy NWS and NOAA Coastwatch)

located just off the coast. Upwelled cold water also may have advected seaward as suggested in a previous observational study (Rosenfeld et al. 1994).

The NCOM SST assimilative forecasts are compared with the Coastwatch SST produced from the AVHRR High Resolution Picture Transmission (HRPT) data and broadcasted continuously by the Polar Orbiting Environmental Satellites (POES) by NOAA's National Environmental Satellite, Data, and Information Service (NES-DIS) (see Fig. 6b). The HRPT data have a resolution of 1.1 km and are mapped to almost full resolution in the production of the CoastWatch AVHRR visible, infrared, and SST images. The AVHRR HRPT SST data are not used in the data assimilation experiments, but the 4-km global area coverage (GAC) SST retrievals from several NOAA satellites are assimilated. The basic observed features are captured by the NCOM forecast as displayed in Fig. 6a for the 18 h SST forecast valid at the satellite observational time (Fig. 6b). These include (1) strong upwelling off Point Ano Neuvo and Point Sur, (2) upwelled water advected southward across the mouth of Monterey Bay that joined with cold water from Point Sur, and (3) warmer offshore water advected toward the mouth of Monterey Bay.

Figure 7 shows a vertical cross section of forecast temperature at 18 h along 37.05 °N at 18Z August 15, 2003. The isopleths of temperature are sloped upward towards the coast, indicating that the upper-layer warm water is pushed offshore and deep cold water is brought to the surface by Ekman transport and

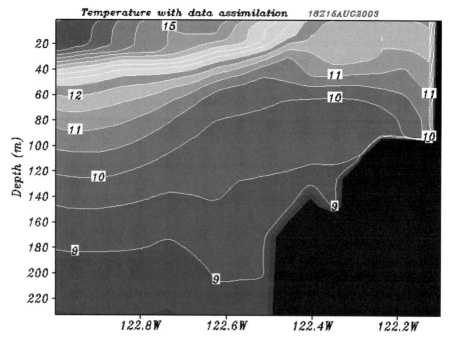

**Fig. 7** Vertical cross section of temperature on 18 Z August 15, 2003 along 37.0°N as a dashed-line shown in Fig. 6

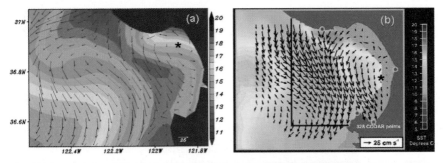

**Fig. 8** 25 h mean from 5Z August 15 to 6Z August 16 2003 for (**a**) NCOM surface temperature and current, (**b**) HF radar surface current with AVHRR-derived SST (Paduan and Lipphardt 2003). The approximated locations of cyclonic centers are marked as "∗"

pumping (Pickett and Paduan 2003). An upwelling front exists between the up-welled and offshore water with a characteristic gradient of 5°C per 100 km across the front.

The NCOM forecast surface current was also compared with the NPS mean HF radar surface currents (Paduan and Lipphardt 2003). The HF radar surface current data were not used in the data assimilation. The comparison of surface current was made for a 25 h mean from 5Z August 15 to 6Z August 16, 2003 during the peak of the upwelling event (Fig. 8). The forecast model shows that cold, upwelled water from Point Ano Nuevo was advected across and into the mouth of Monterey Bay and joined with cold water off Point Sur south of Monterey Bay (Fig. 8a). Both the model and the HF radar show a cyclonic circulation in the bay. However, the size of the cyclonic circulation is smaller in the model and its location is confined within the northern part of the bay. This may be caused by the stronger, southeastward current in the model simulation that advected cold water into the southern part of Monterey Bay. The model results show the warm water offshore in the area of anti-cyclonic circulation to be advected further to the south and closer to the bay. The larger area of cold water in the southern part of Monterey Bay and the stronger warm offshore meander could be due to insufficient model resolutions in both atmospheric and ocean models.

## 5.2 Relaxation Period

An anti-cyclonic meander within the California current moves coastward and cold upwelled water is replaced by warm offshore water during the relaxation period. Figure 9a indicates that warm water occupied the most area with temperatures above 16°C at the surface. Cold water still exists in both upwelling centers; however, the areal extent is considerably reduced as can be seen from the model forecast (Fig. 9a) and from the AVHRR SSTs (Fig. 9b) with the coldest water temperature not less than 14°C. Since there is much less data available for this period as will be seen

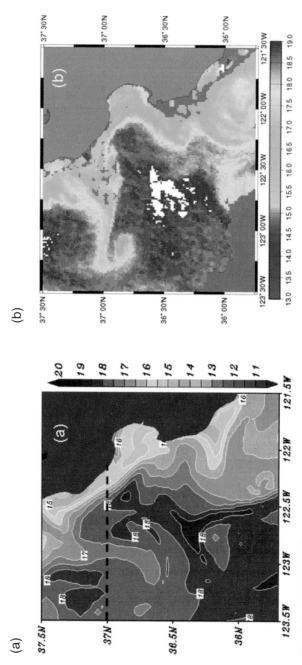

**Fig. 9** (a) NCOM SST from 18 h forecast valid at 18Z August 24, 2003. (b) NOAA POES AVHRR HRPT SST at 1856Z August 24, 2003 (NOAA POES AVHRR, Courtesy NWS and NOAA Coastwatch)

in the validation section, the errors in the model representation of the relaxation episode is most likely due to errors in the atmospheric forcing.

The isopleths of temperature slope downward towards the coast during the relaxation stage (Fig. 10), indicating that offshore warm water is advected to the nearshore. Downwelling forced the upper-layer water downward following the slope of the topography. The upwelling front is located near the coast. Warm water recapped the surface layer in the original upwelling area. There still exist smaller temperature gradients across the front with about a 2.0°C difference.

The 25 h mean NCOM forecast from 5Z August 25 to 6Z August 26, 2003 for a relaxed state is compared to the 25 h mean HF radar observation for the same time period (Fig. 11). Both the model and the data show slightly colder water in the southern part of the bay, a cyclonic circulation inside the bay, and an anti-cyclonic circulation outside the bay. The size and strength of these two circulations are similar in the HF radar analysis. However, the forecast model shows a smaller current speed for the cyclonic circulation inside the bay than for the anti-cyclonic circulation outside the bay. This again could be a result of the coarse horizontal resolution used in the NCOM. The high frequency HF radar can provide significant detail of the surface current in Monterey Bay and allow mesoscale features, like coastal eddies to be resolved with much more accuracy than an array of current meters. In the future, these current data will be used in the assimilation.

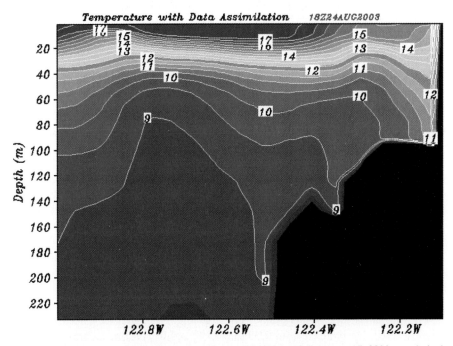

**Fig. 10** Vertical cross section of temperature along 37.0°N on 18 Z August 15, 2003 as a dashed-line shown in Fig. 9

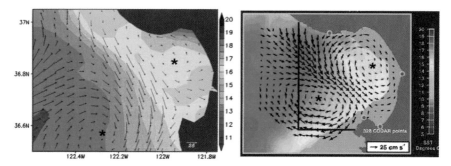

**Fig. 11** 25 h mean from 5Z August 25 to 6Z August 26, 2003 for (**a**) NCOM surface temperature and current, (**b**) HF radar surface current with AVHRR derived SST (Paduan and Lipphardt 2003) The approximated locations of cyclonic centers are marked as "∗"

## 6 Validation

The innovations and residuals for all assimilated observations are saved at the end of each update cycle so that assessment of the impact of the assimilation on the skill of the forecast can be made and the fit of the analysis to specific observations or observing systems can be evaluated.

The innovation and residual root-mean-square error (RMSE) and bias for any analysis or forecast variables are calculated as:

$$RMSE = \sqrt{\frac{1}{N}\{H(x) - y\}^2} \tag{2}$$

$$bias = \frac{1}{N}\{H(x) - y\} \tag{3}$$

where N is the number of observation data used in the analysis and $x$ represents any analysis or forecast variable.

Time series of innovation and residual bias and RMSE for SST and temperature averaged over depth are shown in Fig. 12. Figure 12a and 12b show SST RMS and mean bias errors decreasing with time during the first 3–5 days of the assimilation. After this time the SST innovation errors tend to stabilize suggesting that the model is accepting and retaining information from assimilation of the data. The mean SST innovation RMS error for the entire month is 0.73°C, with a mean bias error of −0.11°C. The analysis consistently reduces forecast errors throughout the assimilation time period. Mean SST residual RMS and bias errors are 0.28°C and −0.01°C, respectively. Similar patterns of reduction in RMS and mean bias errors from the forecast to the analysis are seen for temperature at depth (Fig. 12c, d), although relatively few subsurface observations were available. Nevertheless, forecast RMSE errors at depth are reduced from 0.95 to 0.35°C and forecast mean bias errors are reduced from 0.07 to −0.01°C by the analysis. SST forecast errors tend to be large following periods when few observational data are available for the assimilation

**Fig. 12** (**a**) RMSE and
(**b**) bias for SST innovation
and residual, (**c**) RMSE and
(**d**) bias for temperature
innovation and residual, and
(**e**) number of observation for
SST and temperature

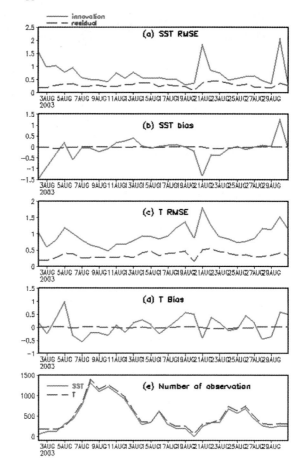

(compare forecast errors on August 22 with data counts on August 20). This suggests that the high surface variability associated with upwelling and relaxation processes in the Monterey Bay will require continuous observations in order to maintain forecast skill at 2 days. Further, the lack of forecast skill during the transition stage between upwelling and relaxation that started on August 20 may also be related to inaccuracies in the ocean model and the atmospheric forcing.

The consistent reduction in RMS error from the forecast to the assimilation throughout the assimilation time period given the changes in observation locations is an indication of a stable analysis/forecast system. This result is further indicated by the zero residual mean bias errors over all update cycles in conjunction with monthly averaged innovation bias errors that are essentially zero. A zero residual mean bias is an expected outcome from a least squares procedure such as optimum interpolation. A non-zero residual mean bias would be an indication of problems in the implementation of the analysis algorithm or in the pre-processing of the observations. A near zero innovation mean bias provides good evidence that, on average, the assimilative ocean model does not have any systematic model errors at the 24 h update cycle forecast period.

The forecast skill is evaluated by comparing the model RMSE at lead forecast times of 24, 48, and 72 h with the RMSE from persisting the previous analysis (nowcast) for the same lead time. The validation time period is from 2 to 31 August with an update assimilation cycle every 24 h, which yields 30 analyses. The forecast and persistence RMSE for SST are summarized from the domain mean values as shown in Fig. 13. The forecast RMSE from non-assimilative run is included and displayed as red solid line. The results show that model forecasts of SST are more skillful than persistence when data assimilation is performed. The SST forecast improvement over persistence is more significant as the forecast lead time increases, indicating that the persistence becomes less important. The forecast RMSEs from non-assimilative run is considerably larger than those from assimilative run at the first day, but with small changes for increased forecast lead time. At day 3, much smaller difference of the SST RMSE between assimilative and non-assimilative due to the impact of the data from the assimilation is lost over time. The error statistics indicate that the forecast skill from data assimilative run in Monterey Bay is about 2 days. The forcing has an impact at day 3 as compared to persistence where forcing changes are not applied.

Results from data assimilative run and non-assimilative run are compared with independent observations from the R/V Point Sur CTD of MBARI (Haddock et al. 2003) on 20 August 2003 as shown in Fig. 14. Figure 14f displays 7 positions (position 4 and 5 are almost overlapped) for providing observed temperature profiles. The cross sections of temperature in Fig. 14a, b, c are plotted from position 1 to 7. The observations show a shallow thermocline caused by previous upwelling processes, resulting in a very strong vertical gradient of temperature in the

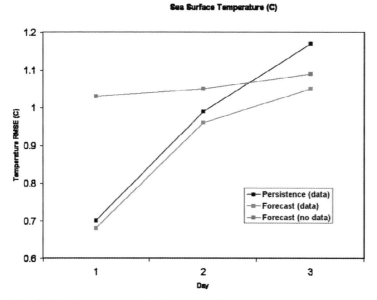

**Fig. 13** The domain-averaged persistence and forecast RMSE of SST for August 2003

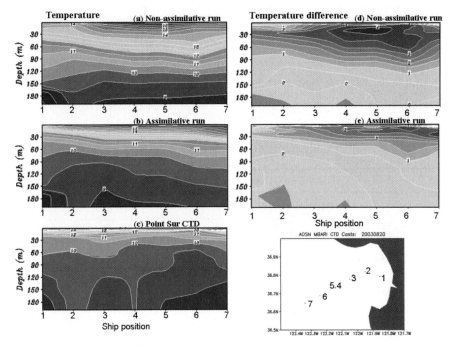

**Fig. 14** Cross section of temperature on August 20, 2003 from (**a**) non-assimilative run, (**b**) assimilative run, and (**c**) R/V Point Sur CTD of MBARI. The differences between NCOM forecasts and independent R/V Point Sur CTD observation are shown in (**d**) for non-assimilative run and (**e**) for assimilative rub. The ship positions are shown in (**f**)

upper surface layer (Fig. 14c). Colder water with a temperature of 10°C is located at about 60 m. There is a much smoother thermocline from the non-assimilative run (Fig. 14a), showing a much smaller vertical gradient of temperature in the upper layer. A better thermocline with a larger vertical gradient of temperature closer to the observation is obtained from the assimilative run (Fig. 14b). The temperature contour of 10°C is about 80 and 20 m deeper from non-assimilative and assimilative runs, respectively, indicating warmer temperatures in the upper layers. The temperature differences between the model results and observations reveal a large model bias in the non-assimilative run (Fig. 14d). Both the bias magnitude and its spatial extent are reduced for the model runs with data assimilation (Fig. 14e).

The mean SST biases for the non-assimilative and assimilative runs are dissected for upwelling and relaxation periods to inspect the forecast skills corresponding to different ocean dynamic processes (Fig. 15). The bias of 24 h forecast is averaged for the time period from August 7 to 19 for the upwelling and from August 20 to 24 for the relaxation. The analysis fields from the data assimilative run with all the available data assimilated are used as the "true" state. NCOM model has very good forecast skill for the upwelling center at Point Ano Nuevo, but much less skillful in the south upwelling center (Fig. 15). This leads to a warmer temperature than observation in the south as shown in Fig. 6. Substantial bias during relaxation period (Fig. 15b) denotes that model is less skillful in response to the transition period from

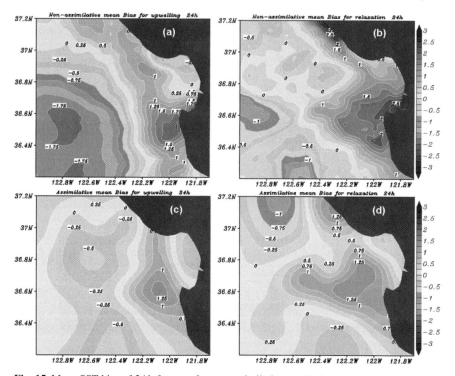

**Fig. 15** Mean SST bias of 24 h forecast for non-assimilative run during (**a**) upwelling period and (**b**) relaxation period, and for assimilative run during (**c**) upwelling period and (**d**) relaxation period

upwelling to relaxation wind regime. The errors have larger horizontal scale during the upwelling (Fig. 15a), indicating ocean response to the larger scale wind forcing, and smaller during the relaxation, indicating ocean dynamics dominates the circulation. The biases for both upwelling and relaxation periods are reduced for assimilative run, especially for the relaxation period with the maximum value decreased from 2.75 to 1.25°C. The assimilation of both operational and AOSN II experimental data gives better initial conditions and reduced forecast errors (Fig. 15b, d).

The mean bias errors and RMSE of temperature based on the observation from moorings m1 and m2 for data assimilative run and non-assimilative run during upwelling and relaxation periods for 24, 48, and 72 h forecasts are displayed in Fig. 16. In general, both mean bias and RMSE are smaller for the upwelling and relaxation periods when data assimilation is performed. The difference is more apparent in the seasonal thermocline due to its large variability and uncertainty and the model could misplace it in the simulation. The errors increase with forecast periods, however, they show relatively small error growth. The bias errors from non-assimilative run are smaller than assimilative run in the surface layer during the upwelling period. This could be due to the model response fairly well to the upwelling favorable winds. Similar to the SST bias, there is a lot worse forecast skill during the relaxation period than the upwelling period due to the transition of driving forcing between the wind and ocean dynamics.

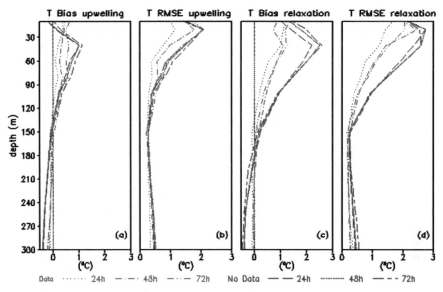

**Fig. 16** Mean bias errors and RMSE of temperature based on the observation from moorings m1 and m2 for 24, 48 and 72 h forecasts during upwelling (**a** and **b**) and relaxation (**c** and **d**) periods. The notation "Data" denotes for data assimilative run and "No Data" for non-assimilative run

# 7 Summary

This paper presents results from an ocean model (NCOM) and a cycling ocean data assimilation system (NCODA) in the Monterey Bay area in conjunction with the AOSN II field campaign. The multivariate analysis of NCODA is cycled with the ocean forecast model NCOM in a sequential, incremental, update cycle. In addition to operational ocean data from the GODAE server, which include remotely sensed SST and SSH and in situ surface and sub-surface observations of temperature and salinity, the assimilated data included high-density aircraft SST data and high-frequency buoy data from the AOSN II field experiment. The ocean forecast model used hourly atmospheric forcing from COAMPS and 3-hourly lateral boundary conditions from Global NCOM. An assimilative run was set up to cycle NCODA and NCOM for the entire month of August 2003, and results are compared with a non-assimilative NCOM run. The Global NCOM nowcast at 00Z August 1 2003 was used to initialize NCOM for the non-assimilative run and for the first forecast (before data assimilation cycle) of the assimilative run. Statistics for simple persistence, forecast skill, and performance measures of the data assimilation are provided to validate and evaluate the NCOM-NCODA cycling system.

Results from the data assimilative run are compared with the NOAA POES AVHRR SSTs and the temporally-averaged HF radar surface currents. Both of these sets of observations were independent of the model assimilation. The assimilative results are comparable with the observations in capturing the detailed coastal features such as upwelling and relaxation processes forced by the atmospheric winds during the late summer time period off Point Ano Nuevo and Point Sur in the vicinity of Monterey Bay. There are significant diurnal fluctuations in these processes. During the upwelling period, the upwelled cold water at the surface off Point Ano Nuevo reaches 11.5°C with a strong upwelling front positioned between the upwelled and offshore water. The cold water transports through the mouth of Monterey Bay and joins with the cold water upwelled from Point Sur, forming a cyclonic circulation inside the Monterey Bay. During the relaxation period, the upwelled cold water diminishes and is replaced with warm water from offshore with temperature greater than 16°C in most of the region. A cyclonic circulation pattern forms inside the bay during the relaxation phase, with an anti-cyclonic circulation outside the bay. These relaxation phase features are well simulated but with a smaller current speed in the inshore cyclone than what is observed.

Diagnosis of the analysis residual with respect to the forecast background error shows that the analysis improves the model initial conditions. The mean innovation RMSE and bias of SST and temperature at depth are reduced from the analysis. The data assimilative forecast is more skillful than the persistence and the non-assimilative run. The data assimilative run is able to simulate a thermocline layer as observed from the R/V Point Sur CTD of MBARI better than the non-assimilative run.

Statistics from forecasts of August 2003 denote that NCOM has more proficient forecast skill during the upwelling period and less skilful during the transition period from upwelling to relaxation wind regime. The biases for both upwelling and relaxation periods are reduced by data assimilation, with a most significant reduction of warm bias from 2.75 to 1.25°C for the relaxation period. The most important improvement in forecast due to the data assimilation is reflected for the seasonal thermocline, where large variability and uncertainty exist due to the strong nonlinearity and turbulence.

The results motivate the need to increase the model resolutions in the future in order to improve the forecast skill of the strength of upwelling, upwelling transport and areal extent of upwelling center. An accurate forecast is very important in the data assimilation system so that the assimilated data can be dynamically incorporated into the model trajectory.

Data assimilation with addition data in the coastal area has show very promising results by reducing bias errors and RMSE in both upwelling and relaxation periods, and in the upper seasonal thermocline layer. Other data collected from the AOSN II field campaign, such as glider data and HF radar data are very useful for providing vertical structure and surface current of upwelling and relaxation. We will explore the assimilation of these data sets in our future work.

**Acknowledgements** The support of the sponsors, the Office of Naval Research, Ocean Modeling and Prediction Program, through program element 7530-07-129 is gratefully acknowledged. Computations were performed on the IBM Cluster power 4+ system at the NAVO DoD Major Shared Resource Center, Stennis Space Center, Mississippi.

# References

Banta RM, Olivier LD, Levinson DH (1993) Evolution of the Monterey Bay sea-breeze layer as observed by pulsed Doppler lidar. J Atmos Sci 50: 3959–3982

Bishop CH, Toth Z (1999) Ensemble transformation and adaptive observations. J Atmos Sci 56: 1748–1765

Bishop CH, Etherton BJ, Majumdar SJ (2001) Adaptive sampling with the ensemble Kalman Filter, I: Theoretical aspects. Mon Weather Rev 129: 420–436

Chavez F (2003a) M1 Mooring Hydrography and Meteorology Data 2002–2003, Autonomuos Ocean Sampling Network (AOSN) 2003 Field Experiment, Monterey Bay Aquarium Research Institute. Retrieved January 2008 from http://aosn.mbari.org

Chavez F (2003b) M2 Mooring Hydrography and Meteorology Data 2003–2004, Autonomuos Ocean Sampling Network (AOSN) 2003 Field Experiment, Monterey Bay Aquarium Research Institute. Retrieved January 2008 from http://aosn.mbari.org

Chelton DB, deSzoeke RA, Schlax G (1998) Geographical variability of the first baroclinic Rossby radius of deformation. J Phys Oceanogr 28: 433–460

Cummings JA (2005) Operational multivariate ocean data assimilation. Q. J. R. Meteorol Soc 131: 3583–3604

Cummings JA (2006) The NRL real-time ocean data quality control system. NRL Technical Note

Daley R (1991) Atmospheric data analysis. Cambridge University Press, Cambridge, UK

Doyle JD, Jiang Q, Chao Y, Farrara J (2008) High-resolution atmospheric modeling over the Monterey Bay during AOSN II. (To appear Deep Sea Research)

Fox DN, Teague WJ, Barron CN, Carnes MR, Lee CM (2002) The modular ocean data assimilation system. J Atmos Ocean Technol 19:240–252

Haddock SHD, Ryan JP, Herren CM, Brewster J, Orrico CM, and Conlin D (2003) Hydrographic and bioluminescent towfish data from the research vessel Point Sur, Aug. 2003, Autonomuos Ocean Sampling Network (AOSN) 2003 Field Experiment, Monterey Bay Aquarium Research Institute. Retrieved January 2008 from http://aosn.mbari.org

Hodur RM (1997) The Naval Research Laboratory's Coupled Ocean/Atmosphere Mesoscale Prediction System (COAMPS). Mon Wea Rev 125: 1414–1430

Hodur RM, Hong X, Doyle JD, Pullen JD, Cummings J, Martin PJ, Rennick MA (2002) The Coupled Ocean/Atmosphere Mesoscale Prediction System (COAMPS). Oceanography 15(1): 88–89

Hong X, and Bishop CH (2005) COAMPS ocean ensemble forecast system. Presented on the 17th Conference on Numerical Weather Prediction, Washington, DC, 1–5 August 2005. http://ams.confex.com/ams/WAFNWP34BC/techprogram/paper_94771.htm. Accessed 15 December 2007

Hong X, and Bishop CH (2006) COAMPS ocean ensemble forecast and adaptive sampling system. Presented on the 2006 Ocean Sciences Meeting, Honolulu, Hawaii, 20-24 February 2006. http://www.agu.org/meetings/os06/cd/. Accessed 15 December 2007

Hong X, Hodur RM, Martin P (2008) Numerical simulation of deep-water convection in the Gulf of Lion. Pure Appl Geophys 164: 2101–2116

Jerlov NG (1976) Marine optics, Elsevier, 231 pp

Kelly KA (1985) The influence of winds and topography on the sea surface temperature patterns over the northern California slope. J Geophys Res 90: 11783–11798

Kondo J (1975) Air-sea bulk transfer coefficients in diabatic conditions. Boundary-Layer Met 9: 91–112

Martin PJ (2000) Description of the Navy Coastal Ocean Model Version 1.0. Naval Research Laboratory, NRL/FR/7322—00-9962, 1–42

Martin PJ, and Hodur RM (2003) Mean COAMPS air-sea fluxes over the Mediterranean during 1999 report. Naval Research Laboratory, Stennis Space Center, Mississippi

Paduan J, and Lipphardt B (2003) Coastal Ocean Dynamics Applications Radar (CODAR) Data 2003, Autonomuos Ocean Sampling Network (AOSN) 2003 Field Experiment, Monterey Bay Aquarium Research Institute. Retrieved January 2008 from http://aosn.mbari.org

Pickett MH, and Paduan JD (2003) Ekman transport and pumping in the California Current based on the U.S. Navy's high-resolution atmospheric model (COAMPS). J Geophys Res 108 (C10): 3327–3337

Pullen J, Doyle JD, Hodur R, Ogston A, Book JW, Perkins H, and Signell R (2003) Coupled ocean-atmosphere nested modeling of the Adriatic Sea during winter and spring 2001. J Geophys Res v108, C10, 3320,doi:10.1029/2003JC001780

Ramp S (2003) Sea surface remote sensing and atmospheric meteorology from the twin otter aircraft, Aug./Sept. 2003, Autonomuos Ocean Sampling Network (AOSN) 2003 Field Experiment, Monterey Bay Aquarium Research Institute. Retrieved January 2008 from http://aosn.mbari.org

Ramp SR, Paduan JD, Shulman I, Kindle J, Bahr FL, Chavez F (2005) Observation of upwelling and relaxation events in the northern Monterey Bay during August 2000. J Geophys Res v110, C07013, doi: 10.1029/2004JC002538

Rosenfeld LK, Schwing FB, Garfield N, and Tracy DE (1994) Bifurcated flow from an upwelling center: a cold water source for Monterey Bay. Cont Shelf Res 14: 931–964

Shulman I, Kindle J, Martin P, deRada S, Doyle J, Penta B, Anderson S, Chavez F, Paduan J, and Ramp S (2007) Modeling of upwelling/relaxation events with the Navy Coastal Ocean Model. J Geophys Res v112, C06023, doi:10.1029/2006JC003946

Tseng Y.-H. and Breaker LC (2007) Nonhydrostatic simulations of the regional circulation in the Monterey Bay area. J Geophys Res v112, C12017, doi:10.1029/2007JC004093

# Comparison of Ensemble-Based Filters for a Simple Model of Ocean Thermohaline Circulation

Sangil Kim

**Abstract** The performance of ensemble-based filters such as Sequential Importance Resampling (SIR) method, Ensemble Kalman Filter (EnKF), and Maximum Entropy Filter (MEF) are compared when applied to an idealized model of ocean thermohaline circulation. The model is a stochastic partial differential equation that exhibits bimodal states and rapid transitions between them. The optimal filtering result against which the methods are tested is obtained by using the SIR filter with $N = 10^4$ for which the method converges. The numerical results reveal advantages and disadvantages of each ensemble-based filter. SIR obtains the optimal result, but requires a large sample size, $N \geq 10^3$. EnKF achieves its best result with relatively small sample size $N = 10^2$, but this best result may not be the optimal solution. MEF with $N = 10^2$ achieves the optimal results and potentially is a better tool for systems that exhibit abrupt state transitions.

## 1 Introduction

Predicting exact states of systems from partially and imperfectly known observations is impossible because of the effects of random noise in the systems or incomplete initial/boundary conditions. It is therefore more reasonable to search for statistical information about the state of the systems such as moments, modes and other statistics. Data assimilation provides a methodology for estimating statistics of systems conditioned on observations. The filtering problem in data assimilation is defined as that of determining the best estimate of the *present* state, given prior observations.

In principle, it is possible to obtain optimal solutions to the filtering problem by calculating the conditional probability density function (pdf) given the observations up to current time (Kushner 1962; Lorenc and Hammon 1988; Stratonovich 1960). That is, between observations, the required conditional pdf can be obtained by solving the Fokker-Planck equation, a parabolic partial differential equation. When

S. Kim (✉)
The College of Oceanic and Atmospheric Sciences, Oregon State University, Corvallis, OR 97331–5503, USA, e-mail: skim@coas.oregonstate.edu

S.K. Park, L. Xu, *Data Assimilation for Atmospheric, Oceanic and Hydrologic Applications*, DOI 10.1007/978-3-540-71056-1_15,
© Springer-Verlag Berlin Heidelberg 2009

observations are available, the conditional density is updated by the Bayes' rule. Such algorithms for the optimal solution are developed for both continuous-time and discrete-time measurements (Campillo et al. 1994; Jazwinski 1970), and the optimal algorithms can be implemented by a numerical discretization of the Fokker-Planck equation to evolve the system statistics (Eyink et al. 2004). However, the algorithms are only feasible for simple models such as one-dimensional stochastic ordinary differential equations (ODEs), the Double-Well Potential Model (DWPM, Miller et al. 1994, 1999), and the 3-variable Lorenz equations (Lorenz63, Lorenz 1963), because of the computational burden of calculating pdfs of realistic spatially-extended systems with many degrees of freedom (Kushner 1967).

Therefore, alternative methods have been developed to approximate the conditional probability density function. Such alternatives are ensemble-based filters such as the Sequential Importance Resampling (SIR) method (Doucet et al. 2001; Gordon et al. 1993; Kitagawa and Gersch 1996), the Ensemble Kalman Filter (EnKF) (Burgers et al. 1998; Evensen 1994a, b), and the Maximum Entropy Filters (MEF) (Eyink and Kim 2006; Kim 2005; Kim et al. 2003). These methods use $N$ realizations, called "samples," that are drawn independently from the given initial distribution and assigned equal weights. The samples are then advanced in time by the evolution equation of the system to approximate unknown states by using the $N$-sample distribution. When observations become available, Bayes' rule is applied either to individual samples or to the $N$-sample distribution, to obtain a new set of realizations. This step is traditionally called the analysis step, and the different methods apply the Bayes' rule at the analysis step in different ways.

Previously, the ensemble-based filters mentioned above have been compared with simple, but highly nonlinear systems (Eyink and Kim 2006; Kim et al. 2003; Miller et al. 1994, 1999) such as low dimensional ODEs, DWPM and Lorenz63. However, the dimensionalities of these models are too small to reveal certain characteristics of the performance of the various ensemble-based filters, SIR, EnKF and MEF, that may be important when they are applied to more realistic models such as General Circulation Models (GCMs). In this study, the ensemble-based filters are evaluated by application to a somewhat more complicated, realistic model, the stochastic Cessi-Young (SCY) equation (Eyink 2005). The SCY model is a partial differential equation (PDE), based on a conceptual model of the Atlantic thermohaline circulation (THC) introduced by Thual and McWilliams (1992), which derived from the depth- and zonally-averaged Boussinesq equations forced by prescribed surface temperatures and salinity fluxes for an ocean with asymptotically small aspect ratio. With the random salinity forcing, the SCY model shows bimodal statistics associated with two distinct steady-forced stable states and state transitions between them, which are analogs of multiple equilibria in more complex models of the THC (Rahmstorf 2003, 1995). Furthermore, SCY has an available closed form of steady state distribution and an exact formula for duration of the stable states (Eyink 2005), which are useful for generation of initial ensemble members, and estimation of the frequency of state transition. We therefore expect that this model is suitable; to illustrate the behavior of various filtering schemes in their ability to estimate statistics such as mean and variance of the filtered state pdfs for a nonlinear, spatially-extended system with abrupt state transitions. A brief overview of the SCY model and description of various ensemble-based filters and their results are given in Sects. 2 and 3, respectively. A summary and remarks are given in Sect. 4.

## 2 Stochastic Cessi-Young Model (SYC)

The SCY model is a stochastic partial differential equation given by Eyink (2005)

$$\partial_t \chi(y,t) = \partial_y^2 \left[ u^2 \chi(t,t) \cdot \{\chi(y,t) - \eta(y)\}^2 - \overline{rf(y)} + \chi(y,t) - \gamma^2 \partial_y^2 \chi(y,t) \right] + \partial_y F(y,t) \tag{1}$$

and boundary conditions are $\chi(\pm\pi) = \partial_y^2 \chi(\pm\pi) = 0$ with constants $\gamma = 0.1$ and $u^2 = 10$. The state variable $\chi(y,t)$ is the meridional gradient of salinity of the ocean water, $y \in [-\pi\pi]$ is latitude from the South Pole $(y = -\pi)$ to the North Pole $(y = \pi)$, the function $\eta(y)$ is a surface thermal gradient defined by $\eta(y) = -\sin(y)$, and $rf(y)$ is defined by

$$\overline{rf(y)} = \begin{cases} \dfrac{2}{3}\left[1 + \dfrac{u^2}{9}\eta^2(y)\right] \cdot \left[1 + \eta^2(y)\right] \cdot \eta(y) & \text{for } -\pi < y < 0, \\[3mm] \dfrac{2}{3}\left[1 + \dfrac{u^2}{9}\eta^2(y)\right] \cdot \eta(y) & \text{for } 0 < y < \pi. \end{cases} \tag{2}$$

The function $F(y,t)$ is a random salinity flux with spacetime white-noise covariance

$$< F(y,t)F(y',t') >= \sigma^2 \cdot \delta(y-y')\delta(t-t'),$$

where $\sigma$ is a given noise strength. The function $F$ represents variability in the surface salinity (freshwater) flux on short, synoptic time scales. Sources of this flux can be freshwater from rainfall or glaciers melting in the high latitudes. Various quantities can be calculated from the salinity gradient $\chi(y,t)$ such as salinity, temperature and current velocity because those variables are enslaved to the state variable in the small aspect ratio limit (Cessi and Young 1992). For its numerical implementation, a semi-implicit method is applied in time, with time increment $dt = 9.37 \times 10^{-5}$ and noise strength $\sigma = 0.08$. The space $y$ is discretized such that the interval $-\pi \leq y \leq \pi$ is represented by uniformly distributed 151 grid points. Thus, the model has 151-state dimension, and the random forcing $F$ at each point is independent. For more detail, see Appendix B in Eyink (2005).

When the random flux $F$ is turned off such that $F = 0$, the system given by Eq. (1) has two stable equilibria (Fig. 1a). These stable states are respectively called the thermally- and salinity-dominated states. Both equilibria are the same in the Southern Hemisphere $(-\pi < y < 0)$, but the patterns differ in the Northern Hemisphere $(0 < y < \pi)$. At high northern latitudes, the strong clockwise circulation

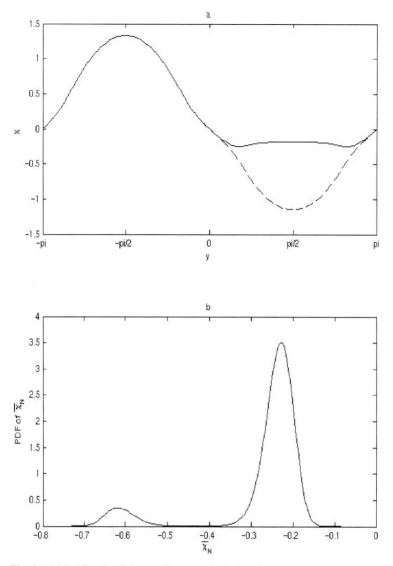

**Fig. 1** (**a**) Meridional salinity gradient $\chi$ on latitude y for the thermally- (*solid line*) and salinity (*dashed lined*) dominated equilibrium states. (**b**) Probability density function of the order parameter $\overline{\chi_N}$ at the North Pole

corresponding to the thermally-dominated state efficiently transports salt to the north pole and smoothes out the equator-to-pole salinity gradient. As a consequence, the north polar water is fresh and cold, and the equatorial waters warm and salty. On the other hand, the circulation of the salinity-dominated state is weak and counter-clockwise, and the high northern latitudes in this state are fresher and colder than those in the strong circulation.

When the random flux is taken into account ($F \neq 0$), the bimodality for the steady state statistics for Eq. (1) can be more clearly seen by defining the salinity quantity $\overline{\chi_N}$ at the North Pole ($y = \pi$) as

$$\overline{\chi_N} = \frac{1}{\pi} \int_0^\pi \chi(y)dy. \tag{3}$$

This quantity $\overline{\chi_N}$ is a so-called order parameter that distinguishes in which of the stable equilibria the system (1) resides at a given time t. The value of the equatorial salinity might be freely defined, but we set it to zero for simplicity. The salinities of the north polar water are then negative values (Fig. 1b). The probability density functions for $\overline{\chi_N}$ shows two modes, one mode at high salinities and the other at lower salinities. The high salinity mode is a state of strong, thermally-dominated circulation, and the low salinity mode is a state of weak, salinity-dominated circulation. The peaks in the modes correspond to the two-equilibrium states seen in Fig. 1a. The bimodal statistics of the model are a consequence of rapid transitions between two stable states, driven by the random fresh water $F$ fluxes at the ocean surface.

To examine the performance of the various ensemble-based filters, we perform an "identical twin" experiment, in which a preliminary run of the model itself is used to generate time histories of a realization as the "truth" that initially starts at the thermally-dominated state, and runs for 500,000 time units. Then, the synthetic observation data are taken at a discrete sampling interval $\Delta t = 8.78$ time units from the truth by adding *i.i.d* white noise $\varepsilon_t$ with mean zero and variance $R_t = 10^{-2}$. For the salinity $\overline{\chi_N}$ at the North Pole ($y = \pi$), the measurement function $h(\chi, t) \in R^q$ ($q = 1$) is

$$h(\chi, t) = \frac{1}{\pi} \int_0^\pi \chi(y, t)dy + \varepsilon_t, \tag{4}$$

where the observational error $\varepsilon_t$ is a normally distributed random error with mean zero and variance $R_t = 10^{-2}$. The main issue in this experiment is whether the approximate filtering methods with reasonable sample sizes $N$ track transitions at the correct time. Initially, the $N$ ensemble members are initially generated by the steady state distribution but with different realizations of the random forcing $F$.

## 3 Ensemble-Based Filters and Results

This section presents brief descriptions of the SIR, EnKF and MEF filtering schemes and the numerical results when the methods are applied to the stochastic Cessi-Young model. Since the SCY model exhibits rapid state transitions between two states, the key issue of interest is the performance of each filtering method in tracking state transitions without a delay as well as correctly estimating the mean and variance.

## 3.1 Sequential Importance Resampling Method

The SIR method is one of Sequentially Monte Carlo (SMC) methods that have received attention recently due to its theoretical appeal and the ease of implementation (Doucet et al. 2001). The method has been proved to converge to the complete non-Gaussian probability density of the state vector conditioned on the measurements in the limit of large sample size, and the convergence rate does not depend on the dimension of the state space (Del Moral 1996). The key implementation of SIR at the analysis step is that for $k = 1, \ldots, N$, the prior weight $w^-(x_k)$ of each member $x_k$ is updated by the exact Bayes rule to obtain the posterior $w^+(x_k)$. That is, when the measurement error density is given by $q(\varepsilon)$ and $y$ is the vector of observed values, the posterior is calculated by the Bayes' rule such that

$$w^+(x_k) = q(y - h(x_k)) \cdot w^-(x_k). \tag{5}$$

According to the updated weights, the previous samples are re-sampled based on their weights for next iterations.

This simple algorithm has been widely used in many fields and shows comparable results with other SMC methods (Arulampalam et al. 2002; Doucet et al. 2001; Van Leeuwen 2003). The main issue of SIR, like other SMC methods, is the "curse of dimensionality"; i.e., the computational complexity grows exponentially as a function of the dimension of the state vector (Daum 2005; Bellman 1961). When available sample size is restricted in high dimensional systems as in oceanography and climatology, SMC methods becomes impractical.

We show the convergent results of SIR with sample size $N = 10^4$ (Fig. 2a). Although the SIR method provides filtering results in terms of conditional pdfs, we only show the corresponding means and mean $\pm$ standard deviations, which are the statistical quantities of greatest interest. The measurements marked by circles are observed in the state of weak, salinity-dominated circulation until the 11th measurement time (t=96.58), and after the 12th measurement time (t=105.36), the observed salinities are in the state of strong, thermally-dominated circulation. The filtering means and standard deviations show that there is a state transition from the weak circulation regime to the strong circulation at the transition time. This can be confirmed by mean profiles of salinity gradient (Fig. 2b). As the sample size is increased further, even more samples make the transition at the transition time and the filtering mean and its standard deviations are smoother, but the qualitative results remain the same. Thus, the result of SIR with $N = 10^4$ can be interpreted as the optimal density, which should be replicated by other filtering methods if they perform properly.

When the sample size is $N = 10^2$, the SIR method fails to track the transition at the 12th measurement time (Fig. 3a). This is because all ensemble members repopulated around the region where consecutive measurements were observed (Kim et al. 2003; Eyink and Kim 2006). And, when the 12th measurement is observed in the other region, there is no sample that has made a transition to the other region. The sample that is the "closest" to the measurement obtains most of the weight and then is repeatedly selected, and all samples collapse to a single sample (Fig. 3c).

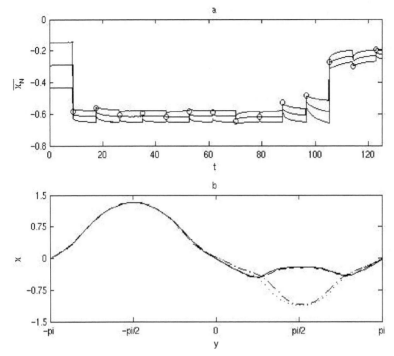

**Fig. 2** Convergent result of SIR with $N = 10^4$. (**a**) Filtering results. The middle solid line is the mean, and the upper and lower lines are mean $\pm$ standard deviation. The circles are measurements at the North Pole. (**b**) Mean profiles of salinity gradient $\chi$ field. The dotted line is $\chi$ field at 10th measurement time, the dotted-dashed is at 11th, the dashed is at 12th and solid is at 13th measurement time

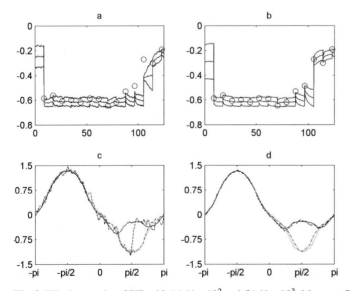

**Fig. 3** Filtering results of SIR with (**a**) $N = 10^2$ and (**b**) $N = 10^3$. Mean profiles of salinity gradient $\chi$ with (**c**) $N = 10^2$ and (**d**) $N = 10^3$. The axes and symbols are the same as in Fig. 2

This problem is known as *sample-impoverishment* or ensemble collapse, and becomes severe when the model exhibits the rapid state transition among states and when the sample size is restricted. Many systematic techniques to solve this problem have been proposed such as the resample-move algorithm (Arulampalam et al. 2002; Doucet et al. 2001), the regularization method (Wand and Jones 1995) and the double density method (Gilks and Berzuini 2001), but they still require exhaustive computational cost for high dimensional systems (Van Leeuwen 2003). After the sample size is increased by a factor 10 to $N = 10^3$, the SIR method shows convergent results for both filtering mean (Fig. 3b), and mean profile of salinity gradient (Fig. 3d).

## 3.2 Ensemble Kalman Filter

When the Kalman filter theory is applied to nonlinear dynamics in which the measurement function $h$ may not be affine, linear approximations such as in Extended Kalman Filter (EKF) are used to make the theory fit the nonlinear problems (Kushner 1967). Evensen (1994b) uses $N$ ensemble members at the prediction step such that the ensemble statistics represent the system dependent statistics and reflect the non-Gaussian density and configuration of the system. At the update step with observation $y$, the EnKF method obtains $N$ new posterior ensemble members $x_k^+, k = 1, \ldots, N$, from the prior member $x_k^-$ by using the Kalman gain matrix such that

$$x_k^+ = x_k^- + \mathbf{K} \cdot \left[ y_k - h(x_k^-) \right],  \tag{6}$$

with $\mathbf{K} = CH^T [HCH^T + R]^{-1}$, and $y_k = y + \varepsilon_k$ (Burgers et al. 1998). Here, $\varepsilon_k$ is an *i.i.d* random noise, C is the covariance of ensemble members $x_k^-$, and the measurement function $h(x)$ is linearly approximated such that $h(x) = Hx + d$. It has been shown that this method yields the optimal filtering solution when a dynamical system is linear with Gaussian statistics. Moreover, it produces very reliable results for many applications even when the system is nonlinear and non-Gaussian. See Evensen (2003) for a thorough introduction for a variety of transformation algorithms and overview of applications.

For the SCY model, the filtering means of EnKF with $N = 10^2$ shows a suboptimal result and delays the state transition by one measurement interval (Fig. 4a). EnKF performs no better than the SIR method with $N = 10^2$. However, the qualitative filtering results of EnKF with both sample sizes $N = 10^2$ and $N = 10^3$ are the same. Thus, increasing sample size does not improve the performance of EnKF such that filtering results with $N = 10^3$ is the same as that of EnKF with $N = 10^2$ except for the smaller fluctuations (Fig. 4a,c). The key ingredient of the quick convergence of EnKF is to presume a good proposal density, a Gaussian distribution that allows representing a system with small sample size. While the filtering mean at the 12th measurement indicates a possible transition, the respective mean profiles of salinity gradient for the EnKF with sample sizes $N = 10^2$ and $N = 10^3$ reveal an inevitable limitation attributable to the Gaussian assumption of the EnKF, such that the mean profiles remain the weak circulation state (Fig. 4c,d). That is, when the model (1)

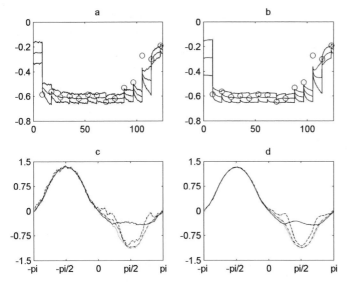

**Fig. 4** Filtering results of EnKF with (**a**) $N = 10^2$ and (**b**) $N = 10^3$. Mean profiles of salinity gradient $\chi$ with (**c**) N $= 10^2$ and (**d**) N$=10^3$. The axes and symbols are the same as in Fig. 2

exhibits transitions between two states, the Gaussian density represents the state to which the transition is observed with only a very small probability, so that the EnKF experiences delayed transitions or totally misses them unless special measures are taken to combine accurate observations with the forecast model to force the system through the transition. It is widely appreciated that the EnKF can fail under these conditions, long time mean residence time and abrupt nonlinear transitions between states. (Eyink and Kim 2006; Kim et al. 2003; Miller et al. 1994).

### 3.3 Maximum Entropy Filter

MEF is a recently introduced filtering method (Eyink and Kim 2006; Kim 2005; Kim et al. 2003) that employs exponential-families for prior distributions

$$P(x,t;\lambda,\Lambda) = \frac{\exp\left[\lambda \cdot \mu + \frac{1}{2}\Lambda : v\right]}{Z(\lambda,\Lambda)} Q(x), \tag{7}$$

where the function $Z$ is a normalization factor, $Q(x)$ is called a reference distribution that represents an estimate of the invariant measure of the system, and the constants $(\mu, v)$ are the first and second moments of the variables $h(x) \in R^q$ that are measured at observation times $t$ such that $\mu = \langle h(x^k) \rangle$ and $v = \langle h(x^k)h^T(x^k) \rangle$, $k = 1,\ldots,N$. The corresponding parameters $\lambda$ and $\Lambda$ which are respectively a $q$-vector and a $q \times q$ symmetric, positive-definite matrix are Lagrange multipliers that enforce the constraint of given values of the moments, and $\Lambda : v = \sum_{i,j=1}^{R^q} \Lambda_{ij} v_{ij}$.

The family (7) consists of "minimum-information distributions," which minimize the relative entropy

$$H(P|Q) = \int p(x) \cdot \ln \left[ \frac{p(x)}{Q(x)} \right] dx,$$

among all distributions $P(x)$ with given first and second moments $\mu$ and $v$ of the measured variables $h(x)$, for fixed $Q(x)$ for which the quotient $P(x)/Q(x)$ makes sense (Cover and Thomas 1991; Jaynes 2003). For practical calculation of the relative entropy $H$, MEF employs the notion of multiple climate regimes in geophysics (Namias 1950; Rex 1950; Stommel 1961) to represent the reference distribution $Q(x)$ with a "Gaussian mixture", a weighted $M$-Gaussian sum denoted by $Q_M(x)$ (Hannachi and Neil 2001; Smyth et al. 1999). Each Gaussian in the mixture model represents a distinctive climate regime of a given model. For SCY, the reference distribution can be a sum of two weighted Gaussians ($M = 2$), which approximate the steady state pdf of SCY (Fig. 1b).

The maximum entropy procedure that minimizes the relative entropy $H$ subject to the moment constraints obtains the probability distribution $P(x)$ by using the method of Lagrange multipliers (Jaynes 2003; Mead and Papanicolaou 1984). The values of the parameters $(\lambda, \Lambda)$ for given moments $(\mu, v)$ are determined by minimizing the Legendre transform

$$L(\mu, v) = \sup_{\lambda, \Lambda} \left\{ \lambda \cdot \mu + \frac{1}{2} \Lambda : v - \log Z(\lambda, \Lambda) \right\}. \tag{8}$$

Once the values of the parameters $(\lambda, \Lambda)$ are obtained, the update procedure of MEF is, then, ($i$) to determine approximate pre-measurement values of the moments $\mu^- = \langle h(x_k^-) \rangle$ and $v^- = \langle h(x_k^-) h^T(x_k^-) \rangle$ from the $N$ samples, ($ii$) to determine fitting parameters $(\lambda^-, \Lambda^-)$ from the minimization (8), ($iii$) to update the parameters to post-measurement values $(\lambda^+, \Lambda^+)$ by the Bayes' rule (5), and then ($iv$) to generate a new ensemble $x_k^+$ by resampling from the posterior distribution $P(x, t; \lambda^+, \Lambda^+)$. One strategy of the efficient sampling procedure for MEF refers to the Appendix in Kim (2005) and Eyink and Kim (2006).

A useful form of MEF is the Mean Field Filter (MFF), which can be formulated by setting $\Lambda = 0$ in the exponential-family (7). Then, the value of $\lambda$ for given $\mu$ can be determined as in MEF. However, the update step is slightly changed. The first steps are essentially the same: ($i'$) to determine approximate pre-measurement values of the moments $\mu^- = \langle h(x_k^-) \rangle$ from the $N$ samples, ($ii'$) to determine fitting parameters $\lambda^-$ from the minimization (8). In the third step ($iii'$), however, the update of $\lambda^-$ to $\lambda^+$ is different and uses a mean-field approximation to the Bayes' rule. In practice, when the observation error statistics are Gaussian with zero mean and covariance $R$, this means that $\lambda^+$ must be determined as the minimum of the following cost function:

$$C(\lambda) = \Gamma(\lambda) \cdot (\lambda - \lambda^-) - \log Z(\lambda) + \frac{1}{2} [\Gamma(\lambda) - y]^T R^{-1} [\Gamma(\lambda) - y], \tag{9}$$

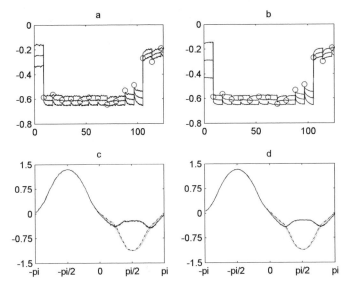

**Fig. 5** Filtering results of MEF with (**a**) $N = 10^2$ and (**b**) $N = 10^3$. Mean profiles of salinity gradient $\chi$ with (**c**) $N = 10^2$ and (**d**) $N = 10^3$. The axes and symbols are the same as in Fig. 5

where $\Gamma(\lambda) = \partial/\partial \lambda Z(\lambda)$ and $y$ is the vector of measurements. Finally, the last step of the update, as before, is ($iv'$) to generate a new ensemble $x_k^+$ by resampling from the posterior distribution $P(x, t; \lambda^+)$.

The filtering results of MEF and MFF with both sample sizes $N = 10^3$ and $N = 10^4$ capture the transition and estimate well its time and duration (Figs. 5a,b, 6a,b). MEF shows almost the same estimation as the optimal result in both mean and

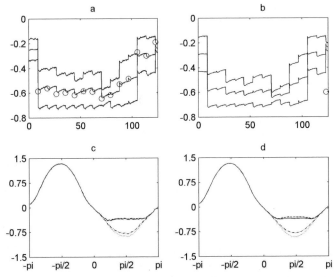

**Fig. 6** Filtering results of MFF with (**a**) $N = 10^2$ and (**b**) $N = 10^3$. Mean profiles of salinity gradient $\chi$ with (**c**) $N = 10^2$ and (**d**) $N = 10^3$. The axes and symbols are the same as in Fig. 2

standard deviation. The corresponding mean profiles of salinity gradient shows that
MEF and MFF correctly generate samples before and after transitions (Figs. 5c, 6c).
On one hand, MFF closely follows observations and also tends to overpredict variances, because the method uses the Bayes' rule only on the mean and keeps the covariances the same as the initial covariances in the reference distribution. However,
the filtering mean of the MFF method indicates the correct state before and after the
transition, and the mean profiles of salinity gradient also agree with those from the
optimal solution (Figs. 5d, 6d). The key success factor of both entropy-based filters'
effective tracking of rapid transitions is the use of mixture models representing all
distinctive states properly, even those with very small probabilities.

## 4 Summary and Remarks

In the numerical experiments with the SCY model, all of the four methods give
the same results as that of the optimal solution before the transition. When there is
a transition from one state to the other, the different filters show different results.
The proven convergent method, SIR does not catch the transition correctly with
$N = 10^2$, simply because of the insufficient sample size $N$. When the number of
sample sizes is increased to $N = 10^3$, SIR gives the optimal solution. The EnKF fails
to track the state correctly at the transition time with both sample sizes, $N = 10^2$
and $N = 10^3$, and lags by one measurement time. However, the EnKF achieves
its best estimation with a small sample size and the results of the EnKF with the
different sample sizes $N = 10^2$ and $N = 10^3$ are the same for the means. In fact,
the same qualitative result can be obtained with $N = 10$ (Kim and Eyink 2008). So,
increasing sample size does not improve the performance of EnKF. However, if we
adjust the measurement error to $R = 10^{-4}$ to provide more accurate measurement
error, EnKF with $N \geq 10$ tracks the transition correctly except for the fluctuation.
Equivalently, by inflating the covariance matrix (Anderson 2007), one might have
better performance of EnKF.

The numerical results of MEF and MFF show far superior results to those of
the standard filters, SIR and EnKF. These two entropy-based filters quickly attain
their convergent results with a small sample size. More importantly, they produce
the correct filtering results in catching state transitions at the correct instant by representing the state properly before and after the transition. The filtering mean of the
MFF method indicates the correct state, but shows a larger standard deviation than
that of the optimal results (Figs. 6a,b) since at analysis steps the method does not
update the covariance, which are the same as the initial covariance in the reference
distribution. The success is attributable to the use of the Gaussian mixture for the
reference distribution. The mixture model properly represents those states that have
low prior but high posterior probabilities, which becomes especially important when
the sample size is restricted. The entropy-based filters are proposed as effective and
efficient filtering methods that can overcome the difficulty of the standard filters
with restricted samples and multi-modal systems.

**Acknowledgements** The author would like to thank Gregory L. Eyink, Roger M. Samelson, and Robert Miller for helpful discussion. The author would also like to thank the anonymous reviewers for many helpful suggestions, which have greatly improved the presentation of this paper. Preparation of the manuscript has been partially supported by NSF Grant #DMC-0113649, and the office of Naval Research, Grant N00014-05-1-891, through the National Ocean Partnership Program.

# References

Anderson JL (2007) An adaptive covariance inflation error correction algorithm for ensemble filters. Tellus 59A:210–224

Arulampalam MS et al (2002) A tutorial on particle filters for online nonlinear/non-Gaussian bayesian tracking. IEEE Trans Signal Process 50(2):174–188

Bellman RE (1961) Adaptive control processes. Princeton University Press, Princeton, NJ

Burgers G et al (1998) Analysis scheme in the ensemble Kalman filter. Mon Wea Rev 126:1719–1724

Campillo F et al (1994) Algorithmes parallèles pour le filtrage non linèare et les èquations aux dèriv'ees partielles stochastiques. Bull Liaison Rech Info Auto 141:21–24

Cessi P, Young WR (1992) Multiple equilibria in two-dimensional thermo-haline circulation. J Fluid Mech 241:291–309

Cover TM, Thomas JA (1991) Elements of information theory. John Wiley & Sons, New York

Daum F (2005) Nonlinear filters: Beyond the Kalman filter. IEEE AES Magazine 20(8):57–69

Doucet A et al (2001) Sequential Monte Carlo in practice. Springer-Verlag, New York

Evensen G (1994a) Inverse methods and data assimilation in nonlinear ocean models. Physica D 77:108–129

Evensen G (1994b) Sequential data assimilation with a nonlinear quasigeostrophic model using Monte Carlo methods to forecast error statistics. J Geophys Res 99(C5):10 143–10 162

Evensen G (2003) The ensemble Kalman filter: Theoretical formulation and practical implementation. Ocean Dyn 53:343–367

Eyink GL (2005) Statistical hydrodynamics of the thermohaline circulation in a two-dimensional model. Tellus A 57:100–115

Eyink GL, Kim S (2006) A maximum entropy method for particle filtering. J Stat Phys 123(5):1071–1128

Eyink GL et al (2004) A mean field approximation in data assimilation for nonlinear dynamics. Physica D 195:347–368

Gilks WR, Berzuini C (2001) Following a moving target – Monte Carlo inference for dynamics Bayesian models. J R Stat Soc Ser B Stat Methodol 63(1):127–146

Gordon N et al (1993) Novel approach to nonlinear/non-Gaussian Bayesian state estimation. IEE Proc Radar Signal Process 140:107–113

Hannachi A, Neil AO (2001) Atmospheric multiple Equilibria and non-Gaussian behaviour in model simulations. Q J R Atmos Sci 127:939–958

Jaynes ET (2003) Probability theory: The logic of science. Cambridge University Press, Cambridge

Jazwinski AH (1970) Stochastic processes and filtering theory. Academic Press, New York

Kim S (2005) Ensemble filtering methods for nonlinear dynamics. Dissertation. University of Arizona

Kim S, Eyink GL (2008) Predicting rapid climate changes. In preparation

Kim S et al (2003) Ensemble filtering for nonlinear dynamics. Mon Wea Rev 131:2586–2594

Kitagawa G, Gersch W (1996) Smoothness priors analysis of time series, volume 116 of Lecture Notes in Statistics. Springer-Verlag, New York

Kushner HJ (1962) On the differential equations satisfied by conditional probability densities of Markov processes, with applications. J SIAM Control Ser A 2:106–119

Kushner HJ (1967) Approximation to optimal nonlinear filters. IEEE Trans Auto Contr 12:546–556

Van Leeuwen PJ (2003) A variance-minimizing filter for large-scale application. Mon Wea Rev 131:2071–2084

Lorenc QC, Hammon O (1988) Objective quality control of observations using Bayesian methods. Theory, and a practical implementation. Q J R Meteorol Soc 114:515–543

Lorenz EN (1963) Deterministic nonperiodic flow. J Atmos Sci 20:130–141

Mead L, Papanicolaou N (1984) Maximum entropy in the problem of moments. J Math Phys 25:2404–2417

Miller RN et al (1994) Advanced data assimilation in strongly nonlinear dynamical systems. J Atmos Sci 51:1037–1056

Miller RN et al. (1999) Data assimilation into nonlinear stochastic models. Tellus 51A:167–194

Del Moral P (1996) Nonlinear filtering: Interacting particle solution. Markov Proc Rel Fields 2:555–579

Namias J (1950) The index cycle and its role in the general circulation. J Meteorl 7:130–139

Rahmstorf S (1995) Bifurcation of the Atlantic thermohaline circulation in response to changes in the hydrological cycle. Nature 378:145–149

Rahmstorf S (2003) The thermohaline circulation: The current climate. Nature 421:699

Rex D (1950) Blocking action in the middle troposphere and its effect upon regional climate. Tellus 2:196–211

Smyth P et al (1999) Multiple regimes in northern hemisphere height fields via mixture model clustering. J Atmos Sci 56:3704–3723

Stommel H (1961) Thermohaline convection with two stable regimes of flow. Tellus 13:224–230

Stratonovich RL (1960) Conditional Markov processes. Theor Prob Appl 5:156–178

Thual O, McWilliams JC (1992) The catastrophe structure of thermohaline convection in a two dimensional fluid model and a comparison with low-order box model. Geophys Astrophys Fluid Dyn 64:67–95

Wand MP, Jones MC (1995) Kernel smoothing. Chapman and Hall, London

# Preconditioning Representer-based Variational Data Assimilation Systems: Application to NAVDAS-AR

Boon S. Chua, Liang Xu, Tom Rosmond and Edward D. Zaron

**Abstract** Assimilation of observations into numerical models has emerged as an essential modeling component in geosciences. This procedure requires the solution of large systems of linear equations. Solving these systems in "real-time" or "near-real-time" in a timely manner is still a computational challenge. This paper shows how new methods in computational linear algebra are used to "speed-up" the representer-based algorithm in a variety of assimilation problems, with particular application to the Naval Research Laboratory (NRL) Atmospheric Variational Data Assimilation System-Accelerated Representer (NAVDAS-AR) system.

## 1 Introduction

The representer-based algorithm is deployed in variational data assimilation systems such as (i) Inverse Ocean Modeling system (IOM) (Chua and Bennett, 2001; Bennett et al., 2008) and (ii) Naval Research Laboratory (NRL) Atmospheric Variational Data Assimilation System-Accelerated Representer (NAVDAS-AR) (Xu et al., 2005; Rosmond and Xu, 2006). Both weak and strong constraint variational assimilation (Sasaki, 1970) may be accomplished with this algorithm, which is formulated in the terms of dual variables (the so-called, "observation space"), the dimension of which is generally much smaller than the corresponding state space (Courtier, 1997). This restriction to the observation space is accomplished by discarding the unobservable degrees of freedom in the system (Bennett, 2002). The power of the representer algorithm as an iterative solver derives from this reduction of degrees of freedom, and results in considerable improvement in solver efficiency as compared to state space algorithms (Zaron, 2006). Despite its computational advantages over other assimilation algorithms, however, solving "real-time", "near-real-time", and "real-world" problems in a timely fashion using this algorithm is still a computational challenge. In this paper we explore some of the most recent methods in computational linear algebra (Simoncini and Szyld, 2007), and implement these methods to "speed-up" the representer-based algorithm.

B.S. Chua (✉)
SAIC, Monterey, CA, USA

S.K. Park, L. Xu, *Data Assimilation for Atmospheric, Oceanic and Hydrologic Applications*, DOI 10.1007/978-3-540-71056-1_16,
© Springer-Verlag Berlin Heidelberg 2009

The IOM system is a Graphical User Interface (GUI)-driven system for configuring and running weak-constraint four-dimensional variational assimilation (W4DVAR) with any model. All of the many model-dependent steps in the optimization algorithms are automatically and correctly generated. The user simply enters the model details via the GUI, and then the master program written in Parametric Fortran (Erwig et al., 2006) will generate the optimization steps as customized Fortran-90 code. If the user supplies the forward nonlinear model coded in Parametric Fortran, the IOM will automatically generate its tangent linear and adjoint models in Fortran-90. The IOM generates the run scripts for partial or complete inversion, as selected by the user via the GUI. In addition to computing the ocean circulation estimates, the IOM also computes dynamical residuals, data residuals, significance tests and posterior error covariances. The IOM has been applied to a level model (free-surface Bryan and Cox), a sigma-coordinate model (ROMS), a finite-element model (ADCIRC) and a spectral-element model (SEOM). Users experiences with it are documented in Muccino et al. 2008. In this paper the methods tested in the IOM system are further extended to an atmospheric variational data assimilation system, NAVDAS-AR.

NAVDAS-AR is an observation space four-dimensional variational data assimilation system based on the accelerated representer algorithm (Xu and Daley, 2000; Chua and Bennett, 2001). It is designed to become the U.S. Navy's next generation operational atmospheric data assimilation system. A wide variety of observations have to be accurately assimilated in a timely fashion. The number of observations assimilated in the current pre-operational NAVDAS-AR tests is around 500,000 during every 6 hour data assimilation window. We anticipate the number of observations to be assimilated will be around 1,000,000 during each 6 hour data assimilation window in the near future. The observations being routinely assimilated are: conventional observations, aircraft observations, feature tracked winds, AMSU-A, SSM/I total precipitable water and wind speeds, scatterometer winds, and Australia synthetic observations. The major computational cost of NAVDAS-AR is the minimization of a weighted least-squares cost function, made up of three computational components: an adjoint model integration, a background error covariance calculation, and a tangent-linear model integration. The cost of these is primarily a function of the model resolution chosen, and surprisingly the cost is only weakly dependent on the number of observations. This is in stark contrast to the NRL Atmospheric Variational Data Assimilation System (NAVDAS), which is the U.S. Navy's current operational three-dimensional assimilation system. The computational cost of NAVDAS is quadratically dependent on the number of observations, a serious liability with anticipated increases in observation volume. Details of the differences between NAVDAS and NAVDAS-AR are discussed in Rosmond and Xu (2006).

To satisfy the operational time constraint, the minimization of the cost function has to be made extremely efficient without degrading the accuracy of the analysis. In this paper we explore the benefits of using an efficient solver and robust preconditioner for NAVDAS-AR.

The outline of this paper is as follows. Section 2 describes three general types of linear algebra problems arising from variational assimilation applications. A linear system solver that is found to be well suited for the applications is discussed in Sect. 3. Section 4 summarizes the essential components of NAVDAS-AR. Section 5 describes the test problems and discusses the test results. Section 6 summarizes the paper.

# 2 Variational Assimilation Problems

There are three general types of linear algebra problems arising from variational assimilation applications. Each has its own characteristics, and therefore has its own computational requirements.

## 2.1 Linear Optimal Problem

In any representer-based variational data assimilation system, the optimal solution is obtained by solving a linear algebraic system. Following the notation in Chua and Bennett (2001), the linear system is given by

$$P\beta = h \tag{1}$$

where P is the stabilized representer matrix of dimension $M \times M$, $h$ is the $M$-dimensional vector of first-guess data misfits (innovations), and $\beta$ is the $M$-dimensional vector of representer coefficients. However, the majority of research publications on solving linear algebraic systems iteratively adopt the following notation:

$$\mathbf{Ax} = \mathbf{b}. \tag{2}$$

The preconditioned system is

$$\mathbf{M}^{-1/2}\mathbf{A}\mathbf{M}^{-1/2}\mathbf{M}^{1/2}\mathbf{x} = \mathbf{M}^{-1/2}\mathbf{b} \tag{3}$$

where the symmetric positive definite matrix $\mathbf{M}$ is a *preconditioner*.

## 2.2 Linear System with Multiple Right-Hand Sides

The diagnostics of assimilated products in a representer-based system, for example significance tests (Bennett et al., 2000; Descroziers and Ivanov, 2001), Kolmogorov-Smirnov (KS) test (Muccino et al., 2004), posterior covariances estimation (Bennett, 2002) and the *adjoint* of the assimilation system itself (Xu et al., 2006) can be obtained by solving a sequence of linear systems in which the coefficient matrix $\mathbf{A}$ is constant, but the right-hand side varies:

$$\mathbf{Ax} = \mathbf{b}_l, \tag{4}$$

where $l = 1, \ldots, L$. The subscript $l$ indicates a sequence of linear systems. In this study we assume the linear systems are solved in sequence, that is the right-hand sides are not available simultaneously. In some applications a right-hand side might be depending on a previously computed right-hand side.

## 2.3 Nonlinear Optimal Problem

In the representer-based variational data assimilation constrained with a non-linear model (e.g. NOGAPS), the optimal solution is obtained by solving for the solution of the nonlinear Euler-Lagrange (EL) problem using the *inner and outer loops* solution strategy (e.g. Chua and Bennett, 2001). In this strategy, the solution is obtained by solving a sequence of linear systems

$$\mathbf{A}^n \mathbf{x}^n = \mathbf{b}^n, \tag{5}$$

where the matrices $\mathbf{A}^n$ vary, the right-hand sides $\mathbf{b}^n$ are arbitrary and $n = 1, \ldots, N$. The superscript $n$ indicates the *outer* iterations.

## 3 Linear System Solver

This section describes a linear system solver for solving the three types of linear algebra problems discussed in the previous section.

## 3.1 Solver

Variational assimilation problems arising in operational systems such as NAVDAS-AR are challenging because the dimensions of the matrix $\mathbf{A}$ are extremely large ($M \sim 10^6$). Normally the construction and subsequent inversion of the full matrix is not feasible. In the representer-based algorithm, the linear algebraic system (2) is solved using an iterative method where the matrix $\mathbf{A}$ is not explicitly formed. Instead, the product of the matrix-vector multiplication $\mathbf{A}\mathbf{y}$, where $\mathbf{y}$ is a trial search vector, is computed by integrating an adjoint model *backward* and a tangent-linear model *forward* (e.g. Chua and Bennett, 2001). A straight application of the standard conjugate gradient (CG) solver (e.g. Golub and Van Loan, 1989) is not chosen here because the number of iterations required for a convergent solution will be large, in addition to the extra iterations required to overcome its rounding errors (Greenbaum and Strakos, 1992). Instead, the flexible conjugate gradient solver of Notay (2000) is chosen because it has been shown to outperform the standard CG solver, and also has shown great flexibility in handling different flavors of preconditioner. The flexible CG solver differs from the standard CG solver in that an explicit re-orthogonalization of the search direction vectors is performed at each iteration. The extra computational cost to form the scalar dot products required in the re-orthogonalization step is relatively small compared to the cost of computing the matrix–vector product $\mathbf{Ay}$. The pseudo code of the solver as described in Notay (2000) is reproduced in Appendix A for completeness.

## 3.2 Preconditioner

In addition to having a good solver, having a good preconditioner is equally important. Constructing a practical and effective preconditioner for the solver is extremely difficult because the dimensions of the matrix $\mathbf{A}$ are extremely large. This factor prohibits the explicit construction of the matrix, and therefore prohibits the use of the well-developed matrix-based preconditioning (Saad, 2003).

In the NAVDAS-AR, there are two levels of preconditioning. The first level employs rescaling of the linear system (2) as follows

$$\mathbf{R}^{-1/2}\mathbf{A}\mathbf{R}^{-1/2}\mathbf{R}^{1/2}\mathbf{x} = \mathbf{R}^{-1/2}\mathbf{b} \tag{6}$$

where $\mathbf{R}$ is the observation error covariance matrix. Because observation errors are assumed to be uncorrelated, only the diagonal of $\mathbf{R}^{1/2}$ is computed, and hereafter, it is incorporated into the measurement operator. This rescaling or change of variables (Courtier, 1997; Xu, Rosmond and Daley, 2005) reduces the condition number of the system by approximately two orders of magnitude. Figure 1 shows the extremal eigenvalues of (6) for a typical case.

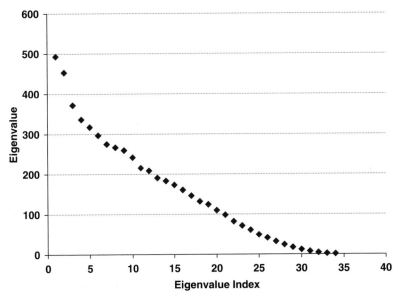

**Fig. 1** Eigenvalue spectrum of the rescaled stabilized representer matrix for a typical data assimilation with 6 hour window

A second level of preconditioning employs a preconditioning technique similar to the technique currently employed in the European Centre for Medium-Range Weather Forecasts (ECMWF) 4DVAR system (Fisher, 1995). It is the so-called spectral preconditioner which is described in Giraud and Gratton (2006):

$$\mathbf{M}^{-1} = \mathbf{I} + \sum_{i=1}^{k} (\alpha/\lambda_i - 1) w_i w_i^T / \|w_i\|^2, \tag{7}$$

where $\lambda_i, i = 1, 2, \ldots, k$ is a set of $k$ eigenvalues, and $w_i, i = 1, 2, \ldots, k$ is a set of corresponding $k$ eigenvectors, and $\alpha$ is a tuning coefficient.

An approximate spectrum of the stabilized representer matrix $\mathbf{A}$ is essential in constructing a good spectral preconditioner. The matrix $\mathbf{A}$ is not computed explicitly in the representer-based algorithm as mentioned before, but we can, as a consequence of the connection between the conjugate gradient and Lanczos algorithms (e.g. Golub and Van Loan, 1989), use the latter algorithm to obtain an approximate spectrum. Let the spectral decomposition of $\mathbf{A}$ be

$$\mathbf{A} = \mathbf{V}\mathbf{\Lambda}\mathbf{V}^T, \tag{8}$$

where $\mathbf{V}$ is orthogonal and $\mathbf{\Lambda}$ is diagonal. At the $k$th conjugate–gradient iterate, an orthonormal matrix $\mathbf{Q}_k$ and a tridiagonal matrix $\mathbf{T}_k$ may be constructed, satisfying

$$\mathbf{A}\mathbf{Q}_k = \mathbf{Q}_k\mathbf{T}_k. \tag{9}$$

The eigenvalues of $\mathbf{T}_k$ are approximations to the eigenvalues of $\mathbf{A}$. The eigenvectors of $\mathbf{T}_k$, when multiplied by $\mathbf{Q}_k$, give *good* approximations of the eigenvectors of $\mathbf{A}$. The construction of the matrices $\mathbf{T}_k$ and $\mathbf{Q}_k$ is given in Appendix B for completeness.

# 4 Technical Summary of NAVDAS-AR System

NAVDAS-AR is an observation space incremental 4D-Var system. It replaces the coupled non-linear Euler-Lagrange equations with a series of linear coupled Euler-Lagrange equations through the so called *inner and outer loops* strategy. It is currently applied to the Navy Operational Global Atmospheric Prediction System (NOGAPS), a global spectral numerical weather prediction model (Hogan and Rosmond, 1991). We use a high horizontal model resolution in the outer loop, a lower horizontal model resolution in the inner loop, and the same vertical model resolution in both the inner and outer loops. The NAVDAS-AR formulation, implementation, and preliminary results were documented in Xu et al. (2005) and Rosmond and Xu (2006). The treatment of model errors in NAVDAS-AR can be found in (Xu et al., 2007). NAVDAS-AR consists of three major components, namely, the pre-processing and quality control, the minimization, and the post-processing. Recently, we also have developed the *adjoint* of system itself (Xu et al., 2006) to

monitor and diagnose the impact of various observations on the short-term forecasts. Both the NAVDAS-AR and its adjoint share the same solver and preconditioner. Since our main focus of this study is to improve the computational efficiency
of the solver using the proposed preconditioner, we only give a brief summary of
the steps involved to minimize the cost function of the system in the following.

**Step 1.** We obtain the four dimensional (4D) background by integrating the fully
nonlinear NOGAPS forecast model with a higher horizontal model resolution (i.e.
the outer loop). Although we typically use a T239 (wave triangular truncation, 239;
nominally equivalent to 0.5° resolution) version of NOGAPS in the horizontal and
30 levels (L30) in the vertical, we use a T119 (1° nom. res.) version of NOGAPS in
the horizontal and 30 levels (L30) in the vertical to get quick turnaround for many
of data assimilation experiments.

**Step 2.** We compute the innovation vector $\mathbf{h} = \mathbf{y} - Hx^b$, where $\mathbf{y}$ is the observation
vector, $x^b$ is the 4D high resolution background produced using the T119L30 version
of NOGAPS, and H is a mapping operator that brings the 4D model background
from model space to the observation space.

**Step 3.** We solve the linear system (or the solver equation) $P\beta = \mathbf{h}$ using flexible
conjugate-gradient solver on a *coarse* (T47, 2.5° nom. res.) horizontal model reso-
lution. Here $P = (\mathbf{HP}^b\mathbf{H}^T + \mathbf{R})$ is the so-called stabilized representer matrix, $\mathbf{P}^b$ and
$\mathbf{R}$ are the 4D background and observation error covariance respectively; $\mathbf{H}$ and $\mathbf{H}^T$
are the tangent linear and the adjoint of the observation operator H, and the obser-
vation space vector $\beta$ is the solution to the solver equation and is also known as the
representer coefficients vector.

**Step 4.** We construct the spectral preconditioner $\mathbf{M}$ using the approximate eigen-
values/eigenvectors of the matrix P computed from a previous conjugate-gradient
iteration. We then apply the preconditioner to various NAVDAS-AR applications
such as the multi-resolution inner loop strategy, the second outer loop with same
inner loop resolution and the adjoint of NAVDAS-AR.

## 5 Experiments and Results

The linear system solver procedure outlined in the previous section are tested in
three different experiments, each for each type of the linear algebra problems previously discussed.

### 5.1 Linear Optimal Problem

In this experiment we compare the convergence rate of the solver for solving the linear optimal problem or so-called the *inner loop* with and without preconditioning.

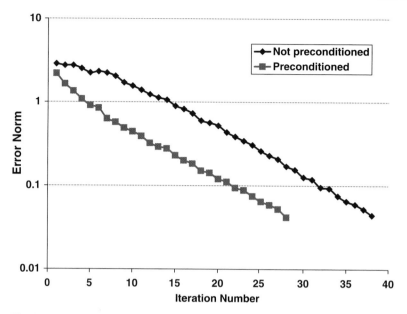

**Fig. 2** This figure compares the convergence rate of the solver for solving the *inner loop* with and without preconditioning. The vertical axis represents the error norm and the horizontal axis represents the iteration number. The blue solid line is the convergence of FCG, and the red solid line is the convergence of FCG preconditioned with the spectral preconditioner

For this comparison, the inner loop model resolution is set to T79L30. However, the spectral preconditioner is constructed using the approximate eigenvalues/eigenvectors of the matrix, based on first running the inner loop iteration at a *coarser* T47L30 model resolution. The numerical results are summarized in Fig. 2, where the vertical axis represents the error norm and the horizontal axis represents the iteration number. The blue solid line represents the convergence of flexible conjugate-gradients (FCG) and the red solid line represents the convergence of FCG preconditioned with the spectral preconditioner. The results shown the preconditioning reduces the number of iterations required to reach a prescribed stopping tolerance by about 25%.

## 5.2 Linear System with Multiple Right-Hand Sides

In this experiment we demonstrate the *recycling* of the approximate eigenvalues/eigenvectors of the matrix constructed from the previous system to reduce the cost of solving the subsequent system in sequence. The test problem for such a system so-called the linear system with multiple right-hand sides is the *forward* and *adjoint* of NAVDAS-AR described in Xu et al. (2006). In this experiment the forward of NAVDAS-AR is first run on a T47L30 model resolution to obtain the *optimal* solution. Subsequently, the adjoint of the system is run with the same model resolution as the forward system. The *adjoint* system is preconditioned with the

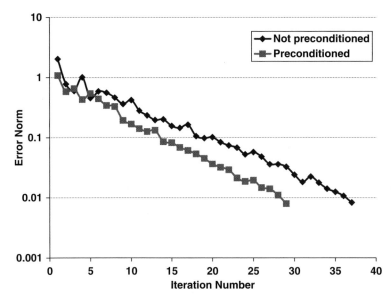

**Fig. 3** This figure shows the convergence of FCG with *recycling* preconditioning. The vertical axis is the error norm and the horizontal axis is the iteration number. The red solid line is the convergence of FCG with preconditioning, and the blue solid line is the convergence of FCG without preconditioning

spectral information gathered during the *forward* system run. The red solid line in the Fig. 3 shows the convergence of FCG with *recycling* preconditioning. The figure clearly illustrates the cost saving of having such preconditioner.

## 5.3 Nonlinear Optimal Problem

As discussed in the previous section, solving the nonlinear Euler-Lagrange (EL) problem is a nonlinear optimal problem. In this section we address a computational strategy used by the representer-based algorithm to solve this type of problem. In this strategy, there are two levels of iteration, the so-called the *outer and inner loops*. The *outer loop* or Picard iteration generates a sequence of linear optimal problems. In NAVDAS-AR the linear optimal problem or *inner loop* is solved using a low-resolution model with a simpler physics package than the full nonlinear model. This is an important saving as the computational cost for solving the inner loop is dominated by the cost of integrating the adjoint of the model *backward* and its tangent-linear model *forward*. Additional saving is made by preconditioning the second outer loop linear system using the first outer loop linear system spectral information. Figure 4 shown the convergence of the FCG with preconditioning (red solid line) and without preconditioning (blue solid line) for solving the inner loop of the second outer iterate. The results shown the number of iterations is reduced substantially for a given stopping criteria with preconditioning.

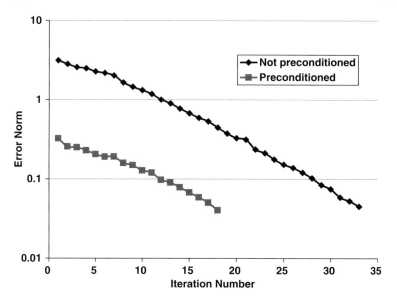

**Fig. 4** This figure demostrates the benefit of using the spectral information of the solver at the first iteration to precondition the inner loop of the second outer loop. The vertical axis represents the error norm and the horizontal axis represents the iteration number. The blue solid line represents the convergence of FCG, and the red solid line represents the convergence of FCG with preconditioning

## 6 Summary

We have described the implementation of a linear system solver for solving a variety of assimilation problems. The solver is applied to an atmospheric variational data assimilation system, the NAVDAS-AR. Computational savings are achieved by using the approximate spectral information of the coefficient matrix computed cheaply on a lower-resolution model, and by recycling of this spectral information. Because our focus is on developing a system for operational prediction, reductions in computation time can be utilized to incorporate more observations, run prognostic simulations at higher resolution, and generally improve the forecasting process. Important questions remain for future research, such as the performance of the recycled spectral preconditioner in the complete, weak constraint, assimilation system. We look forward to continued advances in the methodology of variational assimilation.

**Acknowledgements** The first author gratefully acknowledge the many crucial contributions of Andrew Bennett of Oregon State University (OSU) in developing the IOM system. The IOM system is supported by an National Science Foundation ITR Collaborative Research Medium Award, with subward OCE-121542 to OSU. Support of the Navy sponsors, the Office of Naval Research and PMW 180, under the Rapid Transition Process (RTP), are gratefully acknowledged.

# Appendix 1: Flexible Conjugate-Gradients

The flexible conjugate gradient method as described in Notay (2000) for solving

$$\mathbf{M}^{-1/2}\mathbf{A}\mathbf{M}^{-1/2}\mathbf{M}^{1/2}\mathbf{x} = \mathbf{M}^{-1/2}\mathbf{b} \qquad (10)$$

where $\mathbf{M}$ is a *preconditioner* is represented by the schematic in Fig. 5.

---

$k = 0;\ \mathbf{r}_0 = \mathbf{b} - \mathbf{A}\mathbf{x}_0;\ \omega_0 = \|\mathbf{r}_0\|_2/\|\mathbf{b}\|_2$

while $\omega_k > \omega_{sc}$

    solve $\mathbf{z}_k = \mathbf{M}^{-1}\mathbf{r}_k$

    $k = k + 1$

    if $k = 1$

        $\mathbf{p}_1 = \mathbf{z}_0$

    else

\*        $\mathbf{p}_k = \mathbf{z}_{k-1} - \displaystyle\sum_{l=1}^{k-1}\left(\mathbf{z}_{k-1}^T\mathbf{A}\mathbf{p}_l/\mathbf{p}_l^T\mathbf{A}\mathbf{p}_l\right)\mathbf{p}_l$

    end

    $\sigma_k = \mathbf{p}^T{}_k\mathbf{r}_{k-1}/\mathbf{p}^T{}_k\mathbf{A}\mathbf{p}_k$

    $\mathbf{x}_k = \mathbf{x}_{k-1} + \sigma_k\mathbf{p}_k$

    $\mathbf{r}_k = \mathbf{r}_{k-1} - \sigma_k\mathbf{A}\mathbf{p}_k$

    $\omega_k = \|\mathbf{r}_k\|_2/\|\mathbf{b}\|_2$

end

    $\mathbf{x} = \mathbf{x}_k$

---

**Fig. 5** The schematic of the flexible conjugate gradient (FCG) method for solving a preconditioned linear system as described in Notay (2000). In the schematic, $\mathbf{A}$ is the matrix, and $\mathbf{M}$ is a *preconditioner*. Standard conjugate gradients and FCG methods obtain the identical result in exact arithmetic; however, in finite-precision arithmetic, the FCG method differs by adding an orthogonalization of the search directions (indicated with \*), which stabilizes the method

# Appendix 2: The Lanczos Algorithm

The schematic for computing the orthonormal matrix $\mathbf{Q}_k$ and the tridiagonal matrix $\mathbf{T}_k$ using the Lanczos algorithm at the $k^{th}$ conjugate–gradient iterate is described in Fig. 6.

**Fig. 6** The schematic of the Lanczos algorithms (e.g. Golub and Van Loan, 1989) for estimating the spectrum of the matrix **A**. The schematic shown the computation of the tridiagonal matrix $\mathbf{T}_k$ and the orthonormal matrix $\mathbf{Q}_k$ using on the $k^{th}$ the conjugate–gradient iterate search information

$$\mathbf{Q}_k = [\mathbf{r}_0/\rho_0, \mathbf{r}_1/\rho_1, \ldots, \mathbf{r}_{k-1}/\rho_{k-1}]; \; \rho_i = \|\mathbf{r}_i\|_2$$

$$\Delta_k = diag(\rho_0, \rho_1, \ldots, \rho_{k-1})$$

$$\mathbf{B}_k = \begin{bmatrix} 1 & -\beta_2 & 0 & \cdots & 0 \\ 0 & 1 & -\beta_3 & 0 & \vdots \\ & \ddots & \ddots & \ddots & \\ \vdots & & \ddots & \ddots & -\beta_k \\ 0 & \cdots & & 0 & 1 \end{bmatrix}$$

$$\mathbf{T}_k = \Delta_k^{-1} \mathbf{B}_k^T diag(\mathbf{p}_i^T \mathbf{A} \mathbf{p}_i) \mathbf{B}_k \Delta_k^{-1}$$

# References

Bennett AF (2002) Inverse Modeling of the Ocean and Atmosphere. Cambridge University Press, New York NY

Bennett AF, Chua BS, Harrison DE, McPhaden MJ (2000) Generalized inversion of Tropical Atmosphere-Ocean (TAO) data and a Coupled Model of the Tropical Pacific. Part II: the 1995–1996 La Nina and 1997–1998 El Nino. J Climate 13: 2770–2785

Bennett AF, Chua BS, Pflaum BL, Erwig M, Fu Z, Loft RD, Muccino JC (2008) The Inverse Ocean Modeling System. I: Implementation. J Atmos Ocean Tech 25:1608–1622

Chua BS, Bennett AF (2001) An Inverse Ocean Modeling System. Oc Mod 3: 137–165

Courtier P (1997) Dual Formulation of Four-dimensional Assimilation. Q.J.R. Meteorol Soc 123: 2449–2461

Descroziers G, Ivanov S (2001) Diagnosis and Adaptive Tuning of Observation-error Parameters in a Variational Assimilation. Q.J.R. Meteorol Soc 127: 1433–1452

Erwig M, Fu Z, Pflaum B (2006) Generic Programming in Fortran. ACM SIGPLAN 2006 Workshop on Partial Evaluation and Program Manipulation: 130–139

Fisher M (1998) Minimization Algorithms for Variational Data Assimilation. Proceedings of ECMWF Seminar on Recent Developments in Numerical Methods for Atmospheric Modelling, 7–11 September 1998, pp 364–385

Giraud L, Gratton S (2006) On the Sensitivity of Some Spectral Preconditioners. SIAM. J Matrix Anal Appl 27: 1089–1105

Golub GH, Van Loan CF (1989) Matrix Computations-2nd ed. Johns Hopkins University Press, Baltimore

Greenbaum A, Strakos Z (1992) Predicting the Behavior of Finite Precision Lanczos and Conjugate Gradient Computations. SIAM. J Matrix Anal Appl 13: 121–137

Hogan T, Rosmond T (1991) The Description of the Navy Operational Global Atmospheric Prediction System's Spectral Forecast Model. Mon Wea Rev 119: 1786–1815

Muccino JC, Hubele NF, Bennett AF (2004) Significance Testing for Variational Assimilation. Quart. J Roy Meteor Soc 130: 1815–1838

Muccino JC, Arango HG, Bennett AF, Chua BS, Cornuelle BD, Di Lorenzo E, Ebgert GD, Haivogel D, Levin JC, Luo H, Miller AJ, Moore AM, Zaron ED (2008) The Inverse Ocean Modeling System. II: Applications. J Atmos Ocean Tech 25:1623–1637

Notay Y (2000) Flexible Conjugate Gradients. SIAM. J Sci Comput 22: 1444–1460

Rosmond T, Xu L (2006) Development of NAVDAS-AR: nonlinear formulation and outer loop tests. Tellus 58A: 45–58

Saad Y (2003) Iterative Methods for Sparse Linear Systems-2nd ed. SIAM, Philadelphia PA

Sasaki Y (1970) Some Basic Formalisms in Numerical Variational Analysis. Mon Wea Rev 98: 875–883

Simoncini V, Szyld DB (2007) Recent Computational Developments in Krylov Subspace Methods for Linear Systems. Number. Linear Algebra Appl 14: 1–59

Xu L, Daley R (2000) Data Assimilation with a Barotropically Unstable Shallow Water System using Representer Algorithms. Tellus 54A: 125–137

Xu L, Rosmond T, Daley R (2005) Development of NAVDAS-AR: formulation and initial tests of the linear problem. Tellus 57A: 546–559

Xu L, Langland R, Baker N, Rosmond T (2006) Development of the NRL 4D-Var data assimilation adjoint system. Geophys Res Abs 8: 8773

Xu L, Rosmond T, Goerss J, Chua B (2007) Toward a Weak Constraint Operational 4D-Var System: Application to the Burgers' equation. Meteorologische Zeitschrift 16: 767–776

Zaron ED (2006) A Comparison of Data Assimilation Methods Using a Planetary Geostrophic Model. Mon Wea Rev 134: 1316–1328

# Cycling the Representer Method
# with Nonlinear Models

Hans E. Ngodock, Scott R. Smith and Gregg A. Jacobs

**Abstract** Realistic dynamic systems are often strongly nonlinear, particularly those for the ocean and atmosphere. Applying variational data assimilation to these systems requires the linearization of the nonlinear dynamics about a background state for the cost function minimization, except when the gradient of the cost function can be analytically or explicitly computed. Although there is no unique choice of linearization, the tangent linearization is to be preferred if it can be proven to be numerically stable and accurate. For time intervals extending beyond the scales of nonlinear event development, the tangent linearization cannot be expected to be sufficiently accurate. The variational assimilation would, therefore, not be able to yield a reliable and accurate solution. In this paper, the representer method is used to test this hypothesis with four different nonlinear models. The method can be implemented for successive cycles in order to solve the entire nonlinear problem. By cycling the representer method, it is possible to reduce the assimilation problem into intervals in which the linear theory is able to perform accurately. This study demonstrates that by cycling the representer method, the tangent linearization is sufficiently accurate once adequate assimilation accuracy is achieved in the early cycles. The outer loops that are usually required to contend with the linear assimilation of a nonlinear problem are not required beyond the early cycles because the tangent linear model is sufficiently accurate at this point. The combination of cycling the representer method and limiting the outer loops to one significantly lowers the cost of the overall assimilation problem. In addition, this study shows that weak constraint assimilation is capable of extending the assimilation period beyond the time range of the accuracy of the tangent linear model. That is, the weak constraint assimilation can correct the inaccuracies of the tangent linear model and clearly outperform the strong constraint method.

H.E. Ngodock (✉)
The Naval Research Laboratory, Stennis Space Center, MS 39529,
USA, e-mail: Hans.Ngodock@nrlssc.navy.mil

S.K. Park, L. Xu, *Data Assimilation for Atmospheric, Oceanic and Hydrologic Applications*, DOI 10.1007/978-3-540-71056-1_17,
© Springer-Verlag Berlin Heidelberg 2009

# 1 Introduction

The representer method of Bennett (1992) is a 4D variational assimilation algorithm that relies on the adjoint of the dynamical model and expresses the analyzed solution as a first guess plus a finite linear combination of representer functions, one per datum. The explicit computation and storage of all the representer functions (direct method), however, is not required since the method can be implemented indirectly (Amodei, 1995; Egbert et al., 1994) using the conjugate gradient method (hereafter CGM). A description of the representer methodology is provided in the Appendix. The representer method has earned an established reputation as an advanced data assimilation technique within the past decade, and gained the attention of many potential operational users. Two primary issues, however, need to be addressed prior to implementing the representer method operationally.

The first issue addressed in this paper is the stability and validity of the tangent linear model (hereafter TLM). When the representer method is applied to a nonlinear model, the model must be linearized, preferably using the 1st order approximation of Taylor's expansion. Traditionally, the representer method has been implemented for the assimilation of all observations in the time window considered. As with every other variational data assimilation method with nonlinear dynamics, the representer method necessitates that the TLM and its adjoint be valid and/or stable over the entire assimilation time window. The validity of the TLM is difficult to maintain over a long time period for strongly nonlinear models and complex regions.

The second issue addressed in this paper is the cost of the representer method. The indirect representer method requires the integration of the adjoint and TLM within a CGM that determines the representer coefficients for the minimization of the cost function (see Appendix). This set of representer coefficients is then used to provide a correction to the background. The number of iterations of the CGM (this is referred to as the inner loop) is typically a small fraction of the total number of measurements. For strongly nonlinear systems, outer loops are required. To initialize the outer loop, one would pick a first background solution around which the model is linearized. The best solution (corrected background) obtained from this assimilation would become the background for the next outer loop, and so on until formal convergence (Bennett et al., 1996; Ngodock et al., 2000; Bennett, 2002; Chua and Bennett, 2001; Muccino and Bennett, 2002). This outer loop exacerbates the computational cost of the representer method. In this study the background that serves for linearization is also taken as the first guess.

These two issues have discouraged many potential users of the representer method for operational purposes. It is possible, however, to address these issues and implement the representer method at a reasonable cost for operational applications. Given a time window in which one desires to assimilate observations, it is possible to apply the representer method over cycles of subintervals. The name adopted for this approach is the "cycling representer method" (Xu and Daley, 2000), and its associated solution is called the "cycling solution". The solution that is obtained by assimilating all the observations at once in the original time window will be called the "global solution".

By using the cycling representer method, the assimilation time window is constrained to a period over which the TLM produces an accurate dynamical representation of the nonlinear model. Doing this reduces the need for outer loops. Because the representer method solves a linear assimilation problem, the outer loop is designed to solve the nonlinear Euler-Lagrange conditions associated with the assimilation problem of the nonlinear model. In the global solution problem, the TLM may not be an accurate representation of the dynamical system, and the adjoint would not be an accurate estimate of the derivative of the state with respect to the control variables. If the TLM is an accurate representation of the dynamics, the need for outer loops is removed. In the initial cycles of this assimilation approach, the first guess or background solution may not be accurate and thus outer loops may be required. Once the system is spun up and the TLM is an accurate approximation (thanks to improved background solutions), outer loops may no longer be necessary, thus lowering the computational cost of the assimilation. However, there may be situations in real world applications where a few outer loops would be needed in the current cycle, even though a single outer loop sufficed in previous cycles. An example is a nonlinear ocean response (advection and mixing) to a sudden, stronger than normal, atmospheric forcing, especially in coastal areas with complex bathymetry. The need for additional outer loops may be assessed by the discrepancy between the assimilated solution and the data.

The idea of cycling the representer method was investigated by Xu and Daley (2000) using a 1D linear transport model with synthetic data. In that study, the error covariance of the analyzed solution was updated at the end of each cycle and used as the initial error covariance in the next cycle. Another application of the cycling representer method was performed by Xu and Daley (2002) using a 2D linear unstable barotropic model with no dynamical errors. In this study, the covariance at the end of the cycle was not updated because its computation was too costly to be practical. Updating the covariance requires the evaluation and storage of the representer functions at the final time. These two studies found that updating the covariance at the end of each cycle produced significantly more accurate analyses. However, in these two applications of the cycling representer method, only linear models were used and thus there was no need for a TLM. Most realistic applications are nonlinear and their TLM may not be stable over the time window considered. It is in this context that this study applies the cycling representer method.

There are three clear advantages that one can foresee in this approach: (i) a shorter assimilation window will limit the growth of errors in the TLM, (ii) the background for the next cycle will be improved and, (iii) the overall computational cost will be reduced. It is assumed that the assimilation in the current cycle will improve the estimate of the state at the final time. The ensuing forecast (the solution of the nonlinear model propagated from the final state) is a better background for the next cycle than the corresponding portion of the background used in the global solution. This forecast uses the same forcing as the standalone nonlinear model, although the estimated model error could be ramped to the original external forcing in order to minimize shocks in the model. The latter has not been tested yet.

A good candidate for testing assimilation methods for strongly nonlinear models is the acclaimed Lorenz attractor model (Lorenz, 1963). It has been used to study the behavior of assimilation methods based on sequential filtering and variational techniques: Gauthier (1992), Miller et al. (1994, 1999), Evensen (1997), Evensen and Fario (1997) and Evensen and Van Leeuwen (2000), to cite but a few. This is done with the intent that if an algorithm performs satisfactorily well with this model, then it may be applied to atmospheric and ocean models. This is a necessary but not a sufficient condition.

Although being a strongly nonlinear model, the Lorenz attractor suffers from its low dimension; it has only three scalar prognostic variables. Assimilation experiments with the cycling representer method are presented for the Lorenz attractor (Ngodock et al. 2007a, b) in Sect. 2. Section 3 deals with the second model considered in this study, the one proposed by Lorenz and Emanuel (1998). It is a strongly nonlinear model with 40 scalar prognostic variables. It is called "Lorenz-40" in this paper for the sake of convenience. In Sect. 4, we present the third model in this study: a nonlinear reduced gravity model for an idealized eddy shedding in the Gulf of Mexico by Hurlburt and Thompson (1980). The fourth model is presented in Sect. 5. It is the Navy coastal ocean model (NCOM), a 40-layer primitive equation general circulation model based on the hydrostatic and Boussinesq approximations with a hybrid (terrain-following and z-levels) vertical coordinate. Concluding remarks follow in Sect. 6.

One can clearly notice the progression in this study, as nonlinear models of increasing dimension are considered. In all four applications, the cycling representer method is applied using the full TLM (as opposed to simplified linearizations) and its exact adjoint. In the experiments presented here a significance test is not performed. This would turn the assimilation problem into a search for suitable prior assumptions about errors in the data, initial condition, and dynamical errors, and hence cloud the issue at hand.

## 2 The Lorenz Model

The Lorenz model is a coupled system of 3 nonlinear ordinary differential equations,

$$\frac{dx}{dt} = \sigma(y - x) + q^x,$$
$$\frac{dy}{dt} = \rho x - y - xz + q^y, \tag{1}$$
$$\frac{dz}{dt} = xy - \beta z + q^z,$$

where $x$, $y$ and $z$ are the dependent variables. The commonly used time invariant coefficients are $\sigma = 28$, $\rho = 10$ and $\beta = 8/3$. The model errors are represented by $q^x$, $q^y$ and $q^z$. The initial conditions for Eq. (2) are,

$$x(0) = x_0 + i^x,$$
$$y(0) = y_0 + i^y, \quad (2)$$
$$z(0) = z_0 + i^z,$$

where $x_0 = 1.50887$, $y_0 = -1.531271$ and $z_0 = 25.46091$ are the first guess of the initial conditions. These are the same values that are used in the data assimilation studies by Miller et al. (1994), Evensen (1997), Evensen and Fario (1997), Miller et al. (1999), and Evensen and Van Leeuwen (2000). The initial condition errors are represented by $i^x$, $i^y$ and $i^z$. By setting the model and initial condition errors in Eqs. (1) and (2) to zero, the solution to the Lorenz Attractor is computed for the time interval [0, 20] using the fourth-order Runge-Kutta (RK4) discretization scheme with a time step of $dt = 1/600$ (Fig. 1). This solution is labeled as the true solution, since using time steps smaller than $dt = 1/600$ does not significantly change the solution within the specified time period.

The dimensionless time (t) in the Lorenz model is related to a simplified one-layer atmospheric model time ($\tau$) by $t = \pi^2 H^{-2}(1 + a^2)\kappa\tau$, where $a^2 = 0.5$, $H$ is the depth of the fluid and $\kappa$ is the conductivity. For a fluid depth of 500m and a conductivity of $25 \times 10^{-3}\,\mathrm{m^2 s^{-1}}$, a time unit in the Lorenz model corresponds to

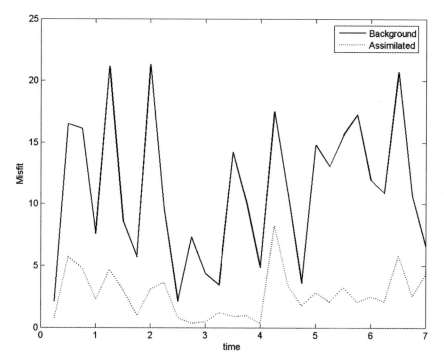

**Fig. 1** RMS misfits between the data and the background (*solid line*) and assimilated (*dotted line*) solutions for the first 7 time units of Fig. 4. This plot reveals that even though the TLM is only reliable for about 0.4 time units the assimilated solution is stable for about 7 time units and is correcting the background towards the data during this time period

7.818 days. The doubling time of the Lorenz attractor is about 1.1 time units, and the tangent linearization is not expected to be stable beyond this time range, which becomes a limiting factor for strong constraint assimilation. It is not so with the weak constraint. The latter is able to assimilate and fit the data beyond the time range of accuracy of the TLM, because the linear perturbation model is not solely driven by initial perturbation, but also by the estimated model error given by the adjoint model.

In the time interval [0, 20] there is a set of $M$ observations $\mathbf{d} \in \mathfrak{R}^M$ such that

$$\mathbf{d} = \mathbf{H}(x, y, z) + \varepsilon \tag{3}$$

where $\mathbf{H}$ is a linear measurement functional (an $M \times 3$ matrix), $\varepsilon \in \mathfrak{R}^M$ is the vector of measurement errors, and $M$ is the number of measurements. The data used for all assimilation experiments are sampled from the true solution with a frequency of 0.25 time units. The measurement error is assumed to be $\varepsilon = 0.002$, and its covariance matrix is assumed to be diagonal. The initial condition error that is used to perturb Eq. (2) is specified to be 10% of the standard deviation of each state variable of the true solution ($i^x = 0.784$, $i^y = 0.897$, and $i^z = 0.870$). The initial condition error covariance ($\mathbf{C}_{ii}$) is simply a $3 \times 3$ diagonal matrix with values equal to the square of the RMS of these initial condition errors. The model error covariance is prescribed as a time correlation function $\exp\left[-\left((t - t')/\tau\right)^2\right]$ multiplied by a $3 \times 3$ stationary covariance matrix

$$\tilde{C}_{qq} = \begin{bmatrix} 1.36 \times 10^{-5} & 5.99 \times 10^{-7} & -1.56 \times 10^{-6} \\ 5.99 \times 10^{-7} & 1.36 \times 10^{-5} & -2.07 \times 10^{-6} \\ -1.56 \times 10^{-6} & -2.07 \times 10^{-6} & 1.36 \times 10^{-5} \end{bmatrix} \tag{4}$$

Even though the time frame of assimilation is far greater than the stability of the TLM, the global solution is able to track the data somewhat for about 7 time units. It can be seen from Fig. 1 that the global solution is able to reduce the prior misfits significantly (even beyond the time range of accuracy of the TLM) before loosing track of the data. Beyond 7 time units, the misfit between assimilated solutions and data grows rapidly and can be attributed to the increasing errors in the TLM approximation. One can therefore conjecture that the error growth in the TLM can be limited by reducing the length of the assimilation window.

The results in Fig. 2 show the RMS error between the truth and the assimilated solution with respect to time for cycle lengths of 1, 2, 5 and 10 time units. It is shown that the RMS error increases with the cycle length. This is to be expected since longer cycles violate the TLM accuracy criterion. In other words, the steady decrease of RMS error with respect to the cycle length indicate that as the latter approaches the TLM accuracy time for the range of perturbations given by the adjoint model, the assimilation algorithm is better able to fit the data.

Results in Fig. 2 are obtained with 4 outer loops in each cycle. However, results with similar accuracy were obtained with 4 outer loops in the first cycle and a single outer loop in subsequent cycles, Ngodock et al. (2007a, b).

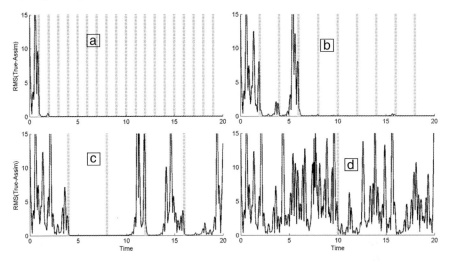

**Fig. 2** RMS of the misfit between assimilated and true solutions using different numbers of cycles: (**a**) 20, (**b**) 10, (**c**) 5, and (**d**) 2 cycles. The cycle boundaries are depicted by vertical dashed lines. By increasing the number of cycles from 2 (**d**) to 20 (**a**), significant improvement in the assimilated solution is achieved

The strong constraint solution (not shown here) is obtained by the same procedure as the weak, except that the model error covariance is set to zero. The weak constraint solution is not only more accurate, but also can afford longer cycles than the strong constraint. The strong constraint is almost confined to the TLM validity time, and needs quite a few cycles to start matching the data. In the experiment with cycles of 2 time units, the weak constraint accurately fits the data after 3 cycles, but the strong constraint never does. When the cycle length is decreased to 1 time unit, the weak constraint fits the data in the second cycle and afterward (Fig. 2a), while the strong constraint starts fitting the data only in the 16th cycle. Strong constraint assimilations will not be carried out with subsequent models.

## 2.1 The Cost

One major reason why the representer method is not widely implemented is the perceived computational cost. The biggest reduction in cost is achieved by limiting the outer loops to one, as was mentioned above. Further gains in computational cost are obtained by cycling the representer method. Assume that the matrix inversion in the indirect method is performed with a cost of $\mathcal{O}(M \log M)$ for computing $M$ representer coefficients, where $M$ is the number of measurements. The cycling approach total cost will be $N_{cy} \times \mathcal{O}(M_{cy} \log M_{cy})$, where $N_{cy}$ is the number of cycles and $M_{cy}$ is the number of measurements within each cycle (assuming that the measurements are uniformly distributed in the assimilation interval). Although $N_{cy} \times M_{cy} = M$, $\log M_{cy}$ gets exponentially smaller with increasing $N_{cy}$, thus decreasing the computational

**Table 1** Computational cost of the global and cycling solution using a single outer loop

|            | Global | 2 cycles | 4 cycles | 5 cycles | 10 cycles | 20 cycles |
|------------|--------|----------|----------|----------|-----------|-----------|
| Time (sec) | 21.37  | 9.33     | 4.49     | 3.59     | 1.90      | 1.09      |

cost as illustrated in Table 1. However, there is a drawback to reducing the cycle length. The data influence is extended beyond the cycle interval only through an improved initial condition for the next cycle. Future data contained in subsequent cycles will not contribute to the assimilation in the current and past cycles. One should keep this in mind, as well as the time decorrelation scale of the model errors, in choosing the appropriate cycle length.

## 3 The Lorenz-40 Model

The Lorenz-40 model (Lorenz and Emanuel, 1998) is a system of 40 coupled non-linear ordinary differential equations designed to represent the time evolution of advection and diffusion of a scalar quantity in one space dimension with periodic boundaries.

$$\frac{dx_i}{dt} = (x_{i+1} - x_{i-2})x_{i-1} - x_i + 8 + q_i, \quad 1 \leq i \leq 40. \tag{5}$$

The model is numerically solved with the 4th-order Runge-Kutta method using a time step of $\Delta t = 0.05$, which corresponds to about 6hr for Atmospheric applications. This model has an estimated fractal dimension of 27.1, and a doubling time of 0.42, given by the leading Lyapunov exponent. It has previously been used to test ensemble-based assimilation schemes by Anderson (2001), Whitaker and Hamill (2002), and Lawson and Hansen (2004).

The assimilation window is [0, 1000]. The data are sampled from a reference solution at every other component and every time step with a variance of $10^{-2}$. The assimilation background uses perturbed initial conditions and forcing. Due to the long time window and the increased chaotic behavior of this model, there is no possibility of computing a global solution; both the TLM and adjoint are unstable. Two cycling assimilations are considered: the first uses 100 cycles of 10 time units and the second uses 10 cycles of 100 time units. Results in Fig. 4 show that the assimilation with a shorter cycle is significantly more accurate. The short-cycle errors decrease rapidly after the first few cycles and never grow again. In contrast, errors in the solution from the longer cycle persist over time, an indication that the global solution would have been unable to match the data.

The cycle lengths of 10 and 100 time units are significantly longer than the doubling time of 0.42 given by the leading Lyapunov exponent. Thus the tangent linearization is not expected to be stable, much less accurate, for any of the cycles. A strong constraint assimilation would therefore fail to fit the data. However, the weak constraint approach is known to be able to fit the data beyond the time limit imposed

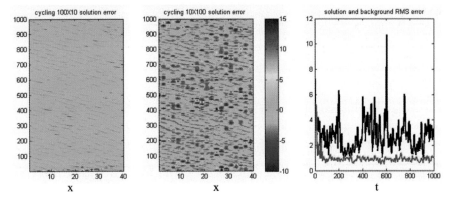

**Fig. 3** The assimilated solution error from the short cycle (*left*) and the long cycle length (*middle*). The right panel shows the RMS of the background (*black line*) and the short cycle assimilated solution (*red line*)

by the linearization stability (mostly because the assimilation is able to minimize the errors in the linearized model), and the results with the Lorenz-40 model shown here in Fig. 3 corroborate that fact.

## 4 The Nonlinear Reduced Gravity Model

A nonlinear reduced gravity (primitive equation) model is used to simulate an idealized eddy shedding off the Loop Current (hereafter LC) in the Gulf of Mexico (hereafter GOM). It is the same as the 1 1/2 layer version of the reduced gravity model introduced by Hurlburt and Thompson (1980). The dynamical equations are:

$$\frac{\partial hu}{\partial t} + \frac{\partial uhu}{\partial x} + \frac{\partial uhv}{\partial y} - fhv + g'h\frac{\partial h}{\partial x} = A_M \left( \frac{\partial^2 hu}{\partial x^2} + \frac{\partial^2 hu}{\partial y^2} \right) + \tau^x - drag_x,$$

$$\frac{\partial hv}{\partial t} + \frac{\partial vhu}{\partial x} + \frac{\partial vhv}{\partial y} + fhu + g'h\frac{\partial h}{\partial y} = A_M \left( \frac{\partial^2 hv}{\partial x^2} + \frac{\partial^2 hv}{\partial y^2} \right) + \tau^y - drag_y, \quad (6)$$

$$\frac{\partial h}{\partial t} + \frac{\partial hu}{\partial x} + \frac{\partial hv}{\partial y} = 0,$$

where $u$ and $v$ are the zonal and meridional components of velocity, $h$ is the layer thickness, $f$ is the Coriolis parameter (here a $\beta$–plane is adopted), $g$ is the acceleration due to gravity, $g'$ is the reduced gravity, $A_M$ is the horizontal eddy diffusivity, computed based on the prescribed Reynolds number Re, the maximum inflow velocity and half the width of the inflow port. The model parameters are listed on Table 2.

Hurlburt and Thompson (1980) showed that it is possible to simulate the eddy shedding by specifying time-invariant transport at the inflow and outflow open

**Table 2** Table of model parameters

| β | $f_0$ | g | $g'$ | Re |
|---|---|---|---|---|
| $2 \times 10^{-11}\,\mathrm{m}^{-1}\mathrm{s}^{-1}$ | $5 \times 10^{-5}\,\mathrm{s}^{-2}$ | $9.806\,\mathrm{ms}^{-2}$ | $0.03\,\mathrm{ms}^{-2}$ | 50.2 |

boundaries (see the model domain in Fig. 4). In this case the wind stress and the bottom drag are neglected. With a transport of 35Sv at inflow and outflow ports, we can simulate an eddy shedding with a period of about 4 months.

The data are sampled from the reference solution according to 8 networks described in Ngodock et al. 2006 (hereafter NG06), with 5cm and 5cm/sec data error for SSH and velocity respectively. The networks are ordered with increasing observation density, with network 8 yielding the most observations. Here the assimilation experiments are carried out for networks 3, 2 and 1 using SSH and velocity data, and for network 3 with only SSH data. The assimilation window is 4 months. In network 3, data are sampled from the reference solution every 200km in each spatial dimension and every 10 days, while networks 2 and 1 sample the reference solution every 300km (in both x and y directions) and every 5 and 10 days respectively. This produces a data density that increases with the network number. The covariances for the data, model and initial errors are the same as in NG06: the data error covariance is assumed diagonal with a variance of $25\,\mathrm{cm}^2$ for SSH and $25\,\mathrm{cm}^2\mathrm{s}^{-2}$ for both components of velocity; the model errors are allowed only in the momentum equations following Jacobs and Ngodock (2003), and have spatial correlation scales 100 km

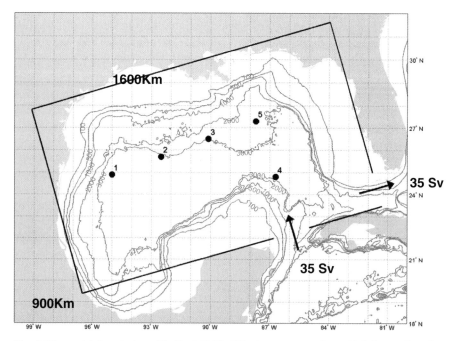

**Fig. 4** The model domain is an idealized Gulf of Mexico representation with inflow and outflow ports. The selected diagnostic locations of the assimilation solution are marked with bullets

in both x and y directions, a standard deviation $10^{-4}\,\mathrm{m^2 s^{-2}}$ (obtained by accounting for a typical wind stress of $0.1\,\mathrm{Nm^{-2}}$ which in turn is divided by a typical density of $1000\,\mathrm{kg\,m^{-3}}$), and a time correlation scale of 10 days. The results from the non-cycling assimilation experiments are available from the experiments reported in the same reference. Only the cycling assimilation experiments are carried out here and compared to the corresponding non-cycling solution obtained with 6 outer loops. It should be noted that the initial error covariance at the beginning of a new cycle is not updated as the posterior error covariance from the previous cycle. This procedure is computationally expensive and is avoided here. The original initial error covariance is used in every cycle. A set of 5 diagnostic stations is used for evaluation in this study. The station locations are shown in Fig. 4. They are selected in such a way that they are common to all the sampling networks; locations 1–3 are distributed along the path of the LCE, location 4 is in the region where the LCE sheds, and location 5 is north of the LCE shedding region.

The first cycling representer assimilation experiments are carried out for network 1 using 4 cycles of 1 month each and 3 outer loops in each cycle. A cycle length of 1 month is chosen to allow (i) a stable and accurate TLM, (ii) time distributed data within each cycle (especially when the data is sampled every 10 days e.g. networks 1 and 3), and (iii) the propagation of the data influence in time through the model dynamics and the model error covariance function. Figure 5 shows the difference between the reference and the assimilated solutions for both the non-cycling and the cycling at the end of each month. This figure shows that although both solutions have comparable discrepancies in velocity and sea surface height with the reference solution at the end of the first month, the discrepancies decrease rapidly in the cycling solution and by the end of the assimilation window they are greatly reduced relative to the non-cycling solution. It is not the case with the non-cycling solution; the discrepancies persist and are mostly located around the region where the LCE sheds from the LC, i.e. where advective nonlinearities are strongest. This indicates that the failure of the non-cycling solution is associated with an inaccurate TLM as suggested in NG06. It is also worth mentioning here that the cycling solution is obtained with 3 outer loops in each cycle, which is half the computational cost of the non-cycling solution computed with 6 outer loops as reported in NG06.

In the second set of cycling representer experiments, data is assimilated for networks 3, 2 and 1 using 4 1-month cycles in two cases: in the first case 3 outer loops are used in each cycle, and in the second case 3 outer loops are used only in the first cycle and 1 outer loop in the remaining cycles. Figure 6 shows the discrepancies to the reference solution computed for the non-cycling and the cycling solutions at the end of the third month for all networks, including an experiment where only SSH data from network 3 is assimilated. This figure shows that the errors in the non-cycling solution are consistent for all networks. One might have expected increasing errors as the data coverage decreases from network 3 to network 1. Such is the case for the non-cycling solution and not for the cycling. It can be hypothesized that the errors in the non-cycling solution are dominated by systematic errors in the TLM approximation. Fortunately, the cycling solution is able to fit the data properly because the growth of TLM errors are inhibited by a limited assimilation interval and a more accurate background provided by the previous cycle nonlinear forecast.

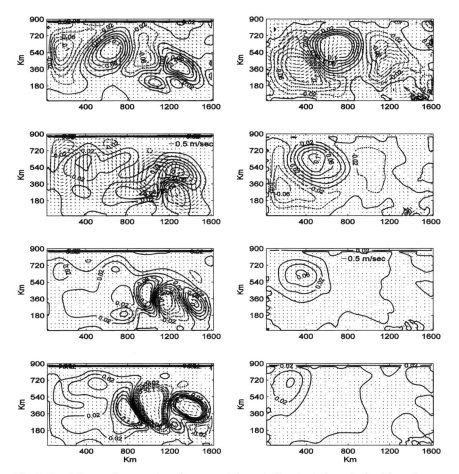

**Fig. 5** The difference between the reference and the assimilated solutions obtained from the non-cycling (*left column*) and the cycling (*right column*) representer algorithms for network 1. The differences are shown at the end the first month (*top row*), second month (*second row*), third month (*third row*) and fourth month (*fourth row*). Arrows represent the velocity and the contour lines represent the sea surface height, with a contour line of 0.01 m (1 cm)

A final experiment is carried out with the assimilation of only SSH data from network 3. As in NG06 for the non-cycling solution, the ability of the cycling algorithm to infer the velocity field through the model dynamics by assimilating only SSH measurements is tested. The non-cycling and the cycling solutions accuracy is evaluated through the rms error to the reference solution at the selected locations. Results in Table 3 show that the non-cycling solution is able to accurately fit the SSH data at all locations (except for location 4 where the rms exceeds 2 standard deviations) and the velocity only at the first two locations. At the remaining and critical locations 3–5, the non-cycling solution miserably fails to correct the velocity components with rms values sometimes exceeding 5–10 standard deviations. In

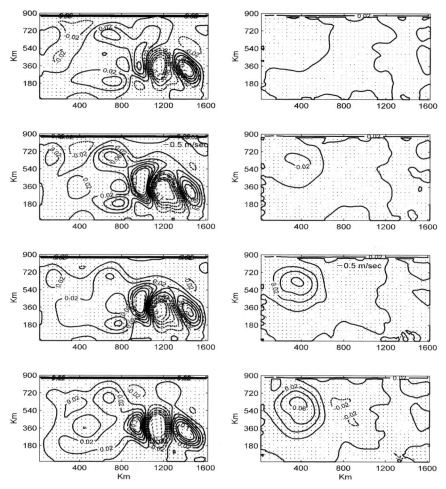

**Fig. 6** Comparison of the difference between the reference and the assimilated solution using the non-cycling (*left column*) and the cycling (*right column*) algorithms at the end of the third month for networks 3 (*first row*), 2 (*second row*), 1 (*third row*) and network 1 with only SSH data assimilated (*last row*)

**Table 3** RMS error of the solutions at the five diagnostic locations for network 3 assimilating only SSH data

| Location | SSH | | U | | V | |
|---|---|---|---|---|---|---|
| | Non-cycling | Cycling | Non-cycling | Cycling | Non-cycling | Cycling |
| 1 | 0.0160 | 0.0619 | 0.0871 | 0.0353 | 0.0307 | 0.0658 |
| 2 | 0.0253 | 0.0330 | 0.0521 | 0.0211 | 0.0416 | 0.0670 |
| 3 | 0.0679 | 0.0173 | 0.1073 | 0.0164 | 0.4772 | 0.0170 |
| 4 | 0.1354 | 0.0060 | 0.1795 | 0.0094 | 0.3671 | 0.0212 |
| 5 | 0.0963 | 0.0075 | 0.2926 | 0.0156 | 0.5844 | 0.0244 |

contrast, the cycling solution accurately fits the SSH data and the inferred velocity accurately matches the non-assimilated velocity data within expected errors.

## 5 The Navy Coastal Ocean Model (NCOM)

NCOM is a free-surface ocean model based on the primitive equations and the hydrostatic, Boussinesq, and incompressible approximations, solved on an Arakawa C-grid with leapfrog time stepping and an Asselin filter. An implicit time stepping is used for the free-surface, and the vertical discretization uses both sigma coordinates (for the upper layers) and z-level coordinates (for the lower layers). Further detailed specifications of NCOM can be found in Barron et al. (2006).

The model domain is shown in Fig. 7 where the 30X34 black dots are spaced 2.5 km apart and represent the center points of the Arakawa C-grid at which sea surface height (SSH), salinity and temperature are solved. This grid resolution requires a 4 minute time-step for numerical stability. In the vertical, there are 40 layers with 19 sigma layers in the upper 137m to resolve the shelf-break. The bathymetry is extracted from a Navy product called DBDB2, which is a global database with 2-min resolution. All of the atmospheric forcing, including wind stress, atmospheric pressure, solar radiation, and surface heat flux, is interpolated from the Navy Operational Global Atmospheric Prediction System (Hogan and Rosmond, 1991), which has a horizontal resolution of 1 degree and is saved in 3 hour increments.

An array of 14 acoustic Doppler current profile (ADCP) moorings was deployed by the Naval Research Laboratory (NRL) for 1 year (May 2004–May 2005) along

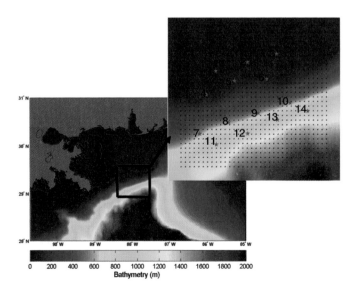

**Fig. 7** The model domain for the Mississippi Bight nested NCOM

the shelf, shelf-break, and slope of the Mississippi Bight (about 100 miles south of Mobile, Alabama). These moorings were spaced about 10–20 km apart and are identified in Fig. 7 as the numbered grey stars. During the time period of this study (the month of June, 2004), the filtered velocity data on the slope (moorings 7–14) exhibits a general transition of the flow field from being predominantly westward to eastward. Also, the flow on the slope had a strong correlation with the wind stress (∼0.8) and was fairly uniform in the along-shelf direction with a slight cross-shelf current towards the shore. In contrast, the circulation on the shelf (moorings 1–6) exhibits a weaker correlation with wind stress (less than 0.6), strong inertial oscillations with a period of about 24 hours, and a substantial velocity shear in the water column. Teague et al. (2006) provides an extensive presentation of this collected data set. The measurements are sampled every 3 hours and at 5 different depths for every mooring. The two velocity components are counted as 2 separate measurements.

As a precursor to the cycling experiments, a long 10-day assimilation experiment was attempted, and the resulting solution misfit (red) is plotted in Fig. 8A. The background misfit (blue) is also plotted for comparison. These misfits are computed as the RMS of the difference between the data and the solution. The assimilation

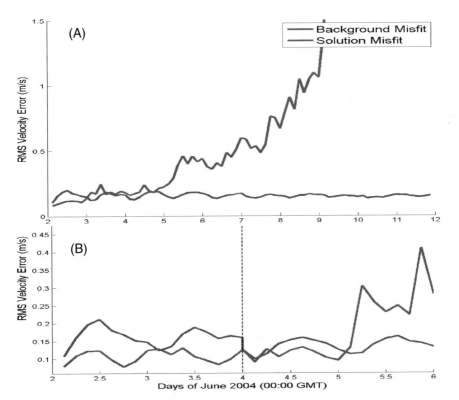

**Fig. 8** RMS misfits of the assimilated solution (*red*) and the background (*blue*) for a 10-day (**A**) and 2-day cycling (**B**) assimilation experiments

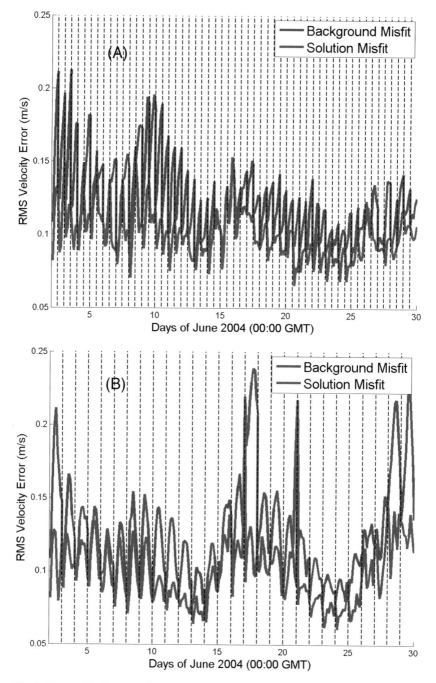

**Fig. 9** Same as Fig. 8, except for 12-hr (**A**) and 24-hr (**B**) cycling lengths

performs fairly well for the first day (which is about the range of TLM accuracy), then the assimilated solution begins to lose its skill, and by the third day it becomes unstable and its errors begin to increase exponentially.

The first cycling experiment performed employs 2-day cycles. The misfit results are displayed in Fig. 8B and reveal that the first cycle did well, but the second cycle began to severely lose skill midway through the cycle. At the end of the second cycle the solution was too poor to provide a sufficient initial condition for the next background forecast (the forecast grew numerically unstable). The dashed black line in this figure represents the break in between cycles and the vertical portion of the blue line along this dashed line is a result of the background being reinitialized to the assimilated solution. It is apparent that a 2-day cycle time period is too long in order to ensure solution accuracy. This falls in line with the time frame of TLM stability.

Two additional assimilation experiments are carried out for a period of 30 days, using 12-hr and 24-hr cycling lengths respectively. Results are shown in Fig. 9, where it is apparent that for the first 12 days of the experiment, the 1-day cycle experiment outperforms the 12-hour cycle experiment. From June 2 to June 14, the solution misfit in the 1-day experiment obtains lower values relative to the 12-hour cycle experiment and the general slope has a steeper downward trend. Also, in the 1-day cycle experiment there is a significant improvement in the background misfit. This is signified by a steep downward trend starting at the middle of each cycle. It is believed that this drastic change in the background misfit is due to the inertial oscillations, which are relatively strong in this region. It appears that the longer 1-day cycles are able to better resolve the inertial oscillations and therefore produce a more accurate solution that better matches the observed flow field. This result illustrates the importance of choosing a cycle time period that is long enough to include the important dynamic features that are prevalent in the region and allow the data to influence as long of a time window as possible.

# 6 Summary and Conclusions

The cycling representer algorithm that was only tested on linear models when initially proposed has now been applied to nonlinear models with increasing complexity and dimensions: the low-dimension Lorenz attractor, the 40-component strongly nonlinear model from Lorenz and Emanuel, the 2-dimension 1.5-layer reduced gravity nonlinear and the 3-dimension 40-layer NCOM models. In each of these models a global assimilation is impractical because the TLM is not stable over the entire assimilation time interval. One may argue that other linearization approaches may be more stable than the TLM. The cycling assimilation method could still be applied should the chosen linearization fail to be stable over the assimilation window considered.

However, the TLM is not the only factor guiding the decision for cycling. The cycled solution has proven to be more accurate than the non-cycling with the linear models to which the algorithm was first applied. Even more so in the context of nonlinear models with limited TLM stability time range. One reason why the cycling algorithm improves the accuracy of the solution is the introduction of new constraints at the beginning of each cycle. These constraints are absent in the non-cycling solution, and thus the cycling solution is more weakly constrained than the non-cycling. Another reason is the immediate improvement of the background in subsequent cycles. This improvement reduces the magnitude of the innovations and thus enables the tangent linear approximation and the assimilation to be more accurate. In contrast, the non-cycling solution has to overcome larger innovations to fit the data, which will require more inner and outer iterations for the process to converge. Finally the computational cost associated with the cycling algorithm is significantly lower than the non-cycling, especially when outer iterations are dropped, as shown in previous studies.

**Acknowledgements** This work was sponsored by the Office of Naval Research (Program Element 0601153N) as part of the project 'Shelf to Slope Energetics and Exchange Dynamics'. This paper is NRL paper contribution number NRL/BC/7320-07-8056.

## Appendix: Solving the Linear EL System Using the Representer Method

Given a background field $\mathbf{x}^f$, the linear EL to be solved is

$$-\frac{d\lambda}{dt} = \left[\frac{d\mathbf{F}(\mathbf{x}^f)}{d\mathbf{x}}\right]^T \lambda - \mathbf{H}^T \mathbf{w}(\mathbf{d} - \mathbf{H}\hat{\mathbf{x}})$$

$$\lambda(T) = 0 \tag{7}$$

and

$$\frac{d\hat{\mathbf{x}}}{dt} = \mathbf{F}(\mathbf{x}^f) + \frac{d\mathbf{F}(\mathbf{x}^f)}{d\mathbf{x}}(\hat{\mathbf{x}} - \mathbf{x}^f) + \mathbf{C}_{qq} \bullet \lambda$$

$$\hat{\mathbf{x}}(0) = \mathbf{x}_0^f + \mathbf{C}_{ii}\lambda(0) \tag{8}$$

The representer expansion for uncoupling (7) and (8) is:

$$\hat{\mathbf{x}}(t) = \mathbf{x}^f(t) + \sum_{m=1}^{M} \alpha_m \mathbf{r}_m(t). \tag{9}$$

Here the background (i.e. the trajectory around which the model is linearized) is also taken as the first guess (the solution that the assimilation will correct). The representer functions $\mathbf{r}_m$, $m = 1, \ldots M$ are computed from

$$-\frac{d\boldsymbol{\lambda}_m}{dt} = \left[\frac{d\mathbf{F}(\mathbf{x}^f)}{d\mathbf{x}}\right]^T \boldsymbol{\lambda}_m - \mathbf{H}^T \delta(t - t_m) \tag{10}$$

$$\boldsymbol{\lambda}(T) = 0$$

and

$$\frac{d\mathbf{r}_m}{dt} = \frac{d\mathbf{F}(\mathbf{x}^f)}{d\mathbf{x}}(\mathbf{r}_m) + \mathbf{C}_{qq} \bullet \boldsymbol{\lambda}_m \tag{11}$$

$$\mathbf{r}_m(0) = \mathbf{C}_{ii}\boldsymbol{\lambda}_m(0)$$

It may be shown (e.g. Bennett, 2002) that the representer coefficients $\alpha_m$ $m = 1,$ ... $M$ in (9) are the solution of the linear system

$$\left[\mathbf{R}^e + \mathbf{w}^{-1}\right]\alpha = \mathbf{d} - \mathbf{H}\mathbf{x}^f \tag{12}$$

where $\mathbf{R}^e$ is the representer matrix, obtained by evaluating the representer functions at the measurements sites, i.e. the mth column of $\mathbf{R}^e$ is $\mathbf{H}\mathbf{r}_m$. In practice, solving (12) does not require the computation of the entire representer matrix. An iterative method such as the conjugate gradient may be invoked, as long as the matrix-vector product on the left hand side of (12) can be computed for any vector in the data space. This is made possible through the indirect representer algorithm (Amodei (1995), Egbert et al. (1994)), which is also used to assemble the right hand side of (9) without the explicit computation and storage of the representer functions. Specifically, given a vector $\mathbf{y}$ in the data space, the product $\mathbf{R}^e\mathbf{y}$ is obtained by solving (10) and (11) with $\mathbf{y}$ replacing the Dirac delta in the right hand side of (10), then applying the observation operator $\mathbf{H}$ to the resulting $\mathbf{r}$. Once the representer coefficients $\alpha$ are obtained, the optimal residuals are computed by solving (10), where the single Dirac delta function is now replaced by the linear combination $\sum_{m=1}^{M} \alpha_m \delta(t - t_m)$. These residuals are then used in the right hand side of (11) to compute the optimal correction to the first guess $\mathbf{x}^f$.

# References

Amodei L (1995) Solution approchée pour un problème d'assimilation de données avec prise en compte de l'erreur du modèle. Comptes Rendus de l'Académie des Sciences 321, Série IIa, 1087–1094

Anderson JL (2001) An ensemble adjustment Kalman filter for data assimilation. Mon Wea Rev 129, 1884–2903

Barron CN, Kara AB, Martin PJ, Rhodes RC, Smedstad LF (2006) Formulation, implementation and examination of vertical coordinate choices in the Global Navy Coastal Ocean Model (NCOM), Ocean Model 11, pp 347–375

Bennett AF (1992) Inverse methods in physical oceanography. Cambridge University Press, New York, 347pp

Bennett AF (2002) Inverse modeling of the ocean and atmosphere. Cambridge University Press, Cambridge, 234pp

Bennett AF, Chua BS, and Leslie LM (1996) Generalized inversion of a global numerical weather prediction model. Meteor Atmos Phys 60, 165–178

Chua BS, Bennett AF (2001) An inverse ocean modeling system. Ocean Model 3, 137–165

Egbert GD, Bennett AF, Foreman MGG (1994) TOPEX/POSEIDON tides estimated using a global inverse method. J Geophys Res 99, 24821–24852

Evensen G (1997) Advanced data assimilation for strongly nonlinear dynamics. Mon Wea Rev 125, 1342–1354

Evensen G, Fario N (1997) Solving for the generalized inverse of the Lorenz model. J Meteor Soc Japan 75, No. 1B, 229–243

Evensen G, Van Leeuwen PJ (2000) An Ensemble Kalman Smoother for nonlinear dynamics. Mon Wea Rev 128, 1852–1867

Gauthier P (1992) Chaos and quadric-dimensional data assimilation: A study based on the Lorenz model. Tellus 44A, 2–17

Hogan TF, Rosmond TE (1991) The description of the navy operational global atmospheric prediction system. Mon Wea Rev 119 (8), pp 1786–1815

Hurlburt HE, Thompson JD (1980) A numerical study of the loop current intrusions and eddy shedding. J Phys Oceanogr 10(10), 1611–1651

Jacobs GA, Ngodock HE (2003): The maintenance of conservative physical laws within data assimilation systems. Mon Wea Rev 131, pp 2595–2607

Lawson WG, Hansen JA (2004) Implications of stochastic and deterministic filters as ensemble-based data assimilation methods in varyingregimes of error growth. Mon Wea Rev 132, 1966–1981

Lorenz E N (1963) Deterministic nonperiodic flow. J Atmos Sci 20, 130–141

Lorenz EN, Emanuel KA (1998) Optimal sites for supplementary weather observations: simulation with a small model. J Atmos Sci 55, 399–414

Miller RN, Ghil M, Gauthiez F (1994) Advanced data assimilation in strongly nonlinear dynamical systems. J Atmos Sci 51, 1037–1056

Miller RN, Carter EF, Blue ST (1999) Data assimilation into nonlinear stochastic models. Tellus 51A, 167–194

Muccino JC, Bennett AF (2002) Generalized inversion of the Korteweg-De Vries equation. Dyn Atmos Oceans 35, 3,227–263

Ngodock HE, Chua BS, Bennett AF (2000) Generalized inversion of a reduced gravity primitive equation ocean model and tropical atmosphere ocean data. Mon Wea Rev 128, 1757–1777

Ngodock HE, Jacobs GA, Chen M (2006) The representer method, the ensemble Kalman filter and the ensemble Kalman smoother: a comparison study using a nonlinear reduced gravity ocean model. Ocean Model 12, pp 378–400

Ngodock HE, Smith SR, Jacobs GA (2007a) Cycling the representer algorithm for variational data assimilation with the Lorenz attractor. Mon Wea Rev 135, 373–386

Ngodock HE, Smith SR, Jacobs GA (2007b) Cycling the representer algorithm for variational data assimilation with a nonlinear reduced gravity ocean model. Ocean Model 19, 3–4, pp 101–111

Teague WJ, Jarosz E, Carnes MR, Mitchell DA, Hogan PJ (2006) Low-frequency current variability observed at the shelfbreak in the northeastern Gulf of Mexico: May–October, 2004, Continental Shelf Res 26, pp 2559–2582

Whitaker JS, Hamill TM (2002) Ensemble data assimilation without perturbed observations. Mon Wea Rev 130, 1913–1924

Xu L, Daley R (2000) Towards a true 4-dimensional data assimilation algorithm: application of a cycling representer algorithm to a simple transport problem. Tellus 52A, 109–128

Xu L, Daley R (2002) Data assimilation with a barotropically unstable shallow water system using representer algorithms. Tellus 54A, 125–137

# Implementation of the Ensemble Kalman Filter into a Northwest Pacific Ocean Circulation Model

Gwang-Ho Seo, Sangil Kim, Byoung-Ju Choi, Yang-Ki Cho and Young-Ho Kim

**Abstract** The Ensemble Kalman Filter (EnKF) was implemented to an ocean circulation modeling system of the Northwest Pacific Ocean. The study area includes the northwestern part of the Pacific Ocean, the East China Sea, the Yellow Sea and the East/Japan Sea. The numerical model used for the system was the Regional Ocean Model System, which is a 3-dimensional primitive-equation ocean circulation model. The performance of EnKF was evaluated by assimilating satellite-observed Sea Surface Temperature (SST) data into the numerical ocean model every 7 day for year 2003. SST data were obtained from 30 fixed points at a time. The number $N$ of ensemble members used in this study was 16. Without localization of covariance matrix, ensemble spread (EnSP) drastically decreased due to rank deficiency and the large correlation between two distant state variables. To resolve the ensemble collapse, localization of covariance matrix was performed and EnSP did not collapse throughout the experiment. Root -mean-square (RMS) error of SST from the assimilative model (RMS error $= 2.2°C$) was smaller than that of the non-assimilative model (RMS error $= 3.2°C$). This work provides promising results that can be further explored in establishing operational ocean prediction systems for the Northwest Pacific including its marginal seas.

## 1 Introduction

The Northwest Pacific Ocean includes four different regions, the northwestern part of the Pacific Ocean, the East China Sea, the Yellow Sea, and the East/Japan Sea (Fig. 1). Each marginal sea has its own distinctive features. The Yellow Sea is tide-dominant and shallow with the average depth of 44 m, whereas the East/Japan Sea is a deep sea more than 3,000 m. Most of the East China Sea is continental shelf. The Kuroshio, one of the world's major western boundary currents, passes the southern area of the East China Sea. After passing the East China Sea, it flows along the southern coast of the Japan and separates from the Japan coast around 40°N.

S. Kim (✉)
The College of Oceanic and Atmospheric Sciences, Oregon State University,
Corvallis OR 97331–5503, USA
e-mail: skim@coas.oregonstate.edu

S.K. Park, L. Xu, *Data Assimilation for Atmospheric, Oceanic and Hydrologic*     341
*Applications*, DOI 10.1007/978-3-540-71056-1_18,
© Springer-Verlag Berlin Heidelberg 2009

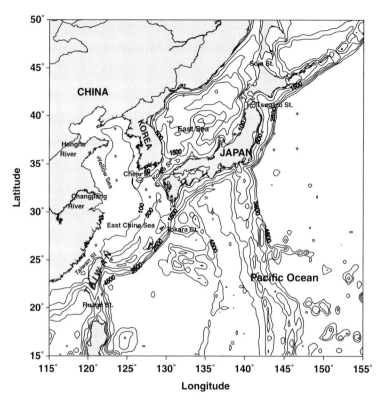

**Fig. 1** Model domain and bottom topography of the northwest Pacific Ocean. Contour lines are water depths. The numbers on the contour plots represent the depth (meter). The 30 diamond marks (♦) are the locations for measurement

The separated Kuroshio flows eastward. The currents transport heat, salt and other suspended materials to and from other seas. Developing a complex current system connecting the multiple regions is extremely difficult due to the distinctiveness of each individual region and the current interaction between the adjacent seas. Yet, it is important to develop such complex current system since it can provide useful information for fisheries, search and rescue, and natural disasters such as typhoons among others.

Numerical simulations of individual Northwest Pacific region have been developed and began to provide reliable representation of the three-dimensional ocean circulations (Guo et al., 2006; Isobe, 1999). This primitive-equation ocean model has given new understandings for the structure and dynamics of ocean circulation of the northwestern part of the Pacific Ocean and individual marginal seas. Recently, a regional ocean modeling system has been developed to cover all regions in a unified grid system to represent the structure and dynamics of the ocean circulation reasonably well (Cho et al., 2008). Furthermore, observing technologies such as satellite observing systems, land-based radars, and in-situ data have greatly advanced to improve our understanding to predict the ocean circulation system. In this context, we attempt to build a successful operational ocean prediction system for the Northwest

Pacific including marginal seas, which counts on combining numerical ocean models and real-time ocean observation data.

The study reported in this chapter is a foundational work in the process of achieving this long-term goal. We first develop a unified grid system for multiple marginal seas of the Northwest Pacific Ocean, which is nested into a global data assimilative ocean circulation model. This is the very first successful attempt to develop and employ the unified grid system for a broad range of the Northwest Pacific Ocean. Up to now, most studies of the Northwest Pacific Ocean region investigate individual marginal seas (Sasajima et al., 2007; Seung et al., 2007; Wantanabe, 2007; Xia et al., 2006) simply due to the absence of the grid system that unifies all marginal seas and northwestern part of the Pacific Ocean. As a next step, data assimilation schemes are applied to the extended numerical model. In particular, we implement the Ensemble Kalman Filter (EnKF) to the unified grid system of the Northwest Pacific Ocean region configured by the Regional Ocean Model System (ROMS) with realistic bathymetry. The performance of EnKF is tested as a preliminary study by assimilating satellite SST observation data into the numerical circulation modeling system. This study will provide better understanding on efficient implementation of EnKF to the numerical ocean circulation model with real observational data.

The remainder of the chapter is organized as follows. Section 2 describes the ocean circulation model and its implementation in a domain representing the Northwest Pacific Ocean region. In Sect. 3, we give a brief outline of EnKF. Finally, a description of numerical results and validation of the EnKF are presented in Sect. 4, followed by concluding remarks in Sect. 5.

## 2 Numerical Setup of a Northwest Pacific Ocean Circulation Model

The Regional Ocean Modeling System was configured for the simulation of Northwest Pacific Ocean circulation (Fig. 1). ROMS is a primitive-equation ocean circulation model widely used in oceanographic community and also used for a variety of studies to investigate the dynamics of ocean circulation. The numerical model is a free-surface and hydrostatic model, and it employs K-Profile Parameterization (KPP) scheme to determine a vertical mixing scheme. The turbulent mixing is parameterized by a non-local bulk Richardson number. The boundary-layer depth is determined at each grid point. Its profile is made to match the similarity theory of turbulence within this boundary layer (Large et al., 1994). The model is also efficient on both single processors and multi-threaded computer architectures (Song and Haidvogel, 1994; Haidvogel et al., 2000; Chassignet et al., 2000; Ezer et al., 2002; Curchitser et al., 2005).

The model domain ranges from 15°N to 50°N and from 115°E to 155°E, and the grid resolution is 0.25° for both Latitude and Longitude. In the vertical, 20 sigma levels are used, which stretch terrain-following coordinates (Song and Haidvogel, 1994). The realistic bathymetry is obtained by interpolating ETOPO-5, with a minimum depth set to 10 m. The model circulation was first spun up for 5 years from 1988 to 1992 with an initial state from the Levitus climatologic data. During the spin-up period, monthly mean winds, solar radiation, air temperature, relative humidity, sea level pressure data for the surface forcing are supplied from Comprehensive Ocean-Atmosphere Data Set (COADS). The initial temperature and salinity are provided from January mean data of World Ocean Atlas (WOA05). The initial currents field is set as geostrophic currents derived from the hydrographic data. After the spin-up period, more realistic surface forcing and boundary data are applied from January 1993 to November 2002. The surface forcing data are obtained from the European Center of Medium Range Weather Forecasting (ECMWF) reanalysis. Since the ECMWF data are available on a spatial resolution of approximately $0.35°$ by $0.35°$, the data have been interpolated to the model grid. Daily mean wind and monthly mean data are used for the air-sea momentum and heat exchanges using a bulk-flux formulation (Fairall et al., 1996). The result from a global data assimilative ocean model, Estimating the Circulation and Climate of the Ocean (ECCO), is used to provide open boundary data. Freshwater discharges from Changjiang and Huanghe River are included in the model, and the discharge is estimated as described in Senjyu et al. (2006). The tidal forcing and the relaxation of temperature and salinity are also applied along the open boundaries.

## 3 Data Assimilation Scheme

A variety of data assimilation (DA) methods have been developed such as optimal interpolation (OI), variational method, and ensemble schemes based on Kalman Filter (Wunsch, 1996; Bennett, 1992, 2002; Anderson, 1996; De Mey, 1997). Here, we will discuss mainly the 4 dimensional variational method (4D-VAR) and EnKF, the two main DA schemes that are most widely used in the community of ocean and atmospheric sciences. The 4D-VAR is a sophisticated DA scheme that can be applied to a highly complex dynamical system and actually has been implemented for operational numerical weather prediction around the world. The 4D-VAR is used to describe the evolution of sensitivity to initial, boundary and parametric conditions backward in time by solving the adjoint equations of the system (Courtier et al., 1993). The EnKF is sequential and straightforward to implement because the algorithm itself is independent of the forecast model, and unlike 4D-VAR, EnKF does not require additional development of the tangent linear and adjoint codes for the nonlinear forecast model. We are currently developing an improved DA method that combines strengths of both 4D-VAR and EnKF for better ocean prediction in the region of the Northwest Pacific Ocean. In the following, EnKF will be discussed more in detail, which is employed in this study and applied to the unified grid system for multiple marginal seas of the Northwest Pacific Ocean.

As described before, EnKF is an ensemble-based, sequential method such that a set of ensemble members is integrated by a given numerical model. When there are available observations, the ensemble statistics approximate the true statistics by Gaussian, which only keeps ensemble mean and covariance. Then, new ensemble members for next iterations are regenerated by using the Kalman gain matrix, which is derived from the assumptions of linearized measurement function and the Gaussian error statistics. Evensen (1994) first proposes EnKF and Burgers et al. (1998) clarifies the perturbation of measurements at the analysis step when observation is available. Later, various alternative algorithms of EnKF have been developed and their successful results are reported in many studies (Houtekamer and Mitchell, 2001; Keppenne and Rienecker, 2001). We refer readers to Evensen (2004) for a more complete overview of the developments and applications of EnKF.

The analysis steps of EnKF discussed here follow Evensen (1994) with the correction of perturbed measurements (Burgers et al., 1998). The equation for the analysis is

$$X_{new}^n = X_{old}^n + K \cdot [y_0^n - HX_{old}^n], \tag{1}$$

Here, $X_{new}^n \in R^p$ is the updated analysis state vector, and $X_{old}^n \in R^p$ is also the prediction state vector from the forward numerical model. The matrix $H$ is the measurement function, and $y_0^n \in R^q$ is an observation vector perturbed with random noises such that $y_0^n = y_0 + \varepsilon^n$, where $y_0 \in R^q$ is measured values from the true state. The random noise $\varepsilon^n \in R^q$ is realization of the assumed observational uncertainty. The difference between the actual measurement and the predicted measurement $y_0^n - HX_{old}^n$ is called the measurement innovation or the residual. The matrix $K$ is what so-called the Kalman gain matrix given by

$$K = CH^T \cdot [HCH^T + R]^{-1}, \tag{2}$$

where $R$ is observation error covariance matrix and $H^T$ is the transpose of the measurement function $H$. The covariance matrix $C$ is calculated as

$$C = \frac{1}{N-1} \sum_{n=1}^{N} (X^n - \mu)(X^n - \mu)^T = \frac{1}{N-1} \sum_{n=1}^{N} X^n (X^n)^T - \mu\mu^T, \tag{3}$$

where $\mu$ is ensemble mean. Then, $CH^T$ and $HCH^T$ can be calculated as in Eq. (3) such that

$$CH^T = \frac{1}{N-1} \sum_{n=1}^{N} (X^n - \mu)(HX^n - H\mu)^T r \quad \text{and}$$

$$HCH^T = \frac{1}{N-1} \sum_{n=1}^{N} (HX^n - H\mu)(HX^n - H\mu)^T. \tag{4}$$

The Kalman gain matrix $K$ in Eq. (2) can be thought as a weighted matrix such that the actual measurement $y_0^n$ gains more weight when the observation error covariance $R$ approaches zero, while the measurement $y_0^n$ loses its weight when the predicted measurement $HX_{old}^n$ is trusted more.

It is worth noting that the forecast covariance matrix $C$ in Eq. (2) is never explicitly calculated, but only appears via the smaller matrices, $CH^T$ and $HCH^T$ of size $p$ by $q$ and $q$ by $q$, respectively, and $q$ is the number of observations and $p$ is size of the state vector. Also, a further computational efficiency can be achieved at the analysis step due to the localized analysis, which will be described in more detail later. The most computational burden of the algorithm is dominated by the integration of ensemble members since the number $q$ of observations is much less than the size $p$ of state vector.

# 4 Assimilation of SST with EnKF

The goal of this experiment is to evaluate the performance of the EnKF when the method is applied to the unified grid system for the Northwest Pacific Ocean region (Fig. 1). The EnKF algorithm described in the previous section is classified as a *stochastic* EnKF. The main issues of implementing the EnKF include the sampling error caused by insufficient ensemble members (Gaspari and Cohn, 1999) and the large correlation among the remote state variables (Houtekamer and Mitchell, 1998; Hamill et al., 2001). The available ensemble size $N$ for the model integration is limited simply because the integration of many ensemble members requires high computational cost. The large correlation of state variables at two distant locations occurs because the covariance matrix does not know the controlling physical dynamics. The covariance localization and covariance inflation will resolve the problems (Gaspari and Cohn, 1999; Anderson and Anderson, 1999) as shown in many studies (Evensen, 2004). Before we exploit these additional techniques, we need to investigate the baseline performance of the stochastic EnKF as it is, which will allow us to assess the impact of the advanced adjustments.

The initial ensemble members could be generated by using statistical analysis of the time-histories of the model run, and the forcing field and open boundary conditions should be perturbed, in order to prevent ensemble spread from collapsing. However, we obtain 16 ensemble members as initial state simply from the snapshots of the integrated forward model during the first spin-up time and the forward model integration period and use them for data assimilation, and also employ the same forcing and open boundary conditions for all ensemble members during model integration. To quantify ensemble spread (EnSP) at each step, the EnSP is defined as

$$EnSP = \sqrt{\frac{1}{p \cdot N} \sum_{i=1}^{N} \sum_{j=1}^{p} [X_{ij} - \overline{X}]}, \tag{5}$$

where $\overline{X}$ is ensemble mean, and $p$ and $N$ are the dimension of state and the number of ensemble, respectively.

As seen in Fig. 2, the EnSP is immediately collapsed at the second analysis, to a relatively small value less than 0.2, which implies that the 16 ensemble members have similar state structures via the Kalman gain matrix $K$ in Eq. (2). Essentially,

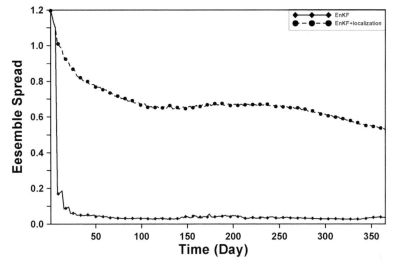

**Fig. 2** Ensemble spread of SST (°C) with and without localization of covariance matrix are indicated with diamond-*solid line* and *circle-dashed lines*, respectively

consecutive observations accelerate the EnSP collapsing due to the small number of sample size $N$ and the large correlation between two distant state variables. The EnSP collapsing results in the filter divergence and the unphysical dynamics. To resolve the ensemble collapse, the covariance localization is adopted as in Gaspari and Cohn (1999). Then, the EnSP decreases from 1.2 to 0.67 after 100 days, but sill retains over 0.6 (Fig. 2, solid line with diamonds).

Figure 3 shows sea surface temperature field of the Northwest Pacific Ocean in July 2003. The satellite SST data are considered as the truth state (or true run) to see the effect of data assimilation although the satellite SST data includes observation error and representation error. For the non-assimilated run, we select a single ensemble member among the 16 ensemble members that is closest to the true state at the initial time, and the ensemble member is integrated for the simulation period without any data assimilation. This free model run without assimilation is called as

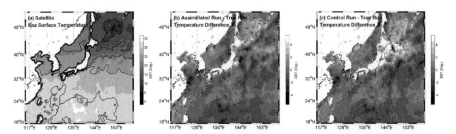

**Fig. 3** Sea surface temperature (°C) of the true and difference fields. Satellite SST data as the true state (*left panel*), difference field between the truth and assimilation run (*center*), and difference field between the truth and control run (*right*)

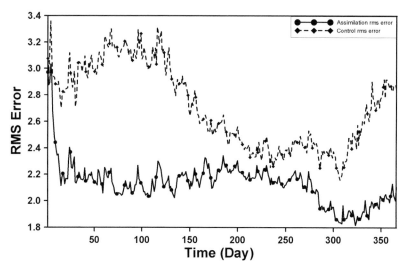

**Fig. 4** Root mean squared (RMS) error of SST in time (day). RMS errors of the assimilation and control are indicated with *solid* and *dashed lines*, respectively

the control run. The SST differences between the assimilated and control runs are compared in July 2003. For the control, the locations of maximum differences from the truth occur along the northern end of the Kuroshio path where Kuroshio meets the Oyashio current to form the Polar Front, and its maximum amplitude for error is about ±8°C. The difference field between the truth and assimilated run shows that EnKF effectively reduced errors in SST. The amplitude of the error of the assimilation run is half as large as that of the non-assimilated, and the averaged root mean squared (RMS) error is reduced from 3.2 to 2.2°C (Fig. 4). Overall, EnKF clearly improves the state of the system, the averaged RMS error for the assimilation run lower than that for the control run over all assimilation period.

## 5 Concluding Remarks and Future Work

We have constructed a unified ocean modeling system for the Northwest Pacific region, which consists of four different regions, the northwestern part of the Pacific Ocean, the East China Sea, the Yellow Sea, and the East/Japan Sea. To this developed unified grid system, a stochastic EnKF is implemented and its performance is evaluated in a simple experiment setup. The assimilation run performs better than the control run over all simulation period such that combining data into the numerical model makes better predictability of the system even with limited observational locations and variables.

In general, EnKF has been successfully implemented in many geophysical fluid dynamical systems with $N = 40$–$100$ ensemble members (Evensen, 2004) although the theoretical ensemble size $N$ for a reliable data assimilation is usually bigger (Van Leeuwen, 1999). Here, we use only 16 ensemble members extracted from the snapshots of the forward model results. For assimilation runs all ensemble members are integrated forward for 7 days with the same forcing and boundary conditions, and then are updated through assimilation of the observation data using the EnKF. As seen in Fig. 2, Sect. 4, a stochastic EnKF shows the EnSP collapsing, small ensemble spread mostly due to the large correlation between two distant state variables after second analysis step. We compensate the large correlation by implementing covariance localization by imposing a fixed correlation length scale on one's estimate of the ensemble spread (Gaspari and Cohn, 1999). This localization also reduces the computational cost of the analysis step. However, in order to avoid the filter divergence and to improve assimilation quality, we still need to employ covariance inflation such as artificially increasing the ensemble spread about the mean (Anderson and Anderson, 1999; Wang and Bishop, 2003; Anderson, 2007).

There is more investigation to carry out particularly; (1) adopting covariance inflation, (2) using more observational location and data such as sea surface height (SSH), shipboard hydrographic and subsurface Argo data, and (3) applying high resolution boundary conditions like the Navy coastal ocean model (NCOM) and the Hybrid coordinate ocean model (HYCOM). In the forthcoming study we will report the development of an improved DA method employing optimally generated initial ensemble members with perturbed forcing and open boundary conditions, the results of assimilating real data to the DA scheme, and the evaluation of the method by comparison between the numerical ocean forecast results and observation data. The significant contributions of the current work include the construction of the unified grid system for the Northwest Pacific including its marginal seas and the first successful attempt to implement ROMS with EnKF for the Northwest Pacific Ocean. We found the effectiveness of EnKF, and believed this study provides a foundation for the development of an ocean forecast system for the Northwest Pacific Ocean.

**Acknowledgements** The authors would also like to thank the anonymous reviewers for many helpful suggestions, which have greatly improved the presentation of this paper. This work was partly supported by the Korean Meteorological Agency (ARGO program) and the Korea Research Foundation Grant funded by the Korean Government (MOEHRD, Basic Research Promotion Fund, KRF-2007-331- C00258).

# References

Anderson JL (1996) A method for producing and evaluating probabilistic forecasts from ensemble model integrations. J Climate 9:1518–1530
Anderson JL, Anderson SL (1999) A Monte Carlo implementation of the nonlinear filtering problem to produce ensemble assimilations and forecasts. Mon Wea Rev 127:2741–2758
Anderson JL (2007) An adaptive covariance inflation error correction algorithm for ensemble filters. Tellus 59A:210–224
Bennett AF (1992) Inverse methods in physical oceanography. Cambridge University Press, Cambridge

Bennett AF (2002) Inverse modeling of the ocean and atmosphere. Cambridge University Press, Cambridge

Burgers G et al (1998) Analysis scheme in the ensemble Kalman filter. Mon Wea Rev 126:1719–1724

Chassignet EP et al (2000) DAMEE–NAB: The base experiments. Dyn Atmospheres Oceans 32:155–184

Cho Y-K et al (2008) Connectivity among straits of the northwest Pacific marginal seas. J Geophys Res Submitted

Curchitser EN et al (2005) Multi-scale modeling of the North Pacific Ocean: Assesment and analysis of simulated basin-scale variability (1996–2003), J Geophys Res 110: C11021, doi:10.1029/2005JC002902

Courtier P et al (1993) Important literature on the use of adjoint, variational methods and the Kalman filter in meteorology. Tellus 45A:342–357

De Mey P (1997) Data assimilation at the oceanic mesoscale: A review. J Meteor Soc Japan 75:415–427

Evensen G (1994) Sequential data assimilation with a nonlinear quasi-geostrophic model using Monte Carlo methods to forecast error statistics. J Geophys Res 99:10143–10162

Evensen G (2004) Sampling strategies and square root analysis schemes for the EnKF. Ocean Dyn 54:539–560

Ezer T et al (2002) Developments in terrain-following ocean models: intercomparisons of numerical aspects. Ocean Model 4:249–267

Fairall CW et al (1996) Bulk parameterization of air-sea fluxes for tropical ocean-global atmosphere coupled-ocean atmosphere response experiment. J Geophys Res 101:3747–3764

Gaspari G, Cohn SE (1999) Construction of correlationfunctions in two and three dimensions. Quart J Roy Meteor Soc 125:723–757

Guo X et al (2006) The Kuroshio onshore intrusion along the shelf break of the East China Sea: the origin of the Tsushima Warm Current. J Phys Oceanogr 36:2205–2231

Haidvogel DB et al (2000) Model evaluation experiments in the North Atlantic Basin: Simulations in nonlinear terrain-following coordinates. Dyn Atmos Oceans 32:239–281

Hamill TM et al (2001) Distance-dependent filtering of background error covariance estimates in an ensemble Kalman filter. Mon Wea Rev 129:2776–2790

Houtekamer PL, Mitchell HL (1998) Data assimilation using an ensemble Kalman filter technique. Mon Wea Rev 126:796–811

Houtekamer, PL, Mitchell HL (2001) A sequential ensemble Kalman filter for atmospheric data assimilation. Mon Wea Rev 129:123–137

Isobe A. (1999) On the origin of the Tsushima warm current and its seasonality. Cont Shelf Res 19:117–133

Keppenne CL, Rienecker MM (2001) Design and implementation of a parallel multivariate ensemble Kalman filter for the Poseidon ocean general circulation model. NASA Tech. Memo-2001-104606, Vol. 21, 35pp

Large WG et al (1994) Oceanic vertical mixing: a review and a model with a nonlocal boundary layer parameterization. Rev Geophys 32:363–403

Sasajima Y et al (2007) Structure of the subsurface counter current beneath the tsushima warm current simulated by an ocean general circulation model. J Oceanogr 63(6):913–926

Senjyu T et al (2006) Interannual salinity variations in the Tsushima Strait and its relation to the Changjiang discharge. J Oceanogr 62:681–692, 2006.09

Seung Y et al (2007) Seasonal characteristics of the Tsushima current in the Tsushima/Korea strait obtained by a fine-resolution numerical model. Cont Shelf Res 27(1):117–133

Song Y, Haidvogel DB (1994) A semi-implicit ocean circulation model using a generalized topography-following coordinate system. J Comp Phys 115(1):228–244

van Leeuwen PJ (1999) Comment on "Data assimilation using an Ensemble Kalman Filter Technique." Mon Wea Rev 127:1374–1377

Wang X, Bishop CH (2003) A comparison of breeding and ensemble transform Kalman filter ensemble forecast schemes. J Atmos Sci 60:1140–1158

Wantanabe M (2007) Simulation of temperature, salinity and suspended matter distributions in-
duced by the discharge into the East China Sea during the 1998 flood of the Yangtze River.
Estuarine Coast Shelf Sci 71(1–2):81–97

Wunsch C (1996) The ocean circulation inverse problem. Cambridge University Press, Cambridge

Xia C et al (2006) Three-dimensional structure of the summertime circulation in the Yellow Sea
from a wave-tide-circulation coupled model. J Geophy Res 111 (C11):Art. No. C11S03

# Particle Filtering in Data Assimilation and Its Application to Estimation of Boundary Condition of Tsunami Simulation Model

Kazuyuki Nakamura, Naoki Hirose, Byung Ho Choi and Tomoyuki Higuchi

**Abstract** We discuss the merit of application of the particle filter compared with the ensemble Kalman filter in data assimilation, as well as its application to tsunami simulation model. The particle filter is one of the ensemble-based methods and is similar to the ensemble Kalman filter that is widely used in sequential data assimilation. We discuss the pros and cons through numerical experiments when the particle filter is used in data assimilation. In next, we review the framework of bottom topography correction based on the tide gauge data. In this procedure, the particle filter was employed to assimilate the tide gauge data, and special localization was used for parameterization. We previously showed the validity of the methods in terms of both attenuation of degeneracy problem and the effectiveness of estimation. We also showed the analysis result of the depth of Yamato Rises in that work. However, the analysis result itself was not sufficiently validated. To validate the analyzed result, we show the result of twin experiment based on artificial bottom topography in this paper. The result fortifies effectiveness of the introduced method for correcting the depth of rise. It also supplements the result of the previous analysis in the Japan Sea.

## 1 Introduction

Bottom topography data set is usually used as fixed boundary condition of the oceanographic simulation. However, they have errors and have been being updated from echo sounder data, the altimetry data, quality control and some interpolation techniques. For example, Smith and Sandwell (1997) found new topography from altimetry data. Incomplete sea depth generates inaccurate results of oceanographic and atmospheric simulations. Bottom topography correction is important to obtain precise geophysical knowledge from simulations.

Data assimilation (DA) is the concept and technique to obtain new knowledge and precise forecast of geophysical simulation models in combination with observed data. Objects of DA are not only to correct physical variables in the simulation

K. Nakamura (✉)
The Institute of Statistical Mathematics, Tokyo 106-8569, Japan
e-mail: nakakazu@ism.ac.jp

S.K. Park, L. Xu, *Data Assimilation for Atmospheric, Oceanic and Hydrologic Applications*, DOI 10.1007/978-3-540-71056-1_19,
© Springer-Verlag Berlin Heidelberg 2009

**Fig. 1** Objectives of DA

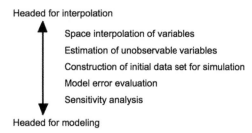

Headed for interpolation

Space interpolation of variables
Estimation of unobservable variables
Construction of initial data set for simulation
Model error evaluation
Sensitivity analysis

Headed for modeling

model and to obtain unknown variables (interpolation and estimation), but also to estimate parameters, to evaluate model error and sensitivity (modeling). Figure 1 shows the summary of various objects in DA. By applying DA, we can precisely assess a forecast accuracy under an appropriate evaluation of errors derived from ambiguity of initial and/or boundary conditions and model error. Hence, using DA to combine incomplete tsunami simulation model and insufficient tide gauge records will lead to new, useful knowledge about tsunamis and tsunami simulation models.

From the viewpoint of DA of tsunami simulation model, tsunami source inversion (Satake 1989; Tanioka and Satake 2001; Tanioka et al. 2006) is also combination of observed data and simulation model. It is an effective method to determine initial conditions of tsunami simulation, which has also uncertainties.

On the other hand, estimation of the bottom topography through tsunami measurements is different in that it is estimation of boundary conditions that is time invariant. Both of them can independently affect the result of tsunami simulation, therefore, they can be computed in parallel though it might be difficult because of mutual interference in the estimation process. We focused on the estimation of boundary condition because it could be localized and find some new geophysical knowledge.

We used ensemble-based method because we aim at assessing probabilities of the estimates. To obtain estimates, we employed the particle filter (PF) (Kitagawa 1996; Gordon et al. 1993; Doucet et al. 2001) because it has good properties in that higher order statistics are preserved and computational burden are small compared with the ensemble Kalman filter (EnKF) (Evensen 1994, 2003). We compared the performance between the PF and the EnKF under the condition that the system is nonlinear.

We also compared the fixed lag smoother of them, the particle smoother (PS) and the ensemble Kalman smoother (EnKS). The result of the experiments suggests the superiority of the PF/PS to the EnKF/EnKS.

On the basis of the result, we applied the PF to the assimilation of the tsunami simulation model. To apply the PF, however, we should manage the problem on the inefficient approximation. We got over this difficulty by parameterizations of the system. We show the way to manage and show the effectiveness of the introduced method by the numerical experiments.

In Sect. 2, we review the state space model and the PF. Numerical experiments of the EnKF/EnKS and the PF/PS are also shown. The formulation of application to tsunami simulation model is shown in Sect. 3. The result of numerical experiment is also shown. Finally we summarize them in Sect. 4.

# 2 State Space Model and the Particle Filter

## 2.1 Nonlinear State Space Model

Nonlinear state space model (SSM) is written in the form

$$\begin{cases} x_t = f_t(x_{t-1}) + v_t \\ y_t = h_t(x_t) + w_t \end{cases},$$

$$v_t \sim N(0, Q_t), \quad w_t \sim N(0, R_t),$$

where $x_t$ is state vector, $y_t$ is observation vector, $t(= \{1, \ldots, T\})$ is time step, $v_t$ is system noise and $w_t$ is observation noise. Once the system is written in this form, we can apply several nonlinear filtering methods explained later. In the context of DA, $x_t$ represents all the variables of the simulation code at time step $t$. $y_t$ represents all the observations at time step $t$. $v_t$ could have various interpretation such as error in physical modeling, discretization or unknown boundary condition. $w_t$ represents mainly measurement error. If $f_t$ and $h_t$ are linear operator, it is easy to estimate $x_t$ conditioned by observations $\{y_1, \ldots, y_t\}$ by means of Kalman filter. In the extended Kalman filter, nonlinear SSM is linearized and applied Kalman filter. However, the stability problem derived from linearization exists. The ensemble Kalman filter originally aims at avoiding this instability and has widely been being used in DA, especially in geophysical research. However, the ensemble Kalman filter has the weak points such as the scalability in the number of samples.

## 2.2 The Particle Filter

The particle filter (PF) is one of the filtering methods mainly used in the statistics and signal processing community. It is the ensemble based method as well as the ensemble Kalman filter (EnKF). The scheme is written in the Fig. 2.

In the ensemble based filtering, estimation of the state $x_t$ is presented by the ensemble set of realizations $\{x_t^{(i)}\}_{i=1}^N$, where N is the number of ensemble members. In the simulation (prediction) step, each ensemble member is used to simulate and obtained simulated ensemble member $x_t^{(i)} = f_t(x_{t-1}^{(i)}) + v_t^{(i)}$, where $v_t^{(i)}$ is generated from some pre-determined distribution. This simulation step of the PF is exactly the same as that of the EnKF. In the analysis step, likelihood $\tilde{\alpha}^{(i)} = p(y_t | x_t^{(i)})$ of each ensemble member $i$ is calculated. Likelihood $\tilde{\alpha}^{(i)}$ represents goodness of fit to observed data.

In the analysis (filtering) step, the set of ensemble members is reconstructed by sampling with replacement. Each sampling probability of ensemble member is determined from the weight

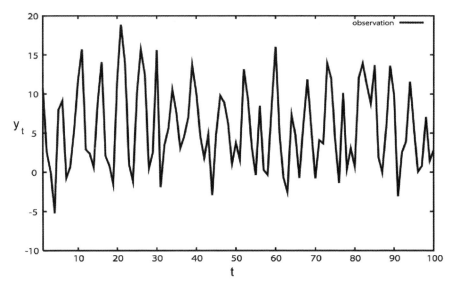

**Fig. 2** Example of observation series. Vertical axis represents observation $y_t$ and horizontal axis represents time $t$. (Reprint from Nakamura et al. (2005))

$$\alpha^{(i)} = \frac{\tilde{\alpha}^{(i)}}{\sum\limits_{i=1}^{N} \tilde{\alpha}^{(i)}}.$$

Therefore, the members that have high likelihood are probable to be duplicated and the members that have low likelihood are probable to be omitted. These processes are conducted in turn.

The PF has good properties in that the ensemble reflects higher order statistics and each ensemble member preserves its physical nature such as continuations. However, it has the critical problem on the Monte Carlo approximation. The problem derives from the fact that "sampling importance resampling technique" (Rubin 1987) used in the PF is too inefficient to obtain effective samples in a high dimensional space. In the statistical and control theory field, the dimension of the system is limited in maximally several tens. Therefore, the problem is not critical in nature. In DA, however, the dimension of the state is too high to obtain good samples. We can consider several approaches to solve the problem. If we do not require online estimates, one potent approach is application of Markov chain Monte Carlo (MCMC) sampling. This method could be more efficient method in that less number of ensemble members would be omitted in ensemble generation. However, we should repeat simulation in series and it is hard to parallelize by algorithmic nature of MCMC. As a result, we should compromise computational speed or preciseness of estimates. We did not apply the MCMC for this reason in the application for tsunami simulation model. We will show the management in Sect. 3 within the framework of the particle filter.

## 2.3 Comparison Between the Ensemble Kalman Filter and the Particle Filter

In this subsection, we review the result of numerical experiments on state estimation and parameter identification in Nakamura et al. (2005). In this work, we used the simulation data $y_t$ that were generated by the following system (Kitagawa 1998)

$$
\begin{cases}
x_t = \dfrac{1}{2}x_{t-1} + \dfrac{25x_{t-1}}{1+x_{t-1}^2} + 8\cos(1.2t) + v_t & (1)\\[3mm]
y_t = \dfrac{x_t^2}{20} + w_t & (2)
\end{cases},
$$

where $x_0 \sim N(0,5)$, $v_t \sim N(0,1)$, $w_t \sim N(0,10)$, and $t = 1,\ldots,T$.

This model is widely used in the statistical field and its properties are well-known (Doucet et al. 2000). We used this model for this reason. The system model (1) can be divided into three parts, the potential part $\frac{1}{2}x_{t-1} + \frac{25x_{t-1}}{1+x_{t-1}^2}$, the oscillatory part $8\cos(1.2t)$ and the small noise part $v_t$. The solutions of the equation $x = \frac{1}{2}x + \frac{25x}{1+x^2}$ are $x = 0$ (unstable equilibrium point) and $x = \pm 7$ (stable equilibrium points). Therefore, the potential part shows the tendency to converge to $x = \pm 7$. On the other hand, the amplitude of the oscillatory part is larger than the distance between a stable equilibrium point and the unstable equilibrium. This means that the state can be changed from the minus equilibrium point to the plus equilibrium point and vice versa. The small noise part provides randomness that causes uncertainty in timing of change of stable equilibrium points. The observation model (2) has square of state variable, which makes it difficult to determine sign of the state. We researched the performance of the filters and the smoothers by the following two experiments.

In the first experiment, the model (1) and (2) were used for estimation of the state $x_t$ and the state was estimated with the PF, the EnKF, the PS and the EnKS. Nonlinear observation was processed according to the method in the previous section. Lag length of the PS and the EnKS was set to be 20. We used the average of ensemble members as estimation variable. In the second experiment, the variance of the system noise $y_t$ was treated as unknown parameter $\sigma_v$ to be estimated and was estimated together with state. The estimation is based on the self-organizing state-space model (Kitagawa 1998) in which the system model (1) is replaced by

$$
\begin{pmatrix} x_t \\ \theta_t \end{pmatrix} = \begin{pmatrix} \dfrac{1}{2}x_{t-1} + \dfrac{25x_{t-1}}{1+x_{t-1}^2} + 8\cos(1.2t) \\[3mm] \theta_{t-1} \end{pmatrix} + \begin{pmatrix} v_t \\ u_t \end{pmatrix},
$$

$$
x_0 \sim N(0,5), \quad v_t \sim N(0,\exp(\theta_t)), \quad u_t \sim N(0,\xi),
$$

where $\xi$ is predetermined constant. In this model, the estimated variance $\sigma_v$ should be $\exp(\theta_t)$. Therefore, we observe whether the self-organized parameter $\theta_t$ converges to the true one.

### 2.3.1 State Estimation

Figure 2 shows the observation series $y_t$ and Fig. 3 shows corresponding state series $x_t$ and estimated state series by the EnKF and the PF. $\hat{x}_t$ represents estimated variable of $x_t$ at time $t$ and is composed by averaging filtered or smoothed ensemble members. It is seen that estimation by the PF is more accurate than by the EnKF.

Table 1 shows the average of the sum of squared difference $\sum\limits_{t=1}^{T} (\hat{x}_t - x_t)^2$ among 100 experiments. Estimation experiments were done 100 times under the condition that the number of ensemble members is 100, 1000 and 2500. It can be seen that estimation by the PF is more accurate than by the EnKF if the number of ensemble members is the same. The same relationship between the PS and the EnKS can be seen. Additionally, the whole computation time on these simulation tests (Table 2) shows that the PF/PS can estimate states more efficiently than the EnKF/EnKS. These results show the superiority of the PF to the EnKF in terms of the accuracy of the state estimation and computational efficiency. The superiority of the PS to

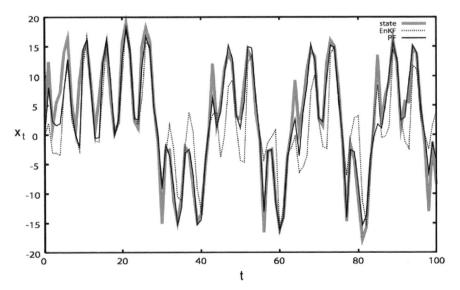

**Fig. 3** State series corresponding to observation series in Fig. 2 and estimated state series by the EnKF and the PF. Vertical axis represents state $x_t$ and horizontal axis represents time $t$. (Reprint from Nakamura et al. (2005))

**Table 1** The sum of squared differences. Estimation experiments were done 100 times and the average of the sums was calculated

|      | PF      | EnKF    | PS     | EnKS    |
|------|---------|---------|--------|---------|
| 100  | 1841.76 | 2853.11 | 567.84 | 1618.73 |
| 1000 | 1710.01 | 2779.09 | 404.90 | 1470.93 |
| 2500 | 1701.90 | 2771.28 | 397.03 | 1447.65 |

**Table 2** Computational time of the test shown in Table 1. Used CPU is Opteron (2.8 GHz)

|              | PF/PS  | EnKF/EnKS |
| ------------ | ------ | --------- |
| CPU time (s) | 155.75 | 215.10    |

the EnKS is also demonstrated. On the other hand, if we apply the PF to real data assimilation, we should solve the problem on the degeneracy originated from finite resampling. In this numerical experiment, it is covered by the richness of the number of the ensemble members compared to the dimension of $x_t$.

### 2.3.2 Parameter Identification

Figure 4 shows the result of the parameter estimation. We can regard the result of numerical integration as accurate estimation. In this experiment, the number of the ensemble members is 100. This result also shows the accuracy of parameter estimation by the PF because the parameters converge to the true one.

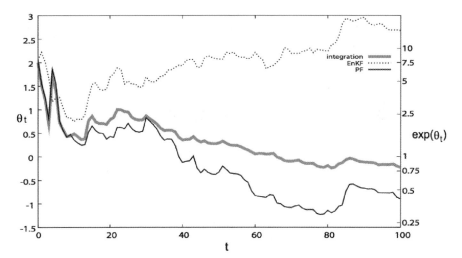

**Fig. 4** Parameter estimation using the EnKF and the PF. Left vertical axis represents $\theta_t$, right vertical axis represents $\exp(\theta_t)$. Horizontal axis represents time $t$. (Reprint from Nakamura et al. (2005))

## 2.4 Discussion

As we noted before, it seems better that we use the PF for sequential data assimilation. However, the inefficiency problem should be resolved to apply the PF to realistic simulation models. The dimension of state vector tends to be large and the likelihood tends to be biased. It results in decrease of variety of ensemble members and bias of estimates.

We can use several approaches to avoid this difficulty. One is to use more sophisticated filtering scheme such as the kernel approximation (Pham 2001), the merging particle filter (Nakano et al. 2007), the ensemble adjustment filter (EAKF) (Anderson 2001) and others. The merit of these approaches is its generality. We do not require any approximations on the structure of simulation models because we only need replacing filter scheme to obtain more sophisticated results. However, we need additional cost in computation when we use them. One of the merits of the PF is its lightweight property. We should discard this property in those approaches.

Another approach is modeling part and to introduce appropriate parameter or model for controlling simulation variables. The advantage of this approach is its ease in implementation and its computational cost. The problems are interpretability of the introduced parameter or model, and inconsistency among simulation models to be applied.

It is also probable that we make observation models more robust (van Leeuwen 2003). For example, Cauchy probability distribution function could be used for observation noise. The problem of this approach is also its interpretability. We employed parameterization in the following because we can parameterize uncertainties naturally.

## 3 Estimation Procedure of Sea Depth

### 3.1 Simulation Model and Observation Data

Simulation model of tsunami is shallow water equations model (Choi and Hong 2001). Target area is discredited spatially and temporally. Leapfrog scheme and C-grid (Arakawa and Lamb 1977) are employed. The number of grid points is $192\backslash$ times 240. On the other hand, assimilated data set is tide gauge data. Tide gauge data is time series of observations of sea level around installation point. In the following Okushiri tsunami case, observation data is measured by Japan Coast Guard, Geographical Survey Institute (Japan), and Sungkyunkwan University (Korea). The sampling time of the records varies from 30 s to 2 min. They have not only tsunami component of the tide gauge data, but also the components of astronomical tide, meteorological tide, harbor oscillation, trend and noise. We should manage these components not oriented from tsunami component. In this analysis, we can apply methods used in time series analysis. To remove astronomical tide, meteorological tide and trend, we can apply high-pass filters or similar functions. We tested time domain method and time-frequency domain method. Time domain method is fitting of one dimensional random walk model. One dimensional random walk model is written by

$$q_t = q_{t-1} + r_t,$$

where $q_t$ denotes the component to be removed and $r_t$ denotes the component to be left. The merit of this method is the ability of flexible estimation of local trend. Tested time-frequency domain method is singular spectrum analysis (Broomhead and King 1986), which is a variant of empirical orthogonal function (EOF) analysis. It can also remove trend component because the frequency domain mainly each component contributes is different and have no aliasing. We confirmed that both methods can be applied, and therefore, we used one dimensional random walk model. Noise component was neglected by taking the diagonal components of the variance $R_t$ large. It can also have the side effect of robustness for the inefficiency problem.

## 3.2 Okushiri Tsunami and Yamato Rises Case

Okushiri tsunami occurred in 1993 is one of the tsunamis occurred in the Japan Sea. This tsunami is observed through tide gauges around the Japan Sea and the records have enough time resolution to use data assimilation. Therefore, we used the records of Okushiri tsunami.

The simulation area is shown in Fig. 5. These areas are discretized spatially and temporally by the scheme described in the previous subsection. The rises called Yamato Rises are in the middle of the Japan Sea. Used data set is DBDB-V (Sandy 1996), JTOPO1 (MIRC 2003), ETOPO2 (Smith and Sandwell 1997) and SKKU (Choi et al. 2002) (see Hirose (2005) for details). Large differences exist in the area among the data set. We targeted this area to be assimilated. On the other hand, the installation points of tide gauge are also shown in Fig. 5.

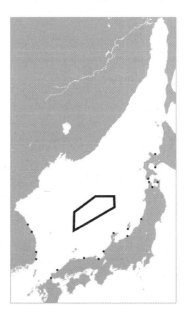

**Fig. 5** Target area and tide gauge installation sites. The points show tide gauge installation sites and the pentagon that is in the middle of the Japan Sea show the target area

To apply sequential DA method, especially the PF, we should parameterize bottom topography. In the work of Nakamura et al. (2007), linear combination of the bottom topography data set is used for parameterization:

$$
\hat{d}_{m,0}^{(i)} = \begin{cases} \sum_{l=1}^{4} \beta_l^{(i)} d_m^l & \text{(target area)} \\ \sum_{l=1}^{4} 0.25 d_m^l & \text{(other area)} \end{cases},
$$

where $d_m^l$ is depth at grid $m$ of the $l$th topography data set and $\beta_l^{(i)}$ is the $i$th weight ensemble of the $l$th topography data set and is distributed as $\beta_l^{(i)} \sim N(0.25, \sigma^2)$. The key in this approach is the form of $\beta_l^{(i)}$ which does not include spatial factor $m$. This parameterization introduces "strong correlation" among estimated points, which enables us to reduce the degree of freedom to obtain estimates. As a result, we can represent estimates by small variations of ensemble members. We used 100 members for this analysis.

Through the introduced framework and analyses, we obtained Yamato Rises, the target area shown in Fig. 5, might be shallower than the average of the data sets by this analysis (see Nakamura et al. 2007). In the next subsection, we fortify the result through the result of the numerical experiment.

## 3.3 Twin Experiment on Artificial Bottom Topography

In the previous work (Nakamura et al. 2006), we conducted twin experiment for similar framework. In that work we corrected bottom topography that is globally biased. However, the setting of the bottom topography estimation in Nakamura et al. (2007) is not the same as the situation in Nakamura et al. (2006). The difference between them is the locality of the analysis area. In Nakamura et al. (2006), whole area of bottom topography is uniformly modified whereas local area of sea bottom is modified in the case of the analysis of Yamato Rises. In the uniform modification case, the degree of the wave deformation is small and the modification of the bathymetry mainly affects arrival time of waves. On the other hand, the local modification causes significant deformation of tsunami wave shape, and therefore, the performance of the assimilation could be affected. To validate the introduced framework for the local modification, we need some test that is similar but simple situation. We test this by twin experiment in artificial bottom topography setting.

The procedure of the experiment is as follows. At first, artificial bottom topography is set that is regarded as true one, and a tsunami is simulated under the setting. We call the bottom topography as "true topography." The observation data set is obtained in this step using observation system. In next, another bottom topography, which has errors about a rise, is set. In this step, error range about bottom topography is also determined. We call it "false topography." At last, we assimilate the observed data to the simulation model that has false topography. We check whether the bottom topography approaches to the true one. The number of used ensemble member is 100.

Figure 6 shows the details of the experiment and result. Left image in each figure in Fig. 6 shows the true topography in simulation area with propagation of tsunami. A source is set in the right middle of the area. Four observation points are set in left coast represented by white dots in Fig. 6. Observation model is linear model:

$$y_t = H_t x_t + w_t,$$

where $H_t$ is a $4 \times n_x$ zero-one matrix. Therefore, the dimension of the observation vector is four. Eight rises can be seen and no rise is in the middle of the simulation area.

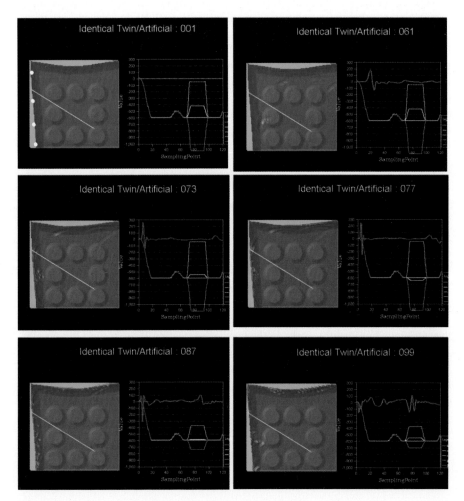

**Fig. 6** Result of the twin experiment for estimation of rise height

As for the bottom topography, we determined that the true topography had no rise in the middle whereas the false topography had rise in the middle. We set distribution on this different area in the following manner. The depth $d_m$ is approximated using the sample set $\{d_m^{(i)}\}_{i=1}^N$, which is generated by

$$d_m^{(i)} = \begin{cases} \max(d_m - c^{(i)}\delta_m, 50) & \text{(false rise area)} \\ d_m & \text{(other area)} \end{cases},$$

where $c^{(i)} \sim N(1.0, 0.5^2)$, $d_m$ is depth at $m$ and $\delta_m$ is height of rise at each grid point $m$. The goal of this experiment is whether the false rise is identified as false one, which means that the middle area is flat. The key in this experiment is also localization of assimilated area. In the Yamato Rises case, assimilated area is localized. We can test effectiveness of localization through this experiment. The result presented in Fig. 6 shows that the estimation procedure works well. Right chart in each figure represents the profile along white line in left image. Red curve represents sea surface height, green curve represents the true bottom topography, blue curve is estimation, orange and purple curves are shallowest and deepest samples from the ensemble. As tsunami arrives at the observation points in the left coast, the center bottom topography starts to approach to true one. The shallowest and deepest samples are also approaching to true one. It shows the degeneracy problem is avoided in that the difference between the deepest sample and the shallowest sample is preserved. The result demonstrates that the introduced approach is effective for correction of local bottom topography, especially, rise area.

# 4 Summary

We reviewed and discussed the performance comparison between the PF/PS and the EnKF/EnKS. The results suggest that the superiority of the PF/PS to the EnKF/EnKS under the condition that the system is nonlinear. We also reviewed the framework of data assimilation to tsunami simulation model with tide gauge data. We also conducted twin experiment of localized assimilation and showed that the introduced procedure works in the case of correcting local bottom topography. Therefore, it could be effective for estimation of the depth of rise. The result fortifies the analysis result of previous work in which we showed Yamato Rises might be shallower than data sets.

**Acknowledgements** The authors would like to thank the Japan Oceanographic Data Center and Geographical Survey Institute of Japan for providing tide gauge data. This research was partially supported by CREST research program of the Japan Science and Technology Agency.

# References

Anderson JL (2001) An ensemble adjustment kalman filter for data assimilation. Mon Wea Rev 129:2884–2903

Arakawa A, Lamb V (1977) Computational design of the basic dynamical processes in the ucla general circulation model. In: J Chang (eds) General circulation models of the atmosphere. Academic Press, New York, pp 174–264

Broomhead DS, King GP (1986) Extracting qualitative dynamics from experimental data. Physica D 20:217–236

Choi B, Hong S (2001) Simulation of prognostic tunamis on the Korean coast. Geophys Res Lett 28:2013–2016

Choi B, Kim K, Eum H (2002) Digital bathymetric and topographic data for neighboring seas of Korea. J Korean Soc Coastal Ocean Eng 14:41–50 (in Korean with English abstract)

Doucet A, De Freitas JFG, Gordon NJ (2001) Sequential Monte Carlo methods in practice, Springer-Verlag, New York

Doucet A, Godsill SJ, Andrieu C (2000) On sequential Monte Carlo sampling methods for Bayesian filtering. Stat Comput 10:197–208

Evensen G (1994) Sequential data assimilation with a non-linear quasi-geostrophic model using Monte Carlo methods to forecast error statistics. J Geophys Res 99(C5):10143–10162

Evensen G (2003) The ensemble Kalman filter: Theoretical formulation and practical implementation. Ocean Dyn 53:343–367

Gordon NJ, Salmond DJ, Smith AFM (1993) Novel approach to nonlinear/non-Gaussian Bayesian state estimation. IEE Proc-F 140(2):107–113

Hirose N (2005) Least-squares estimation of bottom topography using horizontal velocity measurements in the Tsushima/Korea Straits, J Oceanography, 61(4):789–794

Kitagawa G (1996) Monte Carlo filter and smoother for non-Gaussian nonlinear state space model. J Comput Graph Stat 5 (1):1–25

Kitagawa G (1998) Self-organizing state space model. J Am Stat Assoc 93:1203–1215

MIRC (2003) JTOPO1—Northwest Pacific one minute grid data. Marine Information Research Center Japan Hydrographic Association, http://www.mirc.jha.or.jp/products/JTOPO1/, CD-ROM, (in Japanese)

Nakamura K, Ueno G, Higuchi T (2005) Data assimilation: Concept and algorithm. Proc Inst Stat Math 53(2):201–219 (in Japanese with English abstract)

Nakamura K, Higuchi T, Hirose N (2006) Sequential data assimilation: Information fusion of a numerical simulation and large scale observation data. J Comput Sci 12:608–626

Nakamura K, Higuchi T, Hirose N (2007) Modeling for tsunami data assimilation and its application in the Japan Sea. Submitted to Comput Stat Data Anal

Nakano S, Ueno G, Higuchi T (2007) Merging particle filter for sequential data assimilation. Non Proc Geophys 14:395–408

Pham DT (2001) Stochastic methods for sequential data assimilation in strongly nonlinear systems. Mon Wea Rev 129:1194–1207

Rubin DB (1987) A noniterative sampling importance resampling alternative to the data augmentation algorithm for creating a few imputations when fractions of missing information are modest: The SIR algorithm, J Am Stat Assoc 82:543–546

Sandy R (1996) The Navy's bathymetric databases... from the sea. Sea Technol 37:53–56

Satake K (1989) Inversion of tsunami waveforms for the estimation of heterogeneous fault motion of large earthquakes: The 1968 Tokachi-oki and the 1983 Japan Sea earthquakes. J Geophys Res 94:5627–5636

Smith WHF, Sandwell DT (1997) Global sea floor topography from satellite altimetry and ship depth soundings, Science, 277:1956–1962

Van Leeuwen PJ (2003) A variance-minimizing filter for large-scale applications. Mon Wea Rev 131:2071–2084

Tanioka Y, Satake K (2001) Coseismic slip distribution of the 1946 Nankai earthquake and aseismic slips caused by the earthquake. Earth Planets Space 53:235–241

Tanioka Y, Yudhicara, Kususose T, Kathiroli S, Nishimura Y, Iwasaki S, Satake K (2006) Rupture process of the 2004 great Sumatra-Andaman earthquake estimated from tsunami waveforms. Earth Planets Space 58:203–209

# Data Assimilation in Hydrology: Variational Approach

François-Xavier Le Dimet, William Castaings, Pierre Ngnepieba and Baxter Vieux

**Abstract** Predicting the evolution of the components of the water cycle is an important issue both from the scientific and social points of view. The basic problem is to gather all the available information in order to be able to retrieve at best the state of the water cycle. Among some others methods variational methods have a strong potential to achieve this goal. In this paper we will present applications of variational methods to basic problems in hydrology: retrieving the hydrologic state of at a location optimizing the parametrization of hydrologic models and doing sensitivity analysis. The examples will come from surface water as well as underground water. Perspectives of the application of variational methods are discussed.

## 1 Introduction

Predicting the evolution of continental water is an important issue for the anticipation natural hazards like floods and droughts, for the formulation and implementation of policies and regulations in land use planning or water supply. More generally, numerous human activities critically depend on the availability and quality of water resources. Therefore, understanding, predicting and controlling the behavior of the continental water cycle can benefit to many applications of great societal importance.

The continental part of the water cycle is characterised by an important heterogeneity of the environment and by the complexity and variability of the involved hydrological processes. Within the different components of the cycle (atmospheric, land surface and underground), water is present under several phases and fluxes at the interfaces are difficult to quantify. An important part of the cycle takes place below the land surface where geometric properties and flow paths are hardly observable. As a result, understanding and representing streamflow generation processes is still a challenging endeavour.

F.-X. Le Dimet (✉)
Laboratoire Jean-Kuntzmann, Université de Grenoble and INRIA, Grenoble, France
e-mail: Francois-Xavier.Le-

S.K. Park, L. Xu, *Data Assimilation for Atmospheric, Oceanic and Hydrologic Applications*, DOI 10.1007/978-3-540-71056-1_20,
© Springer-Verlag Berlin Heidelberg 2009

The equations governing hydrological processes are non-linear. This has some important consequences on the behavior of the system:

- any situation is unique ("uniqueness of place, action and time", Beven et al., 2001), the equations modeling a geophysical fluids don't have steady state or periodic solution, a consequence is that they have no general properties but only properties in the vicinity of some situation which have to be specified by data.
- there are interactions between the various scales of the flow and, because the numerical models are necessarily finite, the flux of energy and matter between the resolved and unresolved scales has to be parametrized introducing some parameters which are not accessible to direct measurements.
- initial and boundary conditions should be prescribed for the mathematical closure of the system, their observability can be quite limited and they usually have a significant influence on the model simulations.

Briefly said: a model is not sufficient to retrieve the state of a geophysical flow at a given date some additional information must be considered:

- in situ observations representing point scale (soil moisture profile) or integrated information (ex. river discharge)
- static or dynamic remote sensing observations from satellite or airborne sensors
- statistics characterizing the uncertainty affecting quantitative observations (e.g. variance and covariance)
- qualitative information on the behaviour of the system

In the vocabulary of optimal control theory, the parameters resulting from process conceptualization as well as initial and boundary conditions are referred to as control variables, living in a so-called control space. Unfortunately, hydrologic systems, perhaps more so than other geophysical systems, avail very limited knowledge, especially for extreme events, of the control variables due to scarce observations of the system. The objective of Data Assimilation methods is to retrieve the state of a flow by mixing the various sources of information. In a first approximation there are two basic approaches:

- variational methods based on optimal control techniques.
- Stochastic methods based on Kalman filtering

The increasing availability of observation data and the implementation of operational services led to tremendous advances in meteorological and ocean data assimilation. Many of the issues that led these communities to use data assimilation are now at the forefront in hydrology. The theoretical and methodological developments carried out of great interest for various issues related to hydrological modelling.

Variational methods provide a deterministic framework for the theoretical formulation and numerical approximation of numerous problems related to the analysis and control of physical systems, especially those governed by partial differential equations. The mathematical formalism, based on functional analysis and differential calculus, was largely expanded by related disciplines such as optimal control theory and optimization. Sensitivity analysis and nonlinear least squares (data fitting trough state and/or parameter estimation) can be addressed using a unified framework. In particular, the adjoint state method, yielding to an efficient calculation of the derivatives of an objective function with respect to all model inputs, is particularly suited when the dimension of the response function to be analysed (or cost function to be optimized) is small compared to the number of inputs to be prescribed (dimension of the control space). The variational methods have contributed to numerous applications related to the analysis and forecasting of meteorological and oceanographic systems (data assimilation, sensitivity analysis, optimal perturbations for ensemble prediction, adaptive observation strategies ...).

Early applications of the adjoint state method to hydrological systems have been carried out in groundwater hydrology (Chavent, 1974, 1991; Carrera and Neuman, 1986; Sun and Yeh, 1990). The resolution of inverse problems (parameter, state and boundary condition estimation) using non-linear least squares, local sensitivity analysis, where also addressed in this framework in land surface hydrology (Mahfouf, 1991; Callies et al., 1998; Calvet et al., 1998; Bouyssel et al., 1999; Margulis and Entekhabi, 2001; Reichle et al., 2001), in vadose zone hydrology (Ngnepieba et al., 2002), river and floodplain hydraulics (Piasecki and Katopodes, 1997; Sanders and Katopodes, 2000; Yang and LeDimet, 1998; Mazauric, 2003; Belanger and Vincent, 2005; Honnorat et al., 2006) and more recently in catchment hydrology (White et al., 2003; Seo et al., 2003; Castaings et al., 2007).

The purpose of this paper is not to present an exhaustive review of variational data assimilation in hydrology, but to present some illustrative examples of the potential of these methods. The primary building block of any data assimilation system is the model representing the physical processes. For several components of the continental water cycle, there is no consensus of the appropriated model complexity and processes representation. Therefore, after a short presentation of the variational data assimilation framework, an illustrative example is provided in order to emphasize how important it is to take into account the available information for the choice of a model. For some components of the water cycle (e.g. surface runoff and infiltration in the vadoze zone) the efficiency of variational method for the estimation of model parameters is demonstrated. Lastly, for a rainfall-runoff model, it is shown that variational sensitivity analysis provides a local but extensive insight of the relation between model inputs and prognostic variables.

## 2 Variational Data Assimilation Framework

In this section, the basic principle of variational data assimilation is presented and the practical implementation of the approach is discussed.

## 2.1 Principle of VDA

Let us consider that the state of the flow is described by a state variable X depending on time and space, representing the variables of the model (velocity, water content, elevation of the free surface, concentrations in sediment, biological or chemical species. for pollution problems, …). After discretization of the continuous equation the evolution of the flow is governed by the differential system:

$$\begin{cases} \dfrac{dX}{dt} = F\left(X,U\right) \\[2mm] X(0) = V \end{cases} \tag{1}$$

$U$ is some unknown model inputs, e.g. boundary condition, parameters or model error possibly depending on space and time. $V$ represents the uncertain and depends only on space. We assume that, $U$ and $V$ being given, the model has a unique solution between $0$ and $T$ (i.e. the model is deterministic). We denote by $X_{obs}$, the available observations for the diagnostic variables of the model between $0$ and $T$, which for sake of simplicity are supposed continuous in time. The discrepancy between the observation and the state variable is defined by a, so-called, cost-function in the form:

$$J(U,V) = \frac{1}{2} \int\limits_0^T \|CX - X_{obs}\|^2 \, dt + \frac{1}{2} \|U - U_0\|^2 + \frac{1}{2} \|V - V_0\|^2$$

where C is a mapping from the space of the state variable toward the space of observations where the comparison is carried out. The second and the third terms are Tikhonov regularization terms (Tikhonov and Arsenin, 1977) which ensure a well posed inverse problem by introducing a priori information on the control variables U and V. It is important to emphasize that the norms used in the previous equation are on three different spaces. They can take into account the statistical information by introducing the error covariance matrix (observation and background covariance matrices). In the present didactic presentation, only identities will be considered. Although the same value was assigned to the coefficients preceding the different terms of the equation, the coefficients can be used to balance or prioritize the different "objectives" (constants a priori or iterative strategy).

The problem of VDA is the determination of $U^*$ and $V^*$ minimizing $J(U,V)$. As a first approximation (river discharge, soil moisture or concentrations are nonnegative) we have to solve a problem of unconstrained optimization. From the numerical point of view $U^*$ and $V^*$ will be estimated by a descent type algorithm e.g. as the limit of a sequence in the form:

$$\begin{pmatrix} U_{k+1} \\ V_{k+1} \end{pmatrix} = \begin{pmatrix} U_k \\ V_k \end{pmatrix} + \lambda_k D_k$$

where $D_k$ is the direction of descent deduced from the gradient of $J$ and $\lambda_k$ is the step size realizing the minimum of $J$ along the direction $D_k$. The direction of descent $D_k$ can be estimated by BFGS type method (Liu and Nocedal, 1989). For computing the gradient we introduce a so-called adjoint variable P which has the same dimension like the state variable X. It can be demonstrated that if the variable P is solution of the adjoint model:

$$\frac{dP}{dt} + \left[\frac{\partial F}{\partial X}\right]^T P = C^T (CX - X_{obs})$$

$$P(T) = 0 \tag{2}$$

after a backward integration of the adjoint model, the gradient of J is given by:

$$\nabla J = \begin{pmatrix} \nabla_U J \\ \nabla_V J \end{pmatrix} = \begin{pmatrix} -\left[\dfrac{\partial F}{\partial U}\right]^T \cdot P + (U - U_0) \\ -P(0) + (V - U_0) \end{pmatrix} \tag{3}$$

The derivation of the system can be found in Le Dimet and Talagrand (1986) and Lions (1968). It is important to underline that more generally; the so-called adjoint state method enables the calculation of the derivatives of a function J, with respect to the control variables of the model, for a computational cost which is independent for the dimension of the control space. It is precisely the reason why the approach is really suited for the analysis and control of spatially distributed systems. The most general formalism for the deterministic sensitivity analysis of non-linear large scale systems is due to Cacuci (1981a, b).

In the variational data assimilation framework, the model (1) plus the adjoint model (2) with the optimality condition $\nabla J = 0$ constitute the Optimality System (OS). This optimality system contains all the available information and therefore sensitivity studies with respect to the observations must be carried out from the OS rather than from the model (Le Dimet et al., 2002).

## 2.2 Practical Implementation of the Adjoint Model

The practical implementation of the adjoint state method can require substantial efforts but the investment can benefit to many applications (state and parameter trough data assimilation, sensitivity analysis, optimal perturbations for ensemble prediction, adaptive observation strategies ...).

Different paths can be pursued depending if the operations are carried out on the continuous form of the direct model, on its discretized form or directly on the computer code implementing the model. From a numerical point of view, the best representation of the functional to be derived is the associated computer code. Tremendous advances have been made in algorithmic differentiation (Griewank, 2000) and consequently the code based approach is facilitated by the advent of powerful automatic differentiation (AD) engines (see http://www.autodiff.org) such as TAPENADE.

Considering the computer code implementing the direct model (model and objective functional) as a concatenated sequence of instructions, algorithmic differentiation is based on a rigorous application of the chain rule, line by line. The application of the chain rule from the inputs to the outputs of the function is denoted as the forward mode of AD whereas the reverse mode operates from the outputs to the inputs. The reverse mode of AD is the discrete equivalent of the adjoint state method from optimal control theory. The adjoint of a sequence of operations is the reverse sequences of the transposes of each operation. In summary, from the direct model toward the adjoint model there are two basic operations:

– computing the jacobian of the model with respect to the state variable; this operation is not very difficult; the code can be derived statement by statement
– transposing the jacobian; this is the most difficult operation because of multiple dependencies; e.g. for non-linear models the solution of the direct model is needed for the calculation of the adjoint variables (store or/and recompute, checkpointing strategies)

There are some limitations of this approach. The practical ones is on its cost both from the computational point of view and the development of the adjoint requiring an important investment in term of manpower, nevertheless it's important to point out that the adjoint is a generic tools which can be used for several purposes as we will see below? Another difficulty is due to the non-differentiability of the algorithms used in the direct model : physical processes includes thresholds for instance those linked to the changes of phase in the water cycle. From the theoretical point of view, differentiability is necessary to carry out optimizations algorithms permitting to give a numerical solution to the problem. In general if the direct code contains statement (IF-THEN-ELSE), then differentiability is lost. Furthermore the relation between independent and dependent variables should be piecewise differentiable.

There is no consensus on the treatment of on-off switches, several solutions are proposed :

– Use a simplified smooth physics Zupanski (1993), Janiskováa et al. (1999), and Laroche et al. (2002)
– Keep on-off switches in TLM and AM (Zou et al., 1993; Zou, 1997), this strategy is adopted by the Automatic Differentiation. In this case the result of the adjoint model is sub-gradient, many optimization algorithmsare adapted to non-differentiability (see Gilbert and Lemaréchal 1989).
– develop adapted mathematical formalism see e.g. Xu (1996, 2007) and Mu and Wang (2003)

Although models based on partial differential equations form the classical framework for the application of variational methods, most of the time their algorithmic representation does not significantly differ from the one characterising models based on less sophisticated mathematical formulations. The advent of algorithmic differentiation open new trends for the application to other type of models. Representative examples are provided by Lauvernet (2005) or Wu (2005).

It is important to mention that although the approach discretize and derive generally provides the most reliable and accurate derivatives, for some cases it might lead to instability in the solution of the adjoint model (Sirkes and Tziperman, 1997). Therefore, as commented by Sandu et al. (2003)," there is no general rule to decide which approach (discrete vs. continuous) should be used to implement the adjoint model.

# 3 Models and Data (Le Dimet et al. (2004))

Most of the time data are collected and models are designed independently. As emphasized in the Introduction section, in order to carry out a numerical prediction we need a numerical model but also observations (with associated statistical information) characterizing the behavior of the system.

These elements have to be considered as heterogeneous sources of information on a same situation. They have to be combined through a process of Data Assimilation in order to retrieve an initial condition. Subsequently, the prediction is obtained by the numerical integration of the model using the retrieved initial condition.

However, since the prediction is obtained after merging the information contained in models and data, how does the quality of these components impact the quality of the prediction? In other terms:

- What is the impact of an error of the model on the prediction?
- What is the impact of observation error of the prediction?

Improving the quality of the prediction can be achieved by improving both components (i.e. models and data), or by improving the way the information is merged. However, given the quantity and quality of available observations, is it sufficient to improve the model to get a better prediction?

Although the answer to this question might seem trivial, we will see that in some circumstances, if the same set of data is kept, an improvement of the model can lead to a degradation of the forecast.

Usually optimal control methods are used for the assimilation of data, the model is considered as perfect. Nevertheless if some additional term is added and controlled in the model, then it is possible to improve the model and the forecast with a continuous interaction between models and data.

A difficulty comes from the fact that both models and data have errors and these errors will induce errors on data assimilation process then on the prediction.

## 3.1 Sources of Error

Concerning the mathematical/algorithmic model representing the hydrological processes, two main types of errors can be encountered:

- representativity of the governing equations, physical errors due to the approximations used in the model e.g. subscales parametrization, evaluations of the fluxes at the boundary of the model are not perfectly known and modelled by additional terms in the equation.
- numerical errors due to the discretization of the equation and also errors produced by the stopping criterions used in iterative algorithms performed to get a numerical solution (mandatory for nonlinear models).

A priori these errors are independent. Nevertheless, the choice of physical approximation should depends on the size of the mesh of discretization.

Error in observation comes from errors in the direct measurement and errors due to sampling. It worth to point out that a measurement does not have a intrinsic value: the same measurement of wind can be used both in global model or in a very local one. The confidence attached to the value depends of the context and the context is nothing else than the model itself. Moreover, the observation operator $C$ can be very complex and contain another non-linear physical model bounded by the same constraints (model error, uncertain observations for model closure). It is particularly important to acknowledge this feature when assimilating indirect (ex. discharge observations from measurements of water levels) or remote sensing observations (ex. soil moisture from radiometer measurements).

In the assimilation process there will be an interaction between these sources of error that will be transported on the prediction. The problem of accounting for model error in variational data assimilation is a cost effective way has begun to receive more attention in the last decade Derber (1989) and Dalcher and Kalnay (1987) and many references therein. Studies on predictability in meteorological model have shown that the impact of model error on forecast is indeed significant. Dalcher and Kalnay (1987) in their studies extend Lorenz's parametrization by including the effect of growth of errors due to the model deficiencies. Their results lead to the conclusion that, in meteorology, the predictability limit of forecast might be extended by two or three days if the model error were eliminated.

## 3.2 Model Error and Prediction

The discretization of a model both in time and space introduce discretization errors depending on the accuracy of the numerical schemes. Usually a scalar parameter $h$ is associated to the discretization and a majoration of the error is obtained by a numerical analysis of the scheme.

The estimation of the discrepancy between a solution of the model and the observation is carried out in the observations space by using the operator $C$ as defined above. Because $C$ projects the space of the state variable toward the space of observations it depends on the space of the discrete variables.

At first sight we could think that the smaller $h$ is, the better the prediction will be, i.e. if the model is improved then the prediction will be improved. If $h$ get smaller then the model is supposed to improved because of the reduction of the discretization error.

Let see what going with a simple model. We consider the Burgers' equation:

$$\frac{\partial u}{\partial t} + u\frac{\partial u}{\partial x} - v\frac{\partial^2 u}{\partial x^2} = f$$

$$u(0,t) = u(1,t) = 0$$

on the space interval (0,1) and the time interval (0,T), $f$ was chosen such that

$$u(x,t) = t(1 - \exp(2k\pi x)),$$

is the exact solution with the initial condition $u(x,0) = 0$, $x \in (0,1)$ and $k = 2$. The numerical value of the viscosity is set to $v = 0.25$.

The model has been discretized according to a simple uniform finite difference scheme.

A set of observations has been generated from the exact solution on a regular grid associated to the discretization $h = 1/50$. No errors were introduced into these pseudo observations. For different values of $h$ ranging from $h = 1/10$ to $= 1/250$ we carried out a variational data assimilation algorithm. Then the initial condition being retrieved the prediction at time $t = 2T$ obtained by a direct integration of the model. All these numerical experiments were realized with the same set of observations.

The minimization method used in the following experiments is M1QN3 a quasi-Newton algorithm with limited memory developed at INRIA by Gilbert and Lemaréchal (1989).

Figure 1 shows the evolutions of the cost function as a function of the number of the iterations.

This Figure shows the convergence of the data assimilation algorithm for any value of $h$. The convergence is fast and there is not a wide dispersion of the speeds of convergence. In any case we obtain an asymptotic value for the cost function depending on $h$. Table 1 displays the different asymptotic values. The best adjustment is obtained for $h = 1/50$ when the density of observations coincide with the discretization. In Table 1 the asymptotic values of the cost functions are shown.

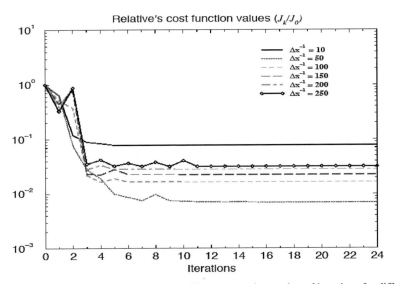

**Fig. 1** Evolution of the cost function with respect to the number of iterations for different values of the discretization parameter $h = \Delta x$

**Table 1** Asymptotic value of the cost function according to the discretization

| $h = \Delta X^{-1}$ | $CJ^*()$ |
|---|---|
| 10 | 7.75777E-002 |
| 50 | 6.90825E-003 |
| 100 | 1.63331E-002 |
| 150 | 2.22294E-002 |
| 200 | 2.80397E-002 |
| 250 | 3.15293E-002 |

The smallest value is obtained for $h = 1/50$, then it increases when the parameter of discretization decreases. Figure 2 displays the prediction obtained at t = 2T, for h = 1/10 it doesn't fit exactly the exact solution, for h = 1/50 we have an almost perfect adjustment, then for smallest value of h the quality of the prediction is worst. In short we can say that when the model is improved (i.e. it has a smaller error, due to discretization, with respect to the «truth») then the prediction is worst. The reason for this fact is mainly due to the operator $C$ mapping the space of the state variable into the space of observations. In the case of a linear model, the exact calculations can be carried out and of course, this not a general result, this example is just a caricature but it demonstrates that the best prediction is not necessarily obtained with the best model. This raises two questions:

– what is the best set of data for a given model?
– what is the best model for a given set of data?

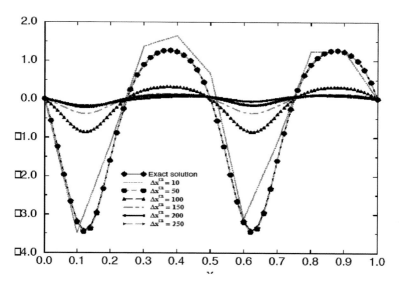

**Fig. 2** Fields predicted at $t = 2T$, according to the discretization

This question is of great importance especially in hydrology when a network has to be designed at a high financial cost for its implementation and its maintenance.

From a mathematical point of view we claim that what has to be estimated is the quality of the OS because it contains all the available information. In the linear case we can obtain an estimation of the condition number of the operators associated to the assimilation. In the non-linear case the problem is open.

# 4 Non-linear Inverse Problems

In the vocabulary of optimal control theory, the parameters resulting from process conceptualization as well as initial and boundary conditions are referred to as control variables, living in a so-called control space. Unfortunately, hydrologic systems, perhaps more so than other geophysical systems, avail very limited knowledge of the control variables due to scarce observations of the system. Although the variational formalism is not restricted to the estimation is this particular type of model input, we will in this section focus on parameter estimation problems (White et al. 2002).

Whatever the complexity of the mathematical representation for hydrological systems, hydrological models contain parameters (ex. diffusivity, roughness, porosity, ...) which are associated to the nature of the environment or to the nature of the flow. These coefficients integrate complex nonlinear phenomena and/or the heterogeneous nature of environment (ex. soil or land cover properties). Consequently they don't have an intrinsic value but this value depends on the scale at which they are considered. A consequence is that these quantities can't be directly measured; they have to be estimated from comparisons between the results of the models and measurements. Variational methods enable an efficient solution of the inverse problem. The general formalism is presented and some applications are described.

## 4.1 General Method for Identification: Steady State Case

We consider a model in the form:

$$F(X,K) = 0$$

where $X$ is the state variable and $K$ some unknown parameter. We assume that $K$ is a function of the coordinate it takes the form:

$$K = \sum_{1}^{N} k_i \phi_i(x)$$

The total domain D is divided into N subdomains $D_i$ on which the global coefficient is homogeneous. $\phi_i$ represents a functional basis for the coefficient $K$. The most common case is when $\phi_i$ is constant on the subdomain $D_i$.

We dispose of observations $X_{obs}$ of the state variable, and we have a-priori evaluation of $K$: $K_0$. As above the observation is supposed to be located on the whole domain; it never happens: observations are pointwise, we use this assumption to simplify notations in a real case, it will be sufficient to change the integrals into sums.

The discrepancy between the solution of the model $X(K)$ and the observation $X_{obs}$ is estimated by the cost function:

$$J(K) = \frac{1}{2} \int_D \|CX(K) - X_{obs}\|^2 dx + \frac{1}{2} \int_D \|K - K_0\|^2 dx$$

The optimal value of $K$: $K^*$ will be the one minimizing the cost function $J$. Both for theoretical reasons (necessary condition for an optimum) and practical ones (running an algorithm of descent) we need to compute the gradient $\nabla J$ of $J$ with respect to $K$. It takes the form of a N-vectors:

$$\nabla J = \left( \frac{\partial J}{\partial ki} \right)_{i=1,N}$$

To do so we have to introduce the adjoint variable $P$ solution of the adjoint model:

$$\left[ \frac{\partial F}{\partial X} \right]^T \cdot P = C^T (CX - X_{obs})$$

Then it comes on the continuous form:

$$\nabla J(K) = - \left[ \frac{\partial F}{\partial K} \right]^T \cdot P + K - K_0$$

Or under the discrete form:

$$\frac{\partial J(K)}{\partial k_i} = - \int_D \left[ \frac{\partial F}{\partial k_i} \right]^T \cdot P dx + k_i - k_{0,i}$$

A major problem is the choice of the basis of function used for the discretization of $K$, it has been to be carried out on physical considerations on the nature of the parameter. The components of $K$ are not necessarily homogeneous they can be associated to heterogeneous physical quantities; a consequence is that a preliminary task before identification will be to nondimensionalize those quantities.

The last term in the cost-function is a regularization term, it helps to make the problem well-posed, furthermore the optimality condition is just a necessary condition, it is not sufficient. If the model is nonlinear then the cost function may not be convex with the possibility of local minima, in this case $K_0$ is also a first guess for the optimization algorithm.

## 4.2 Non-steady Problems

In this case the model is time depending , it is defined on the time interval (0, T), it writes:

$$\frac{dX}{dt} = F(X,K)$$
$$X(0) = X_0$$

The initial condition is supposed to be known. Several cases can be considered:

(i) $K$ depends on time and only on time, it can be written as:

$$K = \sum_{1}^{M} k_i \psi_i(t)$$

(ii) $K$ depends on space and only on space, it can be written as:

$$K = \sum_{1}^{N} k_i \phi_i(x)$$

(iii) $K$ depends both on time and space, it can be written as:

$$K = \sum_{j=1}^{N} \sum_{1}^{M} k_{i,j} \psi_i(t) \phi_j(x)$$

In any case we will have to identify constants.
    The cost function takes the form:

$$J(K) = \frac{1}{2} \int_{0}^{T} \int_{D} \|CX(K) - X_{obs}\|^2 dx + \frac{1}{2} \int_{0}^{T} \int_{D} \|K - K_0\|^2 dxdt$$

The adjoint model will take the form:

$$\frac{dP}{dt} + \left[\frac{\partial F}{\partial X}\right]^T \cdot P = C^T (CX - X_{obs})$$
$$P(T) = 0$$

From where we will deduce the gradient after a backward integration of the adjoint model according to the assumption on $K$ we will find:
    If $K$ is a function of time;

$$\nabla J(K) = \int_{D} \left(-\left[\frac{\partial F}{\partial K}\right]^T \cdot P + K - K_0\right) dx$$

If $K$ is a function of space:

$$\nabla J(K) = \int_0^T \left( -\left[\frac{\partial F}{\partial K}\right]^T \cdot P + K - K_0 \right) dt$$

If $K$ is a function of time and space:

$$\nabla J(K) = \int_0^T \int_D \left( -\left[\frac{\partial F}{\partial K}\right]^T \cdot P + K - K_0 \right) dx dt$$

## 4.3 Calibration of a Rainfall-Runoff Model

### 4.3.1 A Surface Water Model and the Identification Problem

Flood predictions rely on understanding the watershed response to precipitation input. Physically-based models are usually spatially distributed and adopt a bottom up approach to modelling. Considering that infiltration excess is the dominant runoff generation mechanism, the continuity equation over a watershed contains source term for the rain and a sink term for the infiltration. The continuity equation writes:

$$\frac{\partial h}{\partial t} + \frac{\partial (hu)}{\partial x} = R - I \text{ for } (x,t) \in (0,L)X(0,T)$$

where $u = u(x,t)$ is the velocity of the flow and $h = h(x,t)$ the depth of the fluid. $R = R(x,t)$ and $I = I(x,t)$ are rainfall rate and infiltration rate. This equation has two unknowns another equation is needed. The momentum equation is used and simplified by using the kinetic analogy: gradients affecting the momentum are orders of magnitude less than the land surface slope. In this case the momentum equation can be replaced by a relation between velocity, height of the flow, slope and hydraulic roughness in the form:

$$u = \frac{s^{\frac{1}{2}}}{n} h^{\frac{2}{3}}$$

In this relation $s = s(x)$ is the slope and $n = n(x)$ the hydraulic roughness Introducing this relation in the continuity equation we obtain:

$$\frac{\partial h}{\partial t} + \frac{\partial}{\partial x} \left( u = \frac{s^{\frac{1}{2}}}{n} h^{\frac{5}{3}} \right) = R - I \quad \text{for } (x,t) \in (0,L)X(0,T)$$

In this equation rain, infiltration, roughness and slope are supposed to be known everywhere and at any time for $R$ and $I$. These fields are inhomogeneous and therefore we can set an approximate relation in the form:

$$\frac{\partial h}{\partial t} + \beta \frac{\partial}{\partial x} \left( \frac{s^{\frac{1}{2}}}{n} h^{\frac{5}{3}} \right) = \gamma R - \alpha I \quad \text{for } (x,t) \in (0,L)X(0,T)$$

$\alpha, \beta$ and $\gamma$ are unknown empirical parameters. Equipped with initial and boundary conditions the equation as a unique solution when $\alpha, \beta$ and $\gamma$ are given.

The problem will be to get the best estimation of them, if we have an observation of the fields. To do so, in the continuous case, we define a cost function $J(\alpha, \beta, \gamma)$ in the form:

$$2J(\alpha, \beta, \gamma) = \iint\limits_{(0,T)X(O,L)} \|Ch - h_{obs}\|^2 dxdt + (\alpha - \alpha_0)^2 + (\beta - \beta_0)^2 + (\gamma - \gamma_0)^2$$

$C$ is the operator mapping the space of the space variable h into the space of observation. The norm may include the statistical information on the error of observation it takes the form: $\|X\|^2 = (MX, X)$ where $M$ is the inverse of the empirical covariance matrix of the field $X$.

Now the problem can be stated as: «Determine $\alpha, \beta, \gamma$ minimizing J»

### 4.3.2 Optimality System

We need to get the gradient of J with respect to $\alpha, \beta, \gamma$ in order to obtain the Optimality System and to run gradient-based optimization algorithms. The first step consists to compute the Gâteaux derivatives of $J : \hat{J}$ and $h : \hat{h}$ in a direction $\left( \hat{\alpha}, \hat{\beta}, \hat{\gamma} \right)$.

Deriving the model we found that $\hat{h}$ is the solution of:

$$\frac{\partial \hat{h}}{\partial t} + \frac{5}{3}\beta \frac{\partial}{\partial x} \left( \frac{s^{\frac{1}{2}}}{n} h^{\frac{1}{3}} \hat{h} \right) + \hat{\beta} \frac{\partial}{\partial x} \left( \frac{s^{\frac{1}{2}}}{n} h^{\frac{5}{3}} \right) = \hat{\gamma} R - \hat{\alpha} I$$

with homogeneous initial and boundary conditions, and:

$$\hat{J} \left( \alpha, \beta, \gamma, \hat{\alpha}, \hat{\beta}, \hat{\gamma} \right) = \iint\limits_{(0,T)X(O,L)} (C^T (MCh - h_{obs}), \hat{h}) \, dxdt$$

$$+ (\alpha - \alpha_0) \hat{\alpha} + (\beta - \beta_0) \hat{\beta} + (\gamma - \gamma_0) \hat{\gamma}$$

The gradient can be identified by exhibiting the linear dependence of $\hat{J}$ with respect to $\hat{\alpha}, \hat{\beta}, \hat{\gamma}$. To do so we introduce q the adjoint variable to h. Multiplying by q, integrating by parts and after an integration by parts we find that if q is the solution of:

$$\frac{\partial q}{\partial t} + \frac{5}{3}\beta \left( \frac{s^{\frac{1}{2}}}{n} h^{\frac{1}{3}} \right) \frac{\partial q}{\partial x} + = -C^T (MCh - h_{obs})$$

with homogeneous initial and boundary conditions.

Then the gradient is given by:

$$\nabla J(\alpha,\beta,\gamma) = \begin{pmatrix} \nabla J_\alpha \\ \nabla J_\beta \\ \nabla J_\gamma \end{pmatrix} = \begin{pmatrix} -\iint Iqdxdt + \alpha - \alpha_0 \\ \frac{5}{3}\iint \beta\frac{s^{\frac{1}{2}}}{n}h^{\frac{1}{3}}qdxdt + \beta - \beta_0 \\ \iint Rqdxdt + \gamma - \gamma_0 \end{pmatrix}$$

The optimality system is constituted by the model, the adjoint model and the optimality necessary condition:

$$\nabla J(\alpha,\beta,\gamma) = 0$$

### 4.3.3 Numerical Solution

The model has been discretized with a finite element scheme with a leap frog scheme in time. Several values of the parameters of the cost function have been considered (Figs. 3, 4, 5, 6 and 7).

We note that the approximation sequence for parameters appears to converge. However in the case of $\beta$ with $\tau = 0$. When there is no prior information the convergence may not be close to $\beta_0$.

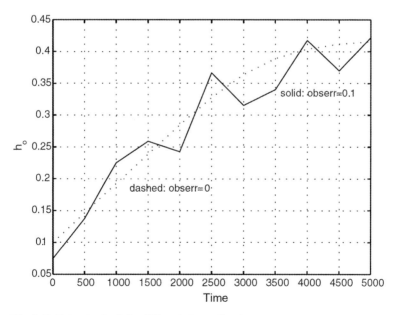

**Fig. 3** Outlet water depth for different observational errors

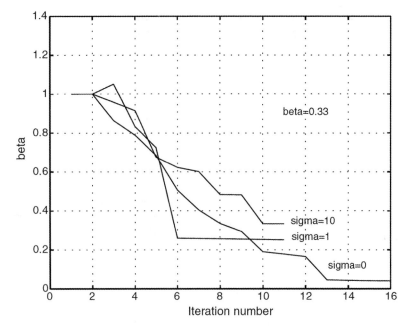

**Fig. 4** Estimated $\beta$ for different values of $\tau$

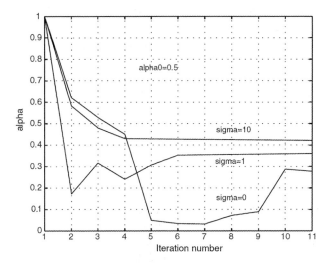

**Fig. 5** Estimated $\alpha$ for different values of $\tau$

## 4.4 Identification for an Infiltration Model

### 4.4.1 The Continuous Problem

Infiltration is an important component of the water cycle. It is a key process for rainfall abstractions and constitutes the driving force of all flow components developing

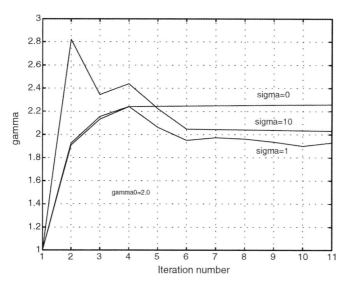

**Fig. 6** Estimated $\gamma$ for different values of $\tau$

below the ground surface (lateral subsurface flow, lateral redistribution of soil moisture, percolation to deep aquifer ...). Infiltration study is a difficult subject because
it requires a good knowledge of the physical and thermodynamical properties of the
soil profile. As we will see later an infiltration model contains several parameters
which can not be directly measured, in the next we will see how, and with simple
measurements on the surface, these parameters can be estimated.

   If $h$ is the relative pressure with respect to the atmospheric pressure measured by an equivalent water height then it is governed by the Richard's equation
(Richards, 1931) which has a unique solution with initial and boundary conditions.
We get a nonlinear partial differential equation of first order with respect to time and
second order with respect to space:.

$$\begin{cases} C(h)\frac{\partial h}{\partial t} = \frac{\partial}{\partial z}\left[K(h)\left(\frac{\partial h}{\partial z} - 1\right)\right] \\ h(0,z) = h_{ini}(z) \\ h(t,0) = h_{surf}(t) \\ h(t,Z) = h_{fond}(t) \end{cases}$$

   In this equation the coefficient C(h) and K(h) are functions of h and they are
defined by:

$$C(h) = \begin{cases} \dfrac{\theta_s(2-n)}{h_g}\left(\dfrac{h}{h_g}\right)^{n-1}\left[1+\left(\dfrac{h}{h_g}\right)^n\right]^{\frac{2}{n}-2} & \text{if } h < 0 \text{ unsaturated case} \\ C(h) = 0 \text{ if } h \geq 0 \text{ saturated case} \end{cases}$$

and

$$K(h) = \begin{cases} K_s \left[ 1 + \left( \dfrac{h}{h_g} \right)^n \right]^{m\left(\frac{2}{n}-1\right)} & \text{if } h < 0 \text{ unsaturated case} \\ K(h) = K_s \text{ if } h \geq 0 \text{ saturated case} \end{cases}$$

In these expressions the different parameters represent:

- $\theta_s$, water content at saturation
- $K_s$, hydraulic conductivity at saturation
- $m$ and $n$ are parameters depending on the soil structure.
- $h_g$ is the inflexion point of the curve:

$$f(x) = \theta_s \left[ 1 + \left( \frac{x}{h_g} \right)^n \right]^{\frac{2}{n}-1}.$$

If all the coefficients were known, with boundaries and initial condition then we can expect a unique solution to the equation. But these coefficients $(\theta_s, K_s, h_g, m, n) = U$ are unknown. The problem is to estimate them from observations. To be more precise at a time $t$, we define the cumulative infiltration by:

$$I_{cal}(t) = \int_0^Z q(t,z)dz,$$

where $Z$ is the depth of the ground and $q$ the infiltration rate . We can also express this quantity by:

$$I_{cal}(t) = \int_0^Z (\theta(t,z) - \theta_{ini})dz,$$

with $\theta$ the water content depending on space and time and $\theta_{ini}$ the initial water content. This quantity can be computed from a solution of the model.

We assume that $M$ observations have been carried out at time $t_i, i = 1, \ldots, M$, an estimator of the discrepancy between the computed solution and the model is given by the cost function $J$ defined by:

$$J(U) = \frac{\Delta t}{2} \sum_{j=0}^M (I_{cal}(t_j) - I_{obs}(t_j))^2$$

$$= \frac{1}{2} \int_0^T \left[ \int_0^Z (\theta(t,z) - \theta_{ini})dz - I_{obs}(t) \right]^2 \cdot \delta(t - t_j)dt,$$

The problem of identification of the unknown coefficient can be stated as:

«Determine $U^* = (\theta_s^*, K_s^*, h_g^*, m^*, n^*)$ minimizing $J$»

The first step is to determine the gradient of J in order to get an optimality condition and to perform an optimization algorithm. To do so we introduce an adjoint Variable $P$ and the adjoint model is derived using the same method as we did above and we obtain the Optimality System (OS), it writes:

$$\begin{cases}
\begin{cases}
C(h)\frac{\partial h}{\partial t} = \frac{\partial}{\partial z}\left[K(h)\left(\frac{\partial h}{\partial z}-1\right)\right], \\
h(0,z) = h_{ini}(z) \\
h(t,0) = h_{aurf}(t) \\
h(t,Z) = h_{fond}(t) \\
-\frac{\partial}{\partial t}(C\cdot P) + \left[\frac{\partial C}{\partial h}\right]\cdot\left(P\cdot\left[\frac{\partial h}{\partial t}\right]\right) - \frac{\partial}{\partial z}\left[K\frac{\partial P}{\partial z}\right] + \left[\frac{\partial K}{\partial h}\right]\cdot\left(\frac{\partial P}{\partial z}\cdot\left[\frac{\partial h}{\partial z}-1\right]\right) = \\
(I_{cal}-I_{obs})\frac{\partial I_{cal}}{\partial h}\delta(t-t_i), i=0,\ldots,(N-1), \\
P(t=T,z) = 0 \\
P(t,z=0) = 0 \\
P(t,z=Z) = 0
\end{cases} \\
\begin{cases}
\nabla J(U) = -\int_0^T\int_0^Z\left(\left[\frac{\partial C}{\partial U}\right]\cdot\left(P\frac{\partial h}{\partial t}\right) - \left[\frac{\partial K}{\partial U}\right]\left(\frac{\partial P}{\partial z}\cdot\left(\frac{\partial h}{\partial z}-1\right)\right)\right)dtdz \\
\qquad + \sum_{i=0}^N\int_0^Z(I_{cal}(t)-I_{obs}(t))\left[\frac{\partial I_{cal}}{\partial U}(t)\right]\delta(t-t_i)dz \\
\qquad + \int_0^T\left[\frac{\partial h_{ini}}{\partial U}\right]\left(K\frac{\partial P}{\partial z}\right)_{|z=Z}dt = 0.
\end{cases}
\end{cases}$$

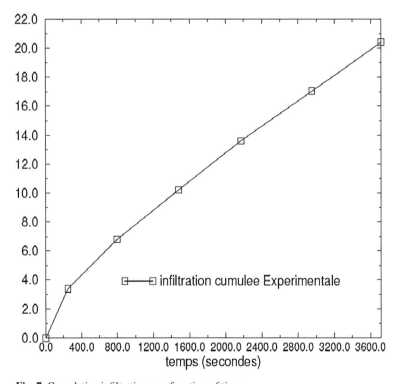

**Fig. 7** Cumulative infiltration as a function of time

## 4.4.2 The Discrete Problem and Numerical Algorithm

The model has to be discretized both in time and space. In this case an Euler explicit scheme has been used for the temporal discretization and a centered finite difference scheme for the discretization in space, therefore at each time step we need to solve a linear system with a tri-diagonal matrix. At it was already pointed out the adjoint of the discretized model is not obtained by a discretization of the adjoint model, we need to derive it directly from the discretized model, and therefore it will include the adjoint operations of the resolution of a linear system.

The algorithm for the estimation of the optimal U can be synthesized as follows:

– Start from a first guess $U_0$
– From iteration $n$ to iteration $n+1$:

- Solve the direct model with the parameters $U_n$
- Solve the adjoint model
- Compute the gradient, then a direction of descent $D_n$
- Estimate $U_{n+1}$ by: $U_{n+1} = U_n + \rho_n D_n$. Where $\rho_n$ is chosen such that $U_{n+1}$ realizes the minimum of $J$ along the direction $D_n$. Therefore, in the nonlinear case, an auxiliary 1-D problem of optimization as to be solved. This problem sounds simple but it requires several evaluations of J and therefore to solve several times the direct model: an expensive operation from the computational point of view.
- Check the optimality condition if the gradient is smaller than some prescribed value then STOP.

## 4.4.3 Checking the Gradient

A bad gradient can inhibit the convergence of the optimization algorithm; therefore a mandatory step is to check its exactness. For some scalar $\alpha$ and some direction $q$, Taylor's formula gives:

$$J(U + \alpha q) = J(U) + \alpha \langle \nabla J(U), q \rangle + \frac{\alpha^2}{2} \langle \nabla^2 J(U)q, q \rangle + o(\alpha^2),$$

If a function $\varsigma$ is defined by:

$$\varsigma(\alpha) = \frac{J(U + \alpha q) - J(U)}{\alpha \cdot \langle \nabla J(U), q \rangle},$$

Then a necessary condition to obtain an exact gradient is:

$$\lim_{\alpha \to 0} \varsigma(\alpha) = 1.$$

**Table 2** Verification of the gradient

| $\alpha$ | $\zeta(\alpha)$ | $\alpha$ | $\zeta(\alpha)$ |
|---|---|---|---|
| $2^{-11}$ | 1.386047463 | $2^{-31}$ | 1.000000345 |
| $2^{-13}$ | 1.095920988 | $2^{-33}$ | 1.000000069 |
| $2^{-15}$ | 1.023945769 | $2^{-35}$ | 0.999999978 |
| $2^{-17}$ | 1.005984309 | $2^{-37}$ | 1.000000029 |
| $2^{-19}$ | 1.001495931 | $2^{-39}$ | 0.999999986 |
| $2^{-21}$ | 1.000373960 | $2^{-41}$ | 0.999999814 |
| $2^{-23}$ | 1.000093475 | $2^{-43}$ | 0.999994405 |
| $2^{-25}$ | 1.000023354 | $2^{-45}$ | 1.000007456 |
| $2^{-27}$ | 1.000005824 | $2^{-47}$ | 0.999929147 |
| $2^{-29}$ | 1.000001441 | $2^{-49}$ | 0.999522488 |

Of course an exhaustive exploration of the directions q cannot be done. A normalized direction with random components can be chosen, then the convergence when $\alpha$ goes to zero tested. In the present case we get the Table 2.

It is seen that, in a first step, the value of $\varsigma$ decreases to 1, then in a second step, for very small values of $\alpha$ the distance to 1 increase, at first sight this is contradictory with Taylor's formulae but it doesn't take into account rounding errors becoming important for small values of $\alpha$.

### 4.4.4 Numerical Example

Some numerical experiments have been carried out in different cases. The effect of observation sampling has been tested. A difficulty is to have an initial estimation (first guess). We obtained the following results for the identification of the coefficients (Table 3).

Numerical results are presented in Figs. 8, 9, 10, 11 and 12. Figure 8 displays the difference between the exact solution and the calculated one, the initial state at the beginning of the iteration is also presented. In Fig. 9. We can see the evolution

**Table 3** Identified values (ident) with their initial condition (init). Three cases were considered, (A) one observation at each time step, (B) one observation every 300 s, (C) one observation every 720 s

|  | A | | B | | C | |
|---|---|---|---|---|---|---|
|  | ident | init | ident | init | ident | init |
| $K_s$ | 5.82E–3 | 1.76E–3 | 4.50E–3 | 1.78E–3 | 4.78E–3 | 1.80E–3 |
| $h_g$ | −17.10 | −16.5 | −17.10 | −17.29 | −17.10 | −17.4 |
| $\theta_s$ | 0.69 | 0.68 | 0.68 | 0.69 | 0.68 | 0.69 |
| $m$ | 6.09 | 6.12 | 6.11 | 6.12 | 6.41 | 6.42 |
| $n$ | 2.47 | 2.65 | 3.19 | 3.2 | 2.65 | 2.63 |

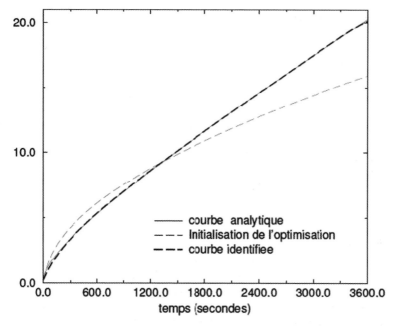

**Fig. 8** Exact solution (*analytic curve*), infiltration with initial value and infiltration at the optimum

of the optimisation of the cost function and of the gradient, this one has a chaotic behavior but it is globally decreasing, the cost function shows a monotone, but non uniform, decrease , this is a typical behavior for the algorithms of variational data assimilation. In Figs. 10, 11, and 12 we can see the results of the optimization and its evolution as a function of the number of iteration. Figure 12 represents cross sections of the cost functions, it shows that this function in not convex because of the non linearity of the model , a consequence is the difficulty for the optimization algorithm to converge toward a global minimum. It also emphasizes the importance of the first guess both for starting the iterative process and also for the regularization of the cost function.

### 4.4.5 A Field Experiment

We have used the observation provided by a field experiment carried out in western part of Cameroon see Fonteh et al. (1998) near the Poneké village in Département des Bamboutos. The superficial ground has a depth between 25 and 90 cm with a weak density (0.8 g/cm$^3$), a strong porosity and a high infiltration rate: 0,68% (Fig. 13).

No a priori knowledge of the coefficients was available. The optimal values found were:

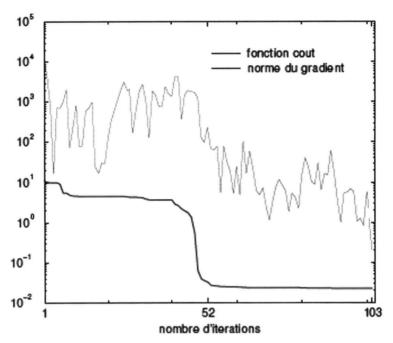

**Fig. 9** Evolution of the cost function and of the gradient when a 720 s sampling of the observations. The cost function has a monotonic evolution; the norm of the gradient shows a more chaotic behavior. In bold the cost function

$$K_s = 3.742707X10^{-5}m/s, h = -0.170999m, \theta_s = 0.648869cm^3/cm^3,$$
$$m = 6.120047, n = 3.20019$$

The identified and observed cumulated infiltrations (Fig. 14) are in good agreement. The norm of the gradient slowly decreases and about 60 iterations are necessary to achieve a minimum. Clearly the cross sections of the cost function show a more regular behavior than in the synthetic case, this is probably due to the nature of the soil. In Fig. 15 we can see that the gradient is, at the beginning, not decreasing, in mean, then after iteration 51 it's rapidly going to zero. This can be explained by a first guess of the iterative process too far from the optimal values (Fig. 16).

## 5 Adjoint Sensitivity Analysis of a Spatially Distributed Hydrological Model

The rainfall-runoff transformation is characterized by the complexity of the involved processes and by the limited observability of the atmospheric forcing, catchment properties and hydrological response. It is therefore essential to understand, analyze

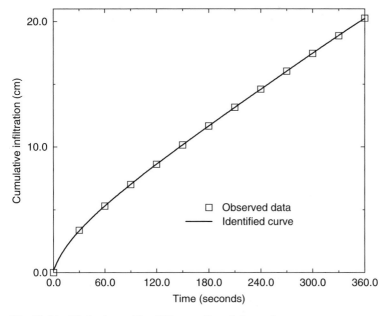

**Fig. 10** Identified values with a 720 s sampling of observations

and reduce the uncertainty inherent to hydrological modelling. The distributed modeling of catchment hydrology offers great potential for understanding and predicting the rainfall-runoff transformation. An exhaustive study is carried out in Estupina-Borrell et al. (2006).

While sampling based sensitivity analysis methods can be limited in handling computer intensive spatially distributed systems (i.e. curse of dimensionality), adjoint sensitivity analysis provide a local but extensive insight of the relation between model inputs and prognostic variables. This paradigm is applied to an event-based distributed rainfall-runoff model adapted to a small watershed on the upper part of the Thoré watershed (Tarn department, South West of France).

## 5.1 Flash Flood Model Description : MARINE Rainfall-Runoff Model

The underlying physics of MARINE flash flood model (Estupina-Borrell et al., 2006) is adapted to events for which infiltration excess dominates the generation of the flood. In the simplified version used for this study, rainfall abstractions are evaluated using the Green Ampt infiltration model and the resulting surface runoff is transferred using the Kinematic Wave Approximation (KWA).

The complex geometry of the watershed described by a uniform grid on which each cell receive water form its upslope neighbours and discharge to a single

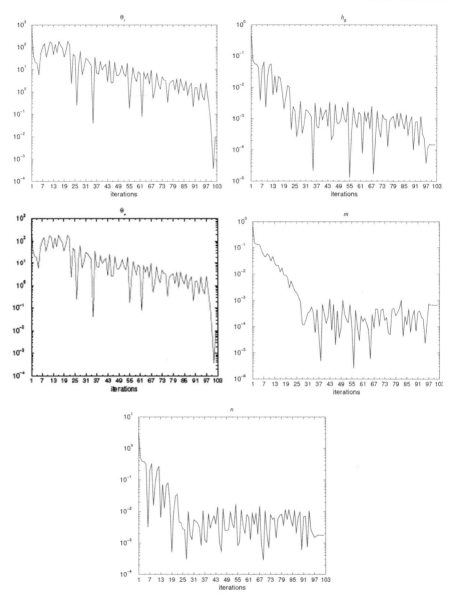

**Fig. 11** Evolution of the five components of the cost function according to the iterations

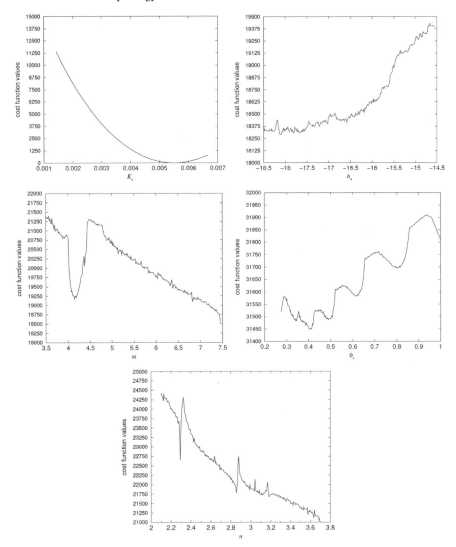

**Fig. 12** Cross sections of the cost function in the vicinity of the solution, it shows its nonlinear character

downslope neighbour (steepest direction). For a one dimensional flow of average velocity $u$ and average depth $h$, the continuity equation can be expressed as:

$$\frac{\partial h}{\partial t} + \frac{\partial uh}{\partial x} = r - i$$

where $r$ the rainfall intensity and $i$ the infiltration rate. Using the KWA approximation which has shown the ability to represent channelized and sheet overland flow, the momentum conservation equation reduces to an equilibrium between the bed

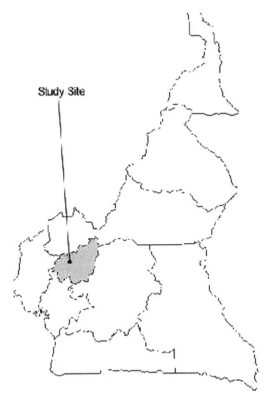

**Fig. 13** Site of the experiment in Cameroon

slope $S_0$ and the friction slope $S_f$. The Manning equation (uniform flow on each grid cell) is used to relate the flow velocity and the flow depth:

$$u = \frac{R^{2/3} S_f^{1/2}}{n} \quad \text{with} \quad R = \frac{hw}{2h+w}$$

where $R$ is the hydraulic radius, $n$ the Manning roughness coefficient and $w$ the elemental flow width. In this simplified version of the model, the flow width is constant (rectangular section) and given the ratio between the width (grid resolution) and the flow depth the hydraulic radius is approximated by the water depth (i.e. $R = h$). As seen above, the resulting equation governing the overland flow is given by:

$$\frac{\partial h}{\partial t} + \frac{S_0^{1/2}}{n} \frac{\partial h^{5/3}}{\partial x} = r - i$$

In the right hand side of the previous equation which represents the excess rainfall, the infiltration rate $i(t)$ is estimated using the Green and Ampt equation, a very classic simplified representation of the infiltration process. For an homogeneous soil

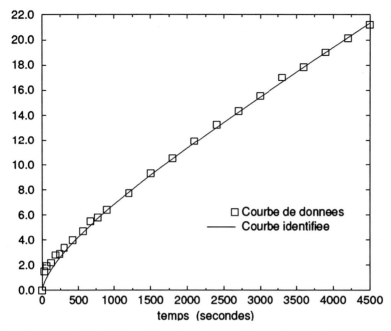

**Fig. 14** Field experiment: observed and identified cumulative infiltration

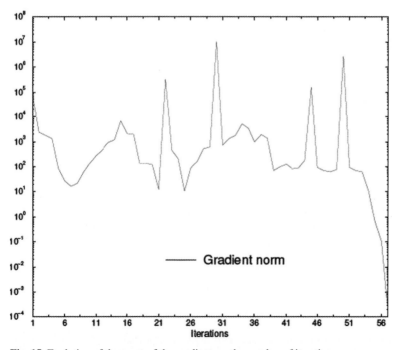

**Fig. 15** Evolution of the norm of the gradient vs. the number of iterations

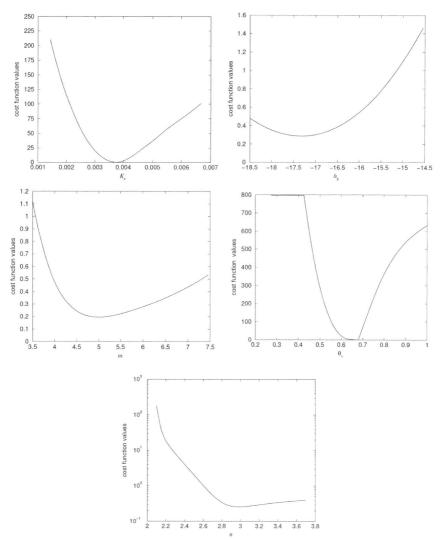

**Fig. 16** Cross sections of the cost variables along the directions of the unknown parameters

column characterized by its hydraulic conductivity $K$ and $\psi$ the soil suction at the downward moving wetting front, the potential infiltration rate is given by

$$i(t) = K \left( \frac{\psi \Delta \theta}{I(t)} + 1 \right) \text{ with } \Delta \theta = \eta (1 - \theta)$$

where $\theta$ is the initial soil moisture content, $\eta$ the soil porosity and $I(t)$ the cumulative infiltration at time $t$. After pounding, the cumulated infiltration at time $t + \Delta t$ can be calculated by the following equation

$$I_{t+\Delta t} - I_t - \psi\Delta\theta \ln \left[ \frac{I_{t+\Delta t} + \psi\Delta\theta}{I_t + \psi\Delta\theta} \right] = K\Delta t$$

which is solved by the Newton's method.

## 5.2 Case Study and Available Data

The previously described model is applied to a very small catchment area ($25\,\text{km}^2$) (Fig. 17) from the upper part of the Thoré basin which was affected by a catastrophic flood event in November 1999. Unfortunately the event was not gauged since all measuring devices were washed away by the flood. Therefore, a priori values (derived from published tables) for the parameters are used for the generation of a reference virtual hydrological reality. For a rainfall forcing estimated from real radar data (from Météo France), the nominal values specified (different spatial variability) produce specific discharges typical for Mediterranean flash flood events.

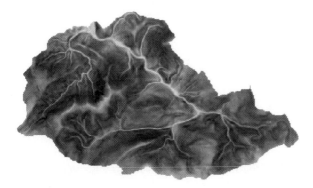

**Fig. 17** Catchment topography

## 5.3 Model Understanding Using Adjoint Sensitivity Analysis

For the distributed modelling of catchment hydrology, parameters are discretized according to the spatial discretization of model state variables. Therefore, different values can be assigned to every single element of the discretization using a priori information. In order to make inverse problem posed for parameter estimation tractable, the calibration usually consists in

– adjusting a few scalar multipliers characterizing an empirical parametrization, most of the time using global and non-smooth optimization techniques (probabilistic and evolutionary methods);

– improving the a priori estimates using a (smooth or non-smooth) local search
  method, i.e. regularizing the inverse problem using a priori information.

Whatever the approach used for the calibration of spatially distributed parameters, an extensive insight in the model behavior would require the analysis of the
influence of the parameters specified for each element of the computational grid on
the simulation results. Depending on the purpose for the sensitivity analysis, it may
not be necessary to average information over the parameter space (global/sampling-
based sensitivity analysis) and local approaches around the behavioral nominal values may prove very informative.

In order to corroborate and improve our understanding of the way the different model parameters control the hydrological response, the tangent linear and adjoint models of MARINE were developed using the direct and reverse modes of the
TAPENADE automatic differentiation engine (Hascoët and Pascual, 2004). For a
given trajectory in the model phase space, the trajectory being defined by the nominal values for the parameters, the initial and boundary conditions, we compute local
sensitivity indices by means of derivative information. The outcome to be analyzed
is a gradient for a scalar response and the entire Jacobian matrix of the transformation for a vectorial response.

## 5.4 Sensitivity of Forecast Aspect to the Model Parameters (Scalar Response)

For the moment, rather than considering the complete flood hydrograph simulated
by the model, we consider two essential aspects of the forecast: the flood volume and
the flood peak. Apart from the classical ranking of parameter values showing that
the friction coefficient mainly drive the flood peak and the hydraulic conductivity
the flood volume, the adjoint of the flash flood model enables a very fruitful spatial-
temporal sensitivity analysis.

During the event, the rainfall intensity is highly variable and the runoff over the
surface leads to flow concentration in the drainage network. The combination of
the previously cited factors will drive an important spatial-temporal variability of
the infiltration capacity. All the cells of the watershed are solicited for infiltration
from both direct rainfall and excess rainfall coming from upstream in the basin
(i.e. *run-on*). Figure 18 exhibits the temporal patterns for the sensitivity of the flood
volume to model parameters. The information provided corresponds to the influence
of instantaneous perturbations of parameter values on the overall response, e.g. the
flood volume. The variability of the rainfall dynamics and the different infiltration
regimes (defined using on/off switches) lead to similar variations for the temporal
increments of adjoint variables. In order to facilitate the interpretation of the results,
the event was divided into four periods, and the results are in agreement with the
infiltration excess overland flow mechanism encoded in the model (i.e. perceptual
model).

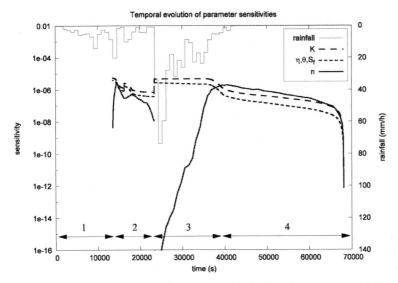

**Fig. 18** Sensitivity of the flood volume, temporal evolution of parameter sensitivities

In fact, during period 1, since the rainfall intensity is much smaller than the potential infiltration capacity, precipitations are directly and totally infiltrated. The period 2 is characterized by the development of run-on processes. Although for this period there is still no flow at the outlet, both soil and friction parameters influence the overall flood volume. When the main rainstorm arrives (period 3), the potential infiltration rate is much smaller than the rainfall intensity and the rising limb of the hydrograph develops. When compared to the previous period, the evolution of parameter sensitivities is very similar. The potential infiltration rate experiments an exponential decay to the hydraulic conductivity. The fact that for an increasing cumulated infiltration the hydraulic conductivity becomes more influent than the other soil parameters is characterized by an increasing gap between sensitivity curves. Moreover, infiltration losses primarily occur through direct infiltration and then run-on processes develop. Although it is not significant at the beginning of the rainstorm, the sensitivity to the friction coefficient is progressively increased. After the rainstorm (period 4), the remaining runoff is concentrated in the drainage network and the friction coefficient dominates the soil parameter for the calculation of infiltration losses.

When considering a single reach the effect of variations of the friction coefficient on the maximum discharge is quite obvious: increasing the roughness will cause a diminution of the flood peak. Over the very complex topography characterizing a catchment, it is much more difficult to anticipate the response. The spatial sensitivity of the peak discharge to the roughness coefficient is provided by Fig. 19. Two colors ramps were used because positive and negatives sensitivities are encountered over the surface of the watershed. In fact, depending on the location, increasing the roughness coefficient can have antagonist effects on the peak discharge. While

**Fig. 19** Sensitivity of the peak discharge, spatial distribution of parameter sensitivities

all sensitivities have the expected sign along the main stream (i.e. negative), some positive sensitivities can counterbalance the overall effect in some concomitant sub-basins. Therefore, when applying scalar multipliers, compensation effects usually occur which are very difficult to identify without such analysis. For example, increasing the nominal of 10% for all roughness coefficient lead to −4.5% on the peak discharge. This variation is −5.9% when only the cells showing a negative sensitivities are modified and +1.5% when the same operation is carried out on the cells featuring positive sensitivities.

## 5.5 Sensitivity of the Simulated Hydrograph to the Model Parameters (Vectorial Response)

In this section, the entire flood hydrograph is analyzed and therefore as underlined previously the sensitivity information consists in a jacobian matrix rather than a gradient vector. The jacobian can be calculated line by line using the adjoint model, column by column using the tangent linear model. When the number of analyzed time steps is much smaller than the number of parameters, the adjoint path is much more efficient. The first line of this jacobian corresponds to the influence of all parameter on the first simulated discharge; the first column corresponds to the influence of the first parameter on the entire hydrograph. While one can propose a physical interpretation for the lines and/or columns of the Jacobian matrix, a very interesting view angle is provided by its singular value decomposition (SVD). This factorization provides singular vectors in the parameter and observations space. The magnitude of the singular values represents the importance of the corresponding singular vectors. Using this information, the parameters really influencing the hydrological response (spatial location) and the measurements really constraining the parameters (temporal location) can be identified.

The visualization of singular vectors in the parameter space for $n$ and $K$ also provide an extensive insight into the model behavior (Fig. 20). The interactions among the two concomitant sub-basins driving the variability of the simulated discharges at

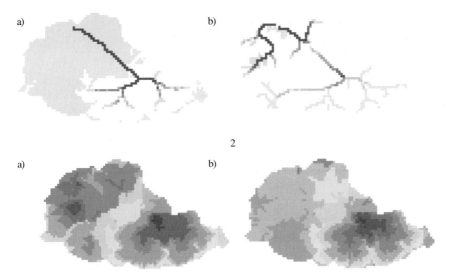

**Fig. 20** Spatial visualization of eigenvectors components in the parameter space for the hydraulic conductivity: (**a**) and (**b**) represent the first and second singular vectors (*red* color ramp for positive components are *gray* for negative)

the outlet of the watershed is clearly characterized. For the roughness coefficient $n$ and the hydraulic conductivity $K$, the first singular vector is dominated by the main stream and its reception basin. For the hydraulic conductivity, the sensitivity magnitude is decreasing with the distance from the principal convergence zone of the network. In an analogous manner, positive components are encountered mainly on the other sub-basin for the second singular vectors. Only some elements, situated very close and/or very far from the outlet, are part of the main sub-basin.

In the experiments presented here, the rainfall forcing is uniform over the watershed. Although the information is local and the model non-linear, the singular vectors in the parameter space are mainly determined by the topography of the watershed rather than the location in the parameter space. The spatial variability of the rainfall forcing influences the different modes but, except if some regions are systematically over exposed, this variability will be smoothed when considering several rainfall events for the calculation of the Jacobian matrix.

The singular spectrum for all the parameters and different forcing conditions (lumped and spatially variable) is given by Fig. 21. The analysis of this figure reveals that the decay of singular values is faster for the roughness coefficient compared to the infiltration parameters. It is also important to note that this gap is reduced when the spatial variability of rainfall is taken into account. In order words, we corroborate the natural reasoning stating that spatially variable rainfall forcing emphasizes the influence of the spatial variability in friction parameters. However, it seems that more complex parameterizations are needed to capture the influence of heterogeneity for the infiltration parameters $K$ and $\theta$. In order to quantify more precisely the number identifiable degrees of freedom in the parametrization, the relative

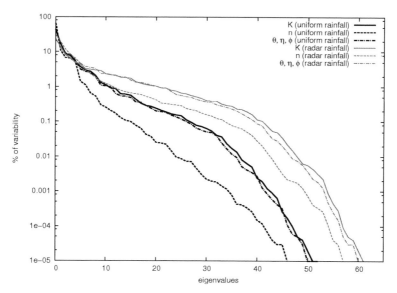

**Fig. 21** Singular values spectrum for lumped and spatially variable rainfall

importance of the parameters and the level of noise in the observations should be taken into account. However, using this approach it is possible to measure the effect of increasing the information content in observations (internal gauging stations, several flood events . . .) on the identifiability of the parameters.

In the experiments carried out in this section, the adjoint of the flash flood model yields very interesting information on the model behavior. In a variational data assimilation framework, the adjoint model is used to estimate the control variables of the original model, minimizing the misfit to observations characterized by a cost function. Therefore, as emphasized before, the optimality system should be considered as a generalized model. The sensitivity of this generalized model requires the derivation of the second order adjoint model.

## 6 Conclusion and Perspectives

Hydrology is a domain of great importance both from the social and scientific point of view. The knowledge we have of the water cycle is partial and imperfect. Improving prediction is a vital necessity for mankind. An improvement can be reached only if all the sources of information are used in a coherent manner. Variational methods have a strong potential of development toward this direction. Variational methods provide tools such as the adjoint model that is useful for many applications : initialization, optimal parametrization of models, sensitivity analysis Variational methods put in light some notions such as the optimality system which is a key point for the integration of heterogeneous sources of information : models, observations, statistics, images. This is a necessity: just one source of information whatever it is can not predict the evolution of the water cycle. For the future, many problems remains open, among them the adequation between data and models: is it necessary to improve models if there is a lack of data? If an accurate prediction, especially in case of natural disaster then it will be necessary to couple hydrological models with atmospheric and oceanic models with their own data such as radar, lidar and other sensors. Prediction makes sense only if an estimation of error is available, stochastic methods sounds more appropriate for this task, nevertheless the development of tools such that second order adjoint (Le Dimet et al., 2002) or sensitivity analysis (Castaings, 2007) permits such an evaluation. A major inconvenient of this approach was its cost, both in term of computation and also of manpower for writing efficient and reliable codes, and consequently their use was difficult for developing countries especially for those vulnerable to climatic hazards during rain seasons. The exponential decreases of the cost of computation now makes prediction affordable for many countries, but a prerequisite is the formation of scientists in this field.

**Acknowledgements** Most of the examples presented in this paper have been developed in the framework of the IDOPT-MOISE: a joint project between INRIA, CNRS, Université Joseph Fourier and Institut National Polytechnique de Grenoble. We would like to thank Yang Junqing,, Cyril Mazauric and Marc Honnorat for their participation in the development of Variational Methods in Data Assimilation. We would like to take advantage of this opportunity to thanks Yoshi Sasaki, from the University of Oklahoma, M.Y. Hussaini from Florida State University, David Furbish from Vanderbilt University for their support.

# References

Belanger E, Vincent A (2005) Data assimilation (4D-VAR) to forecast flood in shallow-waters with sediment erosion. J Hydrol 300(1–4), 114–125

Beven K, Musy A, Higy C (2001) Tribune Libre : L'unicité de lieu, d'action et de temps. Revue des Sciences de l'eau 14(4), 525–533.

Bouyssel F, Cassé V, Pailleux J (1999) Variational surface analysis from screen level atmospheric parameters. Tellus Ser A 51, 453–468

Cacuci D (1981a) Sensitivity theory for nonlinear systems. I. Nonlinear functional analysis approach. J Math Phys 22(12), 2794–2802.

Cacuci D (1981b) Sensitivity theory for nonlinear systems. II. Extensions to additional classes of responses. J Math Phys 22(12), 2803–2812

Callies U, Rhodin A, Eppel DP (1998) A case study on variational soil moisture analysis from atmospheric observations. J Hydrol 212–213, 95–108.

Calvet J, Noilhan J, Bessemoulin P (1998) Retrieving the root-zone soil moisture from surface soil moisture or temperature estimates: A feasibility study based on field measurements. J Appl Meteorol 37(4), 371–386

Carrera J, Neuman S (1986) Estimation of aquifer parameters under transient and steady state conditions: 2. Uniqueness, stability and solution algorithms. Wat Resour Res 2, 211–227

Castaings W, Dartus D, Le Dimet F, Saulnier G (2007) Sensitivity analysis and parameter estimation for the distributed modeling of infiltration excess overland flow. Hydrol Earth Syst Sci Disc 4(1), 363–405

Castaings W (2007) Analyse de sensibilité et estimation de paramètres pour la modélisation hydrologique : potentiel et limitations des méthodes variationnelles. Ph. D. dissertation defended at Univerité Joseph-Fourire, Grenoble, France, November 2007

Chavent G (1974) Identification of Parameters in Distributed Systems, chapter identification of functional parameter in partial differential equations. Heinemann, 367

Chavent G (1991) On the theory and practice of non-linear least square. Adv Wat Resour, 14(2) 55–63

Dalcher A, Kalnay E (1987) Error growth and predictability in operational ECMWF forecasts. Tellus 39A, 474–491

Derber JC (1989) A variational continuous assimilation technique. Mon Wea Rev 117, 2427–2436

Estupina-Borrell V, Dartus D, Ababou R (2006) Flash flooding modelling with Marine hydrological distributed model. Hydrol Earth Syst Sci Disc 3, 3397–3438

Fonteh M, Boukong A, Tankou C (1998) Soil and water management of dry season green Paes pisium sativum, production in the Western highlands of Cameroon. Technical Report, University of Dchang

Gilbert J-C, Lemaréchal C (1989) Some numerical experiments with variable-storage quasi-newton algorithms. Math Program Ser B (45), 407–435

Hascoët L, Pascual V (2004) Tapenade 2.1 User's guide. Technical report .RT-0300. INRIA

Honnorat M, Monnier J, Lai F-X, Le Dimet F (2006), Variational data assimilation for 2D fluvial hydraulics simulation CMWR XVI-computational methods for water resources. Copenhagen, June 2006

Lauvernet C (2005) Assimilation variationelle des observations de télédétection dans les modèles de fonctionnement de la véegétation: utilization du modèle adjoint en et prise en compte des contraintes spatiales. Ph.D. Thesis. University of Grenoble

Le Dimet F-X, Talagrand O (1986) Variational algorithms for analysis and assimilation of meteorological observation. Tellus 38A, 97–110

Le Dimet F, Navon IM, Daescu DN (2002) Second-order information in data assimilation. Mon Wea Rev 130(3), 629–648

Le Dimet F-X, Ngnepieba P, Hussaini MY, Furbish D (2004). Errors and uncertainties for geophysical Fluids. Technical Report, SCRI, Florida State University.

Lions J (1968) Optimal control of systems governed by partial differential equations. Springer-Verlag, Berlin

Liu DC, Nocedal J (1989) On the limited BFGS method for large scale optimization. Math Program 45, 503–528

Mahfouf J (1991) Analysis of soil moisture from near-surface parameters: A feasibility study. J Appl Meteorol 30(11), 1534–1547

Margulis S, Entekhabi D (2001) A coupled land surface-boundary layer model and its adjoint. J Hydrometeorol 2(3), 274–296

Mazauric C (2003) Assimilation de données pour les modèles d'hydraulique fluviale. Estimation de paramètres, analyse de sensibilité et décomposition de domaine. PhD thesis, University Joseph Fourier

Ngnepieba P, Le Dimet F-X, Boukong A, Nguetseng G (2002) Inverse problem formulation for parameters determination using the adjoint method. ARIMA (1) 1, 127–157

Piasecki M, Katopodes N (1997) Control of contaminant releases in rivers and estuaries. Part I: Adjoint sensitivity analysis. J Hydraulic Eng, ASCE 123(6), 486–492

Reichle R, Entekhabi D, McLaughlin D (2001) Downscaling of radio brightness measurements for soil moisture estimation: A four-dimensional variational data assimilation approach. Wat Resour Res 37(9), 2353–2364.

Richards LA (1931) Capillary conduction of liquids through porous mediums. Phys 1, 318–333

Sanders B, Katopodes N (2000) Adjoint sensitivity analysis for shallow-water wave control. J Eng Mech 126(9), 909–919

Sandu A, Daescu D, Carmichael DR (2003) Direct and adjoint sensitivity analysis of chemical kenetics systems with KPP: I -Theory and software tools. Atmos Environ 37, 5083–5096

Seo D, Koren V, Cajina N (2003) Real-time variational assimilation of hydrologic and hydrometeorological data into operational hydrologic forecasting. J Hydrometeorol 4(3), 627–641

Sirkes Z, Tziperman E (1997) Finite difference of adjoint or adjoint of finite Difference. Monthly Weather Review, 125, 3373–3378

Sun N, Yeh W (1990) Coupled inverse problems in groundwater modeling 1. Sensitivity and parameter identification. Wat Resour Res 26(10), 2507–2525

Tikhonov AN, Arsenin VY (1977) Solutions of ill-Posed problems, Wiley, New York

White LW, Vieux BE, Armand D (2002) Surface flow model: Inverse problems and predictions. J Adv Wat Resour 25(3), 317–324

White L, Vieux B, Armand D, Le Dimet F (2003) Estimation of optimal parameters for a surface hydrology model. Adv Wat Resour 26(3), 337–348

Wu L (2005) Variational methods applied to plant functional-structural dynamics: parameter identification, control and data assimilation. D. Thesis. University of Grenoble

Yang J, LeDimet F (1998) Variational data assimilation in the transport of sediment in river. Sci China, Ser D: Earth Sci 41(5), 483–485

# Recent Advances in Land Data Assimilation at the NASA Global Modeling and Assimilation Office

Rolf H. Reichle, Michael G. Bosilovich, Wade T. Crow, Randal D. Koster, Sujay V. Kumar, Sarith P. P. Mahanama and Benjamin F. Zaitchik

**Abstract** Research in land surface data assimilation has grown rapidly over the last decade. We provide a brief overview of key research contributions by the NASA Global Modeling and Assimilation Office (GMAO). The GMAO contributions primarily include the continued development and application of the Ensemble Kalman filter (EnKF) for land data assimilation. In particular, we developed a method to generate perturbation fields that are correlated in space, time, and across variables. The method permits the flexible modeling of errors in land surface models and observations. We also developed an adaptive filtering approach that estimates observation and model error input parameters. A percentile-based scaling method that addresses soil moisture biases in model and observational estimates opened the path to the successful application of land data assimilation to satellite retrievals of surface soil moisture. Assimilation of such data into the ensemble-based GMAO land data assimilation system (GMAO-LDAS) provided superior surface and root zone assimilation products (when validated against in situ measurements and compared to the model estimates or satellite observations alone). Satellite-based terrestrial water storage observations were also successfully assimilated into the GMAO-LDAS. Furthermore, synthetic experiments with the GMAO-LDAS support the design of a future satellite-based soil moisture observing system. Satellite-based land surface temperature (LST) observations were assimilated into a GMAO heritage variational assimilation system outfitted with a bias estimation module that was specifically designed for LST assimilation. The on-going integration of GMAO land assimilation modules into the Land Information System will enable the use of GMAO software with a variety of land models and make it accessible to the research community.

## 1 Motivation

Land surface conditions are intimately connected with the global climate system and have been, through different pathways, associated with predictability of atmospheric variability. Surface and root zone soil moisture control the partitioning of

R.H. Reichle (✉)

Global Modeling and Assimilation Office, NASA Goddard Space Flight Center, Greenbelt, MD, USA, e-mail: rolf.reichle@nasa.gov

S.K. Park, L. Xu, *Data Assimilation for Atmospheric, Oceanic and Hydrologic Applications*, DOI 10.1007/978-3-540-71056-1_21, © Springer-Verlag Berlin Heidelberg 2009

the available energy incident on the land surface into latent and sensible heat fluxes. Through this control, soil moisture impacts local weather parameters, including the boundary layer height and cloud coverage (Betts and Ball 1998). Moreover, root zone soil moisture exhibits memory on weekly to monthly time scales (Entin et al. 2000). Accurate initialization of root zone soil moisture may therefore contribute to enhanced sub-seasonal prediction of mid-latitude summer precipitation over land (Dirmeyer 2003, Koster et al. 2004).

Snow influences climate through its higher albedo and emissivity, its ability to insulate the soil, and its utilization of energy for snowmelt and sublimation. The role of snow in climate variations is well documented in numerous observational and modeling studies. For example, early season snow cover anomalies in Eurasia have been found to modulate wintertime climate variability in northern hemisphere mid-latitudes (Cohen and Entekhabi 1999). Land surface temperature (LST) is at the heart of the surface energy balance and therefore a key variable in weather and climate models. It has a strong influence on the latent and sensible heat fluxes to the atmosphere through which it affects the planetary boundary layer and atmospheric convection. LST also plays an important role in the use of atmospheric remote sensing observations in atmospheric data assimilation systems. Accurate LST specification is therefore critical to improving estimates of the surface water, energy, and radiation balance, which in turn are critical to improving forecast accuracy.

Land surface models driven with observation-based meteorological forcing data (precipitation, radiation, air temperature and humidity, etc.) offer estimates of global land surface conditions (Rodell et al. 2004). Errors in the forcing fields, however, along with the imperfect parameterization of land-atmosphere interactions can lead to considerable drifts in modeled land surface states, both in coupled land-atmosphere models and in uncoupled land models. Satellite remote sensing can provide complementary important information about land surface conditions, including surface soil moisture, snow water equivalent, snow cover, and LST.

Land data assimilation systems combine the modeled land surface fields with observational estimates and produce dynamically consistent, spatially complete and temporally continuous estimates of global land surface conditions based on both sources of information. Besides improving estimates of land surface conditions, land data assimilation has the potential to improve the modeling and prediction of the atmospheric circulation response. The land surface assimilation estimates can be used within atmospheric assimilation systems and also for the initialization of global short-term climate forecasts. Through such use land data assimilation systems offer a unique validation and monitoring perspective because the satellite-based land surface data are continually confronted with independent observations and model estimates. Land data assimilation systems can also be used to establish measurement requirements for future land surface satellite missions.

Operational Numerical Weather Prediction (NWP) systems have long demonstrated that soil moisture and snow conditions impact weather variables in short- and medium-range weather forecasts. Several operational centers initialize their NWP systems based on some form of land surface assimilation. For example, snow observations are assimilated at the Environmental Modeling Center at National Center for Environmental Prediction (NCEP), the Canadian Meteorological Centre (CMC; Brasnett 1999), and the European Centre for Medium-Range Weather Forecasting (ECMWF; Drusch et al. 2004). Furthermore, at ECMWF, the CMC, and the German and French weather services, observations of screen-level air temperature and humidity are used to update soil moisture under certain atmospheric conditions (Mahfouf 1991), leading to improvements in subsequent predictions of screen-level conditions. Even though there are small improvements in weather forecasts in these systems, the quality of the soil moisture estimates is not improved because soil moisture is treated as a tuning parameter (Drusch and Viterbo 2007).

There has been considerable progress in the methodological development and application of land data assimilation algorithms (Andreadis and Lettenmaier 2005, Bosilovich et al. 2007, Crow and Van Loon 2006, Crow and Wood 2003, De Lannoy et al. 2007, Dong et al. 2007, Drusch 2007, Dunne and Entekhabi 2006, Kumar et al. 2008, Margulis et al. 2002, Pan and Wood 2006, Reichle and Koster 2003, 2005, Reichle et al. 2001, 2002a, 2002b, 2007, 2008a, b, Rodell and Houser 2004, Seuffert et al. 2003, Slater and Clark 2006, Walker and Houser 2005, Zaitchik et al. 2008, Zhou et al. 2006). Ensemble-based Kalman filtering and smoothing algorithms have emerged as the most common and most promising method for land data assimilation. Research at the NASA Global Modeling and Assimilation Office (GMAO) contributed significantly to this progress through the development of an ensemble-based GMAO land data assimilation system (hereinafter GMAO-LDAS). In the present paper, we describe the milestones of the system's development and its application to the assimilation of satellite retrievals, along with related GMAO land data assimilation activities.

## 2 The Ensemble-Based GMAO Land Data Assimilation System

In this section, we describe the GMAO-LDAS. The system is based on the ensemble Kalman filter (EnKF, Sect. 2.1.) applied to the NASA Catchment land surface model (Sect. 2.2.). Important aspects of the system are the characterization of model and observation error covariances (Sect. 2.3.) and the adaptive filtering module (Sect. 2.4.).

### 2.1 The Ensemble Kalman Filter

The EnKF is a Monte-Carlo variant of the Kalman filter (Evensen 2003) and works sequentially by performing in turn a model forecast and a data assimilation update. The key feature of the EnKF is that a small ensemble of model trajectories captures the relevant parts of the error structure. Each member of the ensemble experiences perturbed instances of the observed forcing fields (representing errors in the forcing data) and is also subject to randomly generated noise that is added to the model parameters and prognostic variables (representing errors in model physics and parameters). The error covariance matrices that are required for the data assimilation update can then be diagnosed from the spread of the ensemble at the update time. The EnKF was demonstrated for land data assimilation in synthetic studies where it compared well to the weak-constraint variational "representer" method (Reichle et al. 2002a) and favorably to the Extended Kalman filter (Reichle et al. 2002b). Overall, the EnKF is flexible in its treatment of errors in model dynamics and parameters and is very suitable for the modestly nonlinear and intermittent character of land surface processes.

The EnKF works sequentially by performing in turn a model forecast and a data assimilation update. Formally, the forecast step for ensemble member i can be written as

$$x_{t,i}^- = f(x_{t-1,i}^+, q_{t,i}) \tag{1}$$

where $x_{t,i}^-$ and $x_{t-1,i}^+$ are the forecast (denoted with $^-$) and analysis (denoted with $^+$) state vectors at times t and t-1, respectively. The model error (or perturbation vector) is denoted with $q_{t,i}$ and its covariance with $Q_t$. The data assimilation update produces the analyzed state vector $x_{t,i}^+$ at time t and can be written as

$$x_{t,i}^+ = x_{t,i}^- + K_t(y_{t,i} - H_t x_{t,i}^-) \tag{2}$$

where $y_{t,i}$ denotes the observation vector (suitably perturbed) and $H_t$ is the observation operator (which is assumed linear for ease of notation, although this requirement can be relaxed). The Kalman gain matrix $K_t$ is given by

$$K_t = P_t H_t^T (H_t P_t H_t^T + R_t)^{-1} \tag{3}$$

where $P_t$ is the forecast error covariance (diagnosed from the ensemble $x_{t,i}^-$), $R_t$ is the observation error covariance, and superscript T denotes the matrix transpose. Simply put, the Kalman gain $K_t$ represents the relative weights given to the model forecast and the observations based on their respective uncertainties and the error correlations between different elements of the state vector. If the system is linear, if its model and observation error characteristics satisfy certain assumptions (including Gaussian, white, and uncorrelated noise), and if the input error parameters are correctly specified, the Kalman gain of equation (3) is optimal in the sense of minimum estimation error variance. Unfortunately, these optimal conditions are never simultaneously met in land data assimilation applications.

For large modeling domains, the EnKF state vector may contain a large number of variables and the straightforward solution of the (matrix) update equation (3) may not be computationally affordable. Moreover, practical limits on the size of the ensemble typically lead to spurious sample correlations at large separation distances and require a suitable localization of the covariance matrices. Since error correlation lengths in land surface states are typically much shorter than the extent of continental-scale domains, the update equation can be solved in parallel (and thus very efficiently) for subsets of the state vector (Reichle and Koster 2003). In the limit, the EnKF is often applied separately to each land model computational unit, an approximation that is often called "one-dimensional assimilation".

## 2.2 The NASA Catchment Land Surface Model

The land surface model component of the GMAO-LDAS is the NASA Catchment land surface model (hereinafter Catchment model; Koster et al. 2000, Ducharne et al. 2000). The Catchment model's basic computational unit is the hydrological catchment (or watershed). In each catchment, the vertical profile of soil moisture is determined by the equilibrium soil moisture profile from the surface to the water table (defined by a balance of gravity and capillary forces) and by two additional variables that describe deviations from the equilibrium profile in a 1 m root zone layer and in a 2 cm surface layer, respectively. Unlike traditional, layer-based models, the Catchment model includes an explicit treatment of the spatial variation of soil water and water table depth within each hydrological catchment based on the statistics of the catchment topography. Soil heat processes are modeled with a six-layer heat diffusion model. Snow processes are modeled with a three-layer state-of-the-art snow model component. The time step for Catchment model integration is typically 20 minutes.

## 2.3 Error Covariance Modeling

In the GMAO-LDAS, a customized random field generator based on Fast Fourier Transform methods generates time series of cross-correlated perturbation fields. These perturbations are applied to selected meteorological forcing inputs and Catchment model prognostic variables. Collectively, these perturbations allow us to maintain an ensemble of land surface conditions that represents the uncertainty in the land surface states. Depending on the variable, normally distributed additive perturbations or log-normally distributed multiplicative perturbations are applied. The ensemble mean for all perturbations can be constrained to zero for additive perturbations and to one for multiplicative perturbations. Moreover, time series correlations can be imposed via a first-order auto-regressive model (AR(1)) for all fields. Perturbation fields may also be spatially correlated if required by the application (Reichle and Koster 2003, Zaitchik et al. 2008).

Model and forcing errors depend on the application at hand and are extremely difficult to quantify at the global scale. The perturbation parameters used in the soil moisture assimilation study by Reichle et al. (2007; discussed here in Sect. 3.2.), illustrated in Table 1, are largely based on experience. They are supported by earlier studies where model and forcing error parameters were calibrated in twin experiments (Reichle et al. 2002b, Reichle and Koster 2003). For soil moisture, the dominant forcing inputs are precipitation and radiation, and we chose to limit perturbations to these forcing fields. Imperfect model parameters and imperfect physical

**Table 1** Sample parameters for perturbations to meteorological forcing inputs and Catchment model prognostic variables (catchment deficit and surface excess). From Reichle et al. (2007) with minor correction regarding the standard deviations of catchment deficit and surface excess

| Perturbation | Additive (A) or Multiplicative (M)? | Standard deviation | AR(1) time series correlation scale | Cross-correlation with perturbations in | | |
| --- | --- | --- | --- | --- | --- | --- |
| | | | | P | SW | LW |
| Precipitation (P) | M | 0.5 | 1 day | 1.0 | −0.8 | 0.5 |
| Downward shortwave (SW) | M | 0.3 | 1 day | −0.8 | 1.0 | −0.5 |
| Downward longwave (LW) | A | 50 W m$^{-2}$ | 1 day | 0.5 | −0.5 | 1.0 |
| Catchment deficit | A | 0.05 mm | 3 h | 0.0 | 0.0 | 0.0 |
| Surface excess | A | 0.02 mm | 3 h | 0.0 | 0.0 | 0.0 |

parameterizations contribute to model errors. Such errors are represented through direct perturbations to the surface excess and catchment deficit model prognostic variables.

Cross-correlations are only imposed on perturbations of the precipitation and radiation fields. At hourly and daily time scales, the meteorological forcing fields are ultimately based on output from atmospheric modeling and analysis systems and not on direct observations of surface precipitation and radiation. The cross-correlations are therefore motivated by the assumption that the atmospheric forcing fields represent a realistic balance between radiation, clouds, and precipitation. Under that assumption, a positive perturbation to the downward shortwave radiation tends to be associated with negative perturbations to the longwave radiation and the precipitation, and vice versa. The numbers of Table 1 for the imposed cross-correlation coefficients are motivated by an analysis of the cross-correlations between precipitation and radiation in the baseline forcing data sets from the Global Soil Wetness Project 2 (Dirmeyer et al. 2006), and by the assumption that errors behave like the fields themselves.

Generally, data assimilation products are sensitive to input observation and model error variances and, for very poor input error parameters, may even be worse than model estimates without data assimilation. Figure 1 shows an example from a suite of experiments with the GMAO-LDAS in which synthetic surface soil moisture observations are assimilated (Reichle et al. 2008b). Each assimilation experiment has a unique set of input error parameters that leads to a unique pair of scalars: the (space and time) average forecast error variance ($P_0$) and the input observation error variance ($R_0$) for surface soil moisture. We can thus plot two-dimensional surfaces of filter performance as a function of sqrt($P_0$) and sqrt($R_0$). Figure 1a, for example, shows one such surface with the performance measure being the RMSE of surface soil moisture estimates from the (non-adaptive) EnKF. Each of the 30 plus signs in the figure indicates the performance of a 19-year assimilation integration over the entire Red-Arkansas domain.

**Fig. 1** RMSE of surface soil moisture for the (**a**) standard and (**b**) adaptive EnKF as a function of filter input error parameters: (ordinate) forecast and (abscissa) observation error standard deviation. Units are $m^3 m^{-3}$. Each plus sign indicates the result of a 19-year assimilation integration over the entire Red-Arkansas domain. The circled plus sign indicates the experiment that uses the true input error parameters for the assimilation. Thick gray lines indicate RMSE of the open loop (no assimilation) integration. Adapted from (Reichle et al. 2008b)

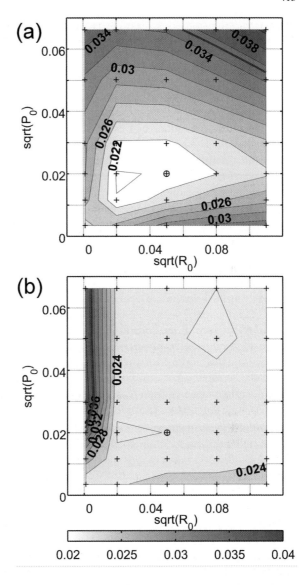

Figure 1a illustrates that the estimation error in surface soil moisture is smallest near the experiment that uses the true model and observation error inputs. The minimum estimation error is around $0.02\,m^3 m^{-3}$, down from the open loop (no assimilation) value of $0.035\,m^3 m^{-3}$. The estimation error increases as the input error parameters deviate from their true values. Figure 1a also indicates where the estimation error surface intersects the open loop error. For grossly overestimated model and observation error variances, the assimilation estimates of surface soil moisture are in fact worse than the open loop estimates. Ultimately, the success of the assimilation (measured through independent validation) suggests whether the selected input error parameters are acceptable.

## 2.4 The Adaptive EnKF

Adaptive filtering methods can assist with the estimation of the filter's input error parameters. The central idea behind adaptive filtering methods is that internal diagnostics of the assimilation system should be consistent with the values that are expected from input parameters provided to the data assimilation system. The most commonly used diagnostics for adaptive filtering are based on the observation-minus-forecast residuals or innovations (computed here as $v_t \equiv E\{y_{t,i} - H_t x_{t,i}^-\}$, where $E\{\cdot\}$ is the ensemble mean operator). For a linear system operating under optimal conditions, the lagged innovations covariance is

$$E[v_t v_{t-k}^T] = \delta_{k,0}(H_t P_t H_t^T + R_t) \tag{4}$$

where $E[\cdot]$ is the expectation operator and $\delta_{k,0}$ is the Kronecker delta. Equation (4) implies that the innovations sequence is uncorrelated in time and that its covariance is equal to the sum of the forecast error covariance $H_t P_t H_t^T$ (in observation space) and the observation error covariance $R_t$. Now recall that the forecast error covariance $P$ depends on the model error covariance $Q$. If the innovations show less spread than expected, the input error covariances ($Q$ and/or $R$) are too large, and vice versa. Such information can be used for adaptive tuning of $Q$ and/or $R$.

Alternative diagnostics are based on the analysis departures $w_t \equiv E\{y_{t,i} - H_t x_{t,i}^+\}$ and the (observation space) analysis increments $u_t \equiv E\{H_t(x_{t,i}^+ - x_{t,i}^-)\}$. For linear systems operating under optimal conditions we have (Desroziers et al. 2005)

$$E[u_t v_t^T] = H_t P_t H_t^T \tag{5}$$

$$E[w_t v_t^T] = R_t \tag{6}$$

Equations (5) and (6) are attractive because they suggest a simple way of estimating the model and observation error covariances separately by tuning the input error parameters such that the output diagnostics on the left-hand-side of (5) and (6) match the right-hand-side error covariances. This approach is the basis of the proof-of-concept example of Desroziers et al. (2005) and it is also the approach currently employed in the GMAO-LDAS. For details of the GMAO implementation consult (Reichle et al. 2008b).

An example of the benefits of the adaptive module is given in Figure 1b (Reichle et al. 2008b). The adaptive estimation of input error parameters leads to improved estimates of surface soil moisture regardless of initial error estimates, except for the case of severe underestimation of the input observation error variance. The poor performance in this special case is due to technicalities in the implementation of the adaptive module and can easily be avoided in applications.

# 3 Examples of Hydrologic Data Assimilation with the GMAO-LDAS

In this section, we present examples of the application of the GMAO-LDAS to hydrologic data assimilation problems, including the assimilation of satellite-based observations of surface soil moisture (Sect. 3.2) and terrestrial water storage (Sect. 3.3). We also show how the GMAO-LDAS can be used to aid with the design of a new soil moisture observing system (Sect. 3.4.) and to investigate the advantages and disadvantages of assimilating different types of snow observations (Sect. 3.5). Prior to discussing the actual assimilation studies, we demonstrate how we deal with the specific problem of model and observational bias in land surface models and observations (Sect. 3.1.).

## 3.1 Soil Moisture Bias and Scaling

Large differences have been identified between the temporal moments of satellite and model soil moisture (Reichle et al. 2004, 2007). Figure 2 shows an example of time-average surface soil moisture (for the top 1–2 cm of the soil column) from the NASA Advanced Microwave Scanning Radiometer for the Earth Observing System (AMSR-E) and a corresponding multi-year integration of the NASA Catchment land surface model. Large differences also exist in the higher temporal moments (standard deviation and skewness, not shown). Similarly striking climatological

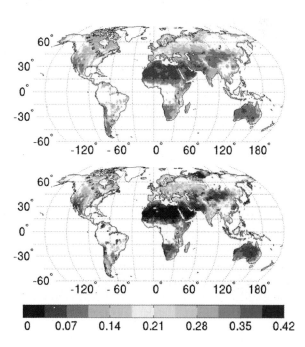

**Fig. 2** Time-average (Jun. 2002–May 2006) surface soil moisture from (*top*) AMSR-E and (*bottom*) the Catchment model ($m^3 m^{-3}$). Adapted from (Reichle et al. 2007)

differences are typical between estimates from different land model integrations (not shown) because land surface models differ markedly in their parameterizations and because surface meteorological forcing data sets have their own unique climatologies. Likewise, different observational products exhibit significant differences (not shown; see (Reichle et al. 2007)) because different retrieval data sets may have different sensor observation depths, different sensor footprints, different sensor cross-over times, and different parameterizations of the retrieval algorithm. At this time, the true climatology of global surface soil moisture can be considered unknown (Reichle et al. 2004).

Because the standard EnKF update Eq. (2) of the assimilation system is only designed to address short-term "random" errors, the climatological differences (or biases) need to be addressed separately within the system. This can be accomplished by scaling the satellite observations to the model's climatology so that the cumulative distribution functions (cdf) of the satellite soil moisture and the model soil moisture match (Reichle and Koster 2004). For weather and climate forecast initialization, knowledge of soil moisture anomalies is, in any case, more important than knowledge of absolute soil moisture. The scaling is performed prior to data assimilation and must be done separately for each observational product because of the differences in the observation characteristics.

For land surface problems, the cdf's are typically estimated for each location from the entire time series available at that location. Note that accurate cdf estimation typically requires a long record of satellite data. Reichle and Koster (2004) demonstrate that using spatial sampling within a suitable neighboring region yields local statistics based on a one-year satellite record that are a good approximation to those that would be derived from a much longer time series. This result increases the usefulness of relatively short satellite data records. Moreover, satisfactory data assimilation results can be obtained without estimating a different cdf for each season (Reichle et al. 2007) even though Drusch et al. (2005) noted that the scaling parameters may vary seasonally.

Alternatively, model bias can be estimated explicitly and dynamically within the cycling data assimilation system. This approach was used by Bosilovich et al. (2007) for LST assimilation (Sect. 4.) and for a field-scale soil moisture problem by De Lannoy et al. (2007). That method's fundamental assumption of unbiased observations, however, is difficult to justify for satellite retrievals of global-scale surface soil moisture.

## 3.2 Assimilation of Satellite Retrievals of Surface Soil Moisture

Reichle et al. (2007) assimilated surface soil moisture retrievals from AMSR-E for June 2002-May 2006 into the GMAO-LDAS. Because of the scaling approach and the basic lack of knowledge regarding soil moisture "truth" (Sect. 3.1), skill cannot usefully be measured in terms of RMS error. Instead, skill is measured in terms of the correlation coefficient R between the time series of the various estimates (expressed as daily anomalies relative to their seasonal climatologies) and in situ

**Table 2** Average time series correlation coefficient R with in situ surface and root zone soil moisture anomalies (sf and rz, respectively) for estimates from satellite (AMSR-E), the Catchment model, and assimilation, all with 95% confidence intervals. N denotes the number of stations available for validation. Adapted from (Reichle et al. 2007)

| Layer | N | Correlation Coefficient R With In Situ Data (dimensionless) for | | |
|---|---|---|---|---|
| | | Satellite | Model | Assim. |
| Sf | 23 | $0.38 \pm 0.02$ | $0.43 \pm 0.02$ | $0.50 \pm 0.02$ |
| Rz | 22 | n/a | $0.40 \pm 0.02$ | $0.46 \pm 0.02$ |

measurements (also expressed as anomalies), averaged over all locations with sufficient in situ data. Table 2 shows that the surface soil moisture from the model ($R = 0.43$) is somewhat more skillful than the satellite retrievals ($R = 0.38$). The skill increases to $R = 0.50$ for the surface soil moisture assimilation product. This increase in skill over that of the satellite or model data alone is highly statistically significant, as indicated by the 95% confidence intervals that are given for each R estimate. Merging the satellite retrievals of surface soil moisture into the model through data assimilation also leads to an increase in skill for the root zone soil moisture anomalies to $R = 0.46$, compared to a model skill of $R = 0.40$. This increase in R is again highly statistically significant. The assimilation system acts to propagate the surface retrieval information into deeper soil layers, giving the retrievals an otherwise unobtainable relevance to such applications as the initialization of weather and seasonal climate forecasts.

The increase in time series correlations with in situ data after assimilation suggests two important things: (i) the satellite and model data contain independent information, and (ii) the assimilation algorithm can successfully combine the independent information into a single, superior data set. This success depends on a large

**Table 3** Evaluation of groundwater estimates from *(Model)* Catchment model integrations without assimilation and *(Assim.)* model integrations with GRACE assimilation against independent groundwater observations. Correlation coefficient R (dimensionless) and RMSE (mm) are calculated with respect to daily average groundwater storage based on observations from 58 piezometers. Bold (italic) fonts indicate a significant increase in R through data assimilation at the 5% (10%) significance level (based on Student's T test). Adapted from (Zaitchik et al. 2008)

| | Catchment Model (No assimilation) | | Grace assimilation | | RMSE Reduction: $1 - \dfrac{RMSE_{Assim.}}{RMSE_{Model}}$ |
|---|---|---|---|---|---|
| | $R_{Model}$ | $RMSE_{Model}$ | $R_{Assim.}$ | $RMSE_{Assim.}$ | |
| Mississippi | 0.59 | 23.5 mm | **0.70** | 18.5 mm | 21% |
| Ohio-Tennessee | 0.78 | 62.8 mm | **0.82** | 40.4 mm | 36% |
| Upper Mississippi | 0.29 | 42.6 mm | 0.27 | 39.6 mm | 7% |
| Red-Arkansas/ Lower Mississippi | 0.69 | 30.9 mm | *0.72* | 26.4 mm | 15% |
| Missouri | 0.41 | 24.5 mm | **0.66** | 19.7 mm | 20% |

number of factors, in particular for root zone soil moisture. The model, for example, must accurately describe the propagation of the surface information into the deeper soil. Also, the assimilation system's model error parameters that co-determine the strength of the coupling between the surface and the root zone must be realistic. The improvement in skill is further limited by the modest skill of the satellite data relative to the model data, and more generally by the large errors in the satellite, model, and in situ data. Note that errors in the in situ data have special relevance for the interpretation of Table 2. Even if the satellite retrievals, the model, and the assimilation system were all perfect, the correlation coefficients could still be much less than 1 due to errors in the in situ data, including errors associated with mismatches in scale because the in situ data are point scale measurements. In other words, the seemingly modest increase in R could be quite large relative to the maximum increase possible given the imperfect validation data, and the impact of the assimilation may be more positive than already implied by the numbers in Table 2.

## 3.3 Terrestrial Water Storage Assimilation

The Gravity Recovery and Climate Experiment (GRACE) satellite mission provides unprecedented observations of variations in terrestrial water storage (TWS), albeit at low spatial ($> 10^5 \text{ km}^2$) and temporal (monthly) resolutions. Depending on topographic and climatologic conditions, TWS variability may be dominated by ground water, soil moisture, surface water, and/or snow. Zaitchik et al. (2008) assimilated terrestrial water storage (TWS) data from GRACE for the Mississippi River basin into the GMAO-LDAS. The Catchment model that is part of the GMAO-LDAS is well-suited for GRACE assimilation because of its implicit representation of the water table and the fact that groundwater is a dynamic component of TWS.

Because of the temporally integrated nature of the GRACE observations, the EnKF component of the GMAO-LDAS was modified to work effectively as an ensemble smoother. GRACE observations were not scaled for assimilation except that the GRACE-derived TWS anomalies were converted to absolute TWS values by adding the corresponding time-mean TWS from a Catchment model simulation. The ensemble perturbations that were added to the model forcing and prognostic variables were generated with a horizontal correlation scale of 2 degrees, which roughly represents error scales in global-scale precipitation fields (Reichle and Koster 2003). For snow-free catchments, assimilation increments were applied to the catchment deficit variable. For snow-covered catchments, positive increments were applied entirely to snow. Negative increments were applied first to snow and then, if all snow was removed, to the catchment deficit.

The GMAO-LDAS separates the contributions of GRACE observations into individual TWS components and downscales the GRACE observations to scales ($\sim 10^3 \text{ km}^2$) typical of global land surface integrations. Assimilation products include catchment-scale groundwater, root zone soil moisture, surface heat fluxes, and runoff. The spatial resolution of the assimilation products is much higher than

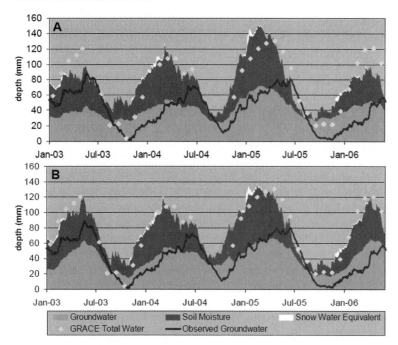

**Fig. 3** Groundwater, soil moisture, and snow water equivalent for the Mississippi river basin for estimates from (**A**) the model without assimilation and (**B**) the GRACE assimilation integration. Also shown are (*solid line*) area averaged daily in situ groundwater observations and (*diamonds*) monthly GRACE-derived TWS anomalies. Note that GRACE assimilation improves agreement of the groundwater estimates with in situ data. GRACE and modeled TWS are adjusted to a common mean, as are observed and modeled groundwater. From (Zaitchik et al. 2008)

that of GRACE observations alone, making the results useful for water resources and forecasting applications. Figure 3 shows that the groundwater time series from the GRACE assimilation integration resembles in situ estimates more closely than model estimates alone. Table 3 quantifies the agreement between estimated and observed groundwater for the Mississippi as a whole and for its four sub-basins and demonstrates that GRACE data assimilation significantly improved estimates of the amplitude and phase of the seasonal cycle of groundwater. In all cases, the GRACE assimilation integration exhibited smaller RMS errors than the model integration without assimilation, with RMS error reductions ranging from 7 to 36%.

Assimilation of GRACE observations also produced improved estimates of hydrologic variability at the sub-observation scale and a small increase in correlation between runoff estimates and gauged river flow in the majority of test watersheds (not shown). Evaluation of the assimilation results at scales finer than the GRACE data revealed no degradation of model performance due to the assimilation of the coarse GRACE observation. The results demonstrate that – through data assimilation – coarse resolution, vertically integrated TWS anomalies from GRACE can be spatially and temporally disaggregated and attributed to different components of the snow-soil-aquifer column in a physically meaningful way. The results emphasize the potential of GRACE observations to improve the accuracy of hydrologic model output, which will benefit water cycle science and water resources applications.

## 3.4 Observing System Design

For the design of new satellite missions it is critical to understand just how uncertain satellite retrievals can be and still be useful. Consider, for example, that a mission assimilation product will have some target accuracy requirement. For a given level of model skill, a specific level of retrieval skill would be needed to bring the merged product to the target accuracy. The required skill level for the retrievals would undoubtedly increase with a decrease in the skill of the raw model product. Knowledge of such retrieval skill requirements, for example, is directly relevant to the planning of the L-band (1.4 GHz) Soil Moisture Active-Passive (SMAP) mission recommended by the National Academy of Sciences for launch in the 2010-2013 timeframe (Space Studies Board 2007).

We designed an Observing System Simulation Experiment (OSSE) that determines the contribution of surface soil moisture retrievals to the skill of land assimilation products (soil moisture and evapotranspiration) as a function of retrieval and land model skill (Reichle et al. 2008a). The OSSE consists of a suite of synthetic data assimilation experiments with the GMAO-LDAS based on integrations of two distinct land models, one representing "truth", and the other representing our flawed ability to model the true processes. Skill is again measured in terms of the correlation coefficient R between the time series of the various estimates (expressed as anomalies relative to their seasonal climatologies) and the assumed (synthetic) truth. The experiment is based on long-term (1981–2000) integrations over the Red-Arkansas river basin in the United States.

Each assimilation experiment is a unique combination of a retrieval dataset (with a certain level of skill, measured in terms of R) and a model scenario (with its own level of skill). We can thus plot two-dimensional surfaces of skill in the data assimilation products as a function of retrieval and model skill. Figure 4a, for example, shows the two dimensional surface corresponding to the surface soil moisture product. As expected, the skill of the assimilation product generally increases with the skill of the model and the skill of the retrievals, for both surface (Fig. 4a) and root zone (Fig. 4b) soil moisture estimates. Except for very low model skill, the contour lines are more closely aligned with lines of constant model skill; that is, the skill of the assimilation product is more sensitive to model skill than to retrieval skill.

Figure 4 also shows skill improvement through data assimilation, defined as the skill of the assimilation product minus the skill of the model estimates (without assimilation). Specifically, Fig. 4c and d show, for a given level of accuracy in the stand-alone model product, how much information can be added to the soil moisture products through assimilation of satellite retrievals of surface soil moisture with a

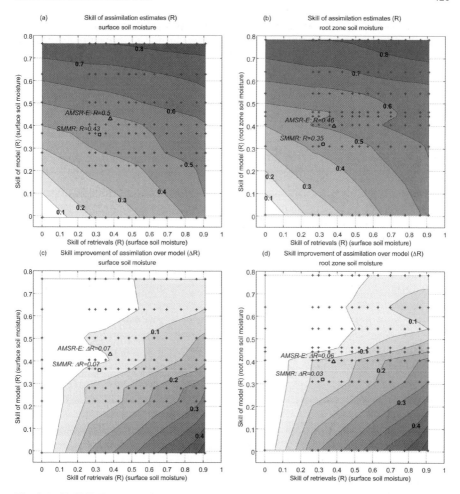

**Fig. 4** (**a, b**) Skill (R) and (**c, d**) skill improvement (ΔR) of assimilation product for (**a, c**) surface and (**b, d**) root zone soil moisture as a function of the (ordinate) model and (abscissa) retrieval skill. Skill improvement is defined as skill of assimilation product minus skill of model estimates. Each plus sign indicates the result of one 19-year assimilation integration over the entire Red-Arkansas domain. Also shown are results from Reichle et al. (2007) for (*triangle*) AMSR-E and (*square*) SMMR. Adapted from (Reichle et al. 2008a)

given uncertainty. Note that the skill of the surface and root zone soil moisture assimilation products always exceeds that of the model. As expected, the improvements in R through assimilation increase with increasing retrieval skill and decrease with increasing model skill. Perhaps most importantly, though, is that even retrievals of low quality contribute some information to the assimilation product, particularly if model skill is modest.

We can also compare the contoured skill levels of Fig. 4 with those of Table 2. From the contours of Fig. 4a we expect that for retrievals with $R = 0.38$ and model estimates with $R = 0.43$, the surface soil moisture assimilation product would have skill of about $R = 0.50$, which is indeed consistent with the AMSR-E result of Table 2. For root zone soil moisture, Table 2 shows that the assimilation of AMSR-E surface soil moisture retrievals also yields improvements, though these improvements fall somewhat short of those suggested by Fig. 4b. Possible explanations include (i) the imperfect translation of information from the surface layer to the root zone in the data assimilation system and (ii) the fact that the in situ data used for validation of the AMSR-E result are themselves far from perfect (unlike the perfectly known truth of the synthetic experiment). Figure 4 also includes the assimilation results of Reichle and Koster (2005) for retrievals from the historic Scanning Multichannel Microwave Radiometer (SMMR), which are similarly consistent with the contours.

A major component of the OSSE is the adaptive component of the GMAO-LDAS. This component estimates experiment-specific input error covariances that enable near-optimal assimilation performance and permit objective comparisons across experiments, as in the surface plots of Fig. 4. Plots of skill improvement similar to Fig. 4 can readily be obtained for estimates of evapotranspiration and other products from the data assimilation system. The general framework permits detailed and comprehensive error budget analyses for data assimilation products. The framework can be used, for example, to study specific trade-offs in sensor design or ancillary data requirements, assessing the impact of each on the quality of the end-product that will be used in science and applications. As the focus on data assimilation products grows in future land surface satellite missions, the OSSE framework presents an important end-to-end tool for mission planning and uncertainty assessment.

## 3.5 Snow Data Assimilation

Observations of fractional snow cover (or snow cover area; SCA) and snow water equivalent (SWE) are available from a variety of remote sensing platforms. SCA observations are typically obtained from visible or infrared satellite sensors at high spatial resolutions and with good accuracy, but are limited to cloud-free conditions. In contrast, passive microwave sensors can provide quantitative observations of SWE under cloudy and nighttime conditions, albeit at coarser spatial resolution and with limited accuracy. Moreover, SWE retrievals are not sensitive to thin snow packs and saturate for very thick snow packs.

We investigated the relative benefits of SCA and SWE observations in a synthetic data assimilation experiment (Kumar et al. 2008). Synthetic "true" snow data from the Catchment model are assimilated into the Noah land surface model over North America for the 2003/2004 snow season. SCA observations are assimilated with the rule-based approach of Rodell and Houser (2004). The rule-based approach for SCA assimilation corrects the SWE fields when there is a mismatch between the observations and the model prediction. If, for example, a SCA observation indicates that snow is absent but the model estimates that snow is present, the snow is removed from the model states. If, on the other hand, a SCA observation indicates that snow is present but snow is absent in the model, then a nominal amount of SWE is added to the model states. Based on the snow density in the model, the snow depth fields are also updated. This empirical approach is used since the SCA observations only provide information on the presence or absence of snow and do not provide a direct measure of snow mass. By contrast, SWE observations are assimilated with the EnKF component of the GMAO-LDAS. Both the SWE and the snow depth fields are corrected using the EnKF update equations based on the sampled ensemble covariance of the observations and the model prognostic variables.

The improvement through data assimilation is measured as the RMS error of the open loop (no assimilation) estimates minus the RMS error of the assimilation estimates, where the error is relative to the (synthetic) truth data. Figure 5 shows the comparison of the time averaged improvement metric for the SWE and snow depth fields from assimilating SCA and SWE observations. The assimilation of SCA produces little change when compared to the open loop integration. By design, assimilation of SCA is only effective when there is a transition between snow-free and snow-covered conditions. Improvements through assimilation of SCA are therefore limited to regions where snow cover changes frequently throughout the season. The SWE fields from the assimilation of SWE observations show positive improve-

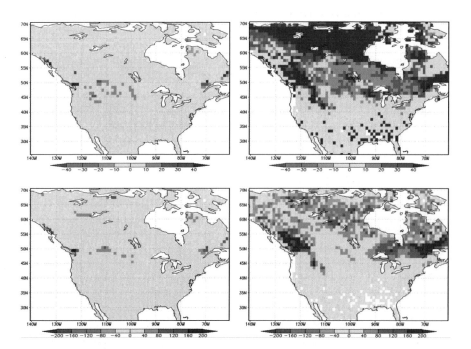

**Fig. 5** Improvements in estimates of (*top*) SWE and (*bottom*) snow depth for assimilation of (*left*) SCA and (*right*) SWE observations. Improvements given as time averaged RMS error of open loop (no assimilation) estimates minus RMS error of assimilation estimates. Blue colors show improved, red tones show deteriorated estimates. Units are mm. Adapted from (Kumar et al. 2008)

ments throughout the domain. The snow depth fields from the assimilation of SWE, however, show negative improvements in large areas, mostly at high latitudes. The discrepancies in the updated snow depth fields in the assimilation estimates is due to a mismatch in the relationship between SWE and the snow density in the Catchment and Noah land surface models. Such a mismatch is representative of likely errors in either model when compared to actual snow fields. The discrepancies may also indicate a lack of snow density information in the assimilated SWE observations.

# 4 Land Surface Temperature Assimilation

Satellite retrievals of LST (also referred to as "skin temperature") are available from a variety of polar orbiting and geostationary platforms. Assimilating such LST retrievals into a land surface model (that is either driven by observed meteorological forcing data or coupled to an atmospheric model) should improve estimates of land surface conditions. Similar to surface soil moisture, however, LST data from retrievals and models typically exhibit very different climatologies for a variety of reasons. For example, LST modeling is fraught with numerical stability problems because in nature the effective heat capacity associated with LST is very small. Land modelers are thus forced to either approximate the heat capacity as zero (and consequently treat LST as an essentially diagnostic variable), or to use a surface temperature prognostic variable that represents more than just a thin layer. LST retrievals from satellite, on the other hand, are strongly affected by the look-angle, a problem that is particularly acute with geostationary satellites. Moreover, satellites observe a radiometric temperature that is difficult to compare to the model's LST because the emissivity of the land surface is not well known. Finally, LST retrievals from different satellite platforms already exhibit different climatologies because sensor characteristics differ. These bias problems are compounded by the strong seasonal and diurnal cycles of LST.

At the GMAO, Bosilovich et al. (2007) developed an algorithm for LST assimilation into a global, coupled land-atmosphere data assimilation system (Bloom et al. 2005). The LST assimilation module is based on a 3D-VAR-like method that uses fixed background error covariances and includes dynamic model bias estimation. An incremental bias correction term is introduced into the model's surface energy budget. In its simplest form, the algorithm estimates and corrects a constant time mean bias for each grid point; additional benefits are attained with a refined version of the algorithm that allows for a correction of the mean diurnal cycle.

The system was tested with the assimilation of LST retrievals from the International Satellite Cloud Climatology Project and validated against independent near-surface air temperature observations. Figure 6 shows that LST assimilation improves estimates of 2m air temperature, both in the mean and the variability. Neglecting the diurnal cycle of the LST bias caused degradation of the diurnal amplitude of background model air temperature in many regions (not shown). In situ measurements of energy fluxes at several locations were used to inspect the surface energy budget

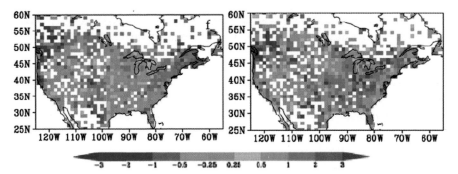

**Fig. 6** Improvements in (*left*) mean and (*right*) amplitude of 2m air temperature through assimilation of ISCCP LST for July 2001. Red colors show improved, blue tones show deteriorated estimates. Units are Kelvin. Adapted from (Bosilovich et al. 2007)

more closely (not shown). In general, sensible heat flux is improved with the surface temperature assimilation, and two locations show a reduction of bias by as much as $30\,Wm^{-2}$. At the Rondonia station in Amazonia, the LST assimilation results in a markedly improved Bowen ratio. At many stations, however, the bias in the monthly latent heat flux increased slightly.

The method of Bosilovich et al. (2007) assumes that the LST observations are unbiased and that the bias is solely due to the model. As an alternative strategy, we applied the rescaling methodology of Sect. 3.1 to LST assimilation with the GMAO-LDAS (Reichle et al. 2006). Preliminary results indicate that not scaling the LST retrievals prior to data assimilation creates serious imbalances in the model-generated mass and energy fluxes and leads to entirely unrealistic land surface fluxes. This problem can be avoided through rescaling.

## 5 Conclusions and Future Directions

Much has been accomplished with the development and application of the GMAO-LDAS over the past few years. The general ensemble-based framework of the system has been established and demonstrated with the assimilation of satellite-based land surface observations. In particular, the assimilation of surface soil moisture retrievals from AMSR-E and SMMR provided improved estimates of surface and root zone soil moisture when compared to the model and the satellite data alone. A precondition for this success was the development of a percentile-based scaling method that addresses biases in modeled and observed global soil moisture fields. The development of an adaptive filtering approach permitted the use of the GMAO-LDAS in a synthetic data assimilation experiment that was developed to aid in the design of future satellite missions for land surface observations. Moreover, a dynamic bias estimation method for LST assimilation was developed and successfully tested in a heritage GMAO assimilation system, leading to improved estimates of near-surface air temperature in the coupled land-atmosphere assimilation system.

The results presented here all reflect the impact of uni-variate assimilation of land surface observations. Errors in the coupled land-atmosphere system are difficult to pin down because they may be related to any number of causes, including errors in precipitation, cloud biases, or errors in land surface parameters. The multi-variate assimilation of LST, soil moisture, and snow observations should lead to more consistent and improved estimates of the land surface water, energy, and radiation budget. Future developments at the GMAO will thus include the transitioning of the LST assimilation to the GMAO-LDAS and the application of the GMAO-LDAS to satellite-based snow observations. Moreover, work is on-going to implement the GMAO-LDAS as an integral component within the next-generation GMAO atmospheric data assimilation system.

Another important development is the gradual implementation of the GMAO data assimilation modules into the Land Information System (LIS, Kumar et al. 2008), a land surface modeling framework that integrates various community land surface models, ground and satellite-based observations, and data management tools within an architecture that allows interoperability of land surface models and parameters, surface meteorological forcing inputs, and observational data. The high performance infrastructure in LIS provides adequate support to conduct assimilation experiments of high computational granularity. Integration of the GMAO assimilation modules into LIS will therefore permit the use of the GMAO software with a variety of land surface models and make the GMAO contributions to land data assimilation development accessible to the research community.

**Acknowledgements** We gratefully acknowledge support from NASA grants NNG05GB61G and NNX08AH36G and from Air Force Weather Agency award F2BBBJ7080G001. We are also indebted to the many collaborators inside and outside the GMAO who contributed to our system.

# References

Andreadis K, Lettenmaier D (2005) Assimilating remotely sensed snow observations into a macroscale hydrology model. Adv Water Resour 29:872–886

Betts AK, Ball JH (1998) FIFE surface climate and site-average dataset 1987–89. *J Atmos Sci* 55:1091–1108

Bloom S, da Silva A, Dee D, Bosilovich M, Chern J-D, Pawson S, Schubert S, Sienkiewicz M, Stajner I, Tan W-W, Wu M-L (2005) Documentation and Validation of the Goddard Earth Observing System (GEOS) Data Assimilation System - Version 4. Technical Report Series on Global Modeling and Data Assimilation, Global Modeling and Assimilation Office, NASA Goddard Space Flight Center, 187 pp, Document number 104606, 26

Bosilovich M, Radakovich J, da Silva A, Todling R, Verter F (2007) Skin temperature analysis and bias correction in a coupled land-atmosphere data assimilation system. J Meteo Soc Japan 85A:205–228

Brasnett B (1999) A global analysis of snow depth for numerical weather prediction. J Appl Meteorol 38:726–740

Cohen J, Entekhabi D (1999) Eurasian snow cover variability and Northern Hemisphere climate predictability. Geophys Res Lett 26:345–348

Crow WT, Van Loon E (2006) Impact of incorrect model error assumptions on the sequential assimilation of remotely sensed surface soil moisture. J Hydrometeorol 7:421–432

Crow WT, Wood EF (2003) The assimilation of remotely sensed soil brightness temperature imagery into a land surface model using ensemble Kalman filtering: A case study based on ESTAR measurements during SGP97. Adv Water Resour 26:137–149

De Lannoy GJM, Reichle RH, Houser PR, Pauwels VRN, Verhoest NEC (2007) Correcting for forecast bias in soil moisture assimilation with the ensemble kalman filter. Water Resour Res 43:W09410. doi:10.1029/2006WR005449

Desroziers G, Berre L, Chapnik B, Poli P (2005) Diagnosis of observation, background and analysis-error statistics in observation space. Quart J R Meteo Soc 131(613):3385–3396. Part C, doi:10.1256/qj.05.108

Dirmeyer PA (2003) The role of the land surface background state in climate predictability. J Hydrometeorol 4:599–610

Dirmeyer PA, Gao X, Zhao M, Guo Z, Oki T, Hanasaki N (2006) GSWP-2: Multimodel analysis and implications for our perception of the land surface. Bull Am Meteor Soc 87:1381–1397

Dong J, Walker JP, Houser PR, Sun C (2007) Scanning Multichannel microwave radiometer snow water equivalent assimilation. J Geophys Res 112:D07108. doi:10.1029/2006JD007209

Drusch M (2007) Initializing numerical weather prediction models with satellite derived surface soil moisture: Data assimilation experiments with ECMWF's Integrated Forecast System and the TMI soil moisture data set. J Geophys Res 112(D3):D03102. doi:10.1029/2006JD007478

Drusch M, Viterbo P (2007) Assimilation of screen-level variables in ECMWF's Integrated Forecasting System: A study on the impact of forecast quality and analyzed soil moisture. Mon Wea Rev 135(2):300–314

Drusch M, Vasiljevic D, Viterbo P (2004) ECMWF's Global snow analysis: Assessment and revision based on satellite observations. J Appl Meteorol 43:1282–1294

Drusch M, Wood EF, Gao H (2005) Observation operators for the direct assimilation of TRMM microwave imager retrieved soil moisture. Geophys Res Lett 32:L15403. doi:10.1029/2005GL023623

Ducharne A, Koster RD, Suarez MJ, Stieglitz M, Kumar P (2000) A catchment-based approach to modeling land surface processes in a general circulation model, 2: Parameter estimation and model demonstration. J Geophys Res 105(20):24823–24838

Dunne S, Entekhabi D (2006) Land surface state and flux estimation using the ensemble Kalman smoother during the Southern Great Plains 1997 field experiment. Water Resour Res 42:W01407. doi:10.1029/2005WR004334

Entin J, Robock A, Vinnikov KY, Hollinger SE, Liu S, Namkhai A (2000) Temporal and spatial scales of observed soil moisture variations in the extratropics. J Geophys Res 105(D9):11865–11877

Evensen G (2003) The ensemble kalman filter: theoretical formulation and practical implementation. Ocean Dynamics 53:343–367. doi:10.1007/s10236-003-0036-9

Koster RD, Suarez MJ, Ducharne A, Stieglitz M, Kumar P (2000) A catchment-based approach to modeling land surface processes in a general circulation model, 1: Model structure. J Geophys Res 105(20):24809–24822

Koster RD, Suarez MJ, Liu P, Jambor U, Berg AA, Kistler M, Reichle RH, Rodell M, Famiglietti J (2004) Realistic initialization of land surface states: Impacts on subseasonal forecast skill. J Hydrometeorol 5:1049–1063

Kumar SV, Reichle RH, Peters-Lidard CD, Koster RD, Zhan X, Crow WT, Eylander JB, Houser PR (2008) A land surface data assimilation framework using the land information system: description and applications. Adv Water Resour. doi:10.1016/j.advwatres.2008.01.013

Mahfouf JF (1991) Analysis of soil moisture from near-surface parameters: A feasibility study. J Appl Meteor 30:506–526

Margulis SA, McLaughlin D, Entekhabi D, Dunne S (2002) Land data assimilation and estimation of soil moisture using measurements from the Southern Great Plains 1997 field experiment. Water Resour Res 38:1299. doi:10.1029/2001WR001114

Pan M, Wood EF (2006) Data assimilation for estimating the terrestrial water budget using a constrained ensemble kalman filter. J Hydrometeorol 7:534–547

Reichle RH, Koster RD (2003) Assessing the impact of horizontal error correlations in background fields on soil moisture estimation. J Hydrometeorol 4(6):1229–1242

Reichle RH, Koster RD (2004) Bias reduction in short records of satellite soil moisture. Geophys Res Lett 31: L19501. doi:10.1029/2004GL020938

Reichle RH, Koster RD (2005) Global assimilation of satellite surface soil moisture retrievals into the NASA Catchment land surface model. Geophys Res Lett 32(2):L02404. doi:10.1029/2004GL021700

Reichle RH, Entekhabi D, McLaughlin DB (2001) Downscaling of radiobrightness measurements for soil moisture estimation: A four-dimensional variational data assimilation approach. Water Resour Res 37(9):2353–2364

Reichle RH, McLaughlin D, Entekhabi D (2002a) Hydrologic data assimilation with the Ensemble Kalman filter. Mon Weather Rev 130(1):103–114

Reichle RH, Walker JP, Koster RD, Houser PR (2002b) Extended versus Ensemble Kalman filtering for land data assimilation. J Hydrometeorol 3(6):728–740

Reichle RH, Koster RD, Dong J, Berg AA (2004) Global soil moisture from satellite observations, land surface models, and ground data: Implications for data assimilation. J Hydrometeorol 5:430–442

Reichle RH, Koster RD, Bosilovich MG, Mahanama S (2006) Biases and scaling in soil moisture and temperature data assimilation. Eos Trans AGU 87(36) Jt Assem Suppl Abstract H22A-03

Reichle RH, Koster RD, Liu P, Mahanama SPP, Njoku EG, Owe M (2007) Comparison and assimilation of global soil moisture retrievals from the Advanced Microwave Scanning Radiometer for the Earth Observing System (AMSR-E) and the Scanning Multichannel Microwave Radiometer (SMMR). J Geophys Res 112:D09108. doi:10.1029/2006JD008033

Reichle RH, Crow WT, Koster RD, Sharif H, Mahanama SPP (2008a) The contribution of soil moisture retrievals to land data assimilation products. Geophys Res Lett 35:L01404, doi:10.1029/2007GL031986

Reichle RH, Crow WT, Keppenne CL (2008b) An adaptive ensemble Kalman filter for soil moisture data assimilation. Wat Resour Res 44:W03423, doi:10.1029/2007WR006357

Rodell M, Houser PR (2004) Updating a land surface model with MODIS-derived snow cover. J Hydrometeorol 5:1064–1075

Rodell M, Houser PR, Jambor U, Gottschalck J, Mitchell K, Meng C-J, Arsenault K, Cosgrove B, Radakovich J, Bosilovich M, Entin JK, Walker JP, Toll DL (2004) The Global Land Data Assimilation System. Bull Am Meteorol Soc 85:381–394. doi:10.1175/BAMS-85-3-381

Seuffert G, Wilker H, Viterbo P, Mahfouf J-F, Drusch M, Calvet J-C (2003) Soil moisture analysis combining screen-level parameters and microwave brightness temperature: A test with field data. Geophys Res Lett 30:1498. doi:10.1029/2063GL017128

Slater A, Clark M (2006) Snow data assimilation via an ensemble Kalman filter. J Hydrometeorol 7:478–493

Space Studies Board (2007) Earth science and applications from space: National imperatives for the next decade and beyond. National Academy of Sciences, Washington, DC., 400 pp

Walker JP, Houser PR (2005) Hydrologic data assimilation. In: Aswathanarayana A (ed) Advances in water science methodologies. AA Balkema, Netherland, p 230

Zaitchik BF, Rodell M, Reichle RH (2008) Assimilation of GRACE terrestrial water storage data into a land surface model: Results for the Mississippi River basin. J Hydrometeorol, 9, 535–548, doi:10.1175/2007JHM951.1, 2008

Zhou Y, McLaughlin D, Entekhabi D (2006) Assessing the Performance of the Ensemble Kalman Filter for Land Surface Data Assimilation. Mon Weather Rev 134:2128–2142

# Assimilation of Soil Moisture and Temperature Data into Land Surface Models: A Survey

Nasim Alavi, Jon S. Warland and Aaron A. Berg

**Abstract** The surface temperature and moisture conditions at the Earth's surface have important controls on land-atmosphere fluxes of energy and water. Over the past decade considerable research has advanced the application of data assimilation systems to ensure the correct specification of these quantities in land surface parameterization schemes. This chapter provides an overview of the primary data assimilation techniques that have evolved for the assimilation of surface temperature and soil moisture into land surface models. We conclude with examination of some of the emerging and current issues for data assimilation including overcoming differences between satellite-retrieved and modeled soil moisture, and strategies that examine assimilation issues given the course spatial resolution obtained with current and near-future satellite sensors.

## 1 Introduction

Surface temperature and moisture at the earth's surface control land-atmosphere fluxes of water and energy and play a critical role in the partitioning of available energy into sensible and latent heat fluxes. Therefore correct specification of these states in land surface models (LSM) has the potential to improve energy and water flux estimates. To correctly update LSMs with observations of the land surface state, recent modeling advances have introduced land data assimilation systems. The application of data assimilation to applied hydrology has progressed rapidly over the past decades. The purpose of this chapter is to present an overview of advances in data assimilation techniques for studies of land surface hydrology and the energy budget. This chapter is organized as follows. In Sect. 2 we review several data assimilation techniques. Application of data assimilation techniques in assimilating land surface temperature into LSMs to improve energy flux terms is discussed in Sect. 3. Section 4 will discuss the assimilation of ground- based or aircraft observation of surface soil moisture into LSMs and its effect in improving evapotranspiration, runoff, soil moisture profile estimates or map soil moisture over large

N. Alavi (✉)
Land Resource Science, University of Guelph, Guelph, ON, Canada, N1G 2W1
e-mail: nalavi@uoguelph.ca

S.K. Park, L. Xu, *Data Assimilation for Atmospheric, Oceanic and Hydrologic Applications*, DOI 10.1007/978-3-540-71056-1_22,
© Springer-Verlag Berlin Heidelberg 2009

areas. Finally, assimilating satellite-measured soil moisture and its limitations are addressed in Sect. 5.

## 2 Data Assimilation Techniques

Data assimilation is the merging of available observations with model outputs to obtain a better estimation of the state given the basic assumption that both the model estimates and observations are imperfect. Observations and model estimates provide different kinds of information and in different time and spatial scales; therefore, when used together, they can provide a level of accuracy that cannot be obtained when used individually. For example, a LSM can provide information on the spatial and temporal variations of near-surface and root-zone soil moisture, but these estimates contain numerous errors due to imperfect model structure, uncertain parameters, error in inputs, and issues with grid size. On the other hand, remote sensing measurements obtain near-surface soil moisture at an instant in time, but do not give the temporal variation or the root zone soil moisture content. Data assimilation techniques have emerged as the primary methods for merging model predictions and remote sensing observations together to obtain more accurate information about the spatial and temporal variations of the soil moisture profile.

Data assimilation techniques have been used widely in meteorology (Sasaki, 1960; Epstein, 1969; Lorenc, 1986; Daley, 1991; Courtier et al. 1993) and oceanography (Ghil and Malanotterizzoli, 1991; Bennett, 1992) for improving prediction. These techniques are now being exploited in hydrological applications. Jackson et al. (1981) and Bernard et al. (1981) were among the first studies that used the direct insertion method to update soil moisture predictions using surface soil moisture observations.

Houser et al. (1998) performed a detailed study of using several alternative assimilation approaches including direct insertion, Newtonian Nudging and statistical Interpolation/correction to update soil surface temperature and moisture in a LSM. Direct insertion replaced the model states values with the observed values directly, while other techniques used statistical techniques to make the update. While the application of these methods is computationally efficient and easy, the updates do not consider uncertainties associated with measurements and the model. In the following sections, more optimal techniques which provide information on uncertainty in the updating procedure will be discussed.

### 2.1 Kalman Filtering

Kalman-filter based methods have been successfully applied in numerous hydrological applications. The Kalman filter is a recursive filter developed by Rudolf Kalman (1960). The filter is used to efficiently estimate states from a dynamic system composed of incomplete and noisy measurements minimizing the mean of the squared error between analysis estimates and the true states. The Kalman filter is a sequential filter which seeks to characterize system states at the current time. It integrates the model forward in time and whenever measurements are available uses them to update the model before the integration continues. The classic Kalman filter is suitable for linear models while for nonlinear models, the extended and ensemble Kalman filters have been developed.

A LSM can be expressed in the following generic dynamic state-space formulation of a stochastic model following notation of Liu and Gupta (2007):

$$x_{k+1} = M_{k+1}(x_k, \theta, u_{k+1}, \eta_{k+1}) \tag{1}$$

$$z_{k+1} = H_{k+1}(x_{k+1}, \theta) + \varepsilon_{k+1} \tag{2}$$

Where $x_k$ and $x_{k+1}$ represent the true system state vectors (e.g., soil moisture and/or temperature) at time $t_k$ and $t_{k+1}$. The nonlinear operator $M_{k+1}$ is equivalent to the model structure and includes model input $u_{k+1}$ and parameter $\theta$ vectors. The observation vector $z_{k+1}$ is related to the model states and parameters through an observation operator $H$. Uncertainties related to errors in the model formulation or the forcing data or parameters are summarized in the model error term $\eta_{k+1}$ with covariance $Q_{k+1}$ while the observation error is summarized by $\varepsilon_{k+1}$, with covariance $R_{k+1}$.

To set up the assimilation system using the above state-space formulation, some assumptions have to be made on the statistics of the two error terms $\eta$ and $\varepsilon$, based on the prior knowledge of the deficiencies in the assimilating system. For example, the mean values of $\eta$ and $\varepsilon$ (i.e., biases) reflect the systematic errors in the modeling and observation systems, while the error covariances $Q_k$ and $R_k$ reflect the uncertainty in the model predictions and observations. One common approach is to assume that the errors are zero-mean white noise sequences with a normal (i.e., Gaussian) probability distribution. In addition, it is often assumed that the model and observation errors are uncorrelated.

In the case of Gaussian model and measurements errors ($\eta$ and $\varepsilon$) and linear model and observation operators (M and H), the data assimilation problem presented in (1) and (2) can easily be solved by the classic Kalman filter (Kalman, 1960) with prediction step (Eqs. (3) and (4)) and an update step (Eqs. (5) and (6)) as follows (Liu and Gupta, 2007):

$$x_{k+1}^- = M_{k+1}(x_k^+, \theta, u_{k+1}) \tag{3}$$

$$P_{k+1}^- = M_{k+1}P_k^+ M_{k+1}^T + Q_{k+1} \tag{4}$$

$$x_{k+1}^+ = x_{k+1}^- + K_{k+1}(z_{k+1} - H_{k+1}x_{k+1}^-) \tag{5}$$

$$P_{k+1}^+ = P_{k+1}^- - K_{k+1}H_{k+1}P_{k+1}^- \tag{6}$$

Where $K$ is Kalman gain matrix

$$K_{k+1} = P_{k+1}^- H_{k+1}^T (H_{k+1}P_{k+1}^- H_{k+1}^T + R_{k+1})^{-1} \tag{7}$$

In the above equations $P$ is the error covariance matrix of the state variables and $M$ and $H$ stand for the linear model and observation operators. The minus and plus superscripts are used to discriminate the states and the error covariance matrix before and after updating, respectively, where $T$ indicates the matrix transpose.

The calculations of (3)–(7) can be repeated at the next time step $k+2$ to assimilate a new observation available at that time; this process can proceed sequentially to assimilate all the available observations. The Kalman gain $(K)$ is determined by the relative magnitudes of the state error covariance $P$ and the observation error covariance $R$ and acts as a weighting factor on the innovation term $(z_{k+1} - H_{k+1}x_{k+1}^-)$. In other words, the larger the observation error covariance, the smaller the Kalman gain, and the smaller the update correction applied to the forecast state vector.

The classic Kalman filter is optimal for linear systems with Gaussian model and measurement errors and therefore it is not applicable in hydrological applications where the model equations are nonlinear. Hence, variations of the Kalman filter have been developed that are applicable to nonlinear problems including the extended Kalman filter (EKF, Jazwinski, 1970) and ensemble Kalman filter (EnKF, Evensen, 1994).

### 2.1.1 Extended Kalman Filter

In the EKF algorithm, linearized forms of the nonlinear model operator $(M)$ and observation operator $(H)$ are generated by using Taylor's series expansion and used in the same equations (3)–(7) for the basic Kalman filter algorithm. However, this may produce instabilities or even divergence due to closure approximation by neglecting the second- and higher-order derivatives of the model (Evensen, 1994). In hydrological application the EKF has been used by Walker et al. (2001a, b), Galantowicz et al. (1999), Reichle et al. (2001b, 2002b), Entekhabi et al. (1994), and Walker and Houser (2001).

### 2.1.2 Ensemble Kalman Filter

The ensemble Kalman filter (EnKF), developed by Evensen (1994) is an extension of Kalman's filter and an alternative to EKF for nonlinear systems. In the EnKF, $n$ ensembles of model states are generated using different initial conditions, forcing, and parameters and propagated forward in time using the nonlinear forward model. At each update time a vector of ensemble observations $(D)$ becomes available and is used to update the model states based on the Kalman gain. The EnKF consists of the following prediction step (equation (8)) and update step (Eq. (9)):

$$x_{k+1}^{-,i} = M_{k+1}(x_k^{+,i}, \theta, u_{k+1}^i) \qquad\qquad i = 1,\ldots,n \qquad\qquad (8)$$

$$x_{k+1}^{+,i} = x_{k+1}^{-,i} + K_{k+1}(d_{k+1}^i - H_{k+1}x_{k+1}^{-,i}) \qquad i = 1,\ldots,n \qquad\qquad (9)$$

where $n$ is the size of the ensemble, the input ensemble $u_{k+1}^i$ is obtained by adding a noise term $\zeta_{k+1}^i$ to the nominal input $u_{k+1}$, i.e., $u_{k+1}^i = u_{k+1} + \zeta_{k+1}^i (\zeta_{k+1}^i \in N(0, U_{k+1})$, and $U_{k+1}$ is the error covariance of $u_{k+1}$. To make the ensemble observation vector (D) at each time of updating, a noise term $\varepsilon_{k+1}^i \in N(0, R_{k+1})$ is added to the nominal observation $z_{k+1}$ as

$$d_{k+1} = z_{k+1} + \varepsilon_{k+1}^i \qquad i = 1, \ldots, n \tag{10}$$

And stored in $D = (d_1, d_2, \ldots \ldots, d_n)$.

The error covariance of state variables $P_{k+1}^-$ can be directly calculated from the ensemble $\left\{ x_{k+1}^{-,i} \right\}$ as expressed in (11)

$$P_{k+1}^- = \frac{1}{N-1} X_{k+1}^- (X_{k+1}^-)^T \tag{11}$$

where $X_{k+1}^- = x_{k+1}^{-,i} - \overline{X_{k+1}^-}$.

To update model states, the traditional updating Eqs. (5) and (6) for the classic Kalman filter are used, with Kalman gain calculated from the error covariances provided by the ensemble (Evensen, 2004).

The advantages of EnKF are: 1) any LSM can be used with EnKF since the equations of the land surface model are not incorporated into the filtering procedure and the model communicates with the filter through its inputs and outputs, 2) EnKF is flexible in the specification of model errors and uncertainties in inputs and measurements can be accounted for 3) it provides information on the accuracy of its estimates, and 4) it is relatively efficient and feasible in large-scale applications. Numerous studies have used EnKF to assimilate soil surface moisture into complex LSMs as discussed in Sects. 4 and 5.

## 2.2 Variational Data Assimilation

The variational data assimilation (VDA) technique merges the noisy measurements with states predicted by a forward model by optimizing a statistical measure over a given time window. The statistical measure minimizes the squared difference between the measured and predicted states considering errors in measurements as well as model estimates. Depending on the spatial and temporal dimensions of the state variable, a VDA method can be one-dimensional (1D-Var), three-dimensional (3D-Var), or four-dimensional (4D-Var).

To construct a variational filter it is assumed that the prior estimate of state variables at time $t_0$ is $x_0$ (with error covariance $Q_0$), the assimilation operates over the time interval $[t_1, t_n]$, with observations $[z_1, z_2, \ldots, z_n]$ available at the $n$ discrete time points $[t_1, t_2, \ldots, t_n]$. A general VDA problem can then be defined as the minimization of a cost function $J$, which represents the aggregated error over the entire assimilation window (assuming that errors at different times are independent and additive) (following notation of Liu and Gupta, 2007):

$$J(x,u,\theta) = \sum_{i=1}^{n} \eta_i^T Q_i^{-1} \eta_i + \sum_{i=1}^{n} (z_i - H_i[x_i])^T R_i^{-1}(z_i - H_i[x_i])$$

$$+ (x_0 - x_0^-)^T Q_0^{-1}(x_0 - x_0^-) + \sum_{i=1}^{n} (u_i - u_i^-)^T C_{uu}^{-1}(u_i - u_i^-)$$

$$+ (\theta - \theta^-)^T C_\theta^{-1}(\theta - \theta^-)$$

$$= J_M + J_O + J_0 + J_u + J_\theta \qquad (12)$$

In Eq. (12), $\eta_i$ represents the model error at $t_i$, $u_i^-$ and $\theta^-$ denote the prior model inputs at $t_i$ and the prior time-invariant parameters respectively. $C_{uu}$ and $C_\theta$ are the time invariant error covariances of inputs and parameters, respectively. The VDA proceeds to minimize $J$, to obtain the least-squares estimates of state variables $x_i$ and input variables $u_i$ for each time point within the assimilation window and the time-invariant parameters $\theta$. In the above general formulation of the cost function, the first term $J_M$ penalizes the difference between the estimated model error vector $\eta_i$ and its prior mean; the second term $J_O$ is used to penalize the differences between model predictions and observations at all time points within the assimilation window; $J_0$, $J_u$, and $J_\theta$ are included to measure the errors associated with the initial conditions, model inputs and parameters, respectively. The errors are weighted by the corresponding error covariance (i.e., $Q$, $R$, $Q_0$, $C_{uu}$, or $C_\theta$) for various sources (e.g., the model, observations, initial conditions, inputs, and parameters).

The VDA which is used in most hydrological studies was formulated based on an adjoint-state variational scheme (Sun 1994) in which the state equation (8) is adjoined to the cost function with a Lagrange multiplier $\lambda$ as follows:

$$J(x,u,\theta) = J_M + J_O + J_0 + J_u + J_\theta + \sum \lambda_i^T [x_i - M_i(x_{i-1}, u_i, \theta)] \qquad (13)$$

The last term in Eq. (13) is called an adjoint physical constraint. The objective of VDA is to find the best estimates of model states and parameters that minimizes $J$. Solution to the variational problem is achieved by minimization (i.e. $\nabla J = 0$) and iteration (Huang and Yang, 1996). In practice for nonlinear, high-dimensional hydrologic applications, solving this optimization problem is very difficult, and often impossible. Consequently, simplifications and approximations are often considered (e.g. neglecting model/parameter errors and/or linearizing the state and observation equations). Even with these simplifications solving a variational problem is not easy, and often a numerical algorithm such as the adjoint model technique is used to obtain solutions in an iterative manner (Huang and Yang, 1996).

Variational techniques are more suitable for linear models. Their application to nonlinear systems requires obtaining the tangent-linear form (using Taylor's series expansion) of the nonlinear model. Instability that arises with this approximation biases the estimate based on this method (Liu and Gupta, 2007).

VDA methods are much less expensive computationally than KF and EKF methods because they do not propagate large error covariance matrices; instead they simultaneously process all data within a given time period (assimilation interval)

(Reichle et al., 2001b). Thus they are preferable for data assimilation in complex systems (e.g., a numerical weather prediction framework). In addition, VDA methods typically provide more optimal estimates than KF and EKF since they use observations inside the assimilation interval all at once. However, the sequential KF methods are more suitable for real-time data assimilation to process observations that arrive continuously in time, while VDA methods can only be run for a finite time interval. KF methods provide error covariance estimates for the prediction, while a VDA method itself does not provide any estimate of the predictive uncertainty. When the assimilation system is nonlinear, both EKF and VDA methods rely on using the tangent-linear models $M$ and $H$ to approximate the state and observation equations while the EnKF can be used efficiently for nonlinear systems. Discussion of VDA applications in land surface parameterization schemes is presented in Sects. 3 and 4.

## 3 Assimilating the Land Surface Temperature

Land surface temperature is a principal control on land-atmosphere fluxes of water and energy. It can be observed from space and aircraft infrared sensors in cloud-free conditions. Assimilation of observed land surface temperature into LSMs can improve energy flux estimates as shown the studies discussed below.

Radiometric observations of the surface temperature were used in a study by Castelli et al. (1999) to estimate the components of the surface energy balance. A force-restore model was combined with the energy balance equation to provide a forward model to predict land-surface temperature. The technique minimized the misfit of surface temperature forecast against observations using a VDA strategy. The approach was tested with the half-hourly observations of radiometric soil surface temperature from the First International Satellite Land Surface Climatology Project Field Experiment (FIFE) across a prairie region in Kansas. Results showed that the estimates of surface heat flux had a root-mean-square error of $20\,\mathrm{W\,m^{-2}}$, when compared to the measurements. This method had the disadvantage of relying on estimates of aerodynamic resistance $r_a$ and values of roughness length; furthermore, canopy resistance was not considered.

Remote sensing data usually have low temporal resolution; if data from low orbit satellites are used as the source of surface soil temperature observations, it is expected that only a few observations per day will be available for assimilation. Therefore, Boni et al. (2001a) applied the same approach as Castelli et al. (1999) over the U.S. Southern Great Plains 1997 (SGP97) hydrologic field experiment and studied the effect of the sparse sampling of surface ground temperature. Surface soil temperature was derived from its relation with soil temperatures at different depths. They showed that observation of surface ground temperature within a $\sim$3h window around the mean time of daily maximum is most effective in the estimation problem. Following the previous study, Boni et al. (2001b) used satellite data from two low Earth orbit platforms (AVHRR on board the NOAA 12 and 14) over two subregions within SGP97 to estimate components of the surface energy balance and land surface control on evaporation. The spatial resolution of the satellite data was 1.1 km and data were available twice daily for each sensor. It was shown that when satellite data are available close to the time of peak of diurnal temperature, the estimation of surface energy balance, ground temperature, and soil moisture or surface control on evaporation were considerably improved.

Caparrini et al. (2003, 2004a) used a VDA scheme to infer two key parameters of the surface energy balance, the dimensionless bulk heat transfer coefficient ($C_H$) and the evaporative fraction ($E_F = (LE)/(LE + H)$, where $H$ and $LE$ are sensible and latent heat fluxes). Caparrini et al. (2003) used remote sensing measurement of land surface temperature from AVHRR and SSM/I sensors over a watershed in Italy for a 18-day period in July 1996, while Caparrini et al. (2004a) tested the approach with radiometric measurements of land surface temperature over the FIFE site. The assimilation of surface temperature focused on estimation of $C_H$ and $E_F$. $C_H$ represents the impact of the land surface characteristic on turbulent transfer and depends on atmospheric stability and surface roughness while $E_F$ shows partitioning between turbulent fluxes and is related to soil moisture control on evaporation. Monthly $C_H$ and daily $E_F$ were obtained by minimizing the root mean square difference between predictions of surface temperature and observations. The results showed that the derived $C_H$ map followed the spatial patterns of the land cover and physiography of the studied basin and $E_F$ variations were consistent with soil moisture variations.

Crow and Kustas (2005) examined this approach over a range of vegetative and hydrological conditions in southern U.S. to demonstrate its limitations for simultaneously retrieving the evaporative fraction and the turbulent-transfer coefficient. Their results showed that the simultaneous retrieval of both parameters is satisfactory over dry and lightly-vegetated areas, but problematic over wet and/or densely-vegetated land surface conditions. They suggested including additional land surface information (e.g leaf area index) in the method for these conditions.

The approach of Caparrini et al. (2003, 2004a) was a single-source model where contributions from soil background to land surface observations were neglected. Caparrini et al. (2004b) applied this approach over a large area within U.S. Great Plains and advanced it in two major new directions. First they extended the VDA system to a multiscale framework where satellite land-surface temperature estimates from several sensors (e.g. AVHRR, SSM/I, GOES) with different spatial (1.1–50 km) and time resolutions (twice daily for AVHRR, SSM/I and half hourly for GOES) were assimilated. Second, they used a two-source model where remotely sensed land surface temperature was considered as a combination of contributions from the canopy and from bare soil surfaces. In this study the contributions of soil and canopy to the land surface temperature and evaporation were estimated separately. The dual-source (DS) energy balance formulation (Kustas et al., 1996) was used in this approach and the turbulent transfer coefficient and evaporative fraction were calculated for soil and canopy separately. Information on LAI provided by remote sensing observations was used to partition land surface temperature into canopy and soil contributions. The spatial pattern of retrieved parameters was consistent with land use maps and seasonal phenology of the study area and to their observed values.

The usefulness of skin temperature in land data assimilation studies is limited by its very short memory (on the order of minutes) due to the very small heat storage it represents (Walker and Houser 2005). In most land surface parameterization schemes, land surface temperature is obtained by solving the energy balance equations; its estimate is therefore subject to errors associated with all the empirical parameters involved in the computation of latent, sensible, snow melt heat fluxes, and net solar radiation. These include soil and vegetation characteristics, and the resistances for momentum, heat and water vapour fluxes through the canopy and over bare ground. Consequently it is difficult to obtain an estimate of model error associated with the land surface temperature. In addition, the observation error of the remotely sensed surface temperature has not been completely understood. Without knowledge of these errors, application of a data assimilation technique can be problematic.

Radakovich et al. (2001), assimilated three-hourly observations of land surface temperature from the International Satellite Cloud Climatology Project (ISCCP) into an Off-line Land-surface Global Assimilation (OLGA). An incremental and diurnal bias correction (IBC and DBC) was included in the assimilation system to account for biased land surface temperature forecasts. The assimilation of ISCCP derived land surface temperature significantly reduced the bias and standard deviation between model predictions and observations. Also, the monthly mean diurnal cycle from the experiment closely matched the diurnal cycle from the observations. Following their approach, Bosilovich et al. (2007) assimilated land surface temperature retrieved from (ISCCP) into a coupled atmosphere land global data assimilation system based on the 3-dimensional variational technique. The incremental bias correction term was introduced in the model's surface energy budget as well as a correction for the mean diurnal cycle. In general the largest biases were improved by a time mean value of bias correction. Accounting for the diurnal cycle of surface temperature bias provided further improvement in mean near surface air temperature and also improved estimation of the amplitude of the air temperature. Sensible heat flux was improved with the surface temperature assimilation. At two locations a reduction of bias by as much as $30\,\mathrm{Wm}^{-2}$ was observed. However, at many stations the monthly latent heat flux bias increased slightly.

# 4 Assimilating Land Surface Moisture

Entekhabi et al. (1994) developed an algorithm to estimate 1-m profile of soil temperature and moisture using remote sensing observations of the top few centimeters of the soil. The technique consisted of a dynamic model that advanced the soil states in time (coupled moisture and heat diffusion equations), a radio-brightness model that computed soil radio-brightness estimates from model states, and an EKF that updated the soil state estimates. The EKF transferred the surface observations to the lower layers considering the correlation of other layer's soil moisture to the surface condition and updated each layer state on an hourly basis. The algorithm was tested

on a basic synthetic example and it was shown that the system is capable of retrieving the profiles below the surface observations. This approach was later tested
with field observations by Galantowicz et al. (1999) in an 8-day field experiment
where the soil moisture was measured by a truck-mounted microwave radiometer
on an hourly basis. The effectiveness of EKF in retrieving 1 m soil state profiles
was shown in short and long time periods with daily updating with field data and
for a 4-month synthetic study with updating every 3 days. The simulation study's
overall retrieval errors were 2.2–8.0% volumetric soil moisture (depending on soil
type knowledge), which suggested that the EKF was efficient in determining the soil
moisture profile.

The ENK methods were adapted by Walker et al. (2001a) to assimilate near surface soil moisture and temperature data into a soil moisture and heat transfer model.
The Kalman filter technique was compared with direct insertion using synthetic
data. They showed that the Kalman filter was superior to direct insertion, with soil
moisture profile retrieval being achieved in 12 h as compared to 8 days or more
(depending on observation depth) for hourly observations. The superiority of the
Kalman filter was due to its ability to adjust the entire profile while direct insertion can only directly change the profile within the observation depth. The Kalman
filter was able to modify the soil moisture below the observation depth using the
correlation between soil moisture in different layers. For application of the Kalman
filter, moisture in the unsaturated zone must be a function of soil moisture of adjacent layers; therefore, the correlations between layers can be used in the covariance
forecasting equations. The Walker et al. (2001a) study also examined the impact of
observation depth and update frequency on soil moisture profile retrieval. For the
ENK filter the updating frequency was more important than the observation depth.

The ability to retrieve the soil moisture profile through assimilation of near-
surface soil moisture measurements (such as would be obtained from remote sensing) has been studied in previous works. However, the majority of assimilation
studies have been limited to synthetic desktop studies and/or short time periods,
such as the 8-day study of Galantowicz et al. (1999). To extend the methodology
to the catchment scale, Walker et al. (2001b) used a simplified soil moisture model
based on an approximation to the Buckingham–Darcy equation. This model was
then used in a 12-month field application with updating at 1-, 5-, 10-, and 20-day
intervals. This study demonstrated that retrieval of the soil moisture profile using
the EKF depends on the adequacy of the forecasting model for prediction of soil
moisture profile dynamics, which is more important than the temporal resolution of
near-surface soil moisture measurements.

Most of the soil moisture assimilation work discussed earlier in this section focused on one-dimensional problems (where only the vertical variations of states and
fluxes were considered) because of computational limitations. A second category of
soil moisture assimilation studies confronts the problem of estimating both horizontal and vertical variations in soil moisture. These studies have been performed by
application of VDA, EKF, or EnKF. Reichle et al. (2001a) used a VDA technique
to assimilate soil moisture into a SVAT scheme where horizontal as well as vertical
variations of soil states and fluxes were simulated. The model coupled moisture and

heat transport where vertical soil moisture and temperature dynamics were modeled with Richards' equation and the force–restore approximation, respectively, while the vegetation layer was treated with diagnostic variables, and fluxes through the canopy were described with a resistance network. The primary model states were soil moisture, soil temperature, and canopy temperature that distributed over time and in three spatial dimensions. The measured variable in this study was synthetically generated brightness temperature. The assimilation algorithm considered both uncertainties in the model and measurements. The results showed that the state estimates obtained with the assimilation algorithm improved model estimation of energy fluxes. A subsequent study by Reichle et al. (2001b) examined the feasibility of estimating large scale soil moisture profiles from L band passive microwave measurements using 4D data assimilation. Estimates of land surface states were derived from airborne passive microwave measurements over U.S. Great Plains, which were related to the states by a radiative transfer model. They also performed a downscaling experiment to downscale brightness images with a resolution of tens of kilometers to scale of a few kilometers. Using high resolution micrometeorological, soil texture, and land cover inputs, the downscaling experiment was performed and soil moisture profiles were retrieved at the finer scale.

EKF has been used in one-dimensional mode by Entekhabi et al. (1994), Galantowicz et al. (1999), and Walker et al. (2001a, b); however, the one-dimensional nature of these studies has restricted the application of this work for evaluation of catchments, where lateral flow and spatial coupling are dominant physical processes. Walker et al. (2002) extended previous work (Walker et al., 2001a, b) to situations where lateral flows are important and spatial coupling of soil moisture must be considered (where there is no correlation between soil moisture of adjacent depths). They applied a simplified EKF covariance forecasting technique to a distributed three-dimensional soil moisture model for retrieval of the soil moisture profile in a 6 ha catchment in Australia using near-surface soil moisture measurements. Surface soil moisture measurements were made using TDR probes on a 20 m by 20 m grid every 2–3 days, to replicate remote-sensing observations. The results showed that when the near-surface soil layer becomes decoupled from the deep soil layer during extended drying periods, the Kalman filter performs poorly and the soil moisture profile cannot be retrieved. This study also showed that, when using the modified EKF, initialization of the model states was not important for adequate retrieval of the soil moisture profile. Moreover, it was demonstrated that the required update frequency is a function of the errors in model physics and forcing data. When model calibration is acceptable and forcing data has a high level of accuracy, the updating interval is relatively unimportant.

Methods such as VDA and EKF have limited capability to deal with model errors, and the necessary linearization approximations can lead to unstable solutions. In a series of studies the EnKF has been adapted to address these issues. Reichle et al. (2002a) examined the feasibility of using the EnKF for soil moisture data assimilation and compared its performance with VDA. In a series of experiments they assimilated synthetically generated L-band microwave data for Great Plains (SGP97) Hydrology Experiment into a SVAT model (Reichle 2001a) and showed that EnKF performed well against the VDA method. In particular the variational approach generally requires an adjoint of the model, which is not usually available and is difficult to derive. The EnKF performance was further compared with the EKF by Reichle et al. (2002b) in a twin experiment for the southern United States. Synthetic observations of near-surface soil moisture were assimilated once every 3 days into the Catchment model. EnKF offered more flexibility in covariance modeling (including horizontal error correlations), which made it a more robust method in soil moisture problems than EKF. This study also found that EnKF, because of its flexibility and performance, is a promising approach for soil moisture initialization problems.

A further application of the EnKF applied to remote sensing data was described in Margulis et al. (2002). In this study airborne L band microwave observations collected over a period of 16 days during the SGP97 field experiment were assimilated into the NOAH model (Chen et al., 1996). The model propagated ensemble replicates between measurement times and a radiative transfer model was used to compute the predicted radio-brightness estimates. Results showed that modeled soil moisture and latent heat flux were in good agreement with ground measurements. Assimilation of brightness temperature data improved the estimated soil moisture at the surface and at depth and consequently the latent heat flux estimation compared to a no-assimilation simulation. Crow and Wood (2003) also used this data set and a version of the EnKF to assimilate the brightness temperature observations into the TOPLATS (TOPMODEL based Land-Atmosphere Transfer scheme by Famiglietti and Wood (1994)). In this application the EnKF was capable of extracting spatial and temporal trends in root-zone (40 cm) soil water content from temperature brightness measurements of surface (5 cm) conditions. Surface state and flux predictions were better than predictions derived from no assimilation modeling.

There have been several recent advances in the development and application of EnKF techniques. Dunne and Entekhabi (2005) argued that soil moisture estimation is a reanalysis-type problem as observations beyond the estimation time are also useful in the estimation. However, the EnKF uses only observations prior to the estimation time to update the current state. To overcome this problem they developed an ensemble moving batch (EnMB) smoother in which all observations within a prescribed window were used to update all of the soil moisture states in a batch. The EnMB takes the EnKF estimates as its first guess; however, in addition to updating the current state, it also updates the best estimates at previous times. This algorithm was applied to NOAH LSM and the results were compared to the EnKF. The approach was evaluated with soil moisture observations synthetically generated from data obtained during SGP97 to simulate the revisit time of a satellite platform. It was shown that smoothing improved the estimates of soil moisture at the surface and at depth. Smoothing also lead to improved surface flux estimation through the dependence of the latent heat flux on root zone soil moisture. A major disadvantage of the EnMB was that the state and observation vectors were augmented to be distributed in time, resulting in a computationally-expensive smoother. Dunne and Entekhabi (2006) found a less expensive approach to ensemble smoothing. Since there is limited memory in soil moisture in the unsaturated zone, rather than use each observation to update all past estimates, the ensemble Kalman smoother can be implemented as a fixed lag smoother in which the observation is only used to update past estimates within a fixed time window. The results showed that the backward propagation of the information from subsequent observations considerably reduced the uncertainty associated with the soil moisture estimates.

In the EnKF method at the update times, it is assumed that the probability density function of the state across the ensemble is Gaussian and therefore the ensemble mean and covariance can be adequate descriptors. In reality the ensemble soil moisture is rarely perfectly Gaussian and during extremely wet or dry conditions it can become particularly skewed. Zhou et al. (2006) examined the efficiency of EnKF in assimilating soil moisture problem in LSMs and showed that, despite the dependency of the filter on normality assumption, it is able to preserve some skewness and multimodality in soil moisture estimates.

Ultimately, data assimilation techniques applied at the point or catchment resolution need to be scaled up for use at the resolution and time steps available from remote sensing. At the continental scale, Walker and Houser (2001) applied an EKF to initialize near-surface soil moisture in the catchment-based land-surface model (Koster et al., 2000). They described this technique as a new method to generate soil moisture initialization that is superior to spinning-up the land-surface model. In this synthetic study, involving assimilation of near-surface soil moisture, errors in forecasting soil moisture profiles as a result of poor initialization were reduced and the resulting evapotranspiration and runoff forecasts were improved. They showed that after 1 month of assimilation, the root mean square error in the profile storage of soil moisture was reduced to 3% vol/vol and after 12 months of assimilation the error was as low as 1% vol/vol. Walker and Houser (2004) used this same EKF framework to address soil moisture satellite mission accuracy, repeat time and spatial resolution requirements through a numerical twin study. Simulated soil moisture profile retrievals were made by assimilating near-surface soil moisture observations with various accuracy (0–10% vol/vol), repeat time (1–30 days), and spatial resolution (0.5–120 arc-min) into the NASA catchment model (Koster et al. 2000). It was found that: (i) near-surface soil moisture observation error must be less than the model forecast error required for a specific application and must be better than 5% v/v accuracy to positively impact soil moisture forecasts, (ii) daily near-surface soil moisture observations achieved the best soil moisture and evapotranspiration forecasts, (iii) near-surface soil moisture observations should have a spatial resolution of around half the model resolution, and (iv) satisfying the spatial resolution and accuracy requirements was much more important than repeat time.

# 5 Assimilating Satellite-Derived Surface Soil Moisture

A number of studies have demonstrated the potential of assimilating satellite-derived soil moisture into LSMs (e.g., Walker and Houser, 2001; Reichle et al., 2001a, b; Crow and Wood, 2003). However, the majority of these studies have been limited to datasets based on either airborne data or synthetic observations from field experiments over limited domains, which do not represent the statistical character-istics of satellite soil moisture retrieval. Application of satellite-retrieved soil mois-ture data in a data assimilation framework is restricted by several problems, which include the limited availability of satellite-retrieved soil moisture, the large system-atic differences between satellite-retrieved soil moisture and in situ observations or model simulations, and the mismatch between the spatial scales of satellite data and the LSMs. Several studies have addressed these problems and suggested different techniques to overcome them as discussed in Sects. 5.1 and 5.2.

## 5.1 Bias Removal and Assimilation of Satellite-Derived Soil Moisture Data

Reichle et al. (2004) examined the compatibility of satellite soil moisture retrievals with modeled soil moisture and ground-based measurements at the global scale. Three independent surface soil moisture datasets for the period 1979–1987, includ-ing global retrieval from SMMR, global soil moisture estimates from NASA catch-ment model, and ground- based measurement in Eurasia and North America were compared. The results showed that there was a significant temporal correlation be-tween these data sets, however the data sets clearly differed in their statistical mo-ments (mean and variance). These differences are due to several factors, including different effective depth of retrieved and modeled soil moisture (Wilker et al., 2006), the effects of subgrid elements (e.g. roads, buildings, small water bodies) within the satellite footprint that are ignored by retrieval algorithms (Gao et al., 2006), and the climatology of the LSM (Koster and Milly, 1997). Therefore, direct as-similation of measured soil moisture (satellite or ground) into a LSM is not ap-propriate. The discrepancy between the datasets demonstrated that some method of rescaling or bias correction are required prior to assimilating observed soil moisture into LSMs. Reichle and Koster (2004) suggested a simple method of bias reduc-tion to match the cumulative distribution function (CDF) of satellite data and model data. Reichle and Koster (2005) and Reichle et al. (2007) showed for AMSR-E and SMMR retrievals and soil moisture estimates from the NASA Catchment model that if satellite-retrieved soil moisture was scaled to the model's climatology with the CDF method, estimates of soil moisture from assimilation were superior to either the satellite or the model alone when compared to in-situ observations. Following their work, Drusch et al. (2005) developed observation operators for the assimila-tion of soil moisture estimates by matching CDFs for the assimilation of soil mois-ture observations into the numerical models. The observation operator is a transfer function that corrects the bias and systematic differences between the retrieved and modeled soil moisture. Results showed that the operators vary by time, space and by land-surface model.

Drusch (2007) incorporated the TRMM Microwave Imager (TMI) surface soil moisture Pathfinder product over the southern United States into an operational forecast system. In this experiment the CDF matching (Drusch et al. 2005) was used to minimize the bias between the modeled and the satellite-derived surface soil moisture. The scaled data then were assimilated into the model through a nudging scheme using 6-hourly increments. Analyzed soil moisture and atmospheric forecasts were compared against independent observations from the Oklahoma Mesoscale Network. Soil moisture analyzed in the nudging experiment compared well with in situ observations. Furthermore, it was shown that the soil moisture analysis influences local weather parameters including the planetary boundary layer height and cloud coverage.

Gao et al. (2007) developed a bivariate statistical approach, based on copula distributions, to reduce bias between satellite observations and model predictions. In this method the conditional probability distribution of model-based soil moisture conditioned on satellite retrieval forms the basis for the soil moisture observation operator.

## 5.2 Downscaling Satellite-Derived Soil Moisture Data

Previous studies (e.g. Walker and Houser, 2001; Margulis et al., 2002; Crow and Wood, 2003) successfully tested the ability of the assimilation of surface brightness temperature observations to correct land-surface model predictions. However, these studies overcame the low spatial resolution of satellite data issues either by using a climatic scale grid size (>50 km) or high resolution airborne soil moisture imagery. The main limitation of the assimilation of passive microwave soil moisture for hydrological applications is the poor resolution obtained with current and near-future microwave based satellite data (50 km).

Since passive microwave remote-sensing signals give an average value of soil moisture over the sensor footprint and can not explain soil moisture variability, capturing the subpixel variability of microwave soil moisture requires that a variety of different downscaling methods be developed to distribute fine-scale soil moisture within the microwave pixel.

To bridge the gap between the resolutions of the surface soil moisture derived from passive microwave data and the input scale required for hydrological applications (about 1 km), a variety of methods have been developed to improve the resolution of passive microwave data. These methods use either statistical correlations between surface soil moisture (or brightness temperature) and fine-scale auxiliary data as in Bindlish and Barros (2002), Kim and Barros (2002), Chauhan et al. (2003), and Reichle et al. (2001b), or distributed physically- based models in conjunction with fine-scale auxiliary data as in Pellenq et al. (2003) and Merlin et al. (2005).

Zhan et al. (2006) conducted an experiment to study the feasibility of retrieving surface soil moisture at a medium spatial scale (10 km) from both coarse-scale (40 km) radiometer brightness temperature and fine-scale (3 km) radar backscatter cross-section observations using the EKF. Results showed that the combined active-passive EKF retrieval algorithm significantly reduced soil moisture error at the medium scale compared to the traditional soil moisture retrieval algorithms.

Merlin et al. (2006a, b) disaggregated SMOS data with the method of Merlin et al. (2005) and assimilated the data into a distributed soil-vegetation-atmosphere transfer scheme (SVAT) with the EnKF. The disaggregation method of Merlin et al. (2005) uses thermal and optical data remotely sensed at a typical resolution of 1 km to estimate the subpixel variability of microwave soil moisture. The performance of the disaggregation assimilation coupling scheme is then assessed in terms of surface soil moisture and latent heat flux predictions. Merlin et al. (2006b) found that the disaggregation improved the assimilation results. The assimilation of the disaggregated microwave soil moisture improved the spatial distribution of surface soil moisture at the observation time. Latent heat flux estimation was subsequently improved.

Alavi (2008) demonstrated that including information about the variability of soil moisture into a Kalman filter based assimilation procedure has the potential to improve latent heat flux estimation by a LSM. In this study 3 different techniques were examined to assimilate the soil moisture data collected from an agricultural field in Ontario into the Canadian Land Surface Scheme (CLASS). The techniques evaluated included 2 versions of ensemble Kalman filter (EnKF), and direct insertion of soil moisture data into the model. The results showed that assimilating field soil moisture variability into CLASS can improve model latent heat flux estimates by up to 14%. Application of EnKF, considering both instrumental error and field variability, resulted in greater improvement in latent heat flux estimates compared to the other two methods. This study showed that assimilation of soil moisture variability into CLASS can result in greater improvement in modeled ET compared to assimilation of the mean of the sampling area.

## 6 Summary

Land surface data assimilation is an efficient method to study land surface states and fluxes. These methods integrate observations into numerical LSMs to provide better estimates of the land-atmosphere system than the raw observations alone. This process is extremely valuable for providing initial conditions for hydrologic or LSM predictions, correcting model predictions, and increasing our understanding of hydrologic system behavior.

Numerous studies have successfully shown that assimilating observed soil surface moisture and temperature data into LSM has the potential to improve the model estimation of surface states and energy fluxes. Depending on the LSM, several data assimilation techniques have been proposed. Variational and extended Kalman filter have limited capability to deal with the model errors, and necessary linearization approximations can lead to unstable solutions. Ensemble Kalman filter and smoothers, while they can be computationally demanding (depending on the size of ensemble) are robust, very flexible and easy to use and are able to accommodate a wide range of model error descriptions; therefore, it they are well suited for hydrological applications.

Preliminary studies on assimilation of observed soil moisture using ground-based or airborne observations of land states into LSMs have shown improvements in the model estimations of soil moisture at the surface and at depth and latent heat flux. These studies prepared a path for using satellite-retrieved soil moisture data in the assimilation framework. However, there are still some limitations in assimilating satellite data into LSMs. The main limitations are: 1) the observed difference between satellite-retrieved and modeled soil moisture, 2) the poor spatial resolution obtained with current and near-future microwave- based satellite data.

Several studies reported the discrepancy between remotely-sensed data and the LSM estimation and differences in their statistical moments and for these reasons direct assimilation of measured soil moisture (satellite or ground) into a LSM is not appropriate. Therefore some method of rescaling or bias correction is required prior to assimilation of observed soil moisture into LSMs.

To overcome the low spatial resolution problem a variety of downscaling or disaggregation techniques have been developed to improve the resolution of passive microwave data. Disaggregation of satellite data can provide valuable information on subpixel soil moisture variability. It has been shown that including this information in the data assimilation procedure improves modeled latent heat flux estimation.

Due to the importance and utility of remote sensing data and our increasing technological capacity, it is expected that the availability of remote sensing data will increase considerably over the next several decades. The usefulness of this information is conditional on our understanding of the accuracy, scale, and calibration of the measurements. In future hydrologic studies, a comprehensive data assimilation framework will be an important component of land surface and hydrologic observation and modeling systems. Land surface data assimilation is still a work in progress, with many open areas for research. Important issues still to be addressed include: (i) improving the LSMs to provide more accurate estimation of land surface states and fluxes (ii) improving our knowledge of observation error and model errors in time and space; (iii) increasing the data assimilation computational efficiency for use in large scale operational applications; (iv) creating multivariate assimilation methods to use multiple observations in different time and spatial scales; (v) enhance data assimilation downscaling; (vi) collecting more appropriate ground-based data for validation of space-borne measurements.

# References

Alavi N (2008) Data assimilation techniques to improve evapotranspiration estimates. Doctoral Dissertation, University of Guelph

Bennett AF (1992) Inverse methods in physical oceanography. Cambridge University Press, Cambridge

Bernard R, et al (1981) Possible use of active microwave remote sensing data for prediction of regional evaporation by numerical simulation of soil water movement in the unsaturated zone. Water Resour Res 17(6):1603–1610

Bindlish, R, Barros AP (2002) Subpixel variability of remotely sensed soil moisture: An intercomparison study of SAR and ESTAR. IEEE Trans Geosci Remote Sens 40:326–337

Boni, G, et al (2001a) Sampling strategies and assimilation of ground temperature for the estimation of surface energy balance components. IEEE Trans Geosci Remote Sens, 39:165–172

Boni, G et al (2001b) Land data assimilation with satellite measurements for the estimation of surface energy balance components and surface control on evaporation. Water Resour Res 37:1713–1722

Bosilovich M, et al (2007) Skin temperature analysis and bias correction in a coupled land-atmosphere data assimilation system. J Meteorol Soc Japan 85A:205–228

Caparrini F, et al (2003) Mapping of land-atmosphere heat fluxes and surface parameters with remote sensing data. Boundary-Layer Meteorol., 107:605–633

Caparrini F, et al (2004a) Estimation of surface turbulent fluxes through assimilation of radiometric surface temperature sequences. J Hydrometeorol 5:145–159

Caparrini F, et al (2004b) Variational estimation of soil and vegetation turbulent transfer and heat flux parameters from sequences of multisensor imagery. Water Resour Res 40:W12515, doi:10.1029/2004WR003358

Castelli F, et al (1999) Estimation of surface heat flux and an index of soil moisture using adjoint-state surface energy balance. Water Resour Res 35:3115–3125

Chauhan NS, et al (2003) Spaceborne soil moisture estimation at high resolution: a microwave-optical/IR synergistic approach. Int J Remote Sens 24:4599–4622

Chen F, et al (1996) Modeling of land-surface evaporation by four schemes and comparison with FIFE observations, J Geophys Res 101:7251–7268

Courtier P, et al (1993) Important literature on the use of adjoint, variational methods and the Kalman filter in meteorology, Tellus Ser A 45:342–357

Crow W T, Wood EF (2003) The assimilation of remotely sensed soil brightness temperature imagery into a land surface model using Ensemble Kalman filtering: a case study based on ESTAR measurements during SGP97. AdvWater Resour 26:137–149.

Crow WT, Kustas WP (2005) Utility of assimilating surface radiometric temperature observations for evaporative fraction and heat transfer coefficient retrieval. Boundary-Layer Meteorol 115:105–130.

Daley R (1991) Atmospheric data analysis. Cambridge University Press, 460 pp.

Drusch M et al (2005) Observation operators for the direct assimilation of TRMM microwave imager retrieved soil moisture. Geophys Res Lett 32:L15403. doi:10.1029/2005GL023623

Drusch M (2007) Initializing numerical weather prediction models with satellite-derived surface soil moisture: Data assimilation experiments with ECMWF's Integrated Forecast System and the TMI soil moisture data set, J Geophys Res 112:D03102. doi:10.1029/2006JD007478.

Dunne S, Entekhabi D (2005) An ensemble-based reanalysis approach to land data assimilation. Water Resour Res 41:W02013. doi:10.1029/2004WR003449.

Dunne S, Entekhabi D (2006) Land surface state and flux estimation using the ensemble Kalman smoother during the Southern Great Plains 1997 field experiment. Water Resour Res 2:W01407. doi:10.1029/2005WR004334

Entekhabi D, et al (1994) Solving the inverse problems for soil-moisture and temperature profiles by sequential assimilation of multifrequency remotely-sensed observations. IEEE Trans.Geosci Remote Sens 32 438–448

Epstein ES (1969) Stochastic dynamic prediction. Tellus 21:739–759

Evensen G (1994) Sequential data assimilation with a nonlinear quasi-geostrophic model using monte-carlo methods to forecast error statistics. J Geophys Res-Oceans 99:10143–10162

Evensen G (2004) Sampling strategies and square root analysis schemes for the EnKF. Ocean Dynamics, 54:539–560

Famiglietti JS, Wood EF (1994) Application of multiscale water and energy-balance models on a tallgrass prairie. Water Resour Res 30:3079–3093

Galantowicz, et al (1999) Tests of sequential data assimilation for retrieving profile soil moisture and temperature from observed L-band radiobrightness. IEEE Trans.Geosci Remote Sens 37:1860–1870

Gao HL, et al (2007) Copula-derived observation operators for assimilating TMI and AMSR-E retrieved soil moisture into land surface models. J Hydrometeorol 8:413–429

Gao H, et al (2006) Using TRMM/TMI to retrieve surface soil moisture over the southern United States from 1998 to 2002. J Hydrometeorol 7:23–38

Ghil M, Malanotterizzoli P (1991) Data assimilation in meteorology and oceanography. Adv Geophys 33:141–266

Houser, et al (1998) Integration of soil moisture remote sensing and hydrologic modeling using data assimilation. Water Resour Res 34:3405–3420

Huang XY, Yang, X (1996) Variational data assimilation with the Lorenz model: High-resolution limited area model (HIRLAM), Tech Rep 26:42 pp., Danish Meteorol. Inst., Copenhagen.

Jackson TJ, et al (1981) Soil-moisture updating and microwave remote-sensing for hydrological simulation. Hydrol Sci Bull.-Bulletin DesSci Hydrologiques, 26:305–319

Jazwinski AH, (1970) Stochastic processes and filtering theory. Academic Press, New York, 376 pp.

Kalman RE (1960) A new approach to linear filtering and prediction problems. Trans AMSE, Ser. D, J. Basic Eng 82:35–45.

Kim GA, Barros P (2002) Downscaling of remotely sensed soil moisture with a modified fractal interpolation method using contraction mapping and ancillary data. Remote Sens Envir 83:400–413

Koster RD, et al (2000) A catchment-based approach to modeling land surface processes in a general circulation model 1. Model structure. J Geophys Res- Atmos 105:24809–24822

Koster RD, Milly PCD (1997) The interplay between transpiration and runoff formulations in land surface schemes used with atmospheric models. J Clim 10:1578–1591

Kustas WP, et al, (1996) Single- and dual-source modeling of surface energy fluxes with radiometric surface temperature. J. Appl Meteorol 35:110–121

Liu Y, Gupta HV (2007) Uncertainty in hydrologic modeling: Toward an integrated data assimilation framework. Water Resour Res 43:W07401. doi:10.1029/2006WR005756

Lorenc A (1986) Analysis methods for numerical weather prediction. Quart J Royal Meteorol Soc 112:1177–1194

Margulis SA, et al (2002) Land data assimilation and estimation of soil moisture using measurements from the Southern Great Plains 1997 Field Experiment. Water Resour Res 38 1299. doi:10.1029/2001WR001114

Merlin O, et al (2005) A combined modeling and multipectral/multiresolution remote sensing approach for disaggregation of surface soil moisture: Application to SMOS configuration. IEEE Trans Geosci Remote Sens 43:2036–2050

Merlin O, et al (2006a) A downscaling method for distributing surface soil moisture within a microwave pixel: Application to the Monsoon '90 data. Remote Sens Envir 101:379–389

Merlin O, et al (2006b) Assimilation of disaggregated microwave soil moisture into a hydrologic model using coarse-scale meteorological data. J Hydrometeorol 7:1308–1322

Pellenq J, et al (2003) A disaggregation scheme for soil moisture based on topography and soil depth. J. Hydrol, 276:112–127.

Radakovich JD, et al (2001) Results from global land-surface data assimilation methods. Proc. AMS 5th Symp. on Integrated Observing Sys. Albuquerque, NM, 14–19 January, 132–134

Reichle RH, et al (2001a) Variational data assimilation of microwave radiobrightness observations for land surface hydrology applications. IEEE Trans Geosci Remote Sens 39:1708–1718

Reichle RH, et al (2001b) Downscaling of radio brightness measurements for soil moisture estimation: A four-dimensional variational data assimilation approach. Water Resour Res 37:2353–2364.

Reichle RH, et al (2002a) Hydrologic data assimilation with the ensemble Kalman filter. Monthly Weath. Rev 130:103–114.

Reichle RH, et al (2002b) Extended versus ensemble Kalman filtering for land data assimilation. J Hydrometeorol 3:728–740.

Reichle RH, Koster RD (2004) Bias reduction in short records of satellite soil moisture. Geophys Res Lett 31:L19501. doi:10.1029/2004GL020938

Reichle RH, et al (2004) Global soil moisture from satellite observations, land surface models, and ground data: Implications for data assimilation. J Hydrometeorol 5:430–442.

Reichle RH, Koster RD (2005) Global assimilation of satellite surface soil moisture retrievals into the NASA Catchment land surface model. Geophys Res Lett 32:L02404.
doi:10.1029/2004GL021700

Reichle RH, et al (2007) Comparison and assimilation of global soil moisture retrievals from
the Advanced Microwave Scanning Radiometer for the Earth Observing System (AMSR-
E) and the Scanning Multichannel Microwave Radiometer (SMMR). J Geophys Res-Atmos
112:D09108,. doi:10.1029/2006JD008033

Sasaki Y (1960) An Objective analysis for determining initial conditions for the primitive equations. Tech Rep 208. Department of oceanography and Meteorology, A & M College of Texas.

Sun NZ (1994) Inverse problems in Groundwater modeling. Kluwer Academic, Dordrecht.

Walker JP, Houser PR (2001) A methodology for initializing soil moisture in a global climate model: Assimilation of near-surface soil moisture observations. J Geophys Res-Atmos
106:11761–11774.

Walker JP, et al (2001a) One-dimensional soil moisture profile retrieval by assimilation of near-
surface measurements: A simplified soil moisture model and field application. J Hydrometeorol
2:356–373.

Walker JP, et al (2001b) One-dimensional soil moisture profile retrieval by assimilation of near-
surface observations: a comparison of retrieval algorithms. Adv Water Resour 24:631–650

Walker JP, et al (2002) Three-dimensional soil moisture profile retrieval by assimilation of near-
surface measurements: Simplified Kalman filter covariance forecasting and field application.
Water Resour Res 38:1301. doi:10.1029/2002WR001545

Walker JP, Houser PR (2004) Requirements of a global near-surface soil moisture satellite mission:
Accuracy, repeat time, and spatial resolution. Adv Water Resour 27:785–801

Walker JP, Houser PR (2005) Hydrologic data assimilation. In: Advances in Water Science
Methodologies, edited by A. Aswathanarayana, 230pp. The Netherlands: A.A. Balkema.

Wilker H, et al (2006) Effects of the near-surface soil moisture profile on the assimilation of L-band
microwave brightness temperature. J Hydrometeorol 7:433–442.

Zhan X, et al (2006) A Method for Retrieving High-Resolution Surface Soil Moisture From Hydros
L-Band Radiometer and Radar Observations. IEEE Trans Geosci Remote Sens 44:1564–1544.

Zhou YH, et al (2006) Assessing the performance of the ensemble Kalman filter for land surface
data assimilation. Monthly Weather Rev 134:2128–2142

# Assimilation of a Satellite-Based Soil Moisture Product into a Two-Layer Water Balance Model for a Global Crop Production Decision Support System

John D. Bolten, Wade T. Crow, Xiwu Zhan, Curt A. Reynolds
and Thomas J. Jackson

**Abstract** Timely and accurate monitoring of global weather anomalies and drought conditions is essential for assessing global crop conditions. Soil moisture observations are particularly important for crop yield fluctuation forecasts provided by the US Department of Agriculture's (USDA) International Production Assessment Division (IPAD) of the Office of Global Analysis (OGA) within the Foreign Agricultural Service (FAS). The current system utilized by IPAD estimates soil moisture from a 2-layer water balance model based on precipitation and temperature data from World Meteorological Organization (WMO) and US Air Force Weather Agency (AFWA). The accuracy of this system is highly dependent on the data sources used; particularly the accuracy, consistency, and spatial and temporal coverage of the land and climatic data input into the models. However, many regions of the globe lack observations at the temporal and spatial resolutions required by IPAD. This study incorporates NASA's soil moisture remote sensing product provided by the EOS Advanced Microwave Scanning Radiometer (AMSR-E) to the U.S. Department of Agriculture Crop Assessment and Data Retrieval (CADRE) decision support system. A quasi-global-scale operational data assimilation system has been designed and implemented to provide CADRE with a daily soil moisture analysis product obtained via the assimilation of AMSR-E surface soil moisture retrievals into the IPAD two-layer soil moisture model. This chapter presents a methodology of data assimilation system design and a brief evaluation of the system performance over the Conterminous United States (CONUS).

## 1 Introduction

Crop yield forecasts affect decisions made by farmers, businesses, and governments by defining global crop production trends and influencing agricultural commodity markets and world food security. The temporal and spatial distribution of agricultural productivity varies widely from region to region and is a major influence on international food production and allocation. Forecasts of crop conditions can also be used to study land cover and land use change, agricultural efficiency, and

J. Bolten (✉)
USDA-ARS Hydrology and Remotes Sensing Lab, Beltsville, MD 20705

S.K. Park, L. Xu, *Data Assimilation for Atmospheric, Oceanic and Hydrologic Applications*, DOI 10.1007/978-3-540-71056-1_23,
© Springer-Verlag Berlin Heidelberg 2009

ecological forecasting. A growing interest in the possible linkages of land use, climate change, and agricultural productivity has increased the need for accurate crop forecasts at various scales. All such issues demonstrate the importance of establishing a reliable and timely means of monitoring not only the current status of global agricultural conditions, but also trends that could lead to the prevention of possible food shortages or signify the impact of regional climate change on agricultural productivity.

The Foreign Agricultural Service (FAS) of the US Department of Agriculture (USDA) is responsible for providing monthly global crop estimates and making predictions of expected commodity market access. These predictions are derived from merging many data sources including satellite and ground observations, and more than 20 years of climatology and crop behavior data over key agricultural areas. The FAS International Production Assessment Division (IPAD) utilizes a wide array of data sources to monitor global food supplies for early warning of food shortages and to provide greater economic security within the agriculture sector. One of the critical tasks of IPAD is to effectively process and integrate large volumes of climatic and land surface data into significant global agricultural assessments such as crop yield calendars and crop hazard algorithms. These assessments are driven by the synergistic merging of meteorological data, land surface models, and crop model forecasts. To most efficiently manage the large amount of data required for this analysis, IPAD has developed a series of analytical tools within a Crop Condition Data Retrieval and Evaluation (CADRE) Data Base Management System (DBMS).

The agricultural forecasting system used by IPAD is primarily driven by observations of precipitation, temperature and surface and sub-surface soil moisture from around the globe. To most efficiently model and monitor crop conditions of specific regions, a combination of station data and remotely sensed observations are used. These parameters are essential to the crop calendar (growth stage) and crop stress (alarm) models within the IPAD DBMS. For example, the crop yield models begin with the assumption of perfect yield, and decrease yield estimates based on crop stresses determined by the amount of estimated available root zone soil moisture, and crop model predictions (Fig. 1). Even though the DBMS receives merged satellite- and station-based precipitation and temperature data to drive these crop models, IPAD relies exclusively on sparse station data for soil moisture observations (Hutchinson et al. 2003). Due to the spatial heterogeneity and dynamic nature of precipitation events and soil wetness, accurate estimation of regional soil moisture dynamics based on sparse ground measurements is difficult (Brubaker and Entekhabi 1996).

This work aims at improving the IPAD surface and sub-surface soil moisture estimates by merging satellite-retrieved soil moisture observations with the current two-layer soil moisture model used within the DBMS. The improved temporal resolution and spatial coverage of the satellite-based EOS Advanced Microwave Scanning Radiometer (AMSR-E) relative to available station-based soil moisture observations is envisaged to provide a better characterization of regional-scale surface wetness and enable more accurate crop monitoring in key agricultural areas. An operational data assimilation system and delivery system has been designed utilizing Ensem-

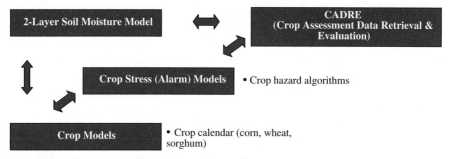

**Fig. 1** Linkage of soil moisture and crop stress models

ble Kalman filtering (EnKF) and calibrated using a suite of synthetic experiments and in situ validation sites. Output from this system is currently being operationally delivered to the IPAD DBMS in near real-time. Figure 2 shows a flow diagram of the integration scheme. This chapter provides an overview of the operational system and discusses the preliminary impact of AMSR-E soil moisture data assimilation on DBMS output. The following sections give an outline of the applied methodology.

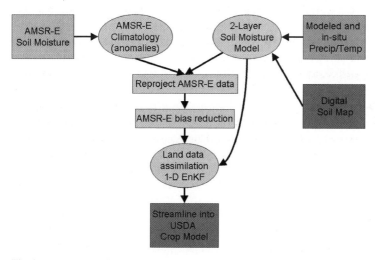

**Fig. 2** Data assimilation system flowchart

## 2 Baseline Datasets

The IPAD DBMS combines approximately seven thousand ground observations from the World Meteorological Organization and climatological estimates provided by the Air Force Weather Agency (AFWA). A major advancement within the IPAD DBMS has been the addition of satellite remote sensing technology within this framework. Satellite remote sensing improves our ability to acquire large-scale meteorological observations at relatively frequent temporal scales. Currently, the DBMS integrates precipitation and surface temperature estimates from coarse-resolution satellite observations provided by the Air Force Weather Agency (AFWA). Within the AFWA Agricultural Meteorology Model (AGRMET), precipitation and temperature observations from the Special Sensor Microwave/ Imager (SSM/I), the Advanced Very High Resolution Radiometer (AVHRR), and geostationary meteorological satellites including the Geostationary Operation Environmental Satellites (GOES) are converted into spatial meteorological information (Cochrane 1981). In addition, global vegetation conditions are monitored using Normalized Difference Vegetation Index (NDVI) and the Enhanced Vegetation Index (EVI) calculations provided by multiple sensors such as the Moderate Resolution Imaging Spectroradiometer (MODIS), Landsat, Advanced Very High Resolution Radiometer (AVHRR), and SPOT-VEGETATION (Huete et al. 2002; Tucker et al. 2005).

The merged products are stored in CADRE as 1/8th mesh grid cells, a stereographic projection with approximately 47 km horizontal grid spacing at 60 degree latitude and added to a baseline data set of historical crop production and climate norms. In this format, key agricultural areas are monitored for abnormal trends in soil moisture to provide an early alert for the onset of agricultural drought that may alter yield potentials or effect international market conditions.

The calculation of soil moisture and vegetation anomalies is fundamental for crop yield fluctuations estimated by CADRE. The departure of regional measurements of precipitation, evapotranspiration, and soil moisture from baseline values are used to signify weather anomalies and force crop hazard alarms. Traditional strategies in drought detection consist mainly of meteorological monitoring through ground or satellite-based observations, and relating these observations to ground-observed drought-affected-area maps (Kogan 1997). In addition, many proxy-based methods use long-term monitoring of energy balance processes to infer changes in regional wetness and/or crop yield (Allen et al. 1998). Drought detection within CADRE is based on a traditional method using temperature and precipitation extremes.

## 3 Two-Layer Water Balance Model

One of the most frequently used indices for regional drought detection is the Palmer Drought Severity Index (PDSI) (Palmer 1965). The PDSI has proven useful because of its ease of application and because it is standardized to facilitate comparisons of drought severity between different regions. The index uses a set of arbitrary rules to determine the intensity and duration of drought conditions based on estimates of soil wetness calculated from a simplified two-layer water balance model. In this manner, moisture deficits are classified based on indexed precipitation, evapotranspiration calculated from temperature extremes, and antecedent soil moisture conditions. Daily amounts of soil moisture withdrawn by evapotranspiration and replenished by precipitation are estimated at each grid cell (1/8th mesh) and adjusted for total water holding capacity based on local soil texture and depth of the soil column.

The IPAD DBMS uses a modified version of the Palmer water balance approach for estimating surface $Ls$, and subsurface $Lu$, soil moisture, as

$$L_s = \min[S_s, (PE - P)], \tag{1}$$

$$L_u = [(PE - P) - L_s]S_u/AWC_u \quad L_u \leq S_u, \tag{2}$$

on a daily timestep, where $PE$ is potential evapotranspiration, $P$ is precipitation, $Ss$ and $Su$ are the antecedent soil moisture values for the surface and second layers, respectively, and $AWC$ is the available water content for the location. The original Palmer model was modified by IPAD to apply the FAO Penman-Monteith equation described by Allen et al. (1998) in replacement of the Thornthwaite (1948) equation to compute PE from station latitude, longitude, elevation and daily temperature extremes. AWC is calculated from soil depth and type using the FAO's (1996) Digital Soil Map of the World (DSMW) AWC as described in (Reynolds et al. 2000). Soil moisture within the two layers is calculated at daily time increments and as depth of water (mm/day).

The top-layer soil moisture is assumed to hold a maximum of 25 mm of available water, and the sub-layer soil moisture is dependent on the soil's water-holding capacity. The sub-layer depth is variable up to a depth of one meter, depending on impermeable soil layers. Acting as an adapted bucket-type water balance model, the first model layer is filled completely by precipitation before the second layer is increased to soil water-holding capacity. When the soil water-holding capacity of both layers is reached, excess precipitation is treated as runoff and is lost from the model. No lateral movement of water is assumed.

IPAD also modified the Palmer model's extraction function to allow a more realistic diffusion of water between the two layers. In the modified form, moisture can be depleted from the lower layer before the surface is completely dry. The modified extraction function allows moisture to be depleted from the surface at the potential evapotranspiration rate until the layer reaches 75% of the surface capacity (or 75% of 25 mm of water). When the surface layer is below 75% capacity, soil moisture is extracted from the lower layer at a fraction of the potential evapotranspiration, calculated as a ratio of water in the sub-layer to the total water-holding capacity. From these assumptions, water holding capacity for both layers normally range from 127 to 203 mm/m of water depending on soil texture (ranging from sand to clay) and soil depth. A global porosity map is used to convert the soil moisture depth units to a volumetric metric compatible with the AMSR-E soil moisture product (volumetric units of $cm^3/cm^3$). Typical CADRE outputs of surface layer soil moisture for the United States from 07/19/2004 to 07/21/2004 are shown in Fig. 3.

## 4 EOS Advanced Microwave Scanning Radiometer

Remote sensing of soil moisture is possible due to the large contrast in the dielectric constant for dry soil and water, and the resulting dielectric properties of soil-water mixtures in the microwave spectrum (Schmugge 1985). These differences can

20040719_20040721

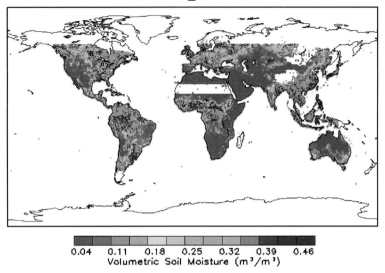

Volumetric Soil Moisture (m³/m³)

**Fig. 3** 3-day composite of the first layer of AFWA surface soil moisture (07/19/2004–07/21/2004)

provide quantitative estimates of soil moisture using physically based expressions derived via the parameterization the land/atmosphere emission pathways and solving the resulting radiative transfer equations (Njoku and Li 1999). Remote sensing at long microwave wavelengths ($\sim$2–30 cm) enables frequent observations of surface soil wetness in areas where point measurements are sparse or infrequently collected. Passive microwave remote sensing is appealing because of the low attenuation of surfaces signals by atmospheric water vapor (cloud transparency), and light vegetation (i.e. vegetation water contents $< 8 \, \text{kg/m}^2$) (Bolten et al. 2003). The validity of soil moisture remote sensing has been well documented (Entekhabi et al. 2004) and is now frequently assimilated into a number of land surface modeling systems (see e.g. Drusch 2007).

Principal to this study is the NASA EOS Advanced Microwave Scanning Radiometer (AMSR-E). AMSR-E is a microwave radiometer launched in 2002 aboard the National Aeronautics and Space Administration (NASA) Aqua satellite. Although there are several operational satellites providing multi-frequency brightness temperature observations, AMSR-E is the first satellite-based remote sensing instrument designed specifically for soil moisture retrieval. The launch of AMSR-E has improved the spatial resolution and frequency range upon earlier satellite-based passive microwave instruments. Since launch, AMSR-E has provided a daily global soil moisture product equivalent for the top 1–3 cm soil depth (Kawanishi et al. 2003).

Aqua follows a sun-synchronous orbit with a descending equatorial crossing at approximately 1330 Local Standard Time and measures vertically and horizontally polarized brightness temperatures at six frequencies: 6.92, 10.65, 18.7, 23.8, 36.5, and 89.0 GHz. Global coverage is possible every 2–3 days. At a fixed incidence

20040719_20040721

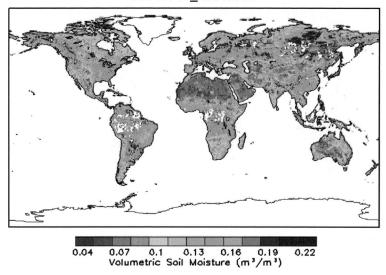

0.04    0.07    0.1    0.13    0.16    0.19    0.22
Volumetric Soil Moisture (m³/m³)

**Fig. 4** 3-day composite of the AMSR-E surface soil moisture (07/19/2004–07/21/2004)

angle of 54.8° and an altitude of 705 km, AMSR-E provides a conically scanning footprint pattern with a swath width of 1445 km. The mean footprint diameter ranges from 56 km at 6.92 GHz to 5 km at 89 GHz. For this system, the AMSR-E Level-3 soil moisture product is used (Njoku 2004). The Level-3 product is a gridded data product of global cylindrical 25 km Equal-Area Scalable Earth Grid (EASE-Grid) cell spacing. Observations of soil moisture are calculated from Polarization Ratios (PR) at 10.7 and 18.7 GHz, and three empirical coefficients used to compute a vegetation/roughness parameter for each grid cell. Deviations from a 18.7 GHz PR baseline value for each grid cell are used to calculate daily soil moisture estimates for each grid cell (Njoku et al. 2003). These soil moisture estimates represent a soil depth comparable to the first layer soil moisture used by the IPAD DBMS. Figure 4 illustrates the daily global coverage of the Level-3 soil moisture product.

## 5 Ensemble Kalman Filtering

Data assimilation is the merging of observations with model integrations for the purpose of reducing uncertainty in model states. Traditional sequential data assimilation techniques use auto-recursive analyses to optimally merge model estimates and state observations. The ensemble Kalman filter (EnKF), for instance, uses a Monte Carlo-based ensemble to propagate the error covariance information required by the Kalman filter update equation (Evensen 1994). The Monte Carlo adaptation within the EnKF provides a nonlinear extension of the standard Kalman filter and has been successfully applied to land surface forecasting problems for a large number of studies. A major benefit of the EnKF is that it can be successfully applied to more highly non-linear models than the standard Kalman filter or the Extended Kalman filter (Reichle et al. 2002).

As in most sequential data assimilation schemes, the application of the filter is divided into forecasting and updating steps. Formally, the forecasting step can be expressed as

$$x^i_{k+1} = f(x^i_k, w_k) \qquad (3)$$

where $f$ is a realization of the (potentially non-linear) forecast model, $x^i_k$ is the model's state vector for the $i$th ensemble member at time $k$ and $w_k$ represents uncertainty (with covariance $Q$) added to each state estimate as it is propagated in time from $k$ to $k+1$ by (3). Data assimilation is based on the availability of some observation $y$ that can be related to $x$ via a known observation operator $H$

$$y^i_k = H_k(x^i_k) + v_k \qquad (4)$$

where $v_k$ represents a random perturbation of observation. Such perturbations are assumed to be Gaussian with a known covariance of $R$. Within the EnKF, the error covariance of both the forecasted observations – or $H_k(x^i_k)$ – ($CM_k$) and the cross-correlation between these observations and each forecasted state variable ($CYM_k$) are calculated by sampling across the ensemble created by the Monte Carlo realization of (3).

The updating step of the EnKF utilizes this error covariance information to optimally update forecasts in response to observations. Such updating is based on the calculation of the Kalman gain as

$$K_k = \frac{CYM_k}{(CM_k + R)} \qquad (5)$$

and the application of the Kalman filter updating equation individually to each realization within the ensemble

$$x^{i,+}_k = x^{i,-}_k + K_k \left[ y^i_k - H_k(x^{i,-}_k) + v^i_k \right]. \qquad (6)$$

where "$-$" and "$+$" notation is used to signify state estimates made before and after updating in response to observations at time $k$. Note that the additive $v^i_k$ term in (6) is required to ensure the proper posterior spread in state estimates after updating (Burgers et al. 1998). The EnKF state estimate at time $k$ is given by simply taking the mean of this updated ensemble.

Our particular implementation of the EnKF integrates soil moisture observations from AMSR-E with the modified Palmer two-layer soil moisture model described in Sect. 3 by applying a 1-dimensional EnKF at daily time-steps when AMSR-E observations are available. The model operator $f$ represents the 2-Layer Palmer introduced in Sect. 3. Since AMSR-E observations are pre-processed into surface soil moisture estimates (assumed to be consistent with the top layer of the Palmer model), our observation operator is simply $H = (1,0)$. However, before AMSR-E soil moisture retrievals can be safely assimilated, the modeled (IPAD) and observed (AMSR-E) data must be scaled to a common climatology to reduce potential biases and differences in dynamic range that commonly exist between modeled and observed soil moisture products (Drusch et al. 2005). For our application, a retrospective analysis of archived AMSR-E and IPAD datasets from June 2002 to June 2007 was performed to establish a representative climatology for both AMSR-E and IPAD soil moisture estimates. Based on these climatologies, a cumulative distribution function (CDF) matching algorithm (see Reichle and Koster 2004) was employed to transform individual AMSR-E retrievals such that their transformed climatology is comparable to that of IPAD surface soil moisture estimates. The climatologically re-scaled AMSR-E data were then applied to an Ensemble Kalman Filter (EnKF) using sequential observations of AMSR-E, AFWA, and WMO climatological data.

# 6 Filter Tuning

A critical element of applying an EnKF is the accurate specification of both the modeling ($Q$) and observation ($R$) error covariances. The misspecification of these error covariance values has been shown to degrade the performance of the EnKF in land data assimilation applications ( Crow and Van Loon 2006). An effective tool for evaluating filter performance is examination of the normalized filter innovations. Normalized filter innovations are defined as the difference between the model background forecast and the observation, normalized by the scalar forecast error and scalar model error

$$v_k = \frac{E\left[y_k^i - H_k(x_k^{i,-})\right]}{\sqrt{(C_M + R)}} \tag{7}$$

where E[.] represents an ensemble averaging operator. A properly working data assimilation system will most efficiently merge the two data sources to reduce state uncertainty, and yield normalized innovations that are serially uncorrelated in time and have a temporal second moment of 1 (Mehra 1971). For this particular EnKF application, the accuracy of remotely sensed soil moisture observations ($R$) varies greatly over different land cover types due to signal attenuation by vegetation and increased scattering over rough terrain. At the wavelengths used by AMSR-E, the accuracy of observed soil moisture is significantly degraded over areas of vegetation water content greater than approximately $10 \, \text{kg/m}^2$ (Jackson and Schmugge 1991; Njoku et al. 2004). These areas of increased uncertainty in our observations must be considered when applied in a data assimilation framework.

To parameterize our implementation of the EnKF, $Q$ was set to an assumed value ($0.04 \, \text{cm}^3/\text{cm}^3$) and $R$ tuned until the resulting innovation time series (7) was serially uncorrelated. Since retrieval error magnitude in remotely sensed soil

20040719_20040721

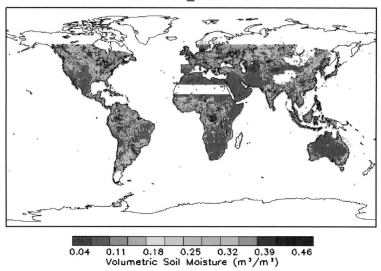

0.04    0.11    0.18    0.25    0.32    0.39    0.46
Volumetric Soil Moisture (m³/m³)

**Fig. 5** 3-day composite of the surface soil moisture output derived from the EnKF (07/19/2004–07/21/2004)

moisture data sets vary strongly as a function of land cover type, such calibration was performed independently for three distinct land cover types: (1) evergreen and deciduous forests (2) wooded grasslands and mixed forests, and (3) bare soil, grasses and open shrub land. The mean value of observation error that satisfied the innovation statistics for each land cover type was then applied to all global areas possessing a similar land cover type. Such tuning ensures that the filter is placing the proper relative weight on model predictions and remote sensing observations when calculating an analysis soil moisture product with minimized error (Crow and Bolten 2007). For the same time period as Figs. 3 and 4, Figure 5 gives an example of the quasi-global scale surface soil moisture output derived from the EnKF described above. Similar imagery is currently being operationally delivered to the USDA IPAD system.

## 7 Preliminary Results

The evaluation of large-scale soil moisture estimates like those in Figs 3, 4, and 5 is greatly complicated by a lack of adequate ground-based soil moisture observations for validation of large-scale soil moisture fields. One of only a few sufficiently data-rich sites suitable for ground-based validation of large-scale soil moisture products is the Walnut Gulch Experimental Watershed near Tombstone, Arizona. The watershed is equipped with an extensive soil moisture and precipitation monitoring network supported by the USDA, the Natural Resource Conservation Service, local Soil Conservation Districts, and local land owners within the watersheds. The approximately $150\,km^2$ watershed is located in a sub-arid environment with predominant vegetation cover consisting of desert shrubs and sparse grassland. Soil textures within the watershed range from alluvial deposits with high gravel content to sandy loam. The combination of low vegetation levels within the region, and the availability of heavily monitored soil moisture and precipitation data concurrent with AMSR-E overpasses makes the Walnut Gulch an ideal location for evaluation of our system.

To demonstrate proof of concept, properly re-scaled AMSR-E soil moisture observations (see Sect. 5) were assimilated into the Palmer model over the Walnut Gulch Watershed for a duration of 20 months at daily time steps. For in situ comparisons within the basin, daily watershed averages of the networked top layer (1–5 cm) soil moisture and precipitation data were calculated. In addition, the mean of the ascending and descending AMSR-E Level 3 data product located within the watershed were used. It is important to note that the in situ soil moisture network sensing depth does not match that of AMSR-E (1–3 cm) or the top layer of the modified Palmer model (2.5 cm) and is included here to demonstrate the temporal soil moisture trend of the top layer. Figure 6 shows a time-series of the baseline IPAD soil moisture product (blue line), the climatologically re-scaled AMSR-E soil moisture product (black line), and the integrated EnKF product from sequential observations

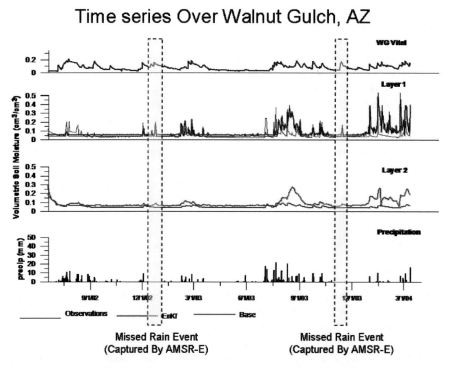

**Fig. 6** Time series of FAS and assimilated products over Walnut Gulch, AZ. The benefit of the EnKF is demonstrated by the precipitation events that were missed by the base run (*blue line*) being observed in both layers of the EnKF run (*red line*)

of AMSR-E and AFWA meteorological input data (red line). Temporal patterns of all three products are validated against the watershed mean in situ soil moisture and rain gauge measurements.

It is evident from Fig. 6 that even though the dynamic ranges of the in situ soil moisture and model forecasts vary, they have a strong correlation with observed precipitation events within the watershed. Comparison of the two layers indicates that the EnKF is able to translate information from AMSR-E to the second layer of the Palmer model, demonstrating the improved performance of the system in areas of precipitation uncertainty. To emphasize this concept, the yellow bars highlight periods when precipitation events were observed by AMSR-E and applied to the EnKF update, yet missed by the AFWA-forced model run. These initial results are promising and imply our EnKF method can offer valuable information to both layers of the integrated two-layer model.

A second evaluation method of the assimilation system was performed over the Conterminous United States (CONUS) area using a data denial strategy. Within this framework, an assessment of the filter performance was accomplished by comparing two model runs – one forced with error-prone precipitation and a second realization forced with a benchmark precipitation product of greater accuracy. Here, we used the TRMM real-time, microwave-only 3B40RT precipitation product as the error-prone precipitation product. This is a real-time product generated solely from satellite observations and without the aid of any ground-based rain rate observations (Huffman et al. 2007). By comparison, our benchmark rain product was the merged AFWA-WMO precipitation product discussed earlier. This product is heavily corrected using ground-based rain gauges measurements and, consequently, is relatively more accurate in data-rich areas than TRMM 3B40RT. The data denial approach is based on three separate model runs over a five-year duration at daily time steps within CONUS. The individual model runs were forced by (1) AFWA-WMO precipitation (i.e. benchmark loop) (2) TRMM 3B40RT precipitation (i.e. open loop) and (3) TRMM 3B40RT precipitation and the EnKF assimilation of AMSR-E soil moisture retrievals (i.e. EnKF loop). In this way, the application of the EnKF to assimilate remotely-sensed soil moisture retrievals in case 3 could be evaluated based on how efficiently it transformed the low-accuracy results in case 2 (generated with the least accurate rainfall product) to match benchmark results in case 1 (generated using the most accurate rainfall product).

Figure 7 shows spatial maps of the forecasted first layer soil moisture over CONUS on 07/21/2004 for all three schemes discussed above (open loop TRMM 3B40RT in Fig. 7a, benchmark AFWA run in Fig. 7c and the AMSR-E/TRMM/EnKF run in Fig. 7d). Also shown is a spatial map of the scaled AMSR-E soil moisture product for the same time period (Fig. 7b). It is apparent when comparing 7a and 7d that the assimilation of AMSR-E soil moisture via the EnKF adds spatial heterogeneity to the open loop case. This emergent heterogeneity allows the EnKF case to better approximate soil moisture patterns in the benchmark case (Fig. 7a) by adding soil moisture in areas (e.g. the south-central and northeastern CONUS) where the TRMM 3B40RT product underestimates antecedent precipitation magnitudes. This demonstrates that the AMSR-E soil moisture retrievals are able to

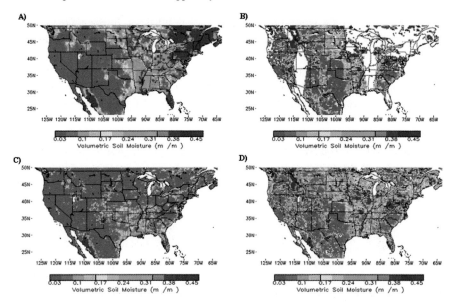

**Fig. 7** Data denial experiment results over CONUS for 07/21/2004. (**A**) Benchmark loop (**B**) re-scaled AMSR-E Soil moisture product (**C**) Open loop and (**D**) EnKF results

effectively compensate modeled soil moisture for the impact of poorly observed rainfall patterns. It is worth noting that over many important – but data poor – agricultural regions of the world, such compensation is critically important due to the lack of ground-based rainfall observations to correct TRMM 3B40RT rainfall estimates.

# 8 Summary and Description of Future Work

IPAD is continually assessing and updating their data base management system to provide timely and accurate estimates of global crop conditions for use in up-to-date commodity intelligence reports. Significant progress has been made in the application of formally integrated models of atmosphere and land-based water cycle to monitor regional land surface processes, namely soil moisture and modeled crops such as corn and soy. This work has demonstrated a currently running data assimilation system that is operating in near-real time to provide the IPAD Crop Condition Data Retrieval and Evaluation (CADRE) Data Base Management System (DBMS) with an additional soil moisture product of improved spatial-temporal scale. The improved spatial structure of wetness patterns demonstrated in Figs. 6 and 7 suggest that the system holds promise for improved characterization of surface wetness conditions at the regional scale and enable more accurate monitoring of soil moisture changes in key agricultural areas.

Future work includes extending our diagnostic calibration of the filter to areas of the globe lacking in sufficient meteorological observations where the IPAD system is more susceptible to agricultural forecasting errors. Further testing will also include additional comparisons to heavily monitored in situ soil moisture ground networks and a more quantitative (and longer term) analysis of data denial results in Fig. 7 to clarify the added impact of AMSR-E soil moisture data assimilation on the accuracy of IPAD soil moisture estimates.

# References

Allen RG, Pereira LS, Raes D, and Smith M (1998) Crop evapotranspiration; guidelines for computing crop water requirements. FAO Irrigation and Drainage Paper 56

Bolten JD, Lakshmi V, Njoku EG (2003) Soil moisture retrieval using the passive/active L- and S-band radar/radiometer. IEEE Trans Geosc Rem Sens 41: 2792–2801

Burgers G, van Leeuwen PJ, Evensen G (1998) Analysis scheme in the Ensemble Kalman filter. Mon Wea Rev126: 1719–1724

Brubaker KL, Entekhabi D (1996) Analysis of feedback mechanisms in land-atmosphere interaction. Water Res Res 325: 1343–1357

Cochrane MA (1981) Soil moisture and Agromet models. Technical Report, USAF Air Weather Service (MAC), USAFETAC, Scott AFB, Illinois, TN-81/001

Crow WT, Van Loon E (2006) The impact of incorrect model error assumptions on the assimilation of remotely sensed surface soil moisture. J Hydr 8(3): 421–431

Crow WT, Bolten JD (2007) Estimating precipitation errors using space borne remote sensing retrievals. doi:10.1029/2007GL029450. Geophys Res Lett 34: L08403

Crow WT, Zhan X (2007) Continental scale evaluation of remotely sensed soil moisture retrievals. Geosci Rems Sens Lett 4(3): 451–455

Drusch M, Wood EF, Gao H (2005) Observation operators for the direct assimilation of TRMM microwave imager retrieved soil moisture. Geophys. Res Lett 32: L15403, doi:10.1029/2005GL023623

Drusch M (2007) Initializing numerical weather prediction models with satellite derived surface soil moisture: Data assimilation experiments with ECMWF's Integrated Forecast System and the TMI soil moisture data set. J Geophys Res 112(D3): D03102, doi:10.1029/2006JD007478

Entekhabi D, Njoku E, Houser P, Spencer M, Doiron T, Kim Y, Smith J, Girard R, Belair S, Crow W, Jackson T, Kerr Y, Kimball J, Koster R, McDonald K, O'Neill P, Pultz T, Running S, Shi J, Wood E, van Zyl J (2004) The Hydrosphere State (Hydros) mission concept: An earth system pathfinder for global mapping of soil moisture and land freeze/thaw. IEEE Trans Geosci Rem Sens 42: 2184–2195

Evensen G (1994) Sequential data assimilation with a nonlinear quasi-geostrophic model using Monte Carlo methods to forecast error statistics. Geophys Res 99: 10143–10162

FAO (1996) The Digitized Soil Map of the World Including Derived Soil Properties. CD-ROM, Food and Agriculture Organization, Rome

Huete A, Didan K, Miura T, Rodriguez EP, Gao X, Ferreira LG (2002) Overview of the radiometric and biophysical performance of the MODIS vegetation indices. Rem Sens Env 83(1–2): 195–213

Hutchinson C, van Leeuwen W, Drake S, Kaupp V, Haithcoat T (2003) Characterization of PECAD's DSS: A zeroth-order assessment and benchmarking preparation. NASA Goddard Space Flight Center, Greenbelt, MD, NASA/PECAD. http://www.asd.ssc.nasa.gov

Huffman GJ, Adler RF, Bolvin DT, Gu G, Nelkin EJ, Bowman KP, Hong Y, Stocker EF, and Wolff DB (2007) The TRMM multi-satellite precipitation analysis: Quasi-global, multi-year, combined sensor precipitation estimates at fine scale. J Hydrometeorol 8: 28–55

Jackson TJ, Schmugge TJ (1991) Vegetation effects on the microwave emission of soils. Rem Sens Env 363: 203–212

Kawanishi T, Sezai T, Ito Y, Imaoka K, Takeshima T, Ishido Y, Shibata A, Miura M, Inahata H, Spencer R (2003) The Advanced Microwave Scanning Radiometer for the Earth Observing System (AMSR-E), NASA's contribution to the EOS for global energy and water cycle studies. IEEE Trans Geosci Rem Sens 41: 184–194

Kogan FN (1997) Global drought watch from space. Bull Am Meteorol Soc 78: 621–636

Mehra RK (1971) On-line identification of linear dynamic systems with applications to Kalman filtering. IEEE Trans Auto Control 16: 12–21

Njoku EG, Li L (1999) Retrieval of land surface parameters using passive microwave measurements at 6–18 GHz. IEEE Trans. Geosci Rem Sens 7(1): 77–93

Njoku EG, Jackson TJ, Lakshmi V, Chan TK, Nghiem SV (2003) Soil moisture retrieval from AMSR-E. IEEE Trans Geosci Rem Sens 41(2): 215–229

Njoku EG (2004) AMSR-E/Aqua daily L3 surface soil moisture, interpretive parameters & QC EASE-Grids. Boulder, CO, USA, National Snow and Ice Data Center. Digital media, March to June

Njoku EG, Chan T, Crosson W, Limaye A (2004) Evaluation of the AMSR-E-E data calibration over land. Ital J Rem Sens 30/31: 19–38

Palmer WC (1965) Meteorological drought, U.S. Weather Bureau Research Paper 45, pp 58

Reichle RH, Koster RD (2004) Bias reducton in short records of satellite soil moisture. Geophys Res Lett 31: L19501, doi10.1029 /2004GL020938

Reichle RH, McLaughlin DB, Entekhabi D (2002) Hydrologic data assimilation with the Ensemble Kalman Filter. Mon Weat Rev 130: 103–114

Reynolds CA, Jackson TJ, Rawls WJ (2000) Estimating Soil Water-Holding Capacities by Linking the FAO Soil Map of the World with Global Pedon Databases and Continuous Pedotransfer Functions. Wat Resours Res, December, 36(12): 3653–3662

Schmugge T (1985) Remote Sensing of Soil Moisture. *Hydrological Forcasting*. M. G. A. a. T. P. Burt. John Wiley and Sons, New York, 604

Thornthwaite CW (1948) An Approach Toward a Rational Classification of Climate. Geograph Rev 38: 55–94

Tucker CJ, Pinzon JE, Brown ME, Slayback D, Pak EW, Mahoney R (2005) An extended AVHRR 8-km NDVI data set compatible with MODIS and SPOT Vegetation NDVI data. Int J Rem Sens 26(20): 4485–4498

# Index

3-D VAR, 25
4DPSAS, 191
4-D VAR, 21, 39
4DVAR, 9, 165, 344
ACARS, 203
Accounting for model error, 374
Accurate data assimilation, 121
Adamas-Bashforth scheme, 235
Adaptive EnKF, 412
Adaptive observation(s), 56, 231, 250, 257
   strategies, 371
Adjoint-based calculations, 187
Adjoint-based targeted observation, 253, 266
Adjoint of data assimilation, 194
Adjoint method(s), 28, 36
Adjoint model, 256, 378
   backwards, 33
Adjoint operator approach, 32
Adjoint sensitivity method, 190
Adjoint state method, 369
Adjoint variable(s), 49, 372
Adjoint weights, 185
AFWA, 447
Agricultural assessments, 448
AIRS, 167
Albedo and emissivity, 406
AMeDAS, 202
Amplification factor, 246
AMSR-E, 172, 413, 449
AMSU-A, 167
Analysis error, 238
   covariance, 30, 74
      matrix, 52
Anti-cyclonic circulation, 245
Anti-cyclonic meander, 281
AOSN II, 269
AQUA AIRS, 190

Arakawa C-grid, 334
Artificial bottom topography, 362
Asian weather systems, 253
Aspect ratio limit, 295
Asselin filter, 224, 334
Assimilation
   algorithm, 326
   interval(s), 81, 204, 216, 324, 327
   of precipitation, 75
   run, 348
   window, 337, 432
Asymptotic values, 375
Atlantic thermohaline circulation, 294
Atmospheric forcing, 283
ATOVS, 203
Attracter, 106
Augmented Lagrangian algorithm, 35
Automatic differentiation (AD), 43, 398
   engine, 398
Auto-recursive analyses, 453
Autoregressive correlation model, 272
AVHRR, 276
Azimuth direction, 159, 204

Background error, 31, 290
   correlation, 183
   covariance, 88, 180, 271
   variances, 272
Background field, 25
Background sensitivity vectors, 181
Backscatter cross-section observations, 441
Backward integration, 379
Baiu front, 199
Balance equation, 9
Baroclinic instability, 3
Baroclinic structures, 245
Bayes rule, 73, 294

Benchmark precipitation product, 458
Best model, 376
Best set of data, 376
BFGS minimization algorithm, 53
Bias correction, 435
Bias errors, 284
Bias of estimates, 360
Bias removal, 440
Bogus vortices, 168
Bottom drag, 330
Bulk heat transfer coefficient, 434

CADRE, 447, 459
California current, 274
Canonical base vectors, 51
CAPE, 222
Catchment
    hydrology, 391
    land surface model, 407
CHAMP, 197
Channel geometry, 132
Checkpointing, 43
CLASS, 442
Clebsch's transformation, 105
Climate hybrid scan tables, 160
Closure assumption, 224
Cloud-physical processes, 104
Cloud-resolving model, 78
CMC, 407
COADS, 344
COAMPS, 270
Coastal topography, 275
CODAR, 270
Cold water, 279
    upwelled, 290
Computational efficiency, 358
Conditional nonlinear optimal perturbation,
    231
Conditional probability density function, 294
Condition number, 311
Conjugate–gradient iterate, 312
Conservation of total enstrophy, 27
Constrained minimization, 34
Continuity equation, 393
Control
    run, 348
    space, 52, 368
    theory, 356
    variable, 89
    vector size, 42
Convective mixing, 118
Convergence, 375
Coriolis effect, 103
Cost function, 37, 188, 378, 389

Coupled land-atmosphere data assimilation,
    422
Covariance
    forecasting equations, 436
    inflation, 346
    localization, 90, 346
Crop yield fluctuation forecasts, 447
Cross-correlated perturbation, 409
Cross correlation, 454
Cross-radial interpolation, 160
Cross-shelf current, 335
Cumulative distribution function, 440
Cumulative infiltration, 386, 396
Cumulus parameterization, 256
CWB, 129
Cycling
    algorithm, 332
    assimilation, 331
    data assimilation system, 414
    experiment, 337
    lengths, 337
    ocean data assimilation, 289
    representer method, 322
Cyclonic circulation, 283
Cyclonic rotation, 114

Data
    holes, 159
    and models, 403
    quality-control, 272
    -sensitive region, 232
Data assimilation, 287
    accurate, 121
    adjoint of, 194
    coupled land-atmosphere, 422
    cycle, 91
    cycling, 414
    ensemble, 67, 70, 86
    land, 424
    land surface, 405
    ocean, 289
    satellite, 164
    sequential, 359
    systems, 194
    techniques, 164, 428
DBMS, 459
Decoupling Euler-Lagrange equations, 38
Degrees of freedom (DOF), 75
Degrees of freedom for signal, 55
Dependent variable, 139
Derivatives of function, 371
Desk calculators, 4
Deterministic framework, 368
Deterministic sensitivity analysis, 371

Diagonal matrix, 206
Diffusive transport approach, 134
Digital Soil Map of the World (DSMW), 451
Dimensionless time, 325
Dipole, 119
Direct insertion, 436
Direct integration, 51
Direction of descent, 370
Direct solution methods, 73
Disaggregation of satellite data, 443
Discretization error, 42
Discretized model, 387
Diurnal timescale, 42
Divergent component, 107
DOF, 93
Doppler radar, 99
Downburst analysis, 121
Downdraft, 101
Drought detection, 450
Dry energy metric, 255
Dual polarization radar, 98
Dual-pol radar, 162
D-value, 200, 211
Dynamical balance of analysis, 75
Dynamical correlation, 267

East China, 257
ECCO, 344
Echo vault, 114
ECMWF, 198, 232
E dimension, 86
Efficient coefficient, 137
Eigenvalues/eigenvectors, 314
Ekman transport and pumping, 280
El Nino, 234
Empirical approach, 421
Empirical covariance matrix, 381
Empirical parameters, 381
EnKF, 220, 293, 341, 407, 449
EnKF/EnKS, 364
Ensemble
    averaging operator, 455
    -based filters, 293
    collapse, 300
    data assimilation, 67, 70, 86
    forecasting technique, 250
    forecast(s), 82, 237
    means, 105
    members, 355
    observations, 430
    perturbation(s), 249
    prediction, 238
    spread, 349
    transform technique, 273

Ensemble Kalman filter (EnKF), 353, 430
Ensemble moving batch (EnMB), 438
EnSP, 341
Entropic balance model, 97, 101, 113
Entropic gradient, 109
Entropic point source, 119
Entropic sink, 102
Entropy
    -based filters, 304
    gradient, 111
    increase, 108
    relative, 302
    Shannon, 85
Environmental flow, 113
Environmental modeling, 143
Environmental wind-shear, 120
EOF, 361
Equator-to-pole salinity gradient, 296
Equilibrium soil moisture profile, 409
Error covariance, 212
    localization, 77
    matrices, 432
Error growth, 288
Estimated model error, 326
Estimation error, 411
ETS, 223
Euler-Lagrange (EL)
    equation, 26
    problem, 315
Evapotranspiration, 427
Explanatory variable, 139
Exponential–families, 301
Extended Kalman filter, 408, 430

FAS, 448
FASTEX, 177, 233, 254
FCG, 315
Filter divergence, 81, 347
Finite element scheme, 382
Finial total energy, 244
Finite resampling, 359
First guess, 338
    path, 209
    refractivity, 206
First order necessary conditions, 45
Fixed lag smoother, 354
Fixed time window, 439
Flash floods, 131
    model, 391
Flexible conjugate gradient
    method, 317
    solver, 310
Flood hydrograph, 400
Flow dependence, 31

FNMOC, 189, 270
FNWC, 10
Force-restore approximation, 437
Forecast covariance matrix, 346
Forecast error covariance, 70
Forecast errors, 186
    reduction in, 193
Forecast skill(s), 237, 285
FORMOSAT–2, 132
Forward operator, 179, 271
Four-dimensional variational (4-Dvar), 236
Free-surface ocean model, 334
Friction
    coefficient, 399
    slope, 394

GA-estimation, 226
Gain matrix, 37
GARP, 23
Gateaux derivative, 48
Gauged river flow, 417
Gauge-radar biases, 157
Gaussian PDF, 74
Gaussian probability density, 298
Gauss's work, 9
Generalized inverse, 55
GENESIS, 200
Genetic algorithm, 220
Geopotential height, 264
Geopotential tendency, 6
Geostrophic wind law, 8
Gibbs Flite Center, 276
Global Band Analysis, 10
Global domain, 90
Global observation
    impact, 187
    network, 192
Global optimization, 225
Global soil moisture fields, 423
Global Soil Wetness Project 2, 410
Global solution, 328
GMAO, 405
GMAO-LDAS, 405
GODAE, 269
GOES–8, 166
GPS, 197
GPS-PWV, 214
GRACE, 416
Gradient
    based minimization, 33
    -based optimization algorithms, 381
    calculation, 260
    of J, 258
    norm of, 390

Graphical methods, 4
Grid projection, 274
Ground-based rainfall observations, 459
Ground water
    estimates, 415
    time series, 417
GSI, 214
GSM, 198
Gulf of Mexico, 329
Gust front, 100

Hamilton matrix, 255
Haurwitz wave, 36
Heat capacity, 108
Heat diffusion model, 409
Heavy precipitation bands, 154
Heavy rainfall, 227
Hessian matrix, 50
Hessian preconditioning, 68
Hessian singular vector (HSV), 53
Hessian/vector, 44
Heterogeneous sources of information, 402
HF radar, 281
High-density aircraft SST, 289
High-dimensional problems, 68
Higher order statistics, 356
High-frequency buoy data, 289
High-pass filter, 226
High surface variability, 285
Horizontal error correlations, 438
Horizontal momentum, 102
Horizontal perturbation vorticities, 118
HRPT, 280
HRQ, 152
Hurricane
    forecasts, 168
    intensity, 170
    track forecast, 170
Hurricane Katrina, 79
Hybrid, 76
    scan reflectivity, 151
    variational scheme, 172
HYCOM, 349
Hydraulic conductivity, 401
    at saturation, 385
Hydrodynamics, 134
Hydrographs
    flood, 400
    simulated, 142
Hydrological applications, 428
Hydrological processes, 367, 373
Hydrologic and landslide modeling, 127
Hydrology, 369

Imperfect model structure, 428
Improved prediction, 80
Incremental algorithm, 55
Incremental formulation, 39
Independent sub-domains, 90
Inertia-gravity wave shock, 27
Infiltration
    losses, 399
    model, 383
    rate, 393
Information
    matrix, 87
    measures, 85
    theory, 85
Infrared radiation pyrometer, 277
Initial analysis, 264
Initial condition, 264
Initial condition errors, 325
Initial total perturbation energy, 244
Inner core data, 169
Inner and outer loops, 310
Inner product, 46
Innovations, 284
    statistics, 31
    vector, 56, 178, 271
Integration by parts, 50
Intense precipitation, 212
Intense rainfall, 202
Intensity of vortex, 107
Interpolation and estimation, 354
Inverse distance weighted method, 133
Inverse Hessian matrix, 52
Inverse problem, 377
IOM, 307
IPADDBMS, 449
IPAD soil moisture estimates, 460
Irrotational flow, 107
ISDC, 199
Iterative method, 339
Iterative power method, 235
Iterative process, 389
Iterative solution methods, 73

Jacobian matrix, 179, 400
Jacobian model, 372
JMA, 3

Kain-Fritsch, 228
Kalman
    filter, 71, 429
    gain matrix, 56, 183, 345
KF scheme, 224
Kinetic analogy, 380
KMA, 228

Korean peninsula, 227
K-Profile Parameterization, 343
Kuroshio, 342
KWA, 391

L2 norm, 237
Lagged innovations covariance, 412
Lagrange multipliers, 26, 105, 301
Lagrangian approach, 32
Lagrangian density, 103
Lambert conformal, 274
Lanczos algorithm, 317
Land-atmosphere fluxes, 433
Land-based radars, 342
Land cover type, 456
Land data assimilation, 424
Land Information System, 424
Landslide
    data, 128
    occurrence, 138
Land surface
    data assimilation, 405
    parameters, 424
Land surface models (LSM), 427
Land surface temperature (LST), 406
La Nina, 234
Larger innovations, 338
Large-scale errors, 181
Large-scale precipitation, 239
Large-scale soil moisture estimates, 456
Large sub-domains, 91
Latent heat flux bias, 435
Lateral subsurface flow, 384
L-band microwave data, 437
Leading Lyapunov exponent, 328
Leading singular vectors, 266
Leap frog scheme, 382
Least squares procedure, 285
LEO, 198
Lidar, 403
Lightning research, 109
Limited observability, 390
Linear analysis, 185
Linear interpolation, 161
Linear optimal problem, 313
Linear sensitivity analysis, 28
Linear system solver, 316
Lithology, 143
Local Analysis and Prediction System, 152
Local domains, 76
    approach, 79
Localization, 364
Local projection operator, 239
Logit model, 138

Lone-wolf approach, 7
Longwave radiation, 410
Loop Current, 329
Lorenz-40 model, 328
Lorenz attractor model, 324
Low-altitude data, 205
Low-dimensional applications, 72
LSM, 442

Manning
    equation, 394
    roughness coefficient, 394
    surface roughness, 135
Master recession curve, 136
Mathematical representation, 377
Matrix-vector multiplication, 185, 310
Maximum posterior probability, 81
Maximum rainfall intensity, 131
MCMC, 356
Mean bias errors, 288
Mean reflectivity, 154
Measurement
    error, 355
    of forecast impact, 194
    function, 345
MEF, 293
Meiyu Front, 253, 263
Mesh of discretization, 374
Meso 4D DAS, 203
Mesoscale, 219
    low, 257, 263
    models, 15
Met office, 54
Micrometeorological, 437
Mid-tropospheric channels, 191
Minimization algorithm, 260
Minimum-information
            distributions, 302
Minimum variance estimation, 29
Mississippi Bight, 335
MLEF, 67
MM5, 165, 221
Model
    bias, 287
    dynamics, complexity of, 94
    error, 40, 68, 323, 431
MODIS, 167
Moist static energy error norm, 190
Monopole rotational center, 120
Monte-Carlo, 71
    -based ensemble, 453
Monterey Bay, 278
Mosaiced hybrid scan reflectivity, 153
Mosaic weighting function, 162

Mudslides, 131
    warnings, 158
Multi-cell structure, 114
Multi-threaded computer architectures, 343
Multivariate assimilation methods, 443

Natural disaster, 403
NAVDAS, 178, 308
NAVDAS-AR, 193, 307
NCEP, 11, 239
    reanalysis, 256
NCODA, 269
NCOM, 269, 324, 349
Negative bias, 201
Neural network, 150
NMC, 10
NOAA, 280
NOGAPS, 312
Non-cycling solution, 332
Non-Gaussian and nonlinear processes, 70
Nonlinear evolution, 241, 249
Nonlinearity, 188, 227
Nonlinear model, 258
Nonlinear ocean response, 323
Nonlinear operator, 48, 74, 429
Nonlinear optimal problem, 315
Nonlinear perturbation, 246
Nonlinear reduced gravity model, 329
Nonlinear state space model (SSM), 355
Non-overlapping sub-domains, 89
Non-precipitation echoes, 150
Non-steady problems, 379
Normalized direction, 388
Normalized filter innovations, 455
Normally distributed random vectors, 41
NORPEX, 233, 254
Northwest Pacific Ocean, 341
Nowcast(ing), 148, 286
NPS, 276
NRL, 189, 307
Numerical prediction, 266
Numerical simulation, 265
Numerical stability, 228
Numerical weather forecast, 254
Numerical Weather Prediction, 21
NWP, 1, 11, 177

Objective analysis, 23
Observation(s)
    adaptive, 56, 231, 250, 257
        strategies, 371
    adjoint-based targeted, 253, 266
    and background sensitivity, 178
    backscatter cross-section, 441

density, 330
density gradient, 182
departures, 33
ensemble, 430
error, 204
error covariance, 180, 208, 313
global
    impact, 187
    network, 192
ground-based rainfall, 459
impact, 186
impact measure, 188
impact statistics, 191
operator, 30, 210, 454
pseudo, 193, 375
quality of, 93
random perturbation of, 454
satellite, impact of, 177
simulated, 77
space, 412
targeted, 164, 253, 267
upper-air, 3
Observational error, 297
Observational noise, 88
Observation System Experiment (OSE), 189
Observing platform, 192
Occultation data, 204, 211
    assimilation, 215
Occultation refractivity, 216
Ocean analysis, 276
Ocean response, 288
Off-diagonal elements, 212
Ohyu, 202
Okushiri tsunami case, 360
Open loop, 421
    estimates, 411
Operational prediction, 316
Operational subjective forecasting, 8
Optimal control, 28
    methods, 373
    theory, 368
    theory approach, 32
Optimal correction, 339
Optimal estimation, 225
Optimal input, 45
Optimal Interpolation (O.I.), 1, 22
Optimality necessary condition, 382
Optimality System (OS), 385
Optimal least-squares estimation, 29
Optimal values, 390
Orthonormal matrix, 318
OSSE, 418
Outer loops, 331
Outer rainband structures, 173

Outflow parameter, 135
Overhang echo pattern, 116
Overpredict variances, 304
Overshooting mechanism, 113
Oyashio current, 348

Parallel processing, 69
Parameter estimation, 219, 359
    problems, 377
Parameterization experiment, 226
Partial differential equation, 295
Partial pressure of water vapor, 207
Particle filter (PF), 354
Passive microwave soil moisture, 441
Path-length, 209
Path Observation Error, 210
Penalty parameter, 35
Percentile-based scaling method, 423
Perceptual model, 398
Percolation to deep aquifer, 384
Perfect model, 236
    assumption, 249
    hypothesis, 39
Persistence, 286
Perturbation (s)
    amplitude, 263
    conditional nonlinear optimal, 231
    cross-correlated, 409
    structure, 248
    wind, 239

Perturbed initial analysis, 265
PF/PS, 364
Phase changes, 108
Phased array radar, 97
PIKAIA, 223
Planetary Froude number, 235
Point Ano Neuvo, 278
Point scale measurements, 416
Polarization Corrected Temperature, 172
Polarized brightness temperatures, 452
Porosity map, 451
Posterior distribution, 303
Potential evapotranspiration, 451
Precipitation
    monitoring network, 456
    radar, 171
    rate, 156
    threshold, 223
    uncertainty, 458
Precipitation distributions, 157
Preconditioner, 309
Preconditioning, 311
Predictability limit, 374

Prediction of tornado, 99
Primitive-equation ocean model, 342
Principles of nature, 13
Prior error covariances, 30
A priori estimates, 398
A-priori evaluation, 378
Prior misfits, 326
Probabilistic and evolutionary methods, 397
Probability
    density function, 439
    forecasting, 238
    of landslide occurrence, 141
    model, 142
Propagation of tsunami, 363
PSAS, 37
Pseudo observations, 193, 375

QPE, 147
QPESUMS, 129
QPF, 221
    skill, 225
Quadratic penalty coefficient, 35
Quadratic polynomial, 23
Quadruple-nested grid system, 274
Quasi-adiabatic, 104
Quasi-geostrophic model, 8, 36
Quasi-Newton minimization, 54
Quasi-steady solution (QSS), 106
QuickSCAT, 166

Radar
    applications, 147
    data, 128
    index, 153
    rainfall, 127
        estimates, 129, 130
    reflectivity, 11, 133
Radiative transfer models, 170
Radiometer brightness temperature, 441
Radius
    of influence, 24
    of maximum winds, 169
Rain-effected data, 168
Rainfall
    abstractions, 383
    distribution, 210
    estimates, 137
    forcing, 397
    intensity, 140, 399
    -runoff model, 369, 380
    -runoff process, 134
    intensity, critical, 141
Random flux, 297
Random noises, 345

Random perturbation of observation, 454
Random search, 222
Random walk model, 361
Range height indicator (RHI), 115
Ray-path, 205
RCWF hybrid scan, 161
Rear-frank downdraft (RFD), 100
Recycled spectral preconditioner, 316
Red-Arkansas domain, 419
Reference
    gauges, 127
    solution, 332
Reflectivity mosaic grid, 151
Refractive index, 199, 208
Regression coefficient, 139
Regularization, 389
Relaxation, 277
    period, 281
Remotely-sensed data, 443
Re-orthogonalization, 311
Representation error, 347
Representer algorithm, 308
Representer-based system, 309
Representer coefficients, 322
Representer coefficients
                vector, 313
Representer functions, 322
Representer matrix, 339
Representer method, 321, 408
Representers technique, 38
Resample-move algorithm, 300
Residual ground clutter, 159
Retrieval
    algorithm, 414
    skill, 419
RFD-gust front, 121
Richard's equation, 384
Robert time filter, 273
ROMS, 343
Root zone soil moisture, 415
Rossby number, 103
Roughness coefficient, 401
Rounding errors, 388
Runoff
    estimates, 417
    forecasts, 439

SACC/JPL, 201
Salinity
    gradient, 298, 304
    mode, 297
Sample-impoverishment, 300
Sample size(s), 303, 347
Sasaki's

dissertation, 6
scientific genealogy, 16
Satellite
-born radiometers, 27
channel, 187
data assimilation, 164
retrievals, 415
observations, impact of, 177
Scaling approach, 414
Scanning
adaptive radar network, 99
footprint pattern, 453
Schur product, 76
SCSMEX, 78
SCY model, 295
Sea-breeze circulation, 278
Sea clutter, 150
Seasonal thermocline, 290
Second derivative, 49
Second order
adjoint methods, 44
adjoint model, 402
information, 46
Sector scans, 160
SEEK filter, 69
Self-organized parameter, 357
Sensible heat flux, 423
Sensitivity
analysis, 44, 179, 255
experiments, 247
gradient, 182, 260, 262, 267
Sequential data assimilation, 359
Sequential filter, 429
Sequentially Monte Carlo, 298
Severe hail, 151
Shallow thermocline, 286
Shelf-break, 335
Signal attenuation, 455
Sign of rotation, 119
Simulated annealing, 220
Simulated hydrographs, 142
Simulated observations, 77
Simulated stream discharge, 140
Single radar Cartesian (SRC), 148
Singular value, 261
Singular value decomposition (SVD), 400
Singular vector, 231, 261, 401
SIR, 293
Skin temperature, 422, 435
Smaller-scale circular vortex, 6
Small perturbation analysis, 106
Small-scale background errors, 181
SMMR, 420
Smoothed ensemble members, 358

Snow
cover area, 420
density, 422
depth fields, 421
Snow-soil-aquifer column, 418
Snow water equivalent (SWE), 420
SOAR, 180
Soil moisture
assimilation, 436
background, 434
content, 396
network, 457
profile estimates, 427
retrieval, 452
retrievals, 458
scheme, 222
suction, 396
Soil Moisture Active-Passive (SMAP), 418
Solar radiation, 273
Spatio-temporal stochastic model, 43
Spectral preconditioner, 314
Spectral projected gradient 2 (SPG2), 234
Squall lines, 15
Square-root EnKF, 72
Square root forecast error covariance, 82
SSM/I, 166
SST gradients, 279
Stabilized representer matrix, 312
Stable equilibrium points, 357
Standard filters, 304
State estimation, 358
Stationary point, 26
Statistical biases, 41
Statistical error covariances, 22
Statistical framework, 72
Statistical information, 373
Statistical interpolation, 29
Steady state pdf, 302
Steepest direction, 393
Steepness of eigenvalue
spectrum, 87
Stopping tolerance, 314
Storage deficit, 135
Stratiform, 156
Stream discharge, 136
Strong constraint, 25, 327
4D-Var, 40
Strong vorticity, 115
Sub-arid environment, 457
Sublimation, 116
Sub-seasonal prediction, 406
Subwatersheds, 130
Successive corrections
method, 24

Supercell, 100
    storm, 111
Super-refractivity, 201
Super-sensitivity, 182
Super Typhoon Paka, 171
Surface energy balance, 433
Surface flux estimation, 438
Surface soil moisture, 413
    assimilation, 419
Surface soil wetness, 452
SVAT, 442
SV method, 250
SWAFS, 270
Symmetric, positive-definite matrix, 301
Systematic error dynamics, 41
System noise, 357

Tangent data point, 200
Tangent linear approximation, 186
Tangent-linear models, 433
Tangent linear model (TLM), 233, 526
Tangent point profiles, 207
Tangent points, 209
TAPENADE, 371
Targeted observation, 164, 253, 267
Taylor's formula, 387
Temporal discretization, 387
Terminal conditions, 51
Terrain regions, complex, 158
Terrestrial water storage (TWS), 413, 416
Thermally-dominated state, 296
Thinning, 211
Thoré basin, 397
Tide gauge, 361
Tikhonov regularization terms, 370
Time distributed data, 331
Time-invariant parameters, 432
TLM, 323
    stability, 338
Topography
    artificial bottom, 362
    coastal, 275
Tornadogenesis, 97
Tornado-like vortex, 111
Torrential rains, 128
Total-energy SVs, 232
Total impact, 192
TPOE, 207
Track and intensity, 163
Track prediction of typhoons, 5
Tridiagonal matrix, 317
TRMM, 165, 197
TRMM Microwave Imager (TMI), 440
Tropical cyclones, 163

evolution, 173
initialization, 169
Tsunami, 360
Turbulent transfer, 434
Twin experiment, 362

Ultra-high resolutions, 98
Unconstrained minimization, 34
Unconstrained optimization, 370
Unified grid system, 346
Unified notation, 21
Uniform refractivity distribution, 206
Unperturbed initial analysis, 258
Unstable dynamics, 80
Update cycle, 284
Update model states, 431
Updraft, 102
Upper-air observations, 3
Upwelling, 277
    front, 283
    period, 279
USDA, 448

Vapor deposition, 110
Variational assimilation applications, 309
Variational data assimilation (VDA), 98,
        321, 431
    framework, 402
    methods, 69
Variational formalism, 34
Variational mechanics, 7
Variational methods, 367
Variational principle, 45
Variational technique, 219
Vegetation anomalies, 450
Verification area, 247
Vertical correlation(s), 208, 216
Vertical gradient of
        temperature, 287
V-J day, 2
Vorticity, 109

W4DVAR, 308
Warm-rain radar, 153
Water balance model, 450
Water content at saturation, 385
Water cycle, 367
Water molecules variety, 101
Water resource managements, 162
Watershed response, 380
Water vapor, 215
    bands, 80
Wave train, 244
Weak constraint, 38, 327, 408

Weak constraint 4D-Var, 40
Weather prediction centers, 12
    *See also* Numerical Weather Prediction
Weather-related disasters, 147
Weather Research and Forecasting (WRF), 67
Wind perturbations, 239
Wind stress, 275, 330
    curl, 275

WRF, 78
WRF-var, 89

Yamato Rises, 353
Yoshikazu Sasaki, 1
Yoshi and Koko, 13

Z-R relationships, 156

Printing: Krips bv, Meppel, The Netherlands
Binding: Stürtz, Würzburg, Germany